Oxidoreduction at the Plasma Membrane: Relation to Growth and Transport

Volume I
Animals

Editors

Frederick L. Crane, Ph.D.
Professor
Department of Biological Sciences
'Purdue University
West Lafayette, Indiana

D. James Morré, Ph.D.
Distinguished Professor
Department of Medicinal Chemistry
Purdue University
West Lafayette, Indiana

Hans E. Löw, Ph.D.
Professor
Department of Endocrinology
Karolinska Hospital
Stockholm
Sweden

CRC

CRC Press
Boca Raton Ann Arbor Boston

Library of Congress Cataloging-in-Publication Data

Oxidoreduction at the plasma membrane: relation to growth and transport/editors. Frederick L. Crane, D. James Morré, Hans E. Löw.

p. cm.

Includes bibliographical references.

Contents: v. 1. Animals.

ISBN 0-8493-6937-1 (v. 1)

1. Oxidoreductases. 2. Cell membranes. 3. Cells--Growth-Regulation. 4. Biological transport. I. Crane, Frederick L. II. Morré, D. James, 1935- . III. Löw, Hans.

[DNLM: 1. Biological Transport. 2. Cell Division. 3. Cell Membrane. 4. Oxidation-Reduction. QU 125 098]

QP602.095 1990

574.87′5--dc20

DNLM/DLC

for Library of Congress 90-1761

CIP

Direct all inquiries to CRC Press, Inc., 2000 Corporate Blvd., N.W., Boca Raton, Florida 33431.

International Standard Book Number 0-8493-6937-1

Library of Congress Card Number 90-1761
Printed in the United States

PREFACE

The objective of this volume is to present a comprehensive coverage of the evidence that oxidoreductase enzymes in the plasma membranes of animal cells play an important role in control of cell growth and to further examine how the oxidoreductase enzymes may influence control of membrane transport. Although isolated experiments in plasma membrane electron transport can be found as far back as 1929, it is only in the last 15 years that a systematic study has begun on the clear determination of the widespread existence of unique plasma membrane oxidoreductase enzymes. As evidence for the presence of plasma membrane redox enzymes solidified, experiments were designed to show potential roles in growth, development, and membrane transport.

At least twenty laboratories around the world have been involved in current studies of plasma membrane oxidoreductase enzymes in animal cells, and even more have worked with plants. The results of these investigations are mostly scattered throughout the literature except for two recent books that contained reports presented at meetings. As a result, it is difficult to evaluate the scope of research which supports the general concept of plasma membrane oxidoreductase enzymes and the evidence which is developing for their contribution to cell function. We hope that these volumes will help to consolidate this basic information. We also expect that these books will serve to emphasize that much more research is needed to understand the inner workings of the electron transport enzymes, how they help to control the cell cycle, where they contribute to transport systems, and their ubiquitous occurrence in living cells. Beyond basic understanding, lies the practical significance which may be revealed in changes in these enzymes in tumor cells and control of the enzymes by antitumor and antimalarial drugs and neuromodulators.

The chapters are organized to present first the basic observation of plasma membrane oxidoreductase activity and the unique properties of these enzyme activities, especially with respect to selective inhibitor and hormone response. The next chapters consider the stimulation of cell growth by transplasma membrane electron transport. This discussion of growth control leads to examination of transmembrane iron and ferric transferrin reduction and its relation to transferrin receptor involvement in growth control. The last chapters consider the effects of plasma membrane electron transport on physiological functions of cells which could be involved in growth control and the relation of the redox system to antitumor drug action and response to neurotransmitters.

The material presented in this book is only half of the story of plasma membrane oxidoreductases. Enzymes of similar character are also present in plant plasma membranes. A discussion of these enzymes and their relation to plant growth will be presented in a second volume.

We very much appreciate the efforts of the contributors who have taken the time to prepare chapters which present a comprehensive view of the subject. We also thank Janet Hollister for skillful help in assembling the final copy and Annika Lindgren and Monica Isaksson for preparation of figure copy.

Frederick L. Crane
D. James Morré
Hans Löw

THE EDITORS

Frederick L. Crane, Ph.D., is a Professor of Biological Sciences in the School of Science at Purdue University in West Lafayette, Indiana.

Dr. Crane obtained his training at the University of Michigan, Ann Arbor, receiving the B.S. degree in chemistry in 1950, followed by the M.S. in 1951, and the Ph.D. in 1953 in botany. From 1953 to 1957 he was a postdoctoral trainee at the Enzyme Institute, University of Wisconsin, where he was Assistant Professor from 1957. In 1959 he was appointed Assistant Professor in the Department of Chemistry, University of Texas, Austin. In 1960, he moved as Associate Professor to the Biology Department of Purdue where he assumed his present position in 1962. He has had visiting appointments at Stockholm University, 1963 to 1964, and at the John Curtin School of Medical Research, Australian National University, Canberra, in 1970/1971 and 1979/1980.

Dr. Crane is a member of the American Society for Biological Chemistry and Molecular Biology, the American Society for Cell Biology, the American Society of Plant Physiologists, the Plant Growth Regulator Society of America, and the Indiana Academy of Science.

In 1961 he was recipient of the American Chemical Society Eli Lilly Medal in Biochemistry. He was awarded a Silver Medal in Biochemistry at the University of Bologna and an Honorary Doctor of Medicine Degree by the Karolinska Institute in Stockholm in 1989.

Dr. Crane is recipient of a Research Career Award from the National Institutes of Health and has served on Research Review Panels at the National Science Foundation and the National Institutes of Health. He was co-chairman of a meeting on Plasmalemma Redox Function in Plants at the 1987 Botanical Congress in Berlin, and was Director of a NATO Advanced Research Workshop on Plasma Membrane Oxidoreductases in Control of Animal and Plant Growth in Cordoba in 1988.

Dr. Crane has served on the editorial board of *Biochimica et Biophysica Acta* and is Review Editor of the *Journal of Bioenergetics and Biomembranes*. He is an author of more than 350 papers and coeditor of two books. His current research interest is the nature and function of plasma membrane oxidoreductases.

D. James Morré, Ph.D. is the Dow Distinguished Professor of Medicinal Chemistry at Purdue University, West Lafayette, Indiana.

Dr. Morré received his B.S. degree from the University of Missouri in 1957, his M.S. from Purdue University in 1959, and the Ph.D. degree in 1963 from the California Institute of Technology in Pasadena, California. He was appointed as an Assistant Professor at Purdue University in 1962. He became an Associate Professor in 1966 and Professor in 1971. From 1976 to 1986 he was Director of the Purdue Cancer Center.

Dr. Morré is a member of the American Society for Cell Biology, the American Society for Biochemistry and Molecular Biology, the American Association for Cancer Research, The American Society for Plant Physiologists, and the American Heart Association.

Dr. Morré has been a recipient of a Senior Investigator Award of the American Heart Association (1976 to 1978), The Lions Cancer Research Award of Purdue University (1978), an Alexander von Humboldt Foundation Senior U.S. Scientist Award of the Federal Republic of Germany (1984), and a Doctorate honoris causa from the University of Geneva in Switzerland (1985). He is presently the 1989 to 1990 Tage Erlander Professor for Sweden as a guest of the Swedish Natural Science Research Council.

Dr. Morré has been the recipient of research grants from the National Institutes of Health, the National Science Foundation, the American Heart Association, and the American Cancer Society. He has published more than 400 papers. His current research interests are membrane biogenesis and cell growth.

Hans Löw, Ph.D., is professor of Experimental Endocrinology in the Department of Endocrinology of the Karolinska Institute in Stockholm.

Dr. Löw was trained at the University of Stockholm and obtained his Ph.D. in cell physiology in 1959 from the Wenner Gren Institute. He was a postdoctoral trainee at the Enzyme Institute, University of Wisconsin during 1960 and 1961 and became an assistant professor the same year. In 1966 he became the head of the laboratory at the Clinic for Endocrinology at the Karolinska Hospital. In 1970 and 1971 he was a visiting scientist at the National Institutes of Health. He became professor of Experimental Endocrinology at the Karolinska Institute in 1980. He has been a visiting investigator in CSIRO Adelaide, South Australia in 1982 and at the Centre de Biochimie, Nice, France in 1985.

Dr. Löw is a member of several Swedish and foreign learned societies and was co-organizer of several symposia related to plasma membrane redox and growth factor action. He is an author of about 200 scientific papers and co-editor of two books. His current research interest includes apart from the mechanism and function of the plasma membrane oxidoreductase reaction, the mode of action of insulin and growth factors with special reference to their function in the diabetic state.

CONTRIBUTORS

Paul Basset, M.D., Ph.D.
LGME Institut de Chimie Biologique
Faculté de Medicine
Strasbourg
France

Alajos Berczi, Ph.D.
Biological Research Center
Biophysical Institute
Hungarian Academy of Science
Szeged, Hungary

C. Buillard
Department of Biochemistry
University of Fribourg
Fribourg, Switzerland

Maria Isabel Burón, Ph.D.
Associate Professor
Department of Cellular Biology
Cordoba
Spain

Rafael García-Cañero, Ph.D.
Chief Associate
Department of Experimental Biochemistry
Madrid
Spain

Frederick L. Crane, Ph.D.
Professor
Department of Biological Sciences
Purdue University
West Lafayette, Indiana

Anthony F. Cutry, Ph.D.
Postdoctoral Fellow
Department of Biochemistry and Cell Biology
SUNY Stony Brook
Stony Brook, NY

Jean-Luc Dreyer, Ph.D.
Professor
Department of Biochemistry
University of Fribourg
Fribourg
Switzerland

K. A. O. Ellem, M.D., Ph.D.
Deputy Director
Department of Cancer Cell Biology
Queensland Institute of Medical Research
Brisbane
Australia

W. Page Faulk, M.D., F.R.C. Path.
Center for Reproduction and Transplantation Immunology
Methodist Hospital of Indiana
Indianapolis, Indiana, and
Adjunct Professor of Medicinal Chemistry
Purdue University
West Lafayette, Indiana

F. Gueggi
Department of Biochemistry
University of Fribourg
Fribourg
Switzerland

Hava Harats, M.D.
Center for Reproduction and Transplantation Immunology
Methodist Hospital of Indiana
Indianapolis, Indiana

Gordon J. Harnadek
Department of Biological Sciences
Wayne State University
Detroit, Michigan

M. Isaksson
Endocrinology Department
Karolinska Institute
Stockholm
Sweden

Visham Jalukar
Department of Biological Sciences
Wayne State University
Detroit, Michigan

G. F. Kay, Ph.D.
Postdoctoral Fellow
Queensland Institute for Medical Research
Brisbane, Queensland
Australia

Patrick Kelley, Ph.D.
Department of Biological Sciences
Wayne State University
Detroit, Michigan

Hans E. Löw, Ph.D.
Professor
Department of Endocrinology
Karolinska Hospital
Stockholm
Sweden

Miguel Angel Medina, Ph.D.
Department of Biochemistry and
Molecular Biology
University of Malaga
Malaga
Spain

D. James Morré, Ph.D.
Distinguished Professor
Dept of Medicinal Chemistry
Purdue University
West Lafayette, Indiana

Plácido Navas, Ph.D.
Professor of Cell Biology
University of Cordoba
Cordoba
Spain

David Njus, Ph.D.
Professor of Biological Sciences
Wayne State University
Detroit, Michigan

Ignacio Núñez de Castro, Ph.D.
Professor
Department of Biochemistry and
Molecular Biology
University of Malaga
Malaga
Spain

Iris L. Sun, Ph.D.
Department of Biology
Purdue University
West Lafayette, Indiana

T. Treichler
Department of Biochemistry
University of Fribourg
Fribourg
Switzerland

Charles E. Wenner, Ph.D.
Associate Chief Cancer Research Scientist
Biological Chemistry Division
Roswell Park Memorial Institute
Buffalo, New York

TABLE OF CONTENTS

Chapter 1

HISTORICAL PERSPECTIVE

F. L. Crane, H. Löw, and D. J. Morré

TABLE OF CONTENTS

I. RECOGNITION OF PLASMA MEMBRANE OXIDOREDUCTASES

The comprehensive study of oxidoreduction reactions in the plasma membrane has been a recent development. It is known now that there are oxidoreductase enzymes on the inner face, the exterior face, and crossing the membrane. Evidence is growing for significant functions for these enzymes.[1-9] Preliminary observations of these oxidoreductases can be traced back to the 1920s, but many of these early observations seem to have faded out of sight before they could stimulate a major investigation into the nature and significance of the oxidoreductases in the plasma membrane.

There are six major lines of investigation which led into the concepts of redox function in the plasma membrane and its functional significance (Table 1). These areas include:

1. Studies of oxidation-reduction reactions of whole cells with impermeable oxidants indicating electron transfer across the plasma membrane: the logical development of this observation would have been study of oxidoreductases in isolated plasma membrane but this was difficult because highly purified plasma membranes were hard to prepare and the oxidoreductase activity in contaminating mitochondrial or endoplastic reticulum membranes overwhelmed any plasma membrance activity. The mammalian erythrocyte membrane was a special case because the absence of internal membranes precluded contamination of plasma membrane preparations by other membranes. This led to early studies of redox functions with these membranes, but the primary emphasis was on known extensions of endoplasmic redox function such as NADH cytochrome c reductase[10] and methemoglobin reduction,[11,12] which represents a useful function of this enzyme in erythrocytes.
2. Stimulation of cell growth by external oxidants: although Brooks' studies in 1947[13] could have opened up this area of research, there seems to have been no further development until evidence for transplasmalemma electron transport stimulated a new study by Ellem and Kay in 1983[14] on using ferricyanide as an external oxidant to stimulate growth.
3. The study of salt respiration in plants led to proposals that plasma membrane redox enzymes could provide energy for ion uptake directly without mediation by ATP.[15] The impetus for these studies was decreased by development of evidence for ATPases to drive transport. New interest developed with evidence for redox effects on amino acid transport. Suggested relations between redox and gastric acid secretion have always been in the background.
4. Evidence that the pigments responsible for response of plants to blue light are located in the plasma membrane:[16] this reaction has specific control function for plant and fungal cells. Questions about plasma membrane purity led to debate about location in the cell.
5. The respiratory burst of leucocytes with consequent hydrogen peroxide production to kill engulfed bacteria led to investigations of the oxidase reaction and evidence for its location in the plasma membrane.[17]
6. External peroxide production in the wall of plant cells has long been related to lignin formation in the cell wall. Peroxidases associated with this peroxidation are bound to the surface of the plasma membrane.[18]

The studies by Voegtlin, Johnson, and Dyer in 1925 were designed to examine a relation between the protoplasmic redox state and cancer,[19] a subject which later attracted others such as Warburg.[20] In their studies, they used both permeable and impermeable oxidoreduction indicator dyes to assess the reduction rate by normal cells and tumor cells. They used a very

TABLE 1
Historical Developments in Plasma Membrane Redox

Dates	Reference events	Tissues, cells	Membranes	Cell division
1920	1923 — 1933, 1948, Keilin, Warburg — respiratory system Green, Lehninger — Mitochrondria as site of energy transfer	1925, Voegtlin et al. — impermeable dye reduction 1925, Barron — transmembrane electron transport 1954, Manyai and Szekely — impermeable ferricyanide introduced as external oxidant		1947, Brooks — dyes increase cell division in eggs
1960	1965, P. Mitchell — chemiosmotic hypothesis	1965, Dormandy and Zarday — ferricyanide reduction and H^+ release by erthrocytes 1969, Mishra and Passow — ferricyanide and erythrocyte	1964, Emmelot — NADH cyt c reductase in erythrocyte PM. 1966, Wallach and Kant — Ehrlich ascites ferricyanide reductase 1966, Zamudio and Canessa — NADH ferricyanide reductase in erythrocyte 1967, Vassilez et al. — NADH dehydrogenase in liver PM	
1970	1971, Morré — membrane flow	1972, Arese et al. — NAD(P)H oxidized by ferricyanide in erythrocytes 1977, Mukherjee and Lynn — NADPH oxidase in plasma membrane oxidizes SH 1978, Morré et al. — cytochemistry, Hatchett Brown 1977, Garcia-Sancho et al. — redox-driven amino acid transport	1970, Fischer — squid axon membrane 1973, Masuda et al. — NADH-cyt c in liver plasma membrane 1973, Frantz, Vigil et al. — liver cell plasma membrane; ferricyanide reduction 1976, Löw and Crane — inhibition and hormones; liver plasma membrane 1977, Gayda — NADH oxidase 1978, Wang and Alaupovic — isolated dehydrogenase 1979, Löw et al. — oriented dehydrogenase 1979, Huang et al. — gel electrophoresis of plasma membrane enzymes	

TABLE 1 (continued)
Historical Developments in Plasma Membrane Redox

Dates	Reference events	Tissues, cells	Membranes	Cell division
1980	1982, Albertsson et al. — phase separation for plasma membrane purification	1981, Clark et al. — ferricyanide reduction by liver cells and perfused liver 1981, Powis et al. — indigo-disulfonate reduction; AKR cells; transformation 1984, Sun — H^+ release with ferricyanide reduction 1986, Navas et al. — NADH oxidized by ferricyanide; HeLa	1980, Goldenberg — SDHA reductase in PM 1983, Morley et al. — diferric transferrin reduction	1980, Barnes and Sato — transferrin required for growth 1983, Kay and Ellem — ferricyanide + cell growth 1984, Sun et al. — ferricyanide and HeLa growth 1985, Sun et al. — redox dyes and growth of HeLa

Dates	Plants: plasma membrane	Blue light effects in plasma membrane	Defense and peroxide formation	External peroxidase and dehydrogenase
1920	1945, Lundegardh — salt respiration 1951, Conway — redox drive in yeast		1954, Morton — xanthine oxidase in milk fat membrane	1957, Crane — cauliflower wall fraction has CN-insensitive NADH oxidase 1959, Forti et al. — NADH oxidase in heavy pea stem particles
1960	1968, Roberston — redox transport			
1970	1977, Polevoi and Salamatova — ferricyanide inhibits auxin-induced O_2 uptake 1977, Chaney et al. — transmembrane iron reduction	1974, Poff and Bulter — blue light effect on plant growth 1977, Brain 1977, Jesaitis — blue light reduction reactions in PM; flavin and cytochrome	1970, Keenan — milk fat membrane is plasma membrane 1975, Briggs et al. — NADH oxidase in plasma membrane neutrophils $O2^-$ 1979, Dewald et al. — NADPH oxidase 1979, Borregaard et al. — redox components and granulomatous disease	1976, Elstner and Heupel — NADH oxidation by wall peroxidase 1977, Gross — peroxide in plant cell wall 1979, Löw et al. — NADH dehydrogenase outside erythrocyte

1980	1981, Novak and Ivankina — PM redox and membrane potential 1981, Craig and Crane — carrot ferricyanide reduction 1984, Sijmons et al. — NADPH oxidation in cytosol 1985, Bienfait — the Turbo reductase 1985, Barr et al. — CN-insensitive NADH oxidase in plant plasma membrane	1981, Widell et al. — blue light effects in pure plasma membrane 1985, Kjellbom et al. — P450	1980, Green et al. — NADPH oxidase is transmembrane 1981, Cross et al. — Low potential cyt b	1981, Cherry — NADH dehydrogenase outside tumor cell 1982, Lin — external oxidase on plant cells 1982, Mader and Amberg-Fischer — external oxidase is peroxidase in plants 1987, Morré et al. — hormone controlled oxidase outside 1985, Barr et al. — cyanide-insensitive NADH oxidase is not peroxidase

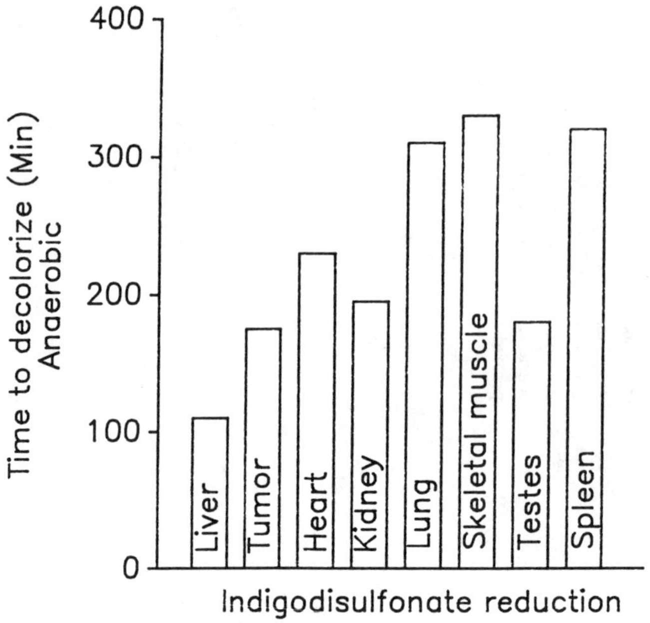

FIGURE 1. Reduction rate of indigo disulfonate exposed to rat tissue rats showing the time taken for the rat to decolorize. (Calculated from data in Reference 19).

simple experimental procedure to determine redox capacity. The time it took to decolorize the dye with a tissue suspension under anaerobiosis was determined and expressed as time to decolorize: in general, the higher the redox potential of the dye, the faster the rate of decolorization (reduction). Both permeable and impermeable dyes were reduced by the cells from all tissues, but the rate of reduction was clearly different from tissue to tissue. In looking at overall dye reduction, including both permeable and impermeable dyes, there was no apparent difference between Flexnor-Jobling tumor tissue and normal liver tissue. If the results of these experiments are considered on the basis that the impermeable dyes measure transplasma membrane electron transport whereas permeable dyes are reduced by many redox systems inside the cell, interesting observations relating to plasma membrane redox are possible. With impermeable indigo disulfonate the reduction rate by Flexnor-Jobling tumor can be half the rate of the one observed with liver (or two times faster than lung). The fact that the impermeable dyes are reduced is in itself good evidence for some type of electron transfer across the plasma membrane and the reduction of indigo disulfonate indicates electron transfer at a relatively low potential (−125 mV) (Figure 1).

Experiments to determine the toxicity of dyes by injecting them into rats gave further evidence for the difference between permeable and impermeable dyes. The impermeable dyes, even though reduced, show very little lethality. They do produce respiratory problems. The permeable dyes were rapidly lethal. This would be consistent with rapid oxidation of cytosolic reductants such as NADH, NADPH, and glutathione, as well as interference with mitochondrial ATP production by the permeable dyes. The impermeable dyes would give a more selective oxidation of cytosolic reductants by electron transfer across the plasma membrane and would not interfere with mitochondrial ATP production by short circuiting the electron transport chain.

The comparison of effects of permeable and impermeable dyes was more fully developed in experiments on starfish eggs by Barron and Hoffman in 1927.[21] They used autooxidizable permeable and nonpermeable dyes to test for effects on oxygen uptake. The permeable dyes stimulate oxygen uptake at low redox potential which is consistent with electron donation by cytosolic redox compounds like NADH or glutathione with redox potentials in the range of −320 to −280 mV. Increased oxygen uptake was only found with high potential impermeable dyes. Gallocyanine (59 mV) gave increased oxygen uptake whereas indigo-disulfonate (−125 mV) had no effect. They came to the clear conclusion that impermeable dye reduction would involve a transplasma membrane electron transport with a component of midpoint potential above −125 mV at the cell surface. It should be noted that restriction of reduction to a redox potential that high would preclude direct reduction by leakage of glutathione or NADH out of the cell, but not ascorbate (+59 mV). The difference between indigo-disulfonate reduction by rat liver cells and lack of reduction by sea urchin eggs may also indicate a difference between external reduction sites in different organisms.

The cytochromes had been rediscovered in 1923—1933 by Keilin[22] and Warburg[23] as components of the major cyanide sensitive respiratory chain. The role of ATP in energy transfer reactions was developed by Lipmann[24] in 1939—1941 and the Krebs cycle was proposed in 1937.[25] Only in 1948 was the combined respiratory system and Krebs cycle, or cyclophorase, proposed by Green[26] and identified with mitochondria by Kennedy and Lehninger.[27] These studies focused the major attention in respiratory activity and especially energy-coupled respiratory activity on the mitochondria. The primary mechanism for energy transfer from both glycolysis and mitochondrial oxidation was recognized as ATP. There was thus no obvious basis to look for redox systems in the plasma membrane to energize any function. The presence of significant membrane-bound redox function not involved in energy transfer, such as the cytochrome b fatty acyl desaturase system and the detoxification system with cytochrome P450, was not developed until 1960s.[28]

In 1945, Lundegardh[29] proposed a special cytochrome system in plant cells that would directly drive ion uptake. To provide direct coupling between the electron transfer and ion transport, this redox system should have been located in the plasma membrane (Figure 2). Without isolated plasma membranes, the identity of the cytochrome changes which responded to addition of salts was difficult to establish.[30] The well-known salt respiration could not clearly be differentiated from mitrochondrial respiration. In 1948, Conway and Brady[31] proposed a plasma membrane redox system as a basis for gastric secretion which would be directly coupled to proton pumping. During this period, until 1968, Robertson[32] also investigated salt respiration and developed further evidence of its unique properties, but a clear definition of a plasma membrane redox system was not possible, primarily because isolation of sufficiently purified plasma membranes was not possible.

In 1954, Manyai and Szekely[33] showed that ferricyanide, which was impermeable to erythrocytes, caused an increase in ATP inside erythrocytes. This implied a possible redox function in the plasma membrane coupled to ATP formation since these cells lack mitochondria. Later studies indicated that the increase in ATP could come from activation of glycolysis by increase in NAD or NADP by oxidation of NADH or NADPH. The oxidation of NADH or NADPH would provide an acceptor to activate glycolytic activity.[34] Only 10 years later, Dormandy and Zarday[35] showed that erythrocytes reduce ferricyanide. They also showed that ferricyanide reduction was directed to proton release, and that this activity was stimulated by insulin. It was at this time (1965) that Mitchell proposed the chemiosmotic hypothesis as the basis for coupling electron transport and transmembrane proton movement to ATP formation in mitochondria[36] (Figure 3). The Mitchell hypothesis was cited by Mishra and Passow in their discussion of possible functions for the plasma membrane ferricyanide reductase when they demonstrated ferricyanide reduction by erythrocytes and associated ATP formation in 1969.[34]

The clearest demonstration of plasma membrane NADH dehydrogenase activity was made by Zamudio and Canessa[37] with erythrocyte membranes. They showed very active NADH ferricyanide reductase activity. Later, Zamudio et al.[38] showed that nitroblue tet-

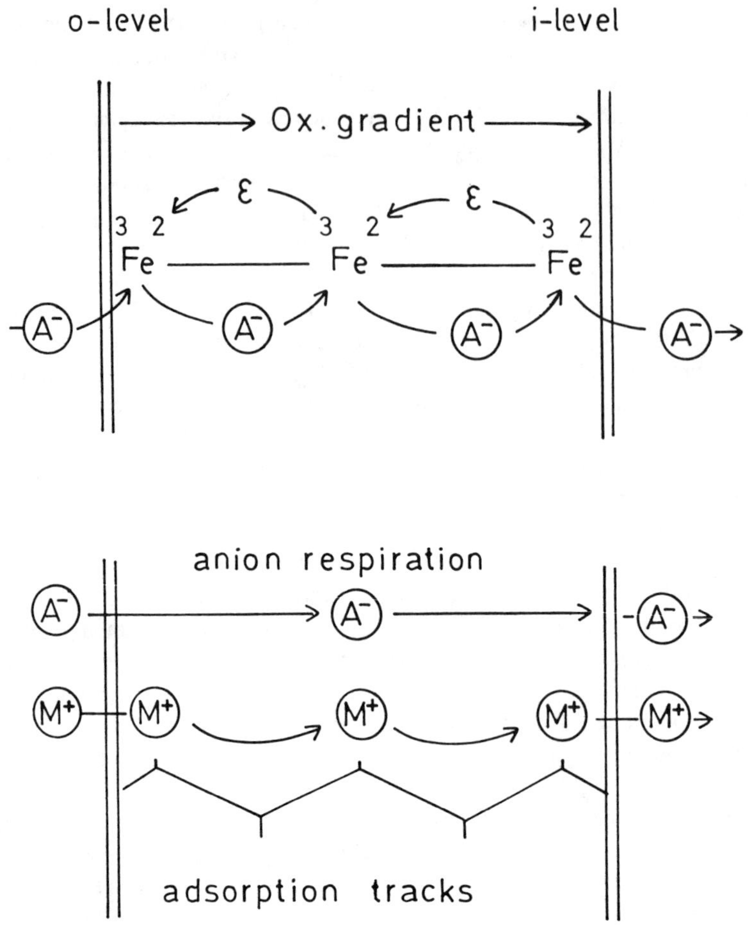

FIGURE 2. Scheme for redox-mediated anion and cation transfer across the plasma membrane from Lundegårdh.[29]

razolium reduction by NADH led to formazan accumulation in the membrane. Wallach and Kamat[39] also reported NADH ferricyanide reductase in Ehrlich ascites plasma membrane (designated as NADH oxidase activity, which was a common name at that time). Prior to this in 1964, Emmelot et al.[40] had reported low NADH cytochrome c reductase activity in rat liver plasma membrane. They concluded that the NADH dehydrogenase activity was intrinsic to the plasma membrane because their preparation had no NADPH cytochrome c reductase and so would not be contaminated with endoplasmic reticulum. This did not rule out possible contamination by outer mitochondrial membrane which may be a problem with early liver plasma membrane preparations.[41] In the late 1960s the membrane flow hypothesis was under development, according to which endomembrane originates in the endoplasmic reticulum, passes through the Golgi, and is incorporated into plasma membrane.[42] Since NADH cytochrome c reductase is a significant and highly active enzyme in the endoplasmic reticulum, interest developed into its possible distribution into the plasma membrane by membrane flow. Early studies by Thines-Sempoux et al.[43] and Fleischer and Fleischer[44] showed no NADH cytochrome c reductase activity in plasma membranes. In the Thines-Sempoux et al.[43,46] work, the purity of the plasma membrane was evaluated by treatment with digitonin, which binds to cholesterol and changes the relative density of the plasma membrane. There was no evidence that any of the NADH cytochrome c reductase moved with the plasma membranes after centrifugation of the digitonin-treated membranes. In

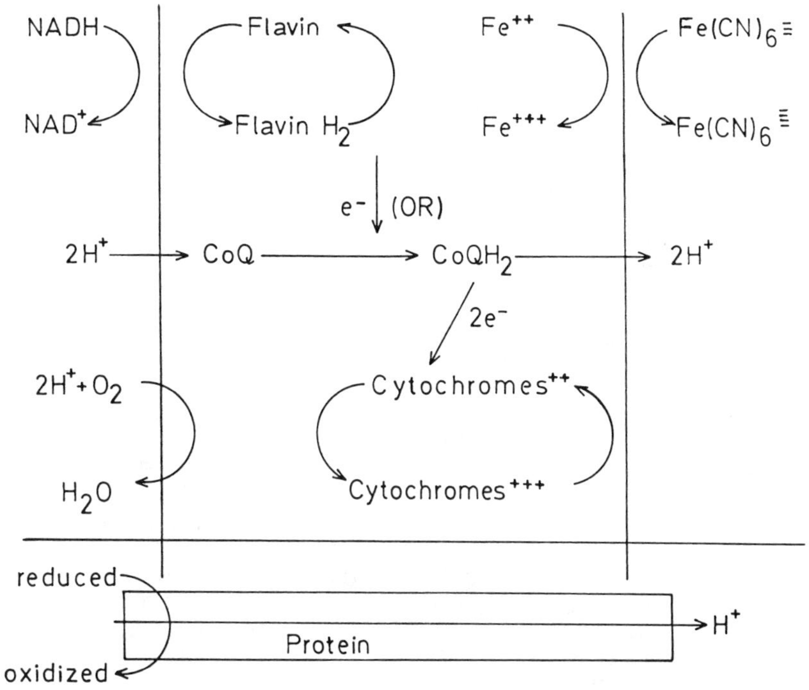

FIGURE 3. Possible pathways for electron flow with accompanying proton movement, based on proposal by Mitchell.[36]

subsequent studies Goldenberg[45] has shown that up to 50% of the NADH cytochrome c reductase in a preparation can be associated with the plasma membrane by the digitonin test (Table 2).

Fleischer et al.[47] also found no significant amount of cytochrome b_5 in plasma membrane preparations which would be consistent with the absence of NADH cytochrome c reductase. Remacle et al.[48] found no reaction of plasma membrane with antibodies to cytochrome b. Thus, for liver plasma membranes there was considerable evidence against the presence of NADH cytochrome b_5 reductase and cytochrome b_5. In 1979, Jarasch et al.[49] made an extensive study of cytochromes in plasma membranes and found evidence for cytochrome b_5 both spectrophotometrically and with antibodies.

The loss of cytochrome b_5 during membrane maturation would still leave the primary dehydrogenase, NADH cytochrome b_5 reductase. In the absence of cytochrome b_5 this activity can be analyzed using ferricyanide as an electron acceptor. The relatively high NADH ferricyanide reductase activity found in plasma membrane preparations could represent the retention of the primary dehydrogenase in the plasma membrane even after the loss of some cytochrome b_5 (Table 2). The NADH dehydrogenase reductase on the erythrocyte plasma membrane was considered to be a remainder of endoplasmic reticulum NADH cytochrome b_5 reductase which had not been released from the membrane during the differentiation of erythrocytes. Both the soluble and membrane-bound forms of the enzyme were active in reduction of methemoglobin back to hemoglobin. They were designated as methemoglobin reductase. Important evidence for the identity of the membrane-bound and soluble enzyme was that both almost completely disappeared in methemoglobinemia.[11,50-52] On the other hand, 25% of the normal ferricyanide reductase activity was found in the membranes isolated from methemoglobinemia patients,[50] which indicated that part of the ferricyanide reductase could represent a different dehydrogenase.[3] NADH ferricyanide reductase was found in liver plasma membranes by Frantz.[53]

TABLE 2
Observations on NADH Ferricyanide Reductase and NADH Cytochrome c Reductase Activity in Purified Plasma Membrane Preparations

Date	Membrane	NADH $Fe(CH)_6$	NADH cyt c	Ref.
1964	Rat liver	—	128	Emmelot et al.[40]
1966	Erythrocyte	570	8	Zamudio and Canessa[37]
1966	Ehrlich ascites	68	—	Wallach and Kamat[39]
1969	Bovine thyroid	290	—	Stanbury et al.[212]
1969	Rat liver	—	0 (corr)	Thines-Sempoux[43]
1970	Rat liver	—	0	Fleischer and Fleischer[44]
1970	Squid retinal axon	1360	74	Fischer et al.[213]
1972	Squid giant axon	1300	3	Marcus et al.[214]
1973	Rat liver	728	80	Frantz[215]
1973	Rat liver	—	120	Raflell and Blomberg[216]
1973	Rat liver	790	116	Masuda et al.[128]
1974	Rat liver	—	139	Fleischer and Kervina[217]
1976	Rat liver	350	145	Crane and Löw[55]
1977	Mouse liver	600	70	Goldenberg[218]
1979	Ehrlich ascites	855	58	Kilberg and Christensen[146]
1982	Rat liver (E)	988 (corr)	128 (corr)	Goldenberg[45]
1982	Rat liver (N)	329 (corr)	65 (corr)	Goldenberg[45]

Note: (corr) indicates value obtained by digitonin correction for endoplasmic reticulum contamination. For Goldenberg (1982) the correction was 49% contamination. E indicates preparation of plasma membrane from endoplasmic reticulum fraction; N indicates purification from nuclear fraction.

The NADH ferricyanide reductase activity in liver plasma membrane was defined histochemically by Morré et al.[53] using the Hatchett Brown procedure to precipitate copper ferrocyanide at the reductase site. Fortunately the plasma membrane enzyme was relatively resistant to mild glutaraldehyde fixation, and heavy staining was found on both plasma and Golgi membranes. Since the staining was prevented by atebrin, the dehydrogenase responsible could be the enzyme identified in isolated plasma membrane.

A different approach to plasma membrane NADH dehydrogenase activity developed from the observations of Löw and Werner[54] starting in 1972 which showed that NADH, but not NAD, inhibited adenylate cyclase in plasma membranes and that Atebrin also inhibited adenylate cyclase. This combination of effects suggested that the adenylate cyclase could be controlled by a flavoprotein which would react with NADH and Atebrin. A test of NADH dehydrogenase activity in plasma membrane showed partial inhibition by Atebrin, whereas the activity in endoplasmic reticulum was much less inhibited (Table 3). This opened the possiblity that the plasma membrane also contained a NADH dehydrogenase different from the NADH cytochrome b_5 reductase proposed by Crane and Löw.[55] Further study of the plasma membrane NADH dehydrogenase activity, especially with 2,6-dichloroindophenol or oxygen as electron acceptors, revealed remarkable sensitivity in both liver and fat cell plasma membrane to hormones such as ACTH, T_3, glucagon, and insulin.[4,56-59] In a series of studies starting in 1977, Mukherjee and Lynn,[60] and Mukherjee and Mukherjee[61] proposed a role for an NADPH oxidase in the plasma membrane which was stimulated by prior insulin treatment of cells. A possible role for H_2O_2 production by this enzyme in control of thiol redox state in the membrane was proposed. The thiol redox changes were also related to control of adenylate cyclase by Mukherjee and Mukherjee in 1981.[60]

TABLE 3
Development of Selective Inhibitors for Plasma Membrane Redox Functions

Date	Electron acceptor: Cyt c	Electron acceptor: Indophenol	Ferricyanide	Transferrin (T) or oxygen (O)	Ref.
1973			Dicoumarol		Frantz[215]
1976	Atebrin, T_3 Azide	Atebrin, T_3 Azide			Crane and Löw[55]
1976				Atebrin (O)	Gayda et al.[58]
1980			Adriamycin		Crane et al.[147]
1981			Actinomycin D		Sun et al.[151]
1984			Cisplatin		Sun and Crane[152]
1985			Bleomycin		Sun and Crane[153]
1988				Chloroquine (T)	Toole-Simms[154]
1988				Chloroquine (T)	Thorstensen and Romslo[86]

In 1977, a direct functional role for a plasma membrane redox system was proposed by Garcia-Sancho et al.[62] when they found evidence for stimulation of amino acid uptake with NADH added to inverted membrane vesicles. This development was important because it provided a clear functional significance to the plasma membrane electron transport. The evidence for an NADH dehydrogenase in the plasma membrane was reinforced by Huang et al.[63] who showed three bands of NADH tetrazolium reductase activity in a nondenaturing gel electrophoresis of detergent-solubilized plasma membrane from rat liver. In 1978, Wang and Alaupovic[64] reported the isolation from erythrocyte membranes of a glycoprotein which had very good NADH ferricyanide reductase activity. This enzyme had no visible prosthetic groups such as flavin or heme, and the molecular weight determined by gel electrophoresis was different from that of the flavoprotein NADH cytochrome b_5 reductase. They proposed that this ferricyanide reductase could be an enzyme with a transmembrane orientation because of the presence of carbohydrate. In 1972, Kant and Steck[10] had shown that the NADH cytochrome c reductase was exclusively exposed on the inside surface of the erythrocyte membrane. No cytochrome c reductase activity was found on the outside of the membrane. The cytochrome b_5 reductase is bound to the membrane by a nonpolar 22 N-terminal amino acid sequence with N-bound myristic acid.[65] The location of both the NADH dehydrogenase and the cytochrome b_5 on the inside of the membrane implied that the reduction of ferricyanide by intact erythrocytes could not be accomplished by the cytochrome b_5 reductase. Therefore another basis for the ferricyanide reduction observed by Mishra and Passow in 1969[34] was needed. Löw et al.[66] in a study of ferricyanide reduction using oriented erythrocyte membranes, showed that in contrast to NADH cytochrome c reductase activity there was more NADH ferricyanide reductase activity revealed in open erythrocyte ghosts (both sides of the membrane exposed) than on sealed vesicles with only the inside or outside of the membrane exposed. There was, of course, NADH ferricyanide reductase activity on the inside of the membrane which would represent the reaction of NADH cytochrome b_5 reductase with ferricyanide. Surprisingly there was also a small amount of NADH ferricyanide reductase on the outside surface, but it could not account for the ferricyanide reduction by open vesicles. Thus it was proposed that a part of the NADH ferricyanide reductase activity was based on a transmembrane-oriented dehydrogenase which would oxidize internal NADH with external ferricyanide. The transplasma membrane NADH ferricyanide reductase would account for ferricyanide reduction by intact erythrocytes. Further support for a transmembrane dehydrogenase in erythrocytes was inherent in the evidence by Arese et al.[67] that cytosolic NAD(P)H in erythrocytes was oxidized when the cells were exposed to ferricyanide as in the 1969 studies of Mishra and Passow.[34] This transmembrane enzyme might be the NADH ferricyanide reductase isolated by Wang and Alaupovic.[64]

Further evidence for the transmembrane dehydrogenase was established, using treatment of intact erythrocytes with diazobenzene sulfonate (DABS) to react with surface sites, which inhibits 30% of the NADH ferricyanide reductase of open vesicles prepared from the treated cells. The DABS treatment had no effect on the NADH cytochrome c reductase activity on the inside of treated cells, but it completely inhibited the cytochrome c reductase when added to open vesicles. Thus the transmembrane redox system can react with external DABS but the NADH cytochrome c reductase inside the cells is not affected.[68,3]

The demonstration of transplasma membrane electron transport using isolated, oriented, membranes was important because it obviates the possibility that the transmembrane reduction by intact cells was based on export of a reducing agent from the cells. In 1979, Orrenger and Roer[60] proposed this alternative as a basis for reduction of external ferricyanide by erythrocytes. They showed that reduction of ferricyanide by erythrocytes was greatly stimulated by dehydroascorbate. They proposed that the dehydroascorbate entered the cell and was reduced to ascorbate in the cytoplasm. The ascorbate would then diffuse out and reduce ferricyanide. Although the transmembrane electron transport enzyme provides an alternative mechanism for transmembrane ferricyanide reduction by erythrocytes, the ascorbate diffusion proposal has not been further tested with erythrocytes. Studies with liver cells later showed that addition of dehydroascorbate, although it increases the rate of ferricyanide reduction, does not produce ascorbate release into the external media.[70] The reduction of ferricyanide by porcine erythrocytes was not stimulated by dehydroascorbate.[68] The basis for dehydroascorbate stimulation remains unknown.

Secretion of phenolic reducing agents had long been considered to be the basis for reduction of ferric ions for uptake into some plant cells.[71] The transmembrane electron transport by root cells to reduce ferric iron and form a ferrous complex with bathophenanthroline disulfonate (BPS) which is impermeable was shown by Chaney et al. in 1972,[72] but excretion of reducing agents remained a possibility and is still under discussion.

A more general occurrence of transplasma membrane electron transport began to develop in 1981. Clark et al.[73] showed that ferricyanide was reduced by liver cells and by perfused liver. It was shown that ferricyanide did not permeate into the cells and that reducing agents were not excreted. In contrast to stimulatory effects that had been observed for NADH indophenol reductase activity with isolated rat liver membranes, the reduction of ferricyanide in the rat liver perfusate was inhibited by glucagon[70] and the β adrenergic agonist ritodrine.[74] Even before these studies, ferricyanide reduction by yeast cells had been studied in November 1979 in A.W. Linnane's laboratory in Melbourne, using the dual beam spectrophotometer to measure reduction directly in the cell suspension.[75] In 1981, Powis et al.[76] used indigodisulfonate as electron acceptor on the outside of AKR cells. This obviated the question of excretion of high potential redox agents such as ascorbate of phenols. The very small amount of cytochrome c reduction observed by Clark et al.[73] or Craig and Crane[77] made excretion of low potential compounds such as glutathione or superoxide unlikely. However, this may not be completely eliminated as a basis for iron reduction in some plants.[78] Higher plant cells also showed transmembrane ferricyanide reduction. With carrot cells the rate of ferricyanide reduction was increased by addition of ethanol and the increase was prevented by pyrazole, which inhibites alcohol dehydrogenase. This indicated that NADH could be the cytosolic electron donor for the transmembrane electron transport.[77] Novak and Ivankina[79,80] also postulated a transplasma membrane electron flow to explain the depolarizing effect observed when nitroblue tetrazolium was reduced by *Elodea* leaves. These studies published in 1981 indicated that the concept of transplasma membrane electron transport could have wide application in both animals and plants.

The functional significance of the transplasma membrane electron transport systems remains to be established, as well as the nature of the natural electron acceptors.

The reduction of iron for uptake by plant cells has been discussed.[9] In many plants a transmembrane reductase for ferric iron chelates is induced in root cells under iron deficiency.

This enzyme has been called the Turbo enzyme by Bienfait,[9,81] and appears to have properties which differentiate it from the transmembrane ferricyanide reductase in iron sufficient roots. For the Turbo reductase, iron chelates secreted by roots or bacteria would be the natural substrate.

In animal cells the transplasma membrane enzyme can act as a diferric transferrin reductase to reduce and release iron from transferrin as shown by Morley et al. in 1983,[82,83] Löw et al.,[84] Sun et al.,[85] and Thorstensen and Romslo in 1988.[86]

With liver and erythrocyte membranes[87,88] and soybean hypocotyl plasma membranes[89] there is evidence that semidehydroascorbate can act as electron acceptor for a plasma membrane dehydrogenase, but the significance of this activity has not been developed.

There is NADH:oxygen oxidoreductase activity in animal[58] plasma membranes. Since oxygen is permeable to membranes, the orientation of the oxidase activity and its relation to the transmembrane NADH ferricyanide reductase is unknown. In studies in M.G. Clark's laboratory in 1981, F. L. Crane[223a] observed that ferricyanide inhibits about 10% of the oxygen uptake by liver cells. Cyanide inhibits 80% of the oxygen uptake by liver cells and ferricyanide inhibits one half of the remaining activity. These observations suggest that 10% of liver cell oxygen uptake is through an electron transport chain which is cyanide insensitive and which ferricyanide can react with at the outer surface of the plasma membrane. Thus the transplasma membrane electron transport chain may use oxygen as its natural electron acceptor. Similar effects of ferricyanide are seen on the cyanide-insensitive oxygen uptake by plant cells.[90]

Studies in 1957 on the fractionation of plant cell membranes gave evidence of a non-mitochondrial cyanide-insensitive NADH oxidase in a fraction heavier than mitochondria which was later shown to contain wall material.[91,92] Clear demonstration that this NADH oxidase was associated with the cell membrane could only come when new methods of plasma membrane preparation, such as two-phase separation and curtin electrophoresis, were developed.[93,89] The nature of the plasma membrane oxidase activities are discussed more extensively in other chapters.

A possible major function for the plasma membrane electron transport became apparent with the demonstration by Ellem and Kay in 1983[14] that ferricyanide could stimulate melanoma cell growth in the absence of fetal calf serum. Prior to this, other oxidants had been shown to stimulate growth, e.g., indigosulfonates by Brooks in 1947,[13] nitroprusside by Zwiller[94] and ferric sulfate by Rudland et al.[95] The general concept that ferric transferrin was essential for cell growth was proposed by Barnes and Sato.[96] The focus of these effects was on cyclic GMP increase by nitroprusside[97] and iron uptake with iron salts and transferrin.[98] Ferricyanide on the other hand is not taken up and does not supply iron to cells[99] so its growth effect could be based on the transmembrane redox system. The ferricyanide effect on growth was soon extended to HeLa[100,101] and other cells.[102,103]

To relate the growth effect to the transmembrane electron transport, it was shown that diverse impermeable oxidants which are reduced by HeLa cells stimulate growth, and those which are not reduced do not stimulate growth.[100]

II. H^+ RELEASE RELATED TO REDOX FUNCTION

One way in which a transmembrane redox system could be important would be to induce or mediate proton transport across the membrane. Although Dormandy and Zarday[35] reported acidification of the external media associated with ferricyanide reduction by erythrocytes, this effect was not examined further for several years. Proton appearance in the external medium was observed during ferricyanide reduction by carrot cells by Craig and Crane in 1981.[77] Investigation of this effect by several laboratories showed that proton release by plant cells did not exceed ferricyanide reduction in most cases.[104] This has been explained by Guern and Ulbrich-Eberius on the basis of proton dissociation by formation of strongly acidic ferrocyanide.[105] The measurment of ferricyanide reduction electrometrically by constant reoxidation of the ferrocyanide formed is still associated with acidification of the media so dissociation of ferrocyanide protons is not the basis for acid formation.[108]

Ferricyanide reduction by animal cells produced proton release into the media which was greater than the ferricyanide reduction,[106] which indicated activation of proton transport by the redox system.[107] When diferric transferrin was used as electron acceptor the proton release was as much as 100 × the rate of electron transfer. This high rate of proton release has been shown to be mediated by the Na/H antiport which is activated by the redox system.[107-109] Although recent evidence indicates that the Na/H antiport activity is not in itself sufficient to explain growth stimulation by the redox system,[110] the change in cytosolic pH mediated by the antiport gave the first approach to a mechanism for redox control. Growth stimulation by external oxidants could therefore be based on activation of the antiport.

In plants elongation growth of postmeristematic cells has been associated with hormone-induced proton release to acidify the periplasmic space within the cell wall.[111] If the redox system is hormone activated and activates proton release by plant cells, it could be a basis for hormone control of proton release leading to subsequent cell elongation. In 1977, Polevoy and Salamatova[112] showed that ferricyanide inhibited auxin-induced oxygen uptake by roots which suggested that auxin activated a plasma membrane electron transport which could be the basis for activation of proton release. Further study of the relation between oxygen uptake and auxin action was developed by Böttger et al.[113,114] They showed that auxin-induced proton release was related to an oxidase with a low affinity for oxygen. Identification of a hormone-stimulated NADH oxidase in isolated plasma membrane from soybean hypocotyl by Morré et al.[89] was consistent with this view. These studies are discussed further in Chapter 2 of this book.

III. BLUE LIGHT-ACTIVATED REDUCTIONS

Short wavelength light (blue) has long been known to have morphogenic effects on plants and to control stomatal cells.[115,116] In 1974, Poff and Butter[117] identified spectral absorbance changes in response to blue light in fungi and other plant tissues which would be consistent with changes in the redox state of flavins and b type cytochromes. The spectrum of the redox response resembles the action spectrum for morphogenic response by plants, which suggests that the redox change might be a basis for the plant response.

The light-induced flavin and cytochrome reduction can also be observed in membrane preparations from cells from light-sensitive tissues.[119,120] Evidence for location of the active components in the plasma membrane was reported by Leong et al.[121,122] but the location remained doubtful until highly purified plasma membranes prepared by phase separation were available.[123] These plasma membrane preparations show blue light absorbance changes under anaerobic conditions which indicate the reduction of a high potential b_{560} cytochrome.[124] There is also evidence that cytochrome P450 is present in these membranes.[125]

IV. REDOX COMPONENTS IN THE PLASMA MEMBRANE

As early as 1967 Vassiletz et al.[126] proposed that rat liver plasma membrane contained iron and copper in addition to cytochromes. Unfortunately, the purity of the preparation was not completely defined. In 1973, Masuda et al.[127,128] also reported the presence of cytochrome b in carefully purified rat liver membranes. In subsequent studies, Jarasch et al.[49] and Bruder et al.[129] clearly demonstrated the presence of b type cytochromes in erythrocyte and rat liver plasma membranes. Bruder et al.[130] also found cytochrome b in bovine milk fat globule membranes which derive from plasma membranes.

In these studies of plasma membrane cytochromes, it is apparent that plasma membrane preparations have significant amounts of P450 in addition to cytochrome b_5. Even more unusual is the high content of P420 relative to P450 found in plasma membranes.[2] The presence of P420 in erythrocyte membranes has been indicated since 1971[136] and has been carefully documented for rat and chicken erythrocyte membranes.[129]

Since erythrocyte membranes have no NADH oxidase activity, in contrast to other plasma membranes, the cytochrome b_5 cannot function as an electron carrier to the fatty acyl coenzyme A desaturase. No other apparent function for the cytochrome b_5, P450, or P420 in plasma membranes has been determined.

Only limited studies have been made of the transition metal composition of plasma membranes since the initial studies by Vassiletz. In 1982, MacKellar and Crane[131] reported significant iron and traces of copper in pig erythrocyte plasma membranes and mouse liver plasma membranes far above the heme content. Iron has also been found in kidney brush border and placenta microvilli membranes.[132,133] Unfortunately there has been no attempt to determine if this iron is of the labile iron sulfur type or in some other form.[134] Characteristic iron sulfur ESR signals have not been found in plasma membranes, but Dreyer and Treichler[135] have reported that iron ESR signals appear when synaptosomal plasma membranes are treated with detergent. Information on the flavin content of plasma membranes is a little more extensive than information about iron. Miner et al.[137] made an extensive study on the flavin content of erythrocyte plasma membranes from several species and found flavin in all membranes with considerable difference between species. Ferricyanide reducatase activity correlates quite well with flavin content except for dog erythrocytes. Flavin has also been reported in kidney brush border[132] and in several plant plasma membranes including yeast and oat coleoptile,[138] corn coleoptile,[120] and soybean hypocotyl.[139] Flavin has also been detected in rat liver plasma membranes by fluorescence measurements.[140]

A remarkably high content of thiol, over 80 nmol per milligram of protein was found in erythrocyte membranes[141,142] but better purification procedures were needed before being examined in other plasma membranes.

Coenzyme Q has also been found in rat liver plasma membranes at concentrations which preclude contamination by other membranes.[143] The presence of a lipophilic quinone would provide a mechanism for transplasma membrane electron and proton transfer if the dehydrogenase can act as a coenzyme Q reductase since the hydroquinone can carry electrons across the lipid bilayer.

V. SELECTIVE INHIBITION OF PLASMA MEMBRANE ELECTRON TRANSPORT

Part of the definition of NADH oxidoreductases was based on selective inhibition as with rotenone inhibition of the NADH dehydrogenase in the mitochondrial inner membrane. An early aid in the definition of a unique electron transport system in the plasma membrane was inhibition by quinacrine (Atebrin). In 1976, Crane and Löw[55] showed that liver plasma membrane NADH indophenol reductase was inhibited 44% by 3 m*M* Atebrin and 63% by 1 μ*M* triiodothyronine T_3, whereas the endoplasmic reticulum activity was insensitive to Atebrin and only 24% inhibited by T_3. From later studies, it is apparent that these membranes had very little contamination by mitochondrial or endoplasmic membranes.[41] The NADH cytochrome c reductase activity of liver plasma membranes was also inhibited 50% by 0.1 *M* azide in contrast to no inhibition in endoplasmic reticulum. Atebrin and T_3 also inhibited the plasma membrane cytochrome c reductase in contrast to no inhibition with endoplasmic reticulum. In 1976, Gayda et al.[58] reported 50% Atebrin inhibition of NADH:oxygen oxidoreductase in rat liver plasma membranes. The same concentration did not inhibit in smooth microsomal membranes and stimulated 50 and 100% with mitochrondria and Golgi membranes. This activity was a cyanide insensitive NADH oxidase. T_3 and β adrenergic catecholamines also specifically stimulated the plasma membrane NADH oxidase with much less or no effect on NADH oxidase in mitochondria, endoplasmic reticulum, or Golgi. The contrasting effects of T_3 and azide on NADH oxidase as compared to NADH indophenol or cytochrome c reductase in the plasma membrane indicated that the redox system is a multicomponent system and not a single simple dehydrogenase.[4] There could be two separate enzymes, one an oxidase and one a reductase. More likely the stimulation of electron flow to oxygen at one site prevented reduction of the other acceptors at another site. Detailed studies of combined activity were never carried out and only recently has the stimulation of dehydrogenase activity under argon been demonstrated.[144] The relationship between specific T_3 binding sites on the plasma membrane and specific T_3 effects on the redox system has not been determined.

Although Atebrin was not an ideal inhibitor because it was active only at m*M* concentrations and the strong yellow color interfered with spectrophotometric or fluorescent assays, it was used in several studies to help define plasma membrane redox function[145,146,129,3] and for a time was the best evidence for a unique electron transport system in the plasma membrane in contrast to endoplasmic reticulum.

A more significant inhibition was discovered during studies of Adriamycin (doxorubicin) inhibition of mitochondrial electron transport. In 1980 it was found that the Adriamycin was a much more effective inhibitor of plasma membrane electron transport compared to mitochondrial electron transport.[147] These studies were complicated by the stimulation of NADH oxidase by Adriamycin, even though ferricyanide reduction showed over 50% inhibition at 10^{-4} *M*. More clear-cut inhibition of NADH dehdrogenase in liver plasma membranes with adriamycin and related anthracycline antibiotics was obtained by preincubation of the drugs with the membrane for at least 3 min before starting the assay. This gave very selective inhibition of rat liver plasma membrane NADH ferricyanide with very low concentrations (10^{-7} *M*) of Adriamycin. The Adriamycin inhibition was also clearly revealed with reduction of ferricyanide by HeLa cells.[148-150] Other antitumor drugs such as actinomycin D,[151] *cis*-diaminodichloroplatinum II,[152] and bleomycin[153] were also found to be effective inhibitors of the transplasma membrane electron transport (Table 3).

Chloroquine, which is even more effective that Atebrin in the treatment of malaria, was also found in 1985 to be an inhibitor of the plasma membrane NADH indophenol reductase.[5] Later it was found to inhibit ferricyanide and diferric transferrin reduction.[154] Chloroquine has long been recognized to inhibit iron uptake and prevent acidification of endosomes or lysosomes.[155]

VI. SEMIDEHYDROASCORBATE REDUCTASE

Semidehydroascorbate reductase activity is well known in endomembranes.[156,157] In 1980, Goldbenberg[158] showed that NADH semidehydroascorbate reductase was present in plasma membranes. It is also present in erythrocyte membranes.[159] Detergents stimulate the activity especially with inside-out membrane vesicles, which indicates that the activity is not entirely dependent on the NADH cytochrome b_5 reductase since the cytochrome b_5 system is exclusively on the inner surface of the membranes. Inhibition by thenoyltrifluoroacetone and BPS also indicates a difference from the cytochrome b_5 system.

A cyanide-insensitive ascorbate oxidase is also present in liver plasma membranes,[147] setting up the possibility that the semidehydroascorbate-ascorbate cycle can act as a link from the dehydrogenase to oxygen.

An external site for the semidehydroascorbate reduction is also indicated by the inhibitory effects of lectins such as *Limax flavus* agglutinin and wheat germ agglutinin.[160]

VII. EXTERNAL REDOX FUNCTION

The oxidation of NADH added to the outside of plant cells has been known for many years.[161,162] A large part of this activity especially in older plant cells is based on oxidation of NADH by peroxidase with the formation of H_2O_2.[163] This activity has been proposed to contribute to lignin formation.[164-166] In 1982, Lin[167] described NADH oxidation by an enzyme on the outside of plant protoplasts which he proposed could control ion transport. Similar function was reported by Mishra et al.[168] Other studies of external NADH oxidation by plant root segments gave the opposite effects to those of Lin in that potassium efflux was increased rather than the uptake.[169,170] The relation of this external NADH oxidase to peroxidase is further amplified by its inhibition with cyanide and catalase and its stimulation by manganese ions and monophenols.[18,171,172]

NADH dehydrogenase using ferricyanide as electron acceptor was found in 1981 on the outer surface of animal cells. On lung tumor cells the activity was not inhibited by superoxide dismutase, cyanide, or removal of oxygen, and cytochrome c reductase was very slow, so it does not depend on formation of superoxide by an NADH oxidase which can then reduce ferricyanide.[173] The same type of activity was found on the outer surface of erythrocyte membrane vesicles. There is no oxidase in erythrocyte membranes, so this activity must be based on an NADH dehydrogenase.[174] There has been little further study of external dehydrogenase activity on animal cells.

With plant cells Morré et al. in 1986 described an external dehydrogenase, other than peroxidase, linked to oxygen and showed that this activity is growth hormone stimulated.[89] The external dehydrogenase may be linked to the same redox chain as the transmembrane dehydrogenase. This oxidase is not inhibited by cyanide so it cannot be peroxidase.[93]

Oxidation of externally added NADH by platelets[175] to generate H_2O_2 has been reported. Cytochemical evidence for NADH-dependent H_2O_2 formation by cerium perhydroxide on the surface of rat small intestine brush border membranes has been described.[176]

Also, Zemkova et al.[177] and Ramasarma showed that NADH oxidation in rat liver plasma membrane produces H_2O_2.[178]

Although H_2O_2 has often been shown to modify functions in cells,[179,180] there has been very little interest in the peroxide-generating oxidase except in plant cells and in bacteriocidal action of leucocytes.

VIII. ANTIBACTERIAL REDOX FUNCTIONS

The respiratory burst which accompanies engulfment of bacteria by neutrophils has been known since 1973 to be associated with the plasma membrane. This increased oxygen uptake leads to formation of peroxide and possibly chloroperoxides which are necessary to kill the phagocytosed bacteria in the vesicles. Early studies of this activity emphasized the action of myeloperoxidase in the generation of peroxide.[181,182] In 1977, Bellavite et al. showed that chick leucocytes which lack peroxidase could generate superoxide.[184] In 1973, Briggs et al.[183] proposed that the peroxide generation was based on the activity of an NADH oxidase in the neutrophil plasma membrane that was localized by cytochemical technique. The recognition of chronic granulomatous disease as a defect in the superoxide generating enzyme provided an important approach to the identification of the enzyme as a plasma membrane oxidase.[185] In 1979, Dewald et al.[186] showed high levels of NADPH oxidase in isolated plasma membranes from activated leucocytes and identified it as the superoxide and peroxide generating enzyme. In 1980, Green et al.[187] proposed that the NADPH oxidase is a transmembrane enzyme which oxidizes internal NADPH to form external (or intraendosomal) superoxide. Membrane potential changes associated with superoxide formation were also recognized at this time (1980). These membrane potential changes were subsequently identified with transmembrane electron flow during NADPH oxidation.[188] In 1981, Cross et al.[189] recognized an unusual low potential cytochrome b_{245} associated with the oxidase enzyme. The presence of a cytochrome b in the leucocyte plasma membrane had been recognized previously.

In 1979, Segal and Jones,[190] Millard et al.,[191] and Borregaard et al.[192] had proposed a three-component redox system including flavoprotein and cytochrome b components. Each component could be nonfunctional and would thus account for more than one type of granulamatous disease.[193]

Purification of NADPH oxidase and dehydrogenases from leucocyte membranes has been described but the exact role of these preparations in the membrane bound NADPH oxidase remains to be established.[194-201] The relation between the NADH dehydrogenase described early in the leucocyte membrane research by Segal and Peters[185] and the NADPH oxidase has not been clarified. It is possible that the NADH dehydrogenase is part of the transplasma membrane redox system found in all cell types. This enzyme would be present even in nonactivated cells[202,203] and could account for the high level of transplasma membrane ferricyanide reduction relative to superoxide dependent cytochrome c reduction observed in nonactivated neutrophils. After activation the cytochrome c reduction is greatly increased and it is inhibited by superoxide dismutase.

The generation of superoxide in plant plasma membranes could serve a similar defensive strategy against invading bacteria. The peroxide-generating peroxidase system has been described in a previous section of this chapter, but it is only recently that peroxide produced in this way has been considered for a role in defense against invasion.[204]

IX. XANTHINE OXIDASE

Xanthine oxidase was recognized to be a membrane-bound enzyme in milk fat globules by Morgon in 1954:[205] the identification of milk fat globule membranes as being derived from apical plasma membranes of milk secreting cells was not accomplished until many years later. In 1970, Keenan et al.[206] showed that the milk fat globule membrane derived from plasma membrane. Jarasch et al.[207] compared the activity in endoplasmic reticulum and plasma membrane to further show the association with plasma membrane. This activity is tightly bound to the membrane[208] and is not extracted by salt and is exposed on the cytoplasmic side of the membrane.[209]

Xanthine oxidase is often found as a soluble enzyme in cellular homogenates and it is not clear why some of it is associated with plasma membranes.[210] It has been suggested that it may be important for superoxide production by macrophages.[211] This suggests that xanthine oxidase may have a defensive role against invading organisms similar to the peroxide producing NADPH oxidase in neutrophil plasma membranes.

X. GASTRIC ACID SECRETION

The idea that proton and chloride ion transport could be based on cytochrome oxidation and reduction was considered for some time in connection with acid secretion in the stomach. These studies have been discussed by Robertson,[15] and studies continued in this area through the 1960s.[219] The concept of ATP-driven proton pumps has very much supplanted the redox-driven activity in recent years.[220,221] There is still some interest in the possible contribution of transport driven by a nonmitochondrial redox system.[222] Further definition of this problem may come from isolation of gastric mucosal plasma membranes and definition of any redox system they may contain.

XI. SUMMARY

The study of plasma membrane oxidoreductase functions and the enzyme systems involved in these functions has a relatively short history compared to mitochondrial or even endoplasmic reticulum oxidations. In this short time a diverse group of oxidoreductase systems have been revealed which either have important well-defined functions or which have the potential to contribute to major systems of cellular control and dynamics. Iron uptake by cells is a vital function which involves movement of only catalytic amounts of iron into the cell. It is now clear that a relatively low rate of electron transport in the plasma membrane is enough to assist in sufficient iron transport to satisfy cellular needs. The activity of these redox systems may be even more restricted when iron uptake is not needed.

The peroxidase-generating system is an example of a redox system which was not obviously related to a plasma membrane function at first, but was looked for in the phagocytic vacuoles. Histochemical studies were necessary to draw attention to the plasma membrane.

New interest in the possible linkage of electron transport and transport across the plasma membrane has developed with more evidence for oxidoreductase enzymes in the plasma membrane. A relation of amino acid transport to plasma membrane electron transport was examined very soon after the plasma membrane redox system was shown to have unique characteristics. There is also a continuing interest in the application to gastric acid secretion.

Other aspects of plasma membrane electron transport and oxidation reduction functions are in the process of development and the final chapters of their history remain to be written. They have potential for major significance in growth and developmental control.

REFERENCES

1. **Löw, H. and Crane, F. L.,** Redox functions in plasma membrane, *Biochim. Biophys. Acta,* 515, 141, 1978.
2. **Goldenberg, H.,** Plasma membrane redox activities, *Biochim. Biophys. Acta,* 694, 203, 1982.
3. **Crane, F. L., Löw H., and Clark, M. G.,** Plasma membrane redox enzymes in *The Enzyme of Biological Membranes,* Martonosi, A. N., Ed., Plenum Press, New York, 1985, 465.
4. **Crane, F. L., Goldenberg, H., Morré, D. J., and Löw, H.,** Dehydrogenases of the plasma membrane in *Subcellular Biochemistry,* Vol. 6, Roodyn, D. B., Eds., Plenum Press, New York, 1979, 345.
5. **Crane, F. L., Sun, I. L., Clark, M. G., Grebing, C., and Löw, H.,** Transplasma-membrane redox systems in growth and development, *Biochim. Biophys. Acta,* 811, 233, 1985.
6. **Lüttge, V. and Clarkson, D. T.,** Mineral nutrition: plasmalemma and tonoplast redox activities, *Prog. Bat.,* 47, 73, 1985.
7. **Møller, I. M. and Lin, W.,** Membrane-bound dehydrogenases in higher plant cells, *Annu. Rev. Plant. Physiol.,* 37, 309, 1985.
8. **Ramirez, J. A., Ed.** , *Redox Function in the Eukaryotic Plasma Membrane,* Consijo Superior Investigaciones Cientificas Publishing Office, Madrid, 1987, 286.
9. **Beinfait, H. F.,** Regulated redox processes at the plasmalemma of plant root cells and their function in iron uptake, *J. Bioenerg. Biomemb.,* 17, 73, 1985.
10. **Kant, J. A. and Steck, T. L.,** Cation impermeable inside-out and right-side out vesicles from human erythrocyte membranes, *Nature,* 240, 26, 1972.
11. **Choury, D., Leroux, A., and Kaplan, J.-C.,** Membrane-bound cytochrome b_5 reductase (methemoglobin reductase) in human erythrocytes: study in normal and methemoglobinemic subjects, *J. Clin. Invest.,* 67, 149, 1981.
12. **Choury, D., Wajeman, H., Boissel, J. P., and Kaplan, J.-C.,** Evidence for endogeneous proteolytic solubilization of human red cell membrane NADH cytochrome b_5 reductase, *FEBS Lett.,* 126, 172, 1981.
13. **Brooks, M. M.,** Activation of eggs by oxidation-reduction indicators, *Science,* 106, 320, 1947.
14. **Ellem, K. A. O. and Kay, G. F.,** Ferricyanide can replace pyruvate to stimulate growth and attachment of serum restricted melanoma cells, *Biochem. Biophys. Res. Commun.,* 112, 183, 1983.
15. **Robertson, R. N.,** *Protons, Electrons, Phosphorylation and Active Transport,* Cambridge University Press, London, 1968.

16. **Briggs, W. R.,** A blue light photoreceptor system in higher plants and fungi, in *Photoreceptors and Plant Development,* Ed., J. De Greef, Antwerpen University Press, Antwerp, 1980, 17.
17. **Prince, R. C. and Gunison, D. E.,** Superoxide production by neutrophils, *Trends Biochem. Sci.,* 12, 86, 1987.
18. **Askerlund, P., Larsson, C., Widell, S., and Møller, I. M.,** NAD(P)H oxidase and peroxidase activities in purified plasma membranes from cauliflower inflorescences, *Physiol. Plant,* 71, 9, 1987.
19. **Voegtlin, C., Johnson, J. M., and Dyer, H. A., Jr.,** Quantitative estimation of the reducing power of normal and cancer tissue, *J. Pharmacol. Exp. Ther.,* 24, 305, 1925.
20. **Warburg, O.,** On the origin of cancer cells, *Science,* 123, 309, 1956.
21. **Barron, E. S. G. and Hoffman, L. A.,** The catalytic effect of dyes on the oxygen consumption of living cells, *J. Gen. Physiol.,* 13, 483, 1929.
22. **Keilin, D.,** *The History of Cell Respiration and Cytochromes,* Cambridge University Press, London, 1966.
23. **Warburg, O.,** Über sauerstoffatmende körnchenous Lebersellen und über Sauerstoffatmung in Berkfeld-Filtraten wässrige Leberextracte, *Arch. Gesch. Physiol.,* 154, 599, 1913.
24. **Lipmann, F.,** Metabolic generation and utilization of phosphate bond energy, *Adv. Enzymol.,* 1, 99, 1941.
25. **Krebs, H. A. and Kornberg, H. L.,** *Energy Transplantation in Living Matter,* Springer-Verlag, Berlin, 1957.
26. **Green, D. E.,** Studies in organized enzyme systems, *Harvey Lect. Ser.,* 52, 177, 1958.
27. **Lehninger, A. L.,** *The Mitochondria: Molecular Basis of Structure and Function,* W. A. Benjamin, New York, 1964.
28. **DePierre, J. W. and Ernster, L.,** Enzyme topology of intracellular membranes, *Annu. Rev. Biochem.,* 46, 201, 1977.
29. **Lundegårdh, H.,** Absorption, transport and exudation of inorganic ions by roots, *Arkiv. Bot.,* 32A, 1, 1945.
30. **Smith, L. and Chance, B.,** Cytochrome in plants, *Annu. Rev. Plant Physiol.,* 9, 449, 1958.
31. **Conway, E. J. and Brady, T. C.,** Source of the hydrogen ions in gastric mucosa, *Nature,* 162, 456, 1948.
32. **Davis, R. E.,** *Metabolic Aspects of Transport Across Cell Membranes,* University of Wisconsin Press, Madison, 1957.
33. **Manyai, S. and Szekely, M.,** Die Wirkung von Natriumfluorid und Monojodessigsäure auf die Glycolyse von menschlichten raten Blutkörpechen, *Acta Physiol. Acad. Sci. Hung.,* 5, 7, 1954.
34. **Mishra, R. K. and Passow, H.,** Induction of intracellular ATP synthesis by extracellular ferricyanide in human blood cells, *J. Memb. Biol.,* 1, 24, 1969.
35. **Dormandy, T. L. and Zarday, Z.,** The mechanism of insulin action: immediate electrophysical effects of insulin on red cells systems, *J. Physiol.,* 180, 684, 1965.
36. **Mitchell, P.,** Chemiosomotic coupling in oxidative and photosynthetic phosphorylation, *Biol. Rev.,* 41, 445, 1965.
37. **Zamudio, I. and Canessa, M.,** Nicotinamide-adenine dinucleotide dehydrogenase activity of human erythrocyte membranes, *Biochim. Biophys. Acta,* 120, 165, 1966.
38. **Zamudio, I., Cellino, M., and Canessa-Fischer, M.,** The relation between membrane structure and the NADH (acceptor oxidoreductase activity of erythrocyte ghosts), *Arch. Biochem. Biophys.,* 129, 336, 1969.
39. **Wallach, D. F. H. and Kamat, V. B.,** Preparation of plasma membrane fragments from mouse ascites tumor cells, in *Methods in Enzymology,* Vol. 8, E. F. Neufeld and V. Ginsberg, Eds., Academic Press, New York, 1966, 164.
40. **Emmelot, P., Bos, C. J., Beneditti, E. L., and Rumke, P.,** Studies on plasma membranes. I. Chemical composition and enzyme content of plasma membranes isolated from rat liver, *Biochim. Biophys. Acta,* 90, 126, 1964.
41. **Goldenberg, H., Crane, F. L., and Morré, D. J.,** NADH oxido-reductase of mouse liver plasma membranes, *J. Biol. Chem.,* 254, 2491, 1979.
42. **Morré, D. J., Franke, W. W., Deumling, B., Nyquist, S. E., and Ovtracht, L.,** Golgi apparatus function in membrane flow and differentiation: origin of plasma membrane from endoplasmic reticulum, in *Biomembranes,* Vol. 2, L. A. Manson, Ed., Plenum Press, New York, 1971, 95.
43. **Thines-Sempoux, D., Amar-Costesec, A., Beaufay, H., and Berthet, J.,** The association of cholesterol, 5′ nucleosidase, and alkaline phosphatase I with a distinct group of microsomal particles, *J. Cell. Biol.,* 43, 189, 1969.
44. **Fleischer, B. and Fleischer, S.,** Preparation and characterization of golgi membranes from rat liver, *Biochim. Biophys. Acta,* 219, 301, 1970.
45. **Goldenberg, H.,** Density gradient fractionation of digitonin treated rat liver plasma membranes and localization of NADH oxidoreductase and B type cytochromes, *Enzyme,* 27, 227, 1982.
46. **Wibo, M., Thines-Sempoux, D., Amar-Costesec, A., Beanfay, H., and Godelaine, D.,** Analytical study of microsomes and isolated subcellular membranes from rat liver. VIII, *J. Cell Biol.,* 89, 456, 1981.

47. **Fleischer, S., Fleischer, B., Azzi, A., and Chance, B.,** Cytochrome b_5 and P-450 in liver cell fractions, *Biochim. Biophys. Acta,* 225, 194, 1971.
48. **Remacle, J., Fowler, S., Beaufay, H., Amar-Costesec, A., and Berthet, J.,** Analytical study of microsomes and isolated subcellular membranes from rat liver VI, *J. Cell Biol.,* 71, 552, 1976.
49. **Jarsch, E.-D., Kartenbeck, J., Bruder, G., Fink, A., Morré, D. J., and Franke, W. W.,** B-type cytochromes in plasma membranes isolated from rat liver in comparison with those of endomenbranes, *J. Cell Biol.,* 80, 37, 1979.
50. **Kitajima, S., Yasukochi, Y., and Minakami, S.,** Purification and properties of human erythrocyte membrane NADH cytochrome b_5 reductase, *Arch. Biochem. Biophys.,* 210, 330, 1981.
51. **Passon, P. G. and Hultquist, D. E.,** Soluble cytochrome b_5 reductase from human erythrocytes, *Biochim. Biophys. Acta,* 275, 62, 1972.
52. **Leroux, A., Torlinski, L., and Kaplan, J. C.,** Soluble and microsomal forms of NADH cytochrome b_5 reductase from human placenta: similarity with NADH methemoglobin reductase from human erythrocytes, *Biochim. Biophys. Acta,* 481, 50, 1977.
53. **Morré, D. J., Vigil, E. L., Frantz, C., Goldenberg, H., and Crane, F. L.,** Cytochemical demonstration of glutaraldehyde resistant NADH ferricyanide oxidoreductase activities in rat liver plasma membranes and Golgi apparatus, *Cytobiologie Z. Exp. Zellforsch.,* 18, 213, 1978.
54. **Löw, H. and Werner, S.,** Effects of reducing and oxidizing agents on the adenylate cyclase activity in adipocyte plasma membranes, *FEBS Lett.,* 65, 96, 1976.
55. **Crane, F. L. and Löw, H.,** NADH oxidation in liver and fat cell plasma membranes, *FEBS Lett.,* 68, 153, 1976.
56. **Löw, H. and Crane, F. L.,** Hormone regulated redox functions in plasma membranes, *FEBS Lett.,* 68, 157, 1976.
57. **Löw, H., Crane, F. L., Grebing, C., Tally, M., and Hall, K.,** Hormone control of plasma membrane oxidation-reduction reactions: effects of insulin, *FEBS Lett.,* 91, 166, 1978.
58. **Gayda, D. P., Crane, F. L., Morré, D. J., and Löw, H.,** Hormone effects on NADH oxidizing enzymes of plasma membranes of rat liver, *Proc. Indiana Acad. Sci.,* 86, 385, 1977.
59. **Goldenberg, H., Crane, F. L., and Morré, D. J.,** Influence of hormones on the NADH dehydrogenase in mouse liver plasma membrane, *Biochem. Biophys. Res. Commun.,* 83, 234, 1978.
60. **Mukherjee, S. P. and Lynn, W. S.,** Reduced nicotinamide adenine dinucleotide phosphate oxidase in adipocyte plasma membrane and its activation by insulin, *Arch. Biochem. Biophys.,* 184, 68, 1977.
61. **Mukherjee, S. P. and Mukherjee, C.,** Role of sulfhydryl oxidation in adipocyte plasma membrane surface in the response of adenylate cyclase to isoproteranol and glucagon, *Biochim. Biophys. Acta,* 677, 339, 1981.
62. **Garcia-Sancho, J., Sanchez, A., Handlogton, M. E., and Christensen, H. N.,** Unexpected additional mode of energization of amino acid transport into Ehrlich cells, *Proc. Nat. Acad. Sci. U.S.A.,* 74, 1488, 1977.
63. **Huang, C. M., Goldenberg, H., Frantz, C., Morré, D. J., Keenan, T. W., and Crane, F. L.,** Comparison of NADH link cytochrome c reductase of endoplasmic reticulum, Golgi apparatus and plasma membrane, *Int. J. Biochem.,* 10, 723, 1979.
64. **Wang, C-S. and Alaupovic, P.,** Isolation and partial characterization of human erythrocyte membrane NADH: (acceptor) oxidoreductase, *J. Supramol. Struct.,* 9, 1, 1978.
65. **Ozales, J., Carr, S. A., and Strittmatter, P.,** Identification of the N-terminal blocking group of NADH cytochrome b_5 reductase as myristic acid and the amino acid sequence of the membrane binding domain, *J. Biol. Chem.,* 259, 13349, 1984.
66. **Löw, H., Crane, F. L., Grebing, C., Hall, K., and Tally, M.,** Metabolic milieu and insulin action, in *Diabetes 1979,* Waldhausl, W. K., Ed., Excerpta Medica, Amsterdam, 1979, 209.
67. **Arese, P., Bosia, A., and Pescamona, G. P.,** Effects of external oxidant on redox state of glutathione and pyridine nucleotide, in *6th Int. Symp. Structur und Function der Erythrocyten,* Rapport, S. and Jung, F., Eds., Akademic-Verlag, Berlin, 1972, 91.
68. **MacKellar, W. C.,** A Transmembrane NADH Dehydrogenase in Porcine Erythrocyte Membranes, Ph.D. thesis, Purdue University, W. Lafayette, IN, 1981.
69. **Orringer, E. P. and Roer, M. E. S.,** An ascorbate mediated transmembrane-reducing system of the human erythrocyte, *J. Clin. Invest.,* 63, 53, 1979.
70. **Clark, M. G., Partick, E., and Crane, F. L.,** Properties and regulation of a transplasma membrane redox system in rat liver, *Biochem. J.,* 204, 795, 1982.
71. **Romheld, V. and Marschner, H.,** Mobilization of iron in the rhizosphere of different plant species, *Adv. Plant Nutr.,* 2, 155, 1986.
72. **Chaney, R. L., Brown, J. C., and Tiffin, L. O.,** Obiligatory reduction of ferric chelates in iron uptake by soybeans, *Plant Physiol.,* 50, 208, 1972.
73. **Clark, M. G., Partick, E. J., Patten, G. S., Crane, F. L., Löw, H., and Grebing, C.,** Evidence for the extracellular reduction of ferricyanide by rat liver, *Biochem. J.,* 200, 565, 1981.

74. **Clark, M. G., Partick, E., and Crane, F. L.,** Effects of catecholamines on the transplasma membrane redox system in rat liver, *Biochem. Int.*, 5, 711, 1982.
75. **Crane, F. L., Roberts, H., Linnane, A. W., and Löw, H.,** Transmembrane ferricyanide reduction by cells of the yeast *Saccharomyces cerevisiae, J. Bioenerg. Biomemb.*, 14, 191, 1982.
76. **Powis, G., Svingen, B. A., and Appel, P.,** Quinone stimulated superoxide formation by subcellular fractions isolated from hepatocyes and other cells, *Mol. Pharmacol.*, 20, 387, 1981.
77. **Craig, T. and Crane, F. L.,** Evidence for a transplasma membrane electron transport system in plant cells, *Proc. Indiana Acad. Sci.*, 90, 150, 1981.
78. **Cakmak, I., Van de Wetering, D. A. M., Marschner, H., and Bienfait, H. F.,** Involvement of the superoxide radical in the extracellular ferric reduction by iron-deficient bean roots, *Plant Physiol.*, 85, 310, 1987.
79. **Ivankina, N. G. and Novak, V. A.,** Localization of redox reactions in the plasmalemma of *Elodea* leaf cells, *Studia Biophys.*, 83, 197, 1981.
80. **Novak, V. A. and Ivankina, N. G.,** Influence of nitro-blue tetrazolium on membrane potential and ion transport in *Elodea, Fysiol. Rasteni*, 30, 1107, 1983.
81. **Bienfait, H. F.,** The turbo reductase in plant plasma membranes, in *Plasma Membrane Oxidoreductase in Control of Animal and Plant Growth,* Crane, F. L., Morré, D. J., and Löw, H., Eds., Plenum Press, New York, 1988, 89.
82. **Morley, C. G. D., Delmastro, D., and Bezkorovainy, A.,** The behavior of transferrin receptors on rat hepatocytes, in *Structure and Function of Iron Storage and Transport Proteins,* Urushizaki, I., Aisen, P., Listowsky, I., and Drysdale, J. W., Eds., Elsevier, Amsterdam, 1983, 356.
83. **Morley, C. G. D., Solberg, L., and Bezkorovainy, A.,** The removal of iron from transferrin by isolated plasma membranes of rat hepatocytes, in *Proteins of Iron Storage and Transport,* Spiky, G., Monteuil, J., Chrichton, R. R., and Mazwier, J., Eds., Elsevier, Amsterdam, 1985, 175.
84. **Löw, H., Sun, I. L., Navas, P., Grebing, C., Crane, F. L., and Morré, D. J.,** Transplasmalemma electron transport in cells is part of the diferric transferrin reductase system, *Biochem. Biophys. Res. Commun.*, 139, 1117, 1986.
85. **Sun, I. L., Navas, P., Crane, F. L., Morré, D. J., and Löw, H.,** NADH diferric transferrin reductase activity in liver plasma membranes, *J. Biol. Chem.*, 262, 15915, 1987.
86. **Thorstensen, K. and Romslo, I.,** Uptake of iron from transferrin by isolated rat hepatocytes, *J. Biol. Chem.*, 263, 8844, 1988.
87. **Goldenberg, H., Grebing, C., and Löw, H.,** NADH monodehydroascorbate reductase in human erythrocyte membranes, *Biochem. Int.*, 6, 1, 1983.
88. **Goldenberg, H.,** Insulin inhibits NADH-semidehydro-ascorbate reductase in rat liver plasma membrane, *Biochem. Biophys. Res. Commun.*, 94, 721, 1980.
89. **Morré, D. J., Navas, P., Penel, C., and Castillo, F. J.,** Auxin stimulated NADH oxidase (semidehydroascorbate reductase) of soybean plasma membrane: role in acidification of cytoplasm?, *Protoplasma,* 133, 195, 1986.
90. **Crane, F. L., Barr, R., Morré, D. J., Brightman, A., Craig, T. A., and Jolivet, A.,** Plasmalemma redox systems in control of growth, in *Current Topics in Plant Biochemistry and Physiology,* Vol. 7, Randall, D. D., Blevins, D. G., and Larson, R. L., Eds., University of Missouri Press, Columbia, MO, 1988, 224.
91. **Crane, F. L.,** Electron transport and cytochromes of subcellular particles from cauliflower buds, *Plant Physiol.*, 32, 619, 1957.
92. **Forti, G., Tua, C., and Tognali, L.,** Fractionation of oxidative particles of pea stem, *Biochim. Biophys. Acta,* 36, 19, 1959.
93. **Barr, R., Sandelius, A. S., Crane, F. L., and Morré, D. J.,** Oxidation of reduced pyridine nucleotides by plasma membranes of soybean hypocotyl, *Biochem. Biophys. Res. Commun.*, 131, 943, 1985.
94. **Zwiller, J., Basset, J., Ulrich, G., and Mandel, P.,** Guanylate cyclase activators and sodium nitroprusside stimulate cell free growth in serum free medium, *Exp. Cell. Res.*, 141, 445, 1982.
95. **Rudland, P. S., Durbin, H., Clingan, D., and Jimenez De Asua, L.,** Iron salts and transferrin are specifically required for division of 3T6 cells, *Biochem. Biophys. Res. Commun.*, 75, 556, 1977.
96. **Barnes, and Sato, G.,** Serum free cell culture, a unifying approach, *Cell,* 22, 649, 1980.
97. **Craven, P. A. and De Rubertis, F. R.,** Requirement for heme in the activation of guanylate cyclase nitric oxide, *Biochim. Biophys. Acta,* 745, 310, 1983.
98. **Ponka, P., Grady, R. W., Wilczynska, B., and Schulman, H. M.,** Iron chelates supply iron for cell growth, *Biochim. Biophys. Acta,* 802, 477, 1984.
99. **Basset, P., Quesneau, Y., and Zwiller, J.,** Iron-induced cell growth: evidence of a transferrin-independent iron transport, *Cancer Res.*, 46, 1644, 1986.
100. **Sun, I. L., Crane, F. L., Löw, H., and Grebing, C.,** Transplasma membrane redox stimulates HeLa cells, *Biochem. Biophys. Res. Commun.*, 125, 649, 1984.

101. **Sun, I. L., Crane, F. L., Grebing, C., and Löw, H.,** Transmembrane redox and control of cell growth, *Exp. Cell. Res.,* 156, 528, 1985.
102. **Crane, F. L., Löw, H., Sun, I. L., Navas, P., and Morré, D. J.,** Redox control of cell growth, in *Redox Functions of the Eukaryotic Plasma Membrane,* Ramirez, J., Ed., Consejo Superior Investigaciones Cientificas, Madrid, 1987, 3.
103. **Garcia-Canero, R.,** Possible role of transplasma membrane ferricyanide reductase in the mitogenic activation of rat liver cells, in *Plasma Membrane Oxidoreductase in Control of Animal and Plant Growth,* Crane, F. L., Morré, D. J., and Löw, H., Eds., Plenum Press, New York, 1988, 27.
104. **Crane, F. L.,** Plasma membrane redox reactions involved in signal transduction, in *Plant Biology,* Boss, W. and Morré, D. J., Eds., Alan R. Liss, New York, 1989, 115.
105. **Guern, J. and Ulbrich-Eberius, C.,** The ferricyanide-driven redox system at the plasmalemma of plant cells: origin of the proton production and reappraisal of the stoichiometry e^-/H^+, in *Plasma Membrane Oxidoreductases in Control of Animal and Plant Growth,* Crane, F. L., Morré, D. J. and Löw, H., Eds., Plenum Press, New York, 1988, 253.
106. **Sun, I. L., Crane, F. L., Grebing, C., and Löw, H.,** Properties of a transplasma membrane electron transport system in HeLa cells, *J. Bioenerg. Biomemb.,* 16, 583, 1984.
107. **Garcia-Canero, R., Diaz-Gil, J. J. and Guerra, M. A.,** Is transmembrane ferricyanide reductase connected with early prereplicative signals in hepatocytes?, in *Redox Functions of the Eukaryotic Plasma Membrane,* Ramirez, J. M., Ed., Consejo Superior Investigaciones Cientificas, Madrid, 1987, 41.
108. **Coleman, J. O. D. and Chalmers, J. D. C.,** Reduction-oxidation activities of plant plasma membranes, in *Plasma Membrane Oxidoreductases in Control of Animal and Plant Growth,* Crane, F. L., Morré, D. S., and Löw, H., Eds., Plenum Press, New York, 1988, 71.
109. **Sun, I. L., Toole-Simms, W., Crane, F. L., Morré, D. J., Löw, H., and Chou, J. Y.,** Reduction of diferric transferrin by SV40 transformed pineal cells stimulates Na/H activity, *Biochim. Biophys. Acta,* 938, 17, 1988.
110. **Grinstein, S., Smith, J. D., Onizuka, R., Chenung, R. K., Gelfand, E. W., and Benedict, S.,** Activation of Na/H exchange and the expression of proto-oncogenes in mitogen and phorbol ester treated lymphocytes, *J. Biol. Chem.,* 263, 8658, 1988.
111. **Cleland, R.,** Cell wall extension, *Annu. Rev. Plant, Physiol.,* 22, 197, 1971.
112. **Polevoy, V. V. and Salamatova, T.,** Auxin, proton pump, and cell trophics, in *Regulation of Cell Membrane Activities in Plants,* Marré, E., and Ciferri, O., Eds., Elsevier, Amsterdam, 1977, 209.
113. **Böttger, M., and Laüthen, H.,** Possible linkage between NADH oxidation and proton secretion in zea *mays* L. roots, *J. Exp. Bot.,* 37, 666, 1986.
114. **Böttger, M., Bigdon, M., and Soll, H.-J.,** Proton translocation in corn coleoptiles, ATPase or redox chain, *Planta,* 163, 373, 1985.
115. **Schmidt, W.,** Blue light physiology, *Bioscience,* 34, 698, 1984.
116. **Senger, H.,** *Blue Light Effects in Biological Systems,* Springer-Verlag, Berlin, 1984.
117. **Poff, K. L. and Butler, W. L.,** Absorbance changes induced by blue light in *Phycomyces blakesleeanus* and *Dictyostelium discoideum, Nature,* 248, 799, 1974.
118. **Munoz, V. and Butler, W. L.,** Photoreceptor pigment for blue light in *Neurospora crassa, Plant Physiol.,* 55, 421, 1975.
119. **Brain, R. D., Freiburg, J. A., Weiss, C. V., and Briggs, W. R.,** Blue-light induced absorbance changes in membrane fractions from corn and *Neurospora, Plant Physiol.,* 59, 948, 1977.
120. **Jesaitis, A. J., Heners, P. R., Hertel, R., and Briggs, W. R.,** Characterization of a membrane fraction containing a b-type cytochrome, *Plant Physiol.,* 59, 941, 1977.
121. **Leong, T.-Y., Vierstra, R. D., and Briggs, W. R.,** A blue light sensitive cytochrome-flavin complex from corn coleoptiles, *Photochem. Photobiol.,* 34, 697, 1981.
122. **Leong, T.-Y., and Briggs, W. R.,** Evidence from studies with acifluorifen for participation of aflavin-cytochrome complex in blue light photoreception for phototropism of oat coleoptiles, *Plant Physiol.,* 70, 875, 1982.
123. **Widell, S., Caubergs, R. J., and Larsson, C.,** Spectral characterization of a light reducible cytochrome in a plasma membrane enriched fraction and in other membranes from cauliflower inflorescences, *Photochem. Photobiol.,* 38, 95, 1983.
124. **Caubergs, R., Asard, H. and De Greef, J. A.,** b-type cytochromes, light and NADH dependent oxidoreductase activities in plant plasma membranes, in *Plasma Membrane Oxidoreductases in Control of Animal and Plant Growth,* Crane, F. L., Morré, D. J., and Löw, H., Eds., Plenum Press, New York, 1988, 273.
125. **Kjellbom, P., Larsson, C., Askerlund, P., Schelin, C., and Widell, S.,** Cytochrome P450/420 in plant plasma membranes: a possible component of the blue light reducible flavoprotein cytochrome complex, *Photochem. Photobiol.,* 42, 779, 1985.
126. **Vassilety, I. M., Derkatchev, E. F., and Neifakh, S. A.,** The electron transfer chain in liver cell plasma membrane, *Exp. Cell. Res.,* 46, 419, 1967.

127. **Masuda, Y., Kuchii, M., Yamamoto, H., and Murano, T.,** Influence of CCl_4 administration on solubilization of the enzyme in the electron transport system in rat liver plasma membranes and microsomes, *Jpn. J. Pharmacol.*, 23, 757, 1973.
128. **Masuda, Y., Kuchii, M., Yamamoto, A., and Murano, T.,** Characterization of the NADH cytochrome c reductase of liver cell membrane in normal and CCl_4 treated rats, *Jpn. J. Pharmacol.*, 23, 653, 1973.
129. **Bruder, G., Bretscher, A., Franke, W. W., and Jarasch, E.-D.,** Plasma membranes from intestinal microvilli and erythrocytes contain cytochrome b_5 and P420, *Biochim. Biophys. Acta*, 600, 739, 1980.
130. **Bruder, G., Fink, A., and Jarasch, E.-D.,** The b-type cytochromes in endoplasmic reticulum of mammary gland epithelium and milk fat globule membranes consist of two components, cytochrome b_5 and P420, *Exp. Cell Res.*, 117, 207, 1978.
131. **Mackellar, W. C. and Crane, F. L.,** Iron and copper in plasma membrane, *J. Bioenerg. Biomemb.*, 14, 241, 1982.
132. **Benevides, J., Garcia, M. L., Valdivieso, F., and Gimenez-Gallego, G.,** Ubiquinone, non heme iron and flavin in the renal brush border membrane, *Arch. Biochem. Biophys.*, 206, 451, 1981.
133. **Bell, S. H., Brown, P. J., Dickson, D. P. E., and Johnson, P. M.,** A mossbauer spectroscopic study of iron location in isolated placental synethiotropoblast microvillious plasma membranes, *Biochim. Biophys. Acta*, 756, 250, 1983.
134. **Tangerås, A., Flatmark, T., Backström, D., and Ehrenberg, A.,** Mitochrondrial iron not bound in heme and iron-sulfur centers, *Biochim. Biophys. Acta*, 589, 162, 1980.
135. **Dreyer, J.-L. and Treichler, T.,** NADH dehydrogenase in synaptic plasma membranes in *Plasma Membrane Oxidoreductases in Control of Animal and Plant Growth*, Crane, F. L., Morré, D. J., and Löw, H.,Eds., Plenum Press, New York, 1988, 406.
136. **Hultquist, D. E., Reed, D. W., Passon, P. G., and Andrews, W. E.,** Purification and properties of S-protein (hemoprotein 559) from human erythrocytes, *Biochim. Biophys. Acta*, 229, 33, 1971.
137. **Miner, C., Lopez-Burillo, S., Garcia-Sancho, J., and Herreros, B.,** Plasma membrane NADH dehydrogenase and Ca^{++}-dependent potassium transport in erythrocytes of several animal species, *Biochim. Biophys. Acta*, 727, 266, 1983.
138. **Ramirez, J. M., Gimenez-Gallego, G., and Serrano, R.,** Electron transfer constituants in plasma membrane fractions of *Avena sativa* and *Saccharomyces cerevisae*, *Plant Sci. Lett.*, 34, 103, 1984.
139. **Barr, R., Sandelius, A. S., Crane, F. L., and Morré, D. J.,** Redox reactions of tonoplast and plasma membranes isolated from soybean hypocolyls by free flow electrophoresi, *Biochim. Biophys. Acta*, 852, 254, 1986.
140. **Nokubo, N., Zs-Nagy, I., Kitani, K., and Ohta, M.,** Characterization of the autofluroescence of rat liver plasma membrane, *Biochim. Biophys. Acta*, 939, 441, 1988.
141. **Haest, C. W. M., Kamp, D., and Duticke, B.** Topology of the membrane sulfhydryl groups in the human erythrocyte. Demonstration of a nonreactive population in intrinsic proteins, *Biochim. Biophys. Acta*, 643, 319, 1981.
142. **Kosower, N. S., Kosower, E. M., Zipser, Z., Faltin, Z., and Shomrat, R.,** Dynamic changes of red cell membrane thiol groups followed by bimanefluorescence labeling, *Biochim. Biophys. Acta*, 640, 748, 1981.
143. **Kalén, A., Norling, B., Appelkvist, E. L., and Dallner, G.,** Ubriguinine synthesis in the microsomal fraction of rat liver, *Biochim. Biophys. Acta*, 926, 70, 1987.
144. **Isaksson, M., Crane, F. L., and Löw, H.,** The transplasma membrane oxidoreductase and its relation cellular proliferation, 13th Eur. Symp., Hormones and Cell Regulation, Mont St.-Odile, France, 1988.
145. **Ramasarma, T., MacKellar, W., and Crane, F. L.,** Nature of the NADH: acceptor oxidoreductase in plasma membranes of mouse liver, *Indian J. Biochem. Biophys.*, 17, 163, 1980.
146. **Kilberg, M. S. and Christensen, H. N.,** Electron transferring enzymes in the plasma membrane of the Ehrlich ascites tumor cell, *Biochemistry*, 18, 1525, 1979.
147. **Crane, F. L., MacKellar, W. C., Morré, D. J., Ramasarma, T.,** Goldenberg, H., Grebing, C., and Löw, H., Adriamycin affects plasma membrane redox functions, *Biochem. Biophys. Res. Commun.*, 93, 746, 1980.
148. **Sun, I. L., Crane, F. L., Löw, H., and Grebing, C.,** Inhibition of plasma membrane NADH dehydrogenase by adriamycin and related anthracycline antibodies, *J. Bioenerg. Biomemb.*, 14, 425, 1984.
149. **Sun, I. L. and Crane, F. L.,** Antitumor drug inhibition of transplasma membrane redox function, *Fed. Proc.*, 41, 737, 1982.
150. **Sun, I. L. and Crane, F. L.,** Effects of anthracyline compounds on transmembrane redox function in cultured HeLa cells, *Proc. Indiana Acad. Sci.*, 93, 267, 1984.
151. **Sun, I. L. and Crane, F. L.,** Transplasmalemma NADH dehydrogenase is inhibited by actinomycin D, *Biochem. Biophys. Res. Commun.*, 101, 68, 1981.
152. **Sun, I. L. and Crane, F. L.,** The antitumor drug cisdiamindichloroplatinum inhibits transplasmalemma electron transport in HeLa cells, *Biochem. Int.*, 9, 299, 1984.

153. **Sun, I. L. and Crane, F. L.,** Bleomycin control of transplasma membrane redox activity and proton movement in HeLa cells, *Biochem. Pharmacol.*, 34, 617, 1985.
154. **Toole-Simms, W.,** Regulation of Proton Release in HeLa Cells by Ferric Reductase, Ph.D. thesis, Purdue University, W. Lafayette, IN, 1988, 157.
155. **Armstrong, N. J. and Morgan, E. H.,** The effect of lysosomatrophic bases and inhibitors of transglutaminase on iron uptake by immature erythroid cells. *Biochim. Biophys. Acta,* 762, 175, 1983.
156. **Morré, D. J., Crane, F. L., Sun, I. L. and Nasas, P.,** The role of ascorbate in biomembrane energetics, *Ann. N.Y. Acad. Sci.*, 498, 153, 1987.
157. **Schultz, H.-U., Gallencamp, H., and Staudinger, H.,** Untersuchungen zum mikrosomalen NADH abhangenigen Elektrontransport, *Hoppe-Seyler's Z. Physiol. Chem.*, 351, 309, 1970.
158. **Goldenberg, H.,** Insulin inhibits NADH-semidehydro-ascorbate reductase in rat liver plasma membrane, *Biochem. Biophys. Res. Commun.*, 94, 721, 1980.
159. **Goldenberg, H., Grebing, C., and Löw, H.,** NADH monodehydroascorbate reductase in human erythrocyte membranes, *Biochem. Int.*, 6, 1, 1983.
160. **Navas, P., Estévez, A., Burón, M. I., Villalba, J. M., and Crane, F. L.,** Cell surface glycoconjugates control the activity of the NADH ascorbate free radical reductase of rat liver plasma membrane, *Biochem. Biophys. Res. Commun.*, 154, 1029, 1988.
161. **Gross, G. G.,** Cell wall bound malate dehydrogenase from horseradish, *Phytochemistry,* 16, 319, 1977.
162. **Elstner, E. F. and Heupel, A.,** Formation of hydrogen peroxide by isolated cell wall from horseradish, *Planta,* 130, 175, 1976.
163. **Mader, M., and Amberg-Fischer, V.,** Role of peroxidase in lignification of tobacco cells: oxidation of NADH and formation of H_2O_2 by cell wall peroxidases, *Plant Physiol.*, 70, 1128, 1982.
164. **Klebanoff, S. J.,** The sulfite activated oxidation of reduced pyridine nucleotides by peroxidase, *Biochim. Biophys. Acta,* 48, 93, 1961.
165. **Halliwell, B.,** Lignin synthesis and the generation of hydrogen peroxide and superoxide by horseradish peroxidase and its estimation by manganese II and phenols, *Planta,* 140, 140, 1978.
166. **Elstner, I. F. and Heupel, A.,** Formation of hydrogen peroxide by isolated walls from horseradish, *Planta,* 130, 175, 1976.
167. **Lin, W.,** Responses of corn root protoplasts to endogenous reduced nicotinamide adenine dinucleotide (NADH): oxygen consumption, ion uptake and membrane potential, *Proc. Nat. Acad. Sci. U.S.A.*, 79, 3779, 1982.
168. **Misra, P. C., Craig, T., and Crane, F. L.,** The link between transport and plasma membrane redox systems in carrot cells, *J. Bioenerg. Biomemb.*, 16, 143, 1984.
169. **Kochian, L. V. and Lucas, W. J.,** Potassium transport in corn roots III perturbation by exogenous NADH and ferricyanide, *Plant Physiol.*, 77, 429, 1985.
170. **Thom, M. and Maretzki, A.,** Evidence for a plasmalemma redox system in sugar cane, *Plant Physiol.*, 77, 873, 1985.
171. **Møller, I. M. and Berczi, A.,** Oxygen consumption by purified plasamalemma vesicles from wheat roots, *FEBS Lett.*, 193, 180, 1985.
172. **Brouwer, K. S. van Valen, T., Day, D. A., and Lambers, H.,** Hydroxamate stimulated O_2 uptake in roots of *Pisum sativum* and *Zea mays,* mediated by a peroxidase, *Plant Physiol.*, 82, 236, 1986.
173. **Cherry, J. M., MacKellar, W., Morré, D. J., Crane, F. L., Jacobson, L. B., and Schirrmacher, W.,** Evidence for a plasma membrane redox system on intact ascites tumor cells with different metastatic capacity, *Biochim. Biophys. Acta,* 634, 11, 1981.
174. **Grebing, C., Crane, F. L., Löw, H., and Hall, K.,** A transmembrane NADH-dehydrogenase in human erythrocyte membranes, *J. Bioenerg. Biomemb.*, 16, 517, 1984.
175. **Finazzi-Agro, A., Menichelli, A., Rersani, M., Biancini, G., and Del Principe, D.,** Hydrogen peroxide release from human blood platelets, *Biochim. Biophys. Acta,* 718, 21, 1982.
176. **Ishikawa, Y. and Mason, H. S.,** Distribution of cytochrome P450 and related redox systems among hepatocyte membranes, in *Oxidases and Related Redox Systems,* Vol. 2, King, T. E., Ed., University Park Press, Baltimore, 1973, 605.
177. **Zemkova, H., Tersinger, J., and Vyskocil, F.,** Hyperpolarization of mouse skeletal muscle plasma membrane-induced by extracellular NADH, *Biochim. Biophys. Acta,* 775, 64, 1984.
178. **Ramasarma, T.,** Generation of hydrogen peroxide in biomembranes, *Biochim. Biophys. Acta,* 694, 69, 1982.
179. **May, J. M. and de Haen, C.,** The insulin like effect of hydrogen peroxide on pathways of lipid synthesis in rat adipocytes, *J. Biol. Chem.*, 254, 9017, 1979.
180. **Czech, M. P., Lawrence, J. C., and Lynn, W. S.,** Evidence for electron transfer reactions involved in the Ca^{2+}-dependent thiol activation of fat cell glucose utilization, *J. Biol. Chem.*, 249, 1001, 1974.
181. **Kakinuma, K. and Chance, B.,** Spectrophotometric studies on NAD(P)H oxidase of leukocytes I. The relation between granule-NAD(P)H oxidase and myeloperoxidase, *Biochim. Biophys. Acta,* 480, 96, 1977.

182. **Murphy, P.,** *The Neutrophil,* Plenum Press, New York, 1976.
183. **Briggs, R. T., Drath, D. B., Karnovsky, M. L., and Karnovsky, M. J.,** Localization of NADH oxidase on the surface of human polymorphonuclear leucocytes by a new cytochemical method, *J. Cell. Biol.,* 67, 566, 1975.
184. **Bellevite, P., Dri, P., Bisiacchi, B., and Patriarca, P.,** Catalase deficiency in myeloperoxidase deficient polymorphonuclear leucocytes from chickens, *FEBS Lett.,* 81, 73, 1977.
185. **Segal, A. W. and Peters, T. J.,** Characterization of the enzyme defect in chronic granulomatous disease, *Lancet,* 1, 1363, 1976.
186. **Dewald, B., Baggiolini, M., Comutte, J. T., and Babior, B. M.,** Subcellular localization of the superoxide forming enzyme in human neutrophils, *J. Clin. Invest.,* 63, 21, 1979.
187. **Green, T. R., Schaeffer, R. E., and Mackler, M. T.,** Orientation of the NADPH dependent superoxide generating oxidoreductase on the outer membrane of human PMNs, *Biochem. Biophys. Res. Commun.,* 94, 262, 1980.
188. **Henderson, L. M., Chappell, J. B., and Jones, O. T. G.,** Internal pH changes associated with the activity of NADPH oxidase of human neutrophils, *Biochem. J.,* 251, 563, 1988.
189. **Cross, A. R., Jones, O. T. G., Harper, A. M., and Segal, A. W.,** Oxidoreduction properties of the cytochrome b found in the plasma membrane fraction of human neutrophils, *Biochem. J.,* 194, 599, 1981.
190. **Segal, A. W. and Jones, O. T. G.,** The subcellular distribution and some properties of the cytochrome b component of the microbiocidal oxidase system of human neutrophils, *Biochem. J.,* 182, 181, 1979.
191. **Millard, J. A., Gerard, K. W., and Schneider, D. L.,** The isolation from rat peritoneal leukocytes of plasma membrane enriched in alkaline phosphatase and a B-type cytochrome, *Biochem. Biophys. Res. Commun.,* 90, 312, 1979.
192. **Borregaard, N., Johansen, K. S., and Esmann, V.,** Quantitation of superoxide production in human polymorphonuclear leukocytes from normals and 3 types of chronic granulomatous disease, *Biochem. Biophys. Res. Commun.,* 90, 214, 1979.
193. **Tauber, A. I., Yoshimoto, S., Tringale, S. M., Mayer, J., and Klitzman, D.,** The human neutrophil, the electron transport chain for killing, in *Redox Function of the Eukaryotic Plasma Membrane,* Ramires, J. M., Ed., Consejo Superior Investigaciones Cientificas, Madris, 1987, 213.
194. **Gabig, T. G. and Lefker, B. A.,** Catalytic properties of the resolved flavoprotein and cytochrome b components of the NADPH dependent O_2-generating oxidase from human neutrophils, *Biochem. Biophys. Res. Commun.,* 118, 430, 1984.
195. **Gabig, T. G.,** The NADPH-dependent O_2^- generating oxidase from human neutrophils, *J. Biol. Chem.,* 258, 6352, 1983.
196. **Tomoto, K., Washida, N., Yukishige, K., Takayama, H., and Koyama, J.,** Electrophoretic isolation of a membrane bound NADPH oxidase from guinea pig polymorphonuclear leukocytes, *Biochim. Biophys. Acta,* 732, 569, 1983.
197. **Bellevite, P., Cross, A. R., Serra, M. C., Davali, A., Jones, O. T. G., and Rossi, P.,** The cytochrome b and flavin content and properties of the O_2^--forming NADPH oxidase solubilized from activated neutrophils, *Biochim. Biophys. Acta,* 746, 40, 1983.
198. **Nisimoto, Y., Wilson, E., Heyl, B. L., and Lambeth, J. D.,** NADH dehydrogenase from bovine neutrophil membranes: purification and properties, *J. Biol. Chem.,* 261, 285, 1986.
199. **Mitchell, R.,** The lethal oxidase of leucocytes, *Trends Biochem. Sci.,* 8, 117, 1983.
200. **Babior, B. M. and Peters, W. A.,** The O_2^- producing enzyme of human neutrophils, *J. Biol. Chem.,* 256, 2321, 1981.
201. **Gabig, T. G., Kipnes, R. S., and Babior, B. M.,** Solubilization of the O_2^- forming activity responsible for the respiratory burst in human neutrophils, *J. Biol. Chem.,* 253, 6663, 1978.
202. **McLoughlin, P., Sun, I. L., and Crane, F. L.,** Membrane redox systems in porcine neutrophils, *Proc. Indiana Acad. Sci.,* 92, 333, 1983.
203. **McLoughlin, P. A., Sun, I. L., and Crane, F. L.,** Adriamycin effects on transplasma membrane redox functions in porcine neutrophils, *Biochim. Biophys. Acta,* 802, 71, 1984.
204. **Choi, H. B. and Doke, N.,** Superoxide anion generation: a response of potato leaves to infection with *Phytophthora infestans, Phytopathology,* 77, 645, 1987.
205. **Morton, R. K.,** The lipoprotein particles in cows milk, *Biochem. J.,* 57, 231, 1954.
206. **Keenan, T. W., Morré, D. J., Olson, D. E., Yunghans, W. N., and Patton, S.,** Biochemical and morphological comparison of plasma membrane and milk fat globule membrane from mammary gland, *J. Cell Biol.,* 44, 80, 1970.
207. **Jarasch, E.-D., Bruder, G., Keenan, T. W., and Franke, W. W.,** Redox constituants in milk fat globule membranes and rought endoplasmic reticulum from lactating mammary gland, *J. Cell Biol.,* 73, 223, 1977.
208. **Briley, M. S. and Eisenthal, R.,** Association of xanthine oxidase with the bovine milk fat globule membrane: nature of the enzyme membrane association, *Biochem. J.,* 147, 417, 1975.
209. **Nielsen, C. S. and Bjerrum, O. J.,** Crossed immunoelectrophoresis of bovine milk fat globule membrane protein solubilized with nonionic detergent, *Biochim. Biophys. Acta,* 466, 496, 1977.

210. **Jarasch, E.-D., Grund, C., Bruder, C., Heid, H. W., Keenan, T. W., and Franke, W. W.,** Localization of xanthine oxidase in mammary gland epithelium and capillary endothelium, *Cell,* 25, 67, 1981.
211. **Tritsch, G. L. and Niswander, P. W.,** Modulation of macrophage superoxide release by purine metabolism, *Life Sci.,* 32, 1359, 1983.
212. **Stanburg, J. B., Wicken, J. V., and Lafferty, M. A.,** Preparation and properties of thyroid cell membranes, *J. Memb. Biol.,* 1, 459, 1969.
213. **Fischer, S., Cellino, M., Zambrano, F., Zampighi, G., Tallez Nagel, M., Marcus, D., and Canessa-Fischer, M.,** The molecular organization of nerve membranes. I. Isolation and characterization of plasma membranes from the retinal axons of the squid, *Arch. Biochem. Biophys.,* 138, 1, 1970.
214. **Canessa-Fischer, M., Zampighi, G. and Fischer, S.,** The molecular organization of nerve membranes. VI. The separation of axolemma from Schwann cell membranes of giant and retinal squid axons by density gradient centrifugation. *J. Memb. Biol.,* 9, 209, 1972.
215. **Frantz, C. E.,** NADH: ferricyanide reductase in rat liver plasma membranes, M.S. thesis, Purdue University, W. Lafayette, IN, 1973, 114.
216. **Raftell, M. and Blomberg, F.,** Membrane fractions from rat hepatoma. III. Immunochemical characterization of detergent soluble membrane phosphatases, electron transport chains and catalase, *Biochim. Biophys. Acta,* 291, 442, 1973.
217. **Fleischer, S. and Kewina, M.,** Subcellular fractionation of rat liver, in *Methods in Enzymology,* Vol. 13, S. Fleischer and L. Packer, Eds., Academic Press, New York, 1974, 6.
218. **Goldenberg, H., Morré, D. J., and Crane, F. L.,** NADH oxidoreductase of liver plasma membranes, *J. Cell Biol.,* 75, 209a, 1977.
219. **Hersey, S. J. and Jobsis, F. F.,** Redox changes in the respiratory chain related to acid secretion by the intact gastric mucosa, *Biochim. Biophys. Acta,* 36, 243, 1969.
220. **Hersey, S. J.,** Interactions between oxidative metabolism and acid secretion in gastric mucosa, *Biochim. Biophys. Acta,* 344, 157, 1974.
221. **Kiddler, G. W.,** III, Theories on gastric acid secretion, *Ann. N.Y. Acad. Sci.,* 341, 259, 1980.
222. **Kidder, G. W.,** III, Azide insensitive cytochrome oxidation in isolated gastric mucosa, *Proc. Fed. Amer. Soc. Exp. Biol. Med.,* 45, 894, 1986.
223. **Crane, F. L.,** unpublished, 1981.

Chapter 2

OXIDOREDUCTASE ENZYMES IN THE PLASMA MEMBRANE

H. Löw, F. L. Crane, D. James Morré, and I. L. Sun

TABLE OF CONTENTS

I. INTRODUCTION

Plasma membranes of animal cells have a variety of oxidoreductases which can oxidize substrates inside or outside of the cell using different biologically significant electron acceptors,[1-3] a series of NADH dehydrogenases, including the NADH cytochrome b_5 reductase which is necessary to desaturation of fatty acids.[4,5] NADH semidehydroascorbate reductase[6] and NADH oxidoreductase activity[7,8] are also present, as well as NADH ferricyanide[9] or iron chelate reductase, which may be part of an NADH diferric transferrin reductase.[10] NADPH may also be used as substrate for the differic transferrin reductase, as well as for a small amount of P450-dependent mixed function oxidase. Thioredoxin reductase in the plasma membrane is also an NADPH-dependent enzyme.[11,18] A DT diaphorase-like enzyme which uses both NADH and NADPH as substrates has also been found in plasma membranes.[12-14] In addition, there are glutathione oxidase[15] and ascorbate oxidase activities. In plasma membranes of neutrophils, there is an inducible NADH oxidase which can generate superoxide to kill bacteria.[16,17,19] These enzymes have specific locations on the membrane (Figure 1) and are involved in several functions of the plasma membrane. For the purposes of this chapter, the enzymes involved in growth and transport are the primary interest.

II. NADH OXIDASE

Liver and fat cell plasma membranes have low levels of NADH oxygen oxidoreductase activity.[7,20,21] The activity is not present in human erythrocyte membranes.[22,23] The activity in liver plasma membranes is greater than the activity in endoplasmic reticulum (microsomes) (Table 1). Since the plasma membrane NADH oxidase is not inhibited by rotenone, antimycin A, or cyanide, and is not stimulated by cytochrome c, it cannot be attributed to contamination by mitochondria. The activity in plasma membrane is stimulated by azide,[7] whereas azide inhibits mitochondrial oxidase. The high activity relative to the endoplasmic reticulum indicates that the activity is not based on contamination by these membranes.[21,24,25]

Atebrin has long been considered as an inhibitor for flavoprotein enzymes,[26] and it has been suggested as an analog inhibitor for pyrroloquinolinoquinone proteins.[27] Atebrin is one of the most selective inhibitors for plasma membrane oxidoreductase activity.[21] Although the concentration required for inhibition is over 1 m*M*, the effects on plasma membrane NADH indophenol reductase (Table 1) or ferricyanide reductase are much greater than on activity in endoplasmic reticulum or mitochondria (Table 1). The selective effect of atebrin is not so apparent for NADH cytochrome c reductase, which suggests that enzyme is the

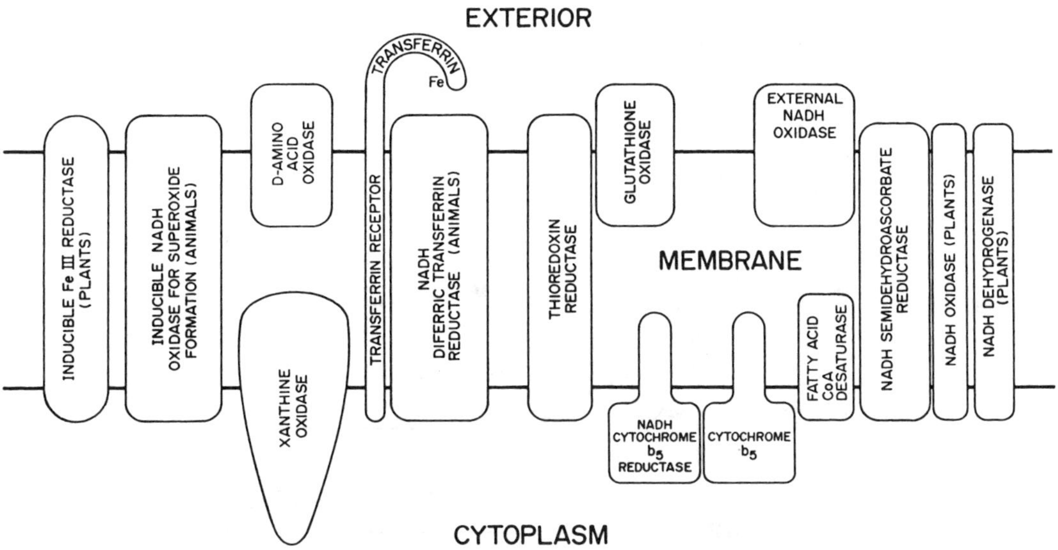

FIGURE 1. Diagram showing the orientation of oxidoreductase enzymes in the plasma membrane.

TABLE 1
NADH Oxidoreductase Activity of Rat Liver Membranes

pH	Activity	Additions	Oxidoreductase activity nmol $min^{-1}mg$ $protein^{-1}$)		
			Plasma membrane	Microsomes	Mitochondria
6.0	NADH → O_2	—	36 ± 16 (6)	6	104 ± 48 (3)
6.0	NADH → O_2	Cytochrome c + KCN	30	—	26
6.0	NADH → O_2	Cytochrome c	44	—	161
6.0	NADH → $Fe(CN)_6$	—	340	2316	4645
7.2	NADH → cytochrome c	—	79	1100	772
6.0	NADH → DCIP	—	68	625	268
6.0	NADH → DCIP	Atebrin 2.4 m*M*	27	610	225
7.0	NADH → DCIP	—	63	310	160
7.0	NADH → cytochrome c	—	91	1030	300
7.0	NADH → cytochrome c	Atebrin 3 m*M*	75	970	240
7.2	NADH → cytochrome c	KCN 1 m*M*	145	1470	510
7.2	NADH → cytochrome c	No CN	71	1370	400
		Azide 0.1 *M*			
7.0	NADH → O_2		14 ± 7 (4)	15	31
7.0	NADPH → DCIP		9 ± 2 (3)	163	62
7.0	NADPH → $Fe(CN)_6$		9	—	87
7.2	NADPH → cytochrome c		12	149	—

same in all membranes. NADH oxidase activity of rat liver plasma membranes and with fat globule membranes is selectively inhibited by millimolar concentrations of Atebrin as compared to NADH oxidase activity in endoplasmic reticulum, Golgi, or mitochondrial membranes (Figure 2).[7]

Azide is an inhibitor of NADH oxidase in mitochondria through its reaction with cytochrome c oxidase. Azide stimulates plasma membrane NADH oxidase at concentrations between 50 and 150 m*M*.[7] The oxidase activity of endoplasmic reticulum and Golgi membranes is not stimulated at these concentrations (Figure 3).

T_3 at 10^{-6} *M* also activates plasma membrane NADH oxidase activity but not oxidase

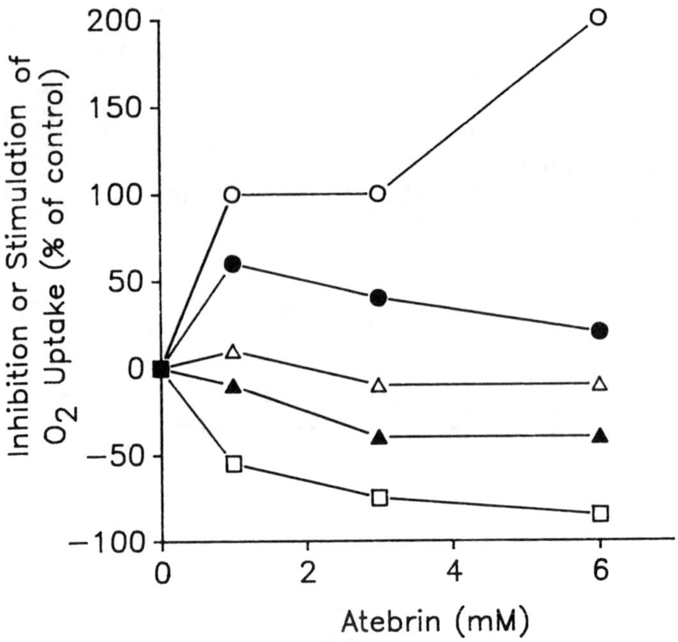

FIGURE 2. The inhibition of stimulation of NADH oxidase activity with various membrane preparations when treated with atebrin. Control activity of membranes (nmol/min/mg protein): ▲-plasma membrane, 14; □-milk fat globular membrane, 24; ●-mitochondria, 13; ○-Golgi apparatus, 10; △-smooth endoplasmic reticulum, 59.

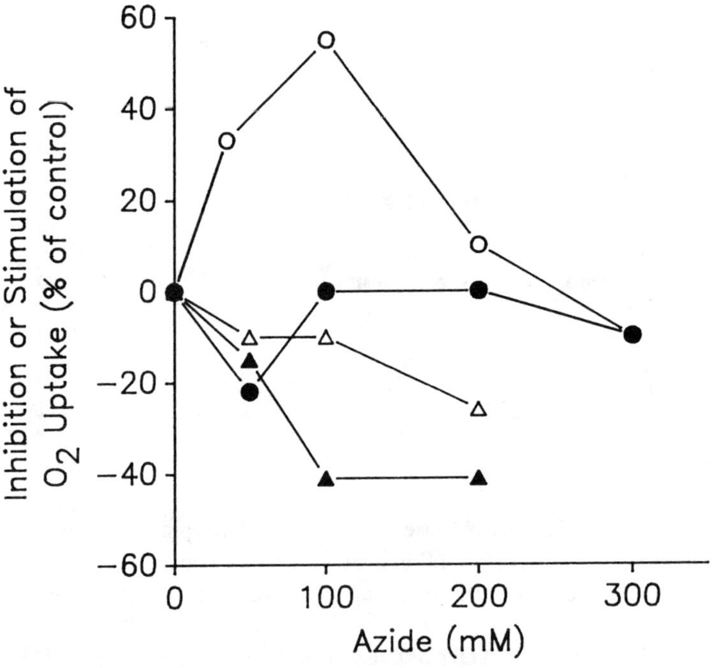

FIGURE 3. The inhibition of stimulation of NADH oxidase activity with various membrane preparations when treated with azide. Control activity of membranes (nmol/min/mg protein): ○-plasma membrane, 25; ▲-milk fat globular membrane, 32; ●-Golgi apparatus, 18; △-smooth endoplasmic reticulum, 38.

TABLE 2
Comparison of T_3 and T_4 Effect on NADH→DCIP in Membranes from Rat Liver

Membrane type		Indophenol reductase (nmol min^{-1} mg^{-1})	% of control
Liver PM	None	63	—
	T_3^a 1.2×10^{-7} *M*	47	75
	T_3 4×10^{-7} *M*	23	37
	T_4^b 2×10^{-7} *M*	56	89
	T_4 4×10^{-7} *M*	56	89
Microsomes	None	310	—
	T_3 1.2×10^{-7} *M*	290	94
	T_3 4×10^{-7} *M*	240	77
	T_4 1.2×10^{-7} *M*	340	110
	T 4×10^{-7} *M*	300	97
Mitochondria	None	160	—
	T_3 1.2×10^{-7} *M*	133	83
	T_3 4×10^{-7} *M*	96	60
	T_4 $\times 10^{-7}$ *M*	150	94

[a] T_3 — L-Triiodothyronine
[b] T_4 — L-Thyroxine

TABLE 3
Oxygen Uptake by Mouse Liver Plasma Membranes with Ascorbic Acid

Additions	Ascorbate oxidase (nmol O_2 min^{-1} mg $protein^{-1}$)
None	146
KCN, 1 m*M*	161
Cytochrome c, 0.03 mg/ml	172
Polylysine (41,000 Da), 0.013 mg/ml	36
L-Thyroxine, 6×10^{-6} *M*	0
L-Triiodothyronine, 6×10^{-6} *M*	43
Tileron[a], 3.6×10^{-4} *M*	36
Diethyldithiocarbanate, 13 μg/ml	0
Atebrin, 5×10^{-3} *M*	10

Note: Assay, 0.07 mg mouse liver plasma membrane in 0.1 K phosphate buffer, pH 7.0; 100 μM sodium ascorbate; oxygen uptake measured with oxygen electrode; 1.5 ml total volume.

[a] Tileron, *N*-(*cis*-2-phenylcyclopentyl) azacyclotridecane-2-imine · HCl.[50]

in endoplasmic reticulum, Golgi or mitochondria (cf. Chapter 5). In contrast, the triiodothyronine inhibits NADH → DCIP reductase in plasma membrane and mitochrondria, whereas it has little effect in endoplasmic reticulum (Table 2). T_4 does not activate plasma membrane NADH oxidase and does not inhibit NADH indophenol reductase, but it does inhibit ascorbic acid oxidase in mouse liver plasma membrane preparations (Table 3).

III. THE TRANSPLASMA MEMBRANE ELECTRON TRANSPORT

The reduction of impermeable oxidants such as ferricyanide by cells was the first indication of a transplasma membrane electron transport system.[28-30] The concept that

TABLE 4
Reduction of Iron Chelates by HeLa Cells

Oxidant	Rate of reduction (nmol min^{-1} gww^{-1})	Ref.
Ferricyanide	120	62
Ferric ammonium citrate	60	55
Diferric transferrin	10	104
Nitriloacetate	25	
Ferric diethylene triamine pentaacetate	12	
Fe EDTA	12	
Ferric pyridoxylisoncotinoyl hydrazone[a]	15	
Fe desferrioxamine	0	

Note: All acceptors present at 0.1 m*M* except Fe EDTA at 0.01 m*M* and ferric ammonium citrate at 15 μ*M*. Reduction measured by rate of ferrous BPS formation using the absorbance difference at 535 to 600 nm in TD tris buffer pH 7.4.[104] Ferricyanide reduction measured without BPS by difference in absorbance change at 420 to 500 nm.[53]

[a] Ferric pyridoxylisoncotinoyl hydrazone[173] kindly provided by Dr. P. Ponka, Lady Davis Institute, McGill University, Montreal.

TABLE 5
Redox Potential of Impermeable Electron Acceptors for the Transmembrane Electron Transport from HeLa Cells

Acceptor	Redox potential	Concentration n*M*	Rate of reduction
Ferricyanide	360	0.1	53
1,2-Naphthoquinone-4-sulfonate	187	0.03	17
FeIII EDTA	94	0.01	12
Hexammine ruthenium III	51	0.1	16
Indigotetrasulfonate	−46	0.1	15
Indigodisulfonate	−125	0.03	10
Anthraquinone-1,5-disulfonate	−174	0.1	2
Anthraquinone-2-sulfonate	−225	0.1	0

Data based on Reference 36.

transplasma membrane electron transport would modify the internal metabolic processes in the cell was recognized in the early studies which showed that ferricyanide reduction by erythrocytes could increase ATP levels.[31,32] With the development of the chemiosmotic concept with proton transport as a basis for ATP synthesis in mitochondria,[33] the transplasma membrane electron transport become a candidate for redox-linked ATP formation.[32,34] As it later developed, the lack of a reversible protonophoric ATP synthase in the erythrocyte plasma membrane made a direct role in ATP synthesis unlikely. Activation of glycolysis offered a different interpretation of the increase in ATP. Transplasma membrane electron transport to ferricyanide from internal NADH would lead to an increase in internal NAD, which would then increase the rate of glycolysis and glycolytic ATP formation.[35] In addition to ferricyanide, other iron complexes can be used as acceptors which will produce external ferrous iron (Table 4), and other nonferric impermeable acceptors can be used as long as their redox potential is over −160 mV (Table 5).[36]

Transplasma membrane electron transport has been found in all cells which have been

TABLE 6
Reduction of Ferricyanide by Cells and Perfused Tissue by Transplasma Membrane Electron Transport

Cell type	Ferricyanide reduction ($nmol\ min^{-1}\ gww^{-1}$) Initial	Slow	Ref.
Primary rat hepatocytes	2300	260	28
Fetal rat hepatocytes	1200	610	51
Fetal rat hepatocytes SV40 transform	260	100	51
Primary rat adipocytes		120	
Primary rat kidney		500	29
Swiss mouse 3T3 fibroblasts	1300	640	
Swiss SV40-transformed 3T3	640	320	
Rat pineal ts A209	1050	460	52
SV40-transformed rat pineal ts A209	379	205	52
HeLa	500	220	53
W138 lung		130	29
Ehrlich ascites (mouse		103	54
K562 leukemic lymphocytes		13	
HL60		98	
L1210		120	55
HepG2		20	
Human erythrocyte		50	32
Pig erythrocyte		50	40
Tetrahymena pyroformis		100	
Porcine neutrophils − SOD		218	42
Porcine neutrophils + SOD 50 units/ml		157	42
Perfused tissue			
Rat liver		220	28
Rat heart		80	56
Rat soleus muscle (whole)		240	29

tested (Table 6). This includes both plants and animals.[29,37] The activity has mostly been determined using ferricyanide as the impermeable external oxidant. Ferricyanide has been tested with various cells to make sure it is impermeable and that no significant reduction occurs within the cell.[28,32] As an alternative, the diffusion of reducing agents out of the cell could also act as a transmembrane electron transport system. All tests for leakage of reducing agents have shown reducing capacity much below the transmembrane electron transport rate.[28,38] Ascorbate has been proposed as a possible transmembrane electron carrier with erythrocytes.[39] Tests with erythrocytes and liver cells have shown that any reductant released, including ascorbate, could not account for the rate of electron transport by intact cells.[28,40] Superoxide is another possible reductant which might escape from cells and reduce external ferricyanide.[41] If superoxide is released, one would expect rapid reduction of cytochrome c as well as ferricyanide by the cells. The addition of superoxide dismutase does not inhibit the transmembrane reduction rate with most cells, which indicates a lack of significant superoxide release.[28] On the other hand, the neutrophils which produce large amounts of superoxide when stimulated by foreign agents do supply a large amount of reducing agent at the outside of the cell, which can be detected as superoxide dismutase-inhibited ferricyanide or cytochrome c reduction.[42,43]

A further indication that excretion of low potential reducing agents, e.g., thiols or ascorbate, is not the basis for transmembrane electron transport, is the fact that cytochrome c is reduced by intact HeLa or liver cells much more slowly than ferricyanide.[28]

The low sensitivity of the transmembrane electron to transport to *p*-chlorophenylmercurisulfonate (PCMBS) shows that the reduction is only partly based on thiol groups exposed

TABLE 7
Inhibition of Transplasma Membrane Electron Transport to Ferricyanide by Cells

Cell type	(Doxorubicin) Inhibitor	Effect on slow rate: Maximum inhibition observed %	Effect on slow rate: Conc of inhibitor required μ*M*	IC_{50}	Ref.
HeLa	Adriamycin	90		0.1 μ*M*	57
	Daunomycin	80		0.2 μ*M*	57
	Cis-diamminodichloroplatinum II	62		0.1 μ*M*	58
	Bleomycin	50		30 μg/ml	59
	PCMBS	40	60 μ*M*	—	53
	PCMB	78	45 μ*M*	—	53
	Iodoacetamide	73	6.6 m*M*	—	53
	Diazobenzensulfonate (DABS)	40	0.2 m*M*		53
Pig erythrocytes	Adriamycin	0			
	Bleomycin	20			
	Actinomycin D	50	7×10^{-6}		60
SV40 ts RLA209-15					
(33°C)	Adriamycin	80		0.9 μ*M*	51
(40°C)	Adriamycin	15			51
Fetal liver cells	Adriamycin	15			51
Porcine neutrophils	NaF	50	5 m*M*		43
Human erythrocyte	PCMBS	100	1 m*M*		61

Note: TsRLA 209-15 is a fetal rat liver cell line transformed with a temperature-sensitive SV40 virus. At 33°C the cells are transformed; at 40°C they are not transformed.

on the cell surface. PCMBS is impermeable to cells, so it reacts only with surface or exported thiols (Table 7). If thioredoxin reductase were involved in transmembrane ferricyanide reduction, then PCMBS inhibition would be expected and may account for the inhibition which is observed.[11,18]

IV. AFFINITY OF SURFACE SITES FOR FERRICYANIDE

The saturation of ferricyanide reduction by cells, with increasing concentration of acceptor, is evidence that the ferricyanide reduction is not based on diffusion of ferricyanide into the cells. Affinity constants for ferricyanide reduction have been determined for some cells (Table 8). With most cells tested, the K_m for slow rate ferricyanide reduction is from 0.02 to 0.08 m*M*. Primary rat hepatocytes and Swiss 3T3 cells have notably higher K_m for ferricyanide at 0.13 to 0.15. A difference this large may indicate that more than one enzyme is involved. Transformation of cells with SV40 virus changes the K_m for external ferricyanide. The transformed cells show a lower affinity for ferricyanide.

V. OXIDATION OF INTERNAL NADH

Reduction of external ferricyanide is accompanied by oxidation of cellular NADH,[44,45] which indicates that the ferricyanide reduction can be based on a transmembrane NADH ferricyanide reductase. Oxidation of internal NADH is observed immediately upon addition of ferricyanide or diferric transferrin to HeLa cells and the decrease reaches a maximum at 15 to 20 min. NAD increases in parallel with the decrease in NADH so the total pyridine nucleotide level in the cells does not change during the short time period. Insulin, which

TABLE 8
Effect of Concentration of Ferricyanide on Reduction by Intact Cells

Cell Type	Ferricyanide K_m m*M* Fast	Slow	Ref.
Pineal			
Nontransformed	0.18	0.048	52
SV40 transformed	0.33	0.074	52
HeLa	0.125	0.024	53,62
Perfused rat heart		0.023	63
Liver cells		0.13	64
Fetal liver cells			
Nontransformed	0.086	0.031	65
SV40 transformed	0.22	0.07	65
Ehrlich ascites (mouse)		0.08	54
Swiss 3T3	0.10	0.15	
Swiss 3T3 SV40	0.045	0.0025	
Yeast cells			
Log phase		0.045	66
Stationary		0.013	66
Carrot cells		0.05	67
HeLa cell growth stimulation (max effect)		0.01	68
FeCN stimulated Na^{22} uptake hepatocytes		0.075	69
Ferricyanide inhbition of iron uptake by liver cells IC_{50}		0.013	70

stimulates the reduction of ferricyanide by HeLa cells, greatly accelerates the oxidation of internal NADH so that most of the change occurs in the first 5 min. The extent of change in NAD is only slightly greater with insulin present.[46] Adriamycin, which inhibits the transmembrane electron transport, inhibits the decline in NADH and increase in NAD levels. In primary liver cells, ferricyanide induces a 33% decrease in NADH in 5 min. With liver growth factor present, the ferricyanide-induced decline is 55%.[45]

Over the longer term, as in a 24- to 48-h incubation period when ferricyanide or transferrin stimulates growth, the total pyridine nucleotide pool increases in the cells. With growth on 0.033 m*M* ferricyanide for 48 h, the total pyridine nucleotides in HeLa cells increase from 210 pmol per mg protein to 590 pmol per mg protein and the NAD/NADH ratio increases from 0.4 to 0.8. The NAD concentration in the cells increases from 60 pmol per mg protein to 260 pmol per mg protein.[46]

The implication in these studies is that oxygen or diferric transferrin in culture media will provide an electron acceptor to effectively increase the NAD concentration in growing cells. Further evidence of this is seen in the effect of retinoic acid on pyridine nucleotide levels in HL60 cells. NAD levels decrease and differentiation of the cells is initiated when retinoic acid is added to the culture media.[47] In a mutant HL60 strain which does not differentiate in the presence of retinoic acid, the NAD + NADH levels stay high and electron transport is not inhibited by retinoic acid. Inhibition of growth by Adriamycin can be partly based on inhibition of NADH oxidation. The Adriamycin analog, daunomycin, also induces a differentiation in HL60 cells at low concentration.[48]

VI. NADH FERRICYANIDE REDUCTASE ACTIVITY IN ISOLATED MEMBRANE

Plasma membranes have enzymes that catalyze NADH ferricyanide reductase activity (Table 1).[1,3,9,11,49] Since many NADH dehydrogenases can reduce ferricyanide, the measurement of transmembrane NADH ferricyanide reductase with isolated membranes has to

TABLE 9
Activity of NADH Ferricyanide Reductase in Oriented Erythrocyte Membranes

	Activity as % of total activity revealed by detergent treatment		
	NADH cytochrome c	Acetylcholine esterase	NADH ferricyanide
Open ghosts	85	107	100
Inside-out vesicles	102	13	57
Right-side-out vesicles	0	103	15

Data from Grebing et al.[71]

be evaluated carefully. The most important consideration is the orientation and closure of membrane vesicles created during the membrane isolation. To observe the transmembrane NADH ferricyanide reductase activity, it is necessary to have both sides of the membrane exposed since both substrate and acceptor are impermeable. NADH dehydrogenases which are exclusively on the inside or on the outside of the membrane will also be available to reduce ferricyanide on open sheets of membrane. Open sheets of membrane will thus show the activity of all orientations of NADH dehydrogenase and it is difficult to determine which part of the activity is transmembranous (Figure 1). Erythrocyte membranes are ideal for evaluating the relative contribution of internal and external dehydrogenases to the overall activity since they can be prepared as open ghosts, inside-out vesicles, and right-side-out vesicles. An evaluation of the ferricyanide reductase activity in human erythrocyte membranes is shown in Table 9. The orientation of the vesicles is determined by assay of marker enzymes. NADH cytochrome c reductase is located exclusively on the inside of the membrane and acetylcholine esterase is exclusively on the outside of the membrane. It is apparent that much NADH ferricyanide reductase is located on the inside surface and is presumed to represent the activity of the NADH cytochrome b_5 reductase. The activity on the outside surface represents an unknown enzyme. In open ghosts the NADH ferricyanide reductase activity is greater than the sum of activities of the internal enzyme and the external enzyme. The additional activity represents about 30% of the total activity and can be attributed to the transmembrane enzyme.

Diazobenzenesulfonate (DABS) is an impermeable agent which will link to amino groups on proteins exposed on the surface of membranes or membrane vesicles. This reagent can be added to intact erythrocytes to test which enzymes are exposed on the outer surface. When added to erythrocyte before isolation of membranes the acetylcholine esterase in the open ghosts is completely lost. NADH cytochrome c reductase is not affected by the DABS treatment, but 30% of the NADH ferricyanide reductase activity is lost. This would be consistent with 30% exposure of the enzyme involved in NADH ferricyanide reducatase activity on the outside of the membrane. Thus, this ferricyanide reductase would be in a position to carry electrons across the membrane.[1,40,71]

NADH ferricyanide reductase activity in membranes isolated from liver cells probably represents contributions of internal, external and transmembrane oxidoreductases to the activity because liver cell membranes are only partly vesiculated when isolated because of the presence of desmosomes.[77] Part of the activity in these membranes should have properties resembling the transplasma membrane ferricyanide reduction observed with whole cells.

Wang and Alaupovic[9] have isolated an NADH ferricyanide reductase from erythrocyte ghosts that is different from NADH cytochrome b_5 reductase. This enzyme is specific for NADH, is inhibited by high concentration (1 m*M*) of *p*-chloromercuribenzoate (PCMB), and contains L-fucose (0.44%), D-galactose (2.05%), D-glucosamine (2.63%), and D-mannose (0.59%). The presence of carbohydrates indicates that this enzyme can be a transmembrane protein and clearly distinguishes it from the NADH cytochrome b_5 reductase.[78] The molecular weight is 40 kDa compared to 32 kDa for the cytochrome b_5 reductase.

VII. INHIBITION OF PLASMA MEMBRANE NADH FERRICYANIDE REDUCTASE

The most specific inhibitors of NADH ferricyanide reductase activity in plasma membranes are the antitumor drugs which inhibit transmembrane ferricyanide reduction by whole cells (Table 10). These include anthracyclines, anthramycin, actinomycin D, *cis*-diaminodichloroplatinum II, and bleomycin.[79] (See also Sun and Crane,[79] Chapter 13). It has been shown that this inhibition by antitumor drugs can be related to inhibition of cell growth.[8,79] Another class of inhibitors includes antimalarial compounds such as Atebrin (quinacrine) and chloroquine.[1,71,80] The effect of chloroquine on NADH ferricyanide reductase of rat liver and rat adipocyte plasma membrane is shown in Figure 4. Inhibition by high concentrations of PCMB may be controlled by the redox state of the electron transport enzymes since preincubation with NADH decreases inhibition.[72,74] This indicates that the location of a thiol site in the membrane protein is altered by reduction.

The concentration of PCMB required for inhibition is much greater than the concentration required for inhibition of the NADH cytochrome b_5 reductase in endoplasmic reticulum, which indicates that the plasma membrane preparation is not grossly contaminated with endoplasmic reticulum and that all cytochrome c reductase in the plasma membrane may not depend on the cytochrome b_5 reductase but may be based on a different enzyme.[81]

Inhibition of activity in open erythrocyte membrane ghosts by PCMBS at 10 μM but not with intact cells, suggests inhibition at a thiol on an internal site on the membrane.

VIII. HISTOCHEMICAL DEMONSTRATION OF NADH FERRICYANIDE REDUCTASE AS A PLASMA MEMBRANE ENZYME

When liver cells are fixed at low concentrations of glutaraldehyde and sectioned, NADH ferricyanide reductase activity remains in the plasma membrane but not in endomembranes. The activity in the sections of membrane can be detected by precipitation of cupric ferrocyanide on the membrane where the reduction of ferricyanide occurs. The histochemical staining reaction known as formation of Hatchett's brown gives direct evidence for the presence of an NADH dehydrogenase with unusual properties in the plasma membrane.[82] Since the NADH dehydrogenase staining reaction is inhibited by Atebrin, the enzyme observed has a property of the transmembrane enzyme. The lack of glutaraldehyde inhibition of NADH ferricyanide reductase activity in the plasma membrane in contrast to inhibition in endoplasmic reticulum is further evidence for existence of a unique dehydrogenase in the plasma membrane.

IX. NADH INDOPHENOL REDUCTASE AND DT DIAPHORASE

Plasma membranes isolated from several tissues have NADH indophenol reductase activity measured by reduction of either dichloroindophenol or trichloroindophenol.[2,12-14,24,74] This activity often shows unusual properties compared to NADH cytochrome c reductase or NADH ferricyanide reductase activity. The largest hormone effects on isolated plasma membrane were seen with this activity.[3,83,84] Indophenol reductase in rat liver plasma membrane has optimum activity at pH 6, and the activity at this pH is strongly inhibited by

TABLE 10
Inhibition of NADH Ferricyanide Reductase Activity of Isolated Plasma Membranes

Membrane source	Inhibitor	Conc.	Maximum % inhibition	Ref.
		Antimalarials		
Mouse liver	Atebrin	0.1 m*M*	16	25
	Atebrin	5 m*M*	50	1
Rat liver	Chloroquine	5 m*M*	62	
	Atebrin	1 m*M*	80	
Human erythrocytes	Suramin	3 μ*M*	40	
		Thiol reagents		
Rat liver	PCMB	5 μ*M*	60	72
Rat liver + NADH	PCMB	5 μ*M*	0	72
Rat liver	NEM	1 m*M*	78	72
Mouse liver	$CdCl_2$	0.1 m*M*	100	25
Ehrlich ascites	PCMB	0.1 m*M*	89	74
	NEM	1 m*M*	35	74
	NADH + PCMB	0.1 m*M*	21	74
Pig erythrocytes	PCMB	10 μ*M*	95	40
	PCMBS	10 μ*M*	90	40
Human erythrocytes	PCMB	2 μ*M*	100	71
		Antitumor drugs		
Mouse liver	Actinomycin D	5 μ*M*	80	60
	Adriamycin	0.1 μ*M*	50	73
	Bleomycin	50 μg/ml	43	59
Pig erythrocytes	*cis*-Platin II	0.1 μ*M*	43	58
	Adriamycin	0.5 μ*M*	60	75
Human erythrocytes	Adriamycin	30 μ*M*	70	75
		Ferrous chelators		
Rat liver	BPS	0.5 m*M*	100	
Pig erythrocytes	BPS	0.85 m*M*	32	40
Human erythrocytes	BPS	3 m*M*	79	71
		Miscellaneous		
Mouse liver	Azide	0.1 *M*	38	25
	Antimycin A	1 μ*M*	0	25
Human erythrocytes	Cloropromazine	0.5 m*M*	100	71
	Lanthanum	20 μ*M*	40	
Pig erythrocytes	Dicumarol	0.3 m*M*	80	40
Rat liver	Dicumarol	0.2 m*M*	65	76

Atebrin (50% at 3 m*M*), triiodothyronine (Table 2), fluoride (50% at 20 m*M*), and chloroquine (60% at $5 \times 10^{-3}M$).

More than one type of NAD(P)H indophenol reductase activity has been reported for astrocyte plasma membranes.[12-14] When NADH indophenol reductase is assayed with intact cells, the activity is completely specific for NADH and is strongly inhibited by dicoumarol.[14] On the other hand, activity in isolated plasma membranes shows equal rates with NADH and NAD(P)H and is relatively insensitive to dicoumarol.[14,40] Evidence is presented in these studies that indophenol does not readily enter cells, which allows the external NADH

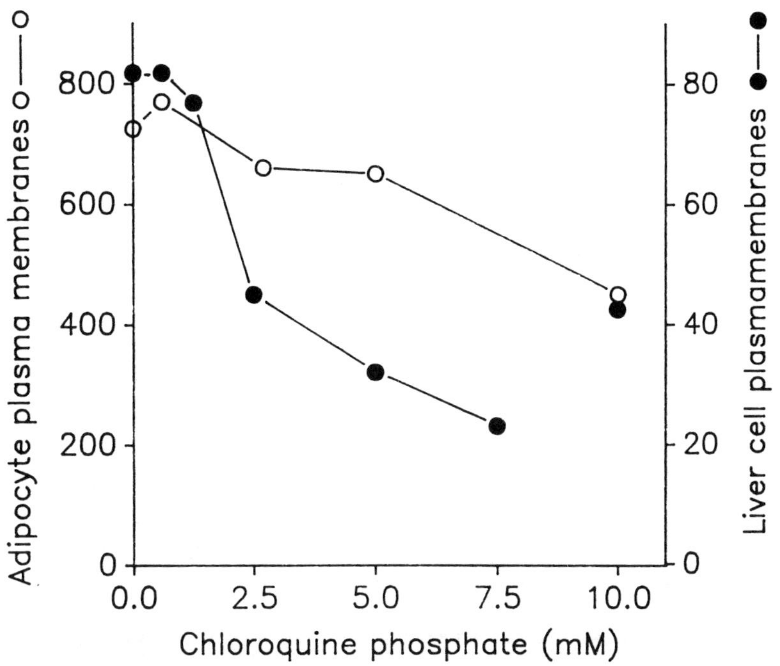

FIGURE 4. Effect of chloroquine on NADH ferricyanide reductase activity of rat liver and adipocyte plasma membranes. Assay as in Reference 71 with 0.05 *M* Na phosphate buffer pH 6.0.

dehydrogenase to be assayed. The equal activity in isolated plasma membrane with both NADH and NADPH resembles the specificity of DT diaphorase, which is a soluble enzyme found in many cells.[12,13] Since this diaphorase activity in the plasma membrane is not inhibited by low concentrations of dicoumarol, it is different from the soluble DT diaphorase.[85] Cell transformation changes the specificity of the enzymes and its response to inhibitors.[88]

X. NADH TETRAZOLIUM REDUCTASE

Rat liver plasma membranes have a low level of NADH MTT tetrazolium reductase activity (16.4 ± 0.9 nmol min^{-1} mg protein $^{-1}$)[5] in 0.1 *M* pH 7.5 Hepes buffer. This activity is inhibited 50% by 2×10^{-6} PCMB in contrast to 90% inhibition of NADH, 3-(4,5-dimethylthiazol-2-yl)-2,5-diphenyltetrazolium bromide MTT reduction in rat liver microsomes by PCMB (H. Crane, unpublished; see also Huang et al.[81]) Activity observed was plasma membrane 16, endoplasmic reticulum 110, and mitochondria 87 nmol min^{-1} per mg $protein^{-1}$. The plasma membrane activity was inhibited 30% by 3×10^{-3} *M* Atebrin, but was not inhibited by 10^{-5} *M* T_3. Activity in mouse liver plasma membranes was 50 nmol min^{-1} per mg protein compared to 137 nmol min^{-1} per mg protein in liver mitochondria.

Gel electrophoresis of rat liver plasma membranes in Triton shows three bands when stained with NADH plus MTT tetrazolium, which indicates up to three enzymes that can react with MTT tetrazolium.[81] In plasma membranes there were two major bands and one minor band. With endoplasmic reticulum there were two bands, and one was much more extensive than the second. At least three dehydrogenases may contribute to the total NADH dehydrogenase activity in plasma membranes, and all of the dehydrogenases in the membrane are not known to react with MTT tetrazolium.

When nitroblue tetrazolium was used to identify NADH and NADPH dehydrogenases

TABLE 11
Comparison of Decavanadate-Stimulated NADH Oxidase and NADH Ferricyanide Reduction in Oriented Erythrocyte Membranes

	Vanadate-stimulated NADH oxidase		NADH ferricyanide reductase		NADH cytochrome c reductase	
Membrane Orientation	**No add**	**+ Triton**	**No add**	**+ Triton**	**No add**	**+ Saponin**
Ghost (open)	440	430	84	76	7.5	8.8
Right-side-out	80	350	9	88	0	6.7
Inside-out	220	360	45	60	4.2	4.2

Note: No add, no addition; activity units nmol min^{-1} mg $protein^{-1}$.

Data from Vijaya et al.[92]

on electrophoretic gels, both adipocyte plasma membrane and the endoplasmic reticulum show the same three peaks of dye reduction with NADH and two peaks for NADPH reduction.[89] It seems that the appearance of three peaks of NADH dehydrogenase activity are often found when extracts from plasma membranes are fractionated.[87]

XI. VANADATE-STIMULATED NADH OXIDATION

Isolated plasma membrane NADH oxidase activity is greatly stimulated in the presence of low concentrations of decavanadate anion[86] (100 μM) $V_{10}O_{28}{}^{6-}$. Other forms of vanadate [V(V)] or vanadyl (V,IV) were much less effective (Table 11). Other endomembranes are able to catalyze the vanadate-stimulated NADH oxidation, so the reaction may be common to many NADH dehydrogenases (Table 10).[90,91] This oxidase activity has engendered considerable controversy as to the mechanism of the reaction and the role of an enzyme in the oxidation.[93,94] The activity is dependent on the presence of oxygen, so the reaction observed does not terminate with decavanadate even though the decavanadate may undergo reversible oxidation reduction during the reaction. The reaction involves formation of superoxide radicals and hydrogen peroxide.[92,95,96] The action of vanadate to increase peroxide production in cells and increase tyrosine kinase activity may well be based on activation of NADH oxidation at the plasma membrane.[97-99] It is clear that vanadate has effects on cellular function similar to insulin[100,101] and can activate the Na^+/H^+ antiport similar to ferricyanide reduction.[102]

XII. DIFERRIC TRANSFERRIN REDUCTION BY CELLS

Cells which have transferrin receptors are able to reduce the iron in diferric transferrin at a slow rate (Table 12). The iron bound in transferrin has a low redox potential of −400 mV,[103] so it would be possible to have reduction by NADH or thiol groups which have a redox potential around −300 mV,[11] but it is not clear if a low redox potential carrier is exposed on the surface of the membrane. Redox probe titration experiments have indicated that the primary reductant site on the exterior of the plasma membrane has a redox potential of −150 mV (Table 5), which would not be suitable for direct reduction of transferrin iron.[36] On the other hand, there are three additional factors to consider in analyzing the observation that transferrin iron can be reduced at the plasma membrane. First, binding of transferrin to its receptor may change the binding affinity for iron and thereby the redox potential. There are no data on the redox potential of iron transferrin bound to transferrin receptors. Second, the assay for transferrin iron reduction is carried out in the presence of

TABLE 12
Diferric Transferrin Reduction by Cells

Cell	Buffer and condition	Rate of ferrous BPS formation (nmol min gww)	Ref.
HeLa	Krebs Henselite, pH 7.4 + 2 m*M* Ca^{2+}	10	55,104
Pineal cell SV40	150 m*M* NaCl, Tris pH 7.4	16.5	105
	250 m*M* sucrose, Tris pH 7.4	7	105
HL60	TD buffer, pH 7.4	4.9	106
Fetal rat liver SV40			
(33°C)	TD buffer, pH 7.4	1.2	107
(40°C)	TD buffer, pH 7.4	5.7	107
		(nmol $min^{-1} 10^{-6}$ cell)	
Human erythrocyte		0	108
Primary hepatocytes	Salts, phosphate, hepes, tricine pH 7.4	0.2	70
Swiss 3T3	Krebs Henselite buffer, pH 7.4	0.25	178
SV40, 3T3 cells	Krebs Henselite buffer, pH 7.4		
Primary hepatocytes	Krebs Henselite buffer + 2 m*M* Ca^{2+}, pH 7.4	0.7	55
K562	TD buffer, pH 7.4	0.1	109
	140 m*M* NaCl, Hepes, pH 7.2	0.1	172

Note: TD buffer is 140 m*M* NaCl, 5 m*M* KCl, 0.7 m*M* Na_2HPO_4, and 25 m*M* Trizma base pH 7.4, 10 μ*M* BPS. Fetal rat liver cells transformed with a temperature-sensitive SV40 virus are transformed at 33°C but have a normal phenotype at 40°C.

a strong ferrous iron chelator, bathophenanthroline sulfonate (BPS), which alters the dissociation equilibrium of the transferrin iron to effectively raise the redox potential by at least 150 mV. An illustration of this effect can be seen with plant cells where ferric pyrophosphate with a redox potential of −140 mV cannot be directly reduced by the surface site on the cell which has a potential of 0 mV.[38] If BPS is present, then the pyrophosphate iron is rapidly reduced by the plant cell. This shows an effect of BPS of increasing the redox potential of the bound iron by 140 mV.

The third factor to consider is the fact that the transplasma membrane electron flow is accompanied by proton release at the cell surface[53,69,70,105,110] which may be based on activation of the Na^+/H^+ antiport.[45,105,110] The release of protons in proximity to the transferrin receptor would decrease the binding affinity of the transferrin for iron and thereby increase the redox potential of the iron as has been suggested by Thorstensen and Romslo.[70] If so, the buffer capacity of the external media should significantly affect the rate of ferric transferrin reduction by cells. Activation of the Na^+/H^+ antiport is dependent on the presence of external sodium ions, which could account for the observation that the presence of external Na^+ increases the rate of diferric transferrin reduction by cells.[105,110] Since NADH diferric transferrin reductase activity is observed with isolated plasma membranes, the activation of the Na^+/H^+ antiport cannot account for all ferric transferrin reductase activity.[10] The first two effects are applicable to the isolated plasma membrane NADH diferric transferrin reduction, since they do not depend on the presence of an intact cell.

Diferric transferrin reduction by HeLa or K562 cells is inhibited by antitumor drugs which inhibit the transplasma membrane ferricyanide reduction so the same electron transport system may be involved for both activities.[79,111,172] Stronger inhibition of transferrin iron reduction by adriamycin suggests that part of the ferricyanide reduction may be through a different adriamycin-insensitive electron transport which cannot react with ferric transferrin (see also Chapter 13).[79,105]

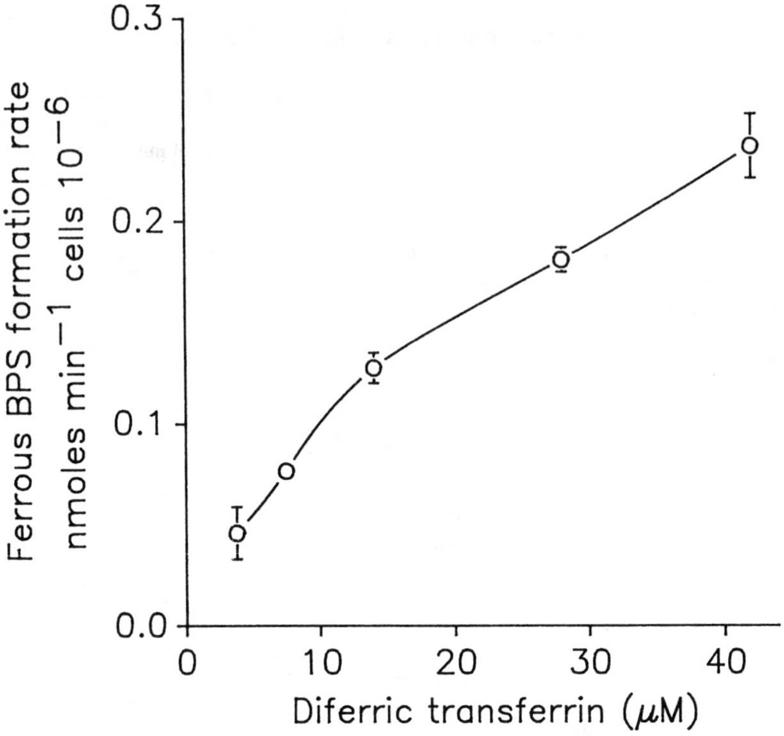

FIGURE 5. Effect of differic transferrin concentration on the rate of ferrous bathophenanthroline sulfonate formation by HeLa S3 cells. Assays in 150 m*M* NaCl, 5 m*M* KCl, 2 m*M* $CaCl_2$, and 1 m*M* $MgCl_2$ with 10 m*M* Hepes pH 7.4; 0.3 × 10^6 cells per ml of assay mixture; 15 μ*M* BPS.

Certain monoclonal antibodies to the transferrin receptor also inhibit the reduction of diferric transferrin by HeLa cells. Those which inhibit include B3/25[112] GB16, GB22.[80] Monoclonal transferrin receptor antibodies 42/6 and OKT9 have little effect. Since 42/6 binds at the high affinity receptor site on the transferrin receptor,[113] the lack of inhibition indicates that the iron transferrin reduction involves a different site. A different site is also indicated by comparison of the K_m for diferric transferrin reduction and the binding constant for the high affinity site on the receptor. Ferric transferrin binding has a K_d below 0.5 μ*M*, whereas ferric transferrin reduction has a K_m at 20 μ*M* (Figure 5).[114,115]

Diferric transferrin reduction by HeLa cells is also inhibited by goat-antihuman transferrin antibody, or rabbit-antihuman transferrin antibody.[80,112] The inhibition of activity by antibody to the transferrin indicates that the iron which is reduced is closely associated with the transferrin and is not free in solution. The antibody to transferrin does not inhibit ferricyanide or ferric citrate reduction.

It should be noted that 42/6 antibody is not a strong inhibitor of iron uptake by fibroblasts, which indicates another site may be important for iron uptake.[116] This may be the redox site, since it also is not inhibited by 42/6.

XIII. STIMULATION OF FERRIC IRON REDUCTION BY DIFERRIC TRANSFERRIN

Ferric salts and ferric chelates can be reduced by the transmembrane electron transport system as measured by formation of ferrous BPS (Table 4). The ferrous BPS is formed on the outside of the cells and remains in the supernatant after centrifugation of the cells. With

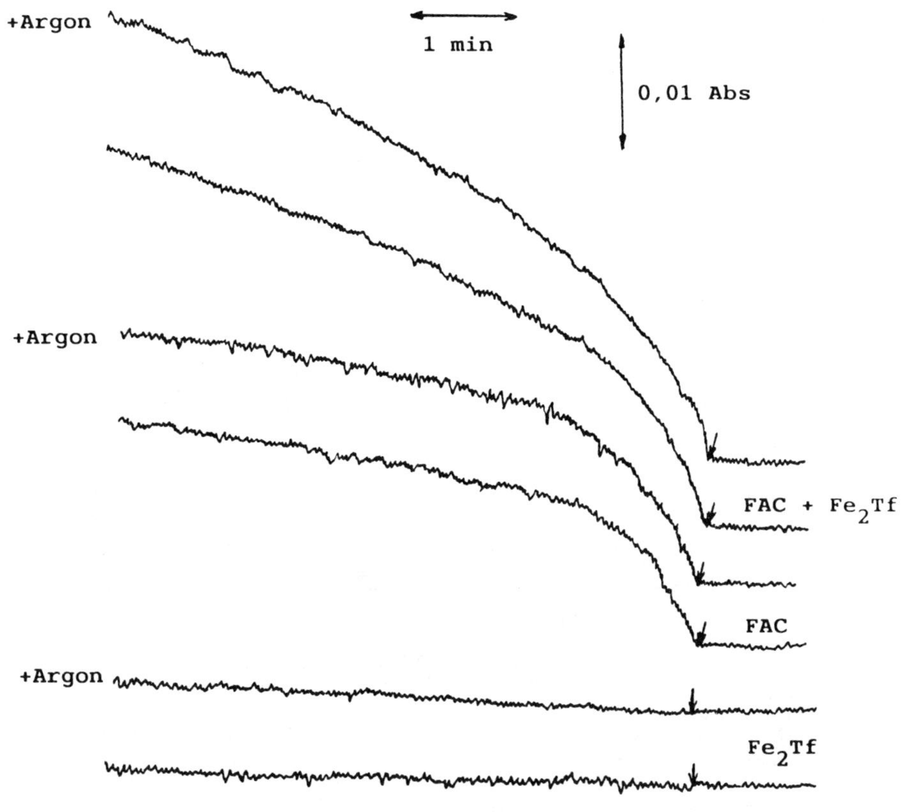

FIGURE 6. Tracing of diferric transferrin stimulation of ferric ammonium citrate (FAC) 7.5 μM with and without differic transferrin (Fe_2Tf) 14 μM reduction by HeLa cells. Assay conditions as for Figure 5 except that 0.5 mM $NaHCO_3$ was also present.

ferric ammonium citrate, the rate of reduction is slightly greater than the reduction of diferric transferrin. Only 7.5 μM ferric ammonium citrate is sufficient to saturate the reduction,[55] which indicates a high affinity for ferric iron at the reduction site. With HeLa cells or SV40-transformed 3T3 cells, the rate of ferrous BPS formation from ferric ammonium citrate is greatly accelerated by addition of diferric transferrin[109] (Figure 6). The additional activity is much greater than the reduction rate for diferric transferrin alone. This greater than additive effect, obtained by adding diferric transferrin to the already saturated ferric ammonium citrate reduction, indicates that the diferric transferrin exerts a synergistic effect on the reduction, which may indicate that binding ferric transferrin to the receptor activates the plasma membrane electron transport. If it is only an activation by binding to the receptor, then gallium transferrin, which also binds to the receptor,[117] should activate ferric ion reduction in ferric ammonium citrate even when it inhibits iron uptake.

Some preparations of diferric transferrin may contain additional loosely bound ferric ions. Preparations with up to 1 mole loosely bound iron per 2 moles of tightly bound iron (each transferrin molecule has two tight binding sites for iron) have been found.[55] The loose iron is easily detected by testing the diferric transferrin preparation with neutral ascorbate in the presence of BPS and measuring ferrous iron formation by increased absorbancy at 535 nm. Ascorbate will only reduce the loosely bound iron and cannot reduce the tightly bound. Preparations of diferric transferrin which contain additional loose iron give a much faster rate of ferrous iron formation than diferric iron without extra iron. The loosely bound iron on the transferrin gives a response like ferric ions in ferric ammonium citrate.

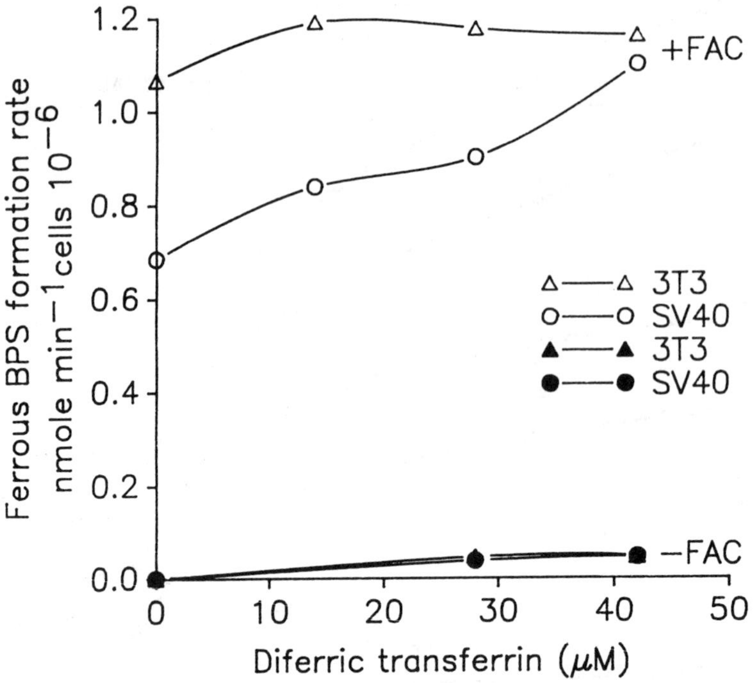

FIGURE 7. Effect of ferric transferrin on the rate of ferrous BPS formation from FAC with untransformed vs. transformed cells (FAC 7.5 μ*M*).

SV40-transformed 3T3 cells show a much greater stimulation of ferric ammonium citrate reduction by diferric transferrin then do untransformed 3T3 cells. The rate of ferric ammonium citrate reduction by SV40-transformed 3T3 cells is much slower than with untransformed 3T3 cells (Figure 7). When diferric transferrin is added, the rate of ferric ammonium citrate reduction increases with the transformed cells until it equals the untransformed cells.[109] There is some stimulation of ferric ammonium citrate reduction by untransformed cells at early growth stages.

Both normal cells at a stage of exponential growth and the transformed cells have more transferrin receptors exposed, so the added rate of ferric ammonium citrate reduction may depend on the exposure of transferrin receptors.[116,118] This effect of stimulating transmembrane electron transport by diferric transferrin is consistent with previous evidence that the binding of diferric transferrin to its receptor activates electron transport. The relationship between the redox enzyme at the cell surface and the transferrin receptor is not clear. Mutagenesis of the receptor has shown that phosphorylation of serine in the cytosolic domain of the receptor does not control endocytosis of the receptor. Deletion of 36 amino acids on the cytosolic domain does prevent endocytosis and iron uptake by 70% at 20 μg/ml ferric transferrin. These results indicate that part of the cytosolic domain is required for endocytosis, and that endocytosis is required for iron uptake.[119] It is not known if the deletion of amino acids of the cytosolic domain of the transferrin receptor had any effect on transplasma membrane electron transport.

XIV. EFFECTS OF APOTRANSFERRIN

If diferric transferrin reduction requires binding of transferrin at the transferrin receptor, then it is generally expected that apotransferrin would not interfere with the reduction reaction at pH 7.0 or above.[120] Apotransferrin does not bind to the transferrin receptor at pH above

7.0, whereas diferric transferrin does. On the other hand, apotransferrin does bind to the receptor at pH below 7.0. Apotransferrin also rapidly binds any free iron in solution or in low affinity chelates when the pH is above 7.0. Therefore, addition of diferric transferrin would remove ferric ions from ferric ammonium citrate, and the reduction rate would approach the rate for diferric transferrin rather than for ferric chelate. Addition of apotransferrin decreases the rate of ferric ammonium citrate reduction until it reaches the rate found with diferric transferrin (with HeLa from 0.51 to 0.12 nmol min^{-1} cells 10^{-6} at 7 μM apotransferrin with 7 μM ferric ammonium citrate).[55,104] Since 3.5 μM apotransferrin should take up all the ferric iron where the rate is 0.22, then the extra apotransferrin may inhibit the diferric transferrin reduction (Table 13).

HeLa cells have an unusually high binding affinity for apotransferrin at pH 7.4. Apotransferrin binding at alkaline pH is not found with reticulocytes. Since HeLa cells can bind apotransferrin at pH 7.4, the inhibition of diferric transferrin reduction at high concentration of apotransferrin may be based on apotransferrin binding to the receptor.[104] If the diferric transferrin contains loosely bound iron, then apotransferrin will bind the extra iron and cause decrease in activity to that found with diferric transferrin. At pH 5.7 apotransferrin binds to the receptor and also partially inhibits reduction of iron from added ferric transferrin. Thus, the apotransferrin binding to the receptor does give an inhibitory effect even for free iron reduction.[55]

Another effect of apotransferrin may be to decrease the extent of saturation of iron on the diferric transferrin to decrease the rate of reduction. Ferrous iron formation from transferrin which is only 2/3 saturated with iron is slower that with diferric transferrin. Thus, if apotransferrin is added to diferric transferrin, there may be a distribution of iron from the saturated to the unsaturated sites to decrease the average saturation of the transferrin with iron and thus decrease the reduction rate.

Apotransferrin inhibition of ferric transferrin reduction can be composed of three components. First, binding of any free ferric iron will decrease the activity down to the level of tightly bound iron reduction. Second, the apotransferrin can bind to the transferrin receptor at pH 7.4 on some cells which bind apotransferrin at alkaline pH, like HeLa, to compete with the iron transferrin. Third, the apotransferrin may slowly desaturate the saturated transferrin to lower the reduction rate. Desferrioxamine has very little effect on diferric transferrin reduction or ferricyanide reduction,[70] so iron exchange does not cause much inhibition. Since it does inhibit FAC reduction, any reduction which remains with diferric transferrin would be through tight binding sites. Since ferric desferrioxamine has a redox potential at −430 mV,[121] the failure to reduce desferrioxamine iron clearly shows that the transplasma membrane electron transport cannot reduce low potential compounds (Table 12). This is in contrast to the ability of iron-deficient plant cells to reduce iron desferrioxamine.[122]

With liver cells there is evidence that reduction and uptake of iron at high diferric transferrin concentrations involve prior release of ferric iron from the transferrin before it is reduced and taken up by the cell. Desferrioxamine, which has high affinity for ferric iron, inhibits iron uptake by cells but does not inhibit uptake of transferrin.[70] The desferrioxamine effect is interpreted on the basis that iron is released from transferrin at the cell surface by local acidification (e.g., Na^+/H^+ antiport), and the ferric iron is bound to chelate sites on the cell surface where it can be reduced for uptake.[70,123] Desferrioxamine would compete for the ferric receptor on the cell surface and would be unavailable for reduction because of its low redox potential (−430 mV). Inhibition of iron uptake by HepG2 cells is inhibited by 2 m*M* primaquin, an antimalarial similar to chloroquine that inhibits the plasma membrane redox system,[124] which indicates that the redox system may control iron uptake.

XV. EFFECT OF OXYGEN ON DIFERRIC TRANSFERRIN REDUCTION

When transferrin reacts with ferrous iron, the iron is tightly bound to the transferrin in the ferric form. Oxygen is necessary to oxidize the ferrous iron to ferric upon binding. Transferrin increases the rate of ferrous iron oxidation in solution, so the binding reaction is effectively a ferrous oxidase reaction.[125,126]

$$2H^+ + {}^1/_2O_2 + \text{apotransferrin} + 2Fe^{2+} = 2Fe^{3+}\ \text{transferrin} + H_2O$$

If iron bound to transferrin is reduced to the ferrous form, or if it is released and reduced, the apotransferrin formed will immediately tend to bind and reoxidize the iron. Thus, the formation of a ferrous complex, such as with bathophenanthroline sulfonate or α-dipyridyl, should be in competition for the ferrous iron with apotransferrin when oxygen is present. Absence of oxygen would favor formation of the ferrous chelates.

When diferric transferrin or ferric ammonium citrate plus diferric transferrin are used as substrates for iron reduction by cells with bathophenanthroline sulfonate present to measure ferrous iron formation, the ferrous BPS formation is increased when oxygen is removed and the reaction run under argon (Figure 6).[55,109] This increased rate in absence of oxygen is consistent with reduction of transferrin iron and competition between reoxidation and ferrous chelate formation when oxygen is present. The reduction of ferric ammonium citrate alone, in the absence of transferrin, is not increased much under argon, so the link for electron flow to oxygen is in the diferric transferrin and the transferrin receptor.[109] A more direct indication of the oxygen effect is seen during reduction of diferric transferrin by NADH with isolated liver membranes. This reaction is measured directly by following decrease of ferric transferrin using the diferric transferrin absorbance at 435 nm without other chelators present. When oxygen is present no reduction is observed, but under argon the diferric transferrin absorbance is decreased.[10]

Oxygen has also been shown to affect iron uptake from transferrin by liver cells[127-129] but not by reticulocytes. High levels of oxygen inhibit iron uptake. Transferrin uptake by endocytosis or receptor binding is not inhibited.[70,127] High concentration of oxygen (300 μM) would favor retention of iron by the transferrin, which would increasc endocytosis of the iron attached to the transferrin. Since high oxygen concentration decreases iron uptake 90%, prior reduction of iron at the surface rather than endocytosis appears to be the favored route for liver cells with 1 μM or above ferric transferrin.[70,130] At lower concentrations of ferric transferrin which saturate the tight binding sites on the transferrin receptor, endocytosis appears to predominate. Reduction of diferric transferrin does not reach maximum activity until 40 μM, which is far above the affinity of the tight binding site on the receptor (0.1 to 0.5 μM) (Figure 5). Lower affinity binding sites must be involved in diferric transferrin reduction and the major part of iron uptake by liver cells and fibroblasts.[70,116,130] Since the concentration of diferric transferrin in serum is much higher than required for specific high affinity binding, reduction of diferric transferrin can be expected to participate in iron uptake by liver.

On the other hand, the rate of diferric transferrin reduction by liver cells is much greater than the rate of iron uptake. Other cells show a similar discrepancy;[131,132] therefore much of the reduction of iron must be compensated by reoxidation and rebinding to transferrin at the surface. For example, HepG2 cells take up iron from ferric transferrin at 4 pmol g wet wt^{-1} min^{-1},[124] whereas they reduce diferric transferrin iron at 30 nmol min^{-1} gww^{-1}. This allows a transplasma membrane electron transport to oxygen through a reduction-reoxidation cycle of transferrin without equivalent iron uptake.

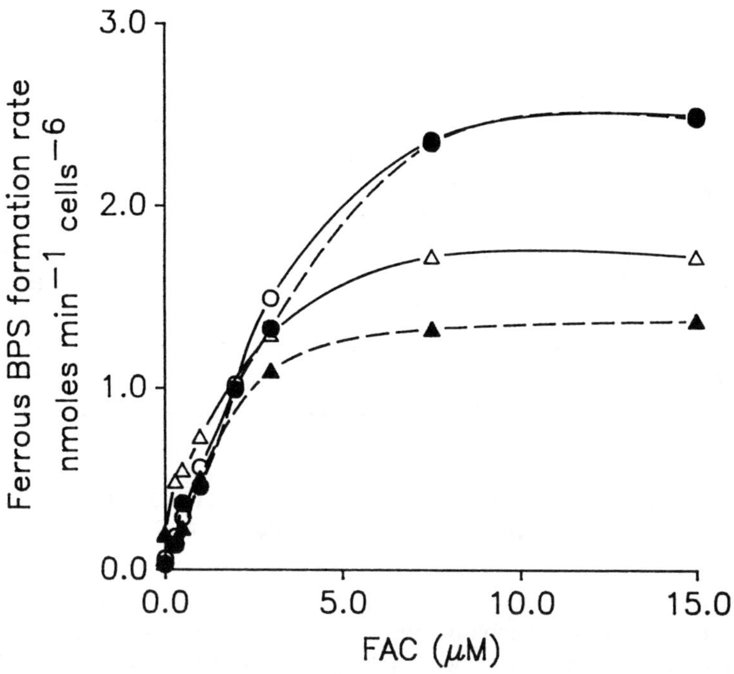

FIGURE 8. Comparison of oxygen (argon) effect on SV3T3 cells vs. 3T3 cells diferric transferrin reduction: SV 3T3 cells ▲ in air, △ in argon; 3T3 cells ● in air, ○ in argon.

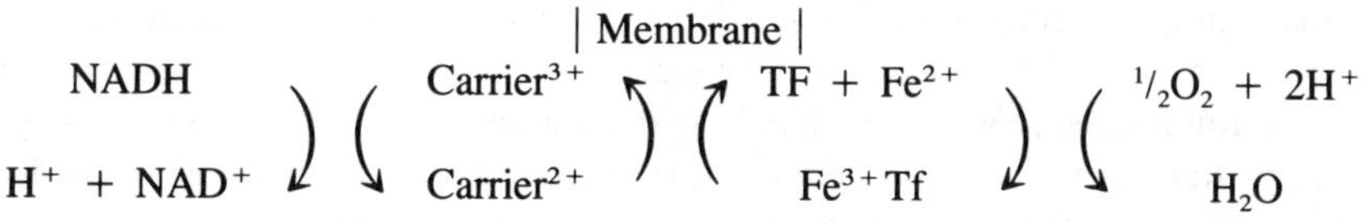

(Carrier indicates the electron transport system in the membrane.)

It should be noted that high oxygen (300 μ*M*) also inhibited ferricyanide reduction by liver cells. This suggests that the oxygen reduction site on the plasma membrane has a relatively high K_m for oxygen.[70] Since bicarbonate is necessary for rebinding of ferric iron and transferrin, the presence of bicarbonate in the media should inhibit ferrous BPS formation.

In contrast to HeLa or primary liver cells, oxygen has little effect on the reduction of iron when ferric ammonium citrate and diferric transferrin are present with 3T3 cells. The SV40-transformed 3T3 cells show a stimulation of iron reduction with ferric ammonium citrate and diferric transferrin. The implication is that the transformed 3T3 cells have a diferric transferrin-induced link to oxygen which is not available to nontransformed 3T3 cells (Figure 8). The increased interaction with oxygen in the transformed cells may be based on the threefold greater exposure of transferrin receptors on the surface of transformed compared to nontransformed cells.

XVI. THE CYTOSOLIC ELECTRON DONOR FOR TRANSMEMBRANE ELECTRON TRANSPORT

The substrates in the cytoplasm which are the most likely donors for transmembrane electron transport are NADH, NADPH, and glutathione. There is evidence that NADH can be the principal electron donor. Internal NADH is oxidized in cells which are exposed to ferricyanide or diferric transferrin. Inhibition of electron flow across the membrane by adriamycin inhibits the oxidation of internal NADH. Electron flow across the membrane can be based on NADH generated from glycolysis. Further evidence for this comes from inhibition of electron transport by HeLa cells by inhibitors of glycolysis.[53] With primary liver cells it was not possible to stimulate transmembrane ferricyanide reduction by adding glucose or other glycolytic substrates except for dihydroxyacetone. This lack of response by liver cells may be based on retention of high concentrations of stored substrates in the liver cell after isolation.[128] Isolated Ehrlich ascites cells, on the other hand, do respond to glucose, lactate, and glutamine which stimulate transmembrane electron transport after 10 to 20 min incubation.[133]

As pointed out by Kruger and Böttger,[134] the electron transport across the plasma membrane will oxidize all of the NADH in the cytoplasm in a few seconds, so the longer term rate of transmembrane electron transport ultimately depends on the rate of glycolysis or of hexose monophosphate shunt.

XVII. NAD(P)H DIFERRIC TRANSFERRIN REDUCTASE IN ISOLATED MEMBRANES

Futher evidence for transmembrane electron transport from cytosolic NAD(P)H to diferric transferrin derives from the demonstration that isolated plasma membranes have NAD(P)H diferric transferrin reductase. This activity has been described in plasma membranes from hepatocytes and HL60 cells.[10,47,123] The properties of the oxidoreductase activity in plasma membrane preparations resemble those of the activity observed with whole cells.

NADH diferric transferrin reductase activity of isolated plasma membranes can be assayed by three different methods.[10] First, the decrease in diferric transferrin can be followed by decrease in absorbancy at 465 nm, which is the wavelength maximum of the diferric transferrin (Figure 9). This assay must be carried out under anaerobic conditions because the ferrous iron formed reacts immediately with apotransferrin in the presence of oxygen to reform ferric transferrin. Second, the formation of ferrous ions can be measured by formation of ferrous bathophenanthroline chelate which is measured by increase in absorbance at 535 nm (Figure 9). This assay will also measure reduction of any loosely bound iron associated with the transferrin, and may be decreased by reoxidation of the ferrous iron by transferrin plus oxygen. Third, stimulation of NADH oxidase acitivity of plasma membrane by diferric transferrin (Figure 9). This assay will also measure NADH oxygen oxidoreductase in the plasma membrane (cf. Chapter 3). The rate of NADH oxidation without diferric transferrin is subtracted from the rate of NADH decrease (measured at 340 nm) found after addition of diferric transferrin. All three procedures have been used to measure NADH diferric transferrin reductase.

The properties of the plasma membrane NADH ferricyanide and diferric transferrin reductase activity in plasma membranes are similar to the activity observed using intact cells. The major problem for assay of an enzyme of transmembrane orientation is that NADH dehydrogenase is located only on the inside of the membrane, and the ferricyanide and diferric transferrin reduction occurs on the outer side of the membrane. Therefore, maximum activity can only be seen with fully open nonvesiculated plasma membrane. Liver plasma membranes are advantageous for study of transmembrane oxidoreductase activity with impermeable donors and acceptors because they are obtained with extensive nonvesicular structure because of desmosomes. Other plasma membranes as well as endomembranes undergo extensive vesicularization during preparation, which prevents measurement of transmembrane activity.[135] To attain maximum activity, the membrane vesicles may require opening by detergent or other means. CHAPS detergent has been used to get maximum activity from membrane vesicles.[10]

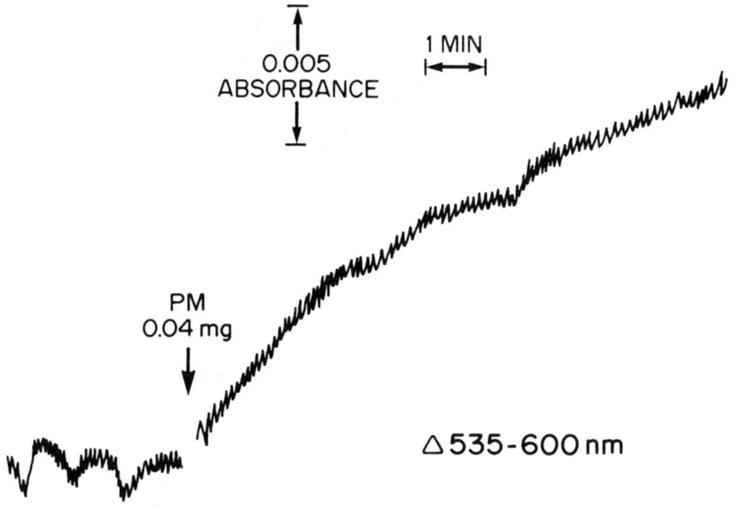

A

FIGURE 9. Three assays for NADH-diferric transferrin reduction by isolated liver plasma membrane. (A) Reduction of diferric transferrin measured by formation of ferrous BPS with plasma membrane, 3.4 μ*M* diferric transferrin, and 25 μ*M* NADH in 2.8 ml of TD buffer, pH 7.4. The absorbance measured at 600 nm was subtracted from absorbance at 535 nm with the dual-beam spectrophotometer. PM, plasma membrane. (B) Oxidation of NADH by liver plasma membrane in the presence of diferric transferrin. Reaction in 2.8 ml of TD buffer, pH 7.4, 1 m*M* KCN, and 25 μ*M* NADH. The reaction was started by the addition of 3.4 μ*M* diferric transferrin. Note that the NADH oxidase rate prior to transferrin addition is subtracted from the rate of NADH oxidation after transferrin addition. Absorbance measured at 340 minus 430 nm. PM, plasma membrane. (C) Reduction of diferric transferrin in the presence of 17 μ*M* diferric transferrin, 50 μ*M* NADH, and rat liver plasma membrane in TD buffer in an argon atmosphere. The absorbance at 600 nm was subtracted from the absorbance at 465 nm with the dual-beam spectrophotometer. PM, plasma membrane, Fe_2T, diferric transferrin. TD buffer contains 140 m*M* NaCl, 5 m*M* KCl, 0.7 m*M* Na_2HPO_4, and 25 m*M* TRIZMA base pH 7.4. (Figure continued on next page.)

The affinity for diferric transferrin for the isolated membrane is similar to the affinity of the reduction site on the plasma membrane of the cell (33 μ*M*). This low affinity is not consistent with reduction at the high affinity diferric transferrin binding site (K_d = 0.5 μ*M*).

The second ($Fe(BPS)_3$ formation) and third assay (NADH oxidation) procedure for NADH diferric transferrin reductase will also measure NADH ferric chelate reductase activity. The problem of extra iron bound to the diferric transferrin must therefore be considered. Apotransferrin inhibition of diferric transferrin reductase activity appears to provide an excellent control on reduction of loosely bound iron or iron chelates because it converts the iron into tightly bound transferrin iron (10^{22} M^{-1}) very rapidly.[136] The amount of apotransferrin used should ideally only be slightly greater than stoichiometric to the free iron since excess apotransferrin may inhibit by competing for the low affinity binding site where reduction occurs or may decrease the degree of iron saturation of the ferric transferrin.

In the study on NADH diferric transferrin reduction by liver plasma membrane,[10] the direct (first) assay gave a slow rate of 5 nmol min^{-1} mg $protein^{-1}$. In some experiments, often with different preparations of diferric transferrin, the BPS assay and the NADH oxidation assay gave higher rates, up to 29 nmol min^{-1} mg^{-1}. These higher rates were inhibited by apotransferrin at two times the stoichiometric level to diferric transferrin to decrease the rate to between 8 and 10 nmol min^{-1} mg^{-1}. Therefore, rates of diferric transferrin reduction over 10 nmol min^{-1} mg^{-1} appear to be based on NADH ferric chelate

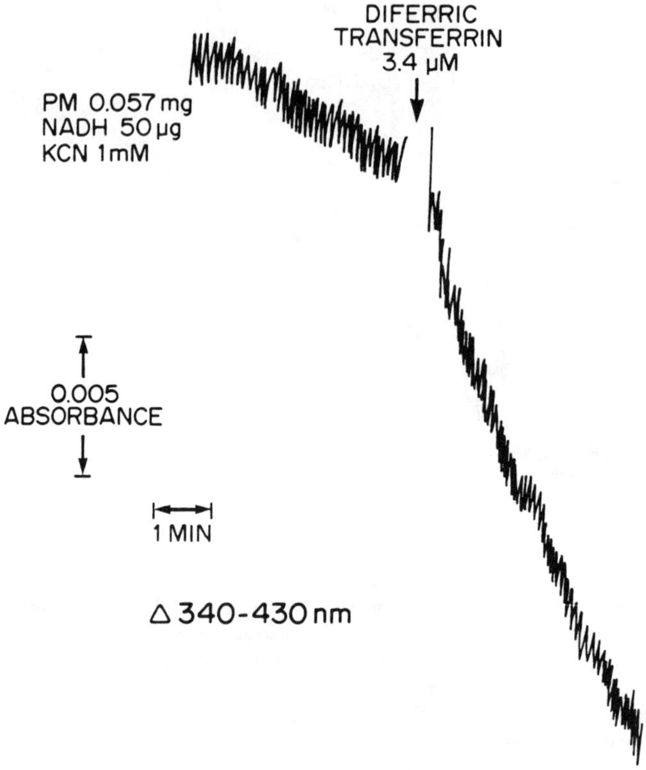

FIGURE 9B.

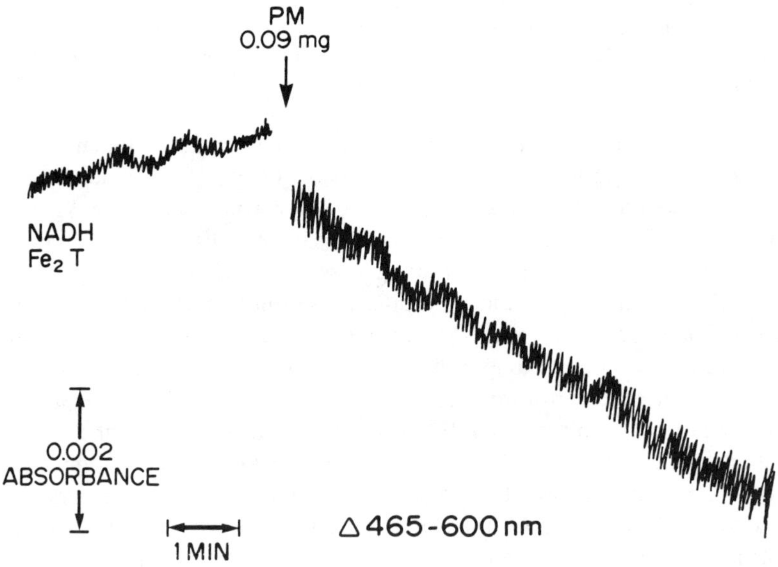

FIGURE 9C.

TABLE 13
Effect of Apotransferrin on Ferric Ammonium Citrate Reduction by Cells and NADH Ferric Ammonium Citrate Reduction by Isolated Membranes

Cell or Membrane	FAC μM	Apotransferrin μM	$Fe(BPS)_3$ formation (nmol min^{-1} cell 10^{-6})	Ref.
HeLa S3	7.5	0	0.41	55
	7.5	7	0.04	55
CCl-39	7.5	0	0.41	55
	7.5	14	0.09	55
			(nmol min^{-1} mg $prot^{-1}$)	
Liver plasma membrane	0[a]	0	29	10
	0	17	10	10
Placenta membranes	7.5	0	+ + +	
	7.5	2	+	
	7.5	4	0	

Note: Apotransferrin will bind 2 iron per mol, so complete inhibition of free iron reduction should occur at 4 μ*M* apotransferrin when 7.5 μM FAC is present. (FAC is ferric ammonium citrate.)

[a] For liver plasma membrane, 17 μ*M* diferric transferrin was the substrate, and the apotransferrin inhibition can be a measure of the amount of loosely bound iron associated with the transferrin. No more than 1 mol loose iron has been found in any diferric transferrin preparation.[55]

reductase activity (Table 13). It is noteworthy that all apparent NADH diferric transferrin reductase activity found in endomembrane fractions was completely inhibited by apotransferrin.[10] This is not unexpected since the endoplasmic reticulum and other endomembranes form sealed vesicles during isolation, so it would be unlikely to find transmembrane electron transport to diferric transferrin, and any transferrin receptors in these vesicles would be exposed inside the vesicles. The NADH ferric reductase activity in the surface of the vesicle is most likely based on NADH cytochrome b_5 reductase which is exposed on the outer surface.[137] Preparations of membranes from placenta show strong apparent NADH diferric transferrin reductase which is completely inhibited by apotransferrin (Table 13). Since these membranes are not highly purified plasma membrane and are more vesiculated than liver membranes, the lack of true diferric transferrin reductase activity is no surprise even though the high concentration of transferrin receptors would suggest such activity. The inhibition with apotransferrin of ferric chelate reductase activity by these membranes is a good illustration of the effectiveness of apotransferrin, an inhibitor of ferric iron reduction by other dehydrogenases (Table 13). The inhibition by apotransferrin is stoichiometric for the amount of ferric ammonium citrate in the assay mixture on the basis of two high affinity binding sites for iron on each transferrin molecule. The implication is that the apotransferrin has effectively removed all weakly chelated iron and that the lack of activity after all the iron has been incorporated into the transferrin shows that these membranes cannot carry out reduction of iron in diferric transferrin. In contrast, the reduction of ferric ammonium citrate by CC139 cells or HeLa cells in the presence of two times the stoichiometric concentration of apotransferrin continues at a significant rate which would represent diferric transferrin reductase activity (Table 13).[55]

The NADH diferric transferrin reductase of isolated liver plasma membranes is inhibited by the antitumor drugs adriamycin and *cis*-diamminedichloroplatinum at 1 μ*M*, which is consistent with inhibition of diferric transferrin reduction by cells.[79,111] Atebrin (30 μ*M*) and chloroquine (100 μ*M*) also inhibit the NADH diferric transferrin reductase because they inhibit NADH ferricyanide reductase by membranes or diferric transferrin reduction by cells.[10,80] Retinoic acid, which inhibits diferric transferrin reductase activity with HL60 cells and NADH diferric transferrin reductase of HL60 cell membranes,[10,47] also inhibits the activity in liver plasma membranes.[138]

The NADH oxidase activity in liver plasma membrane is not inhibited by 1 m*M* KCN. The NADH oxidase activity induced by diferric transferrin is also not inhibited by cyanide. It is not clear at this time if the basal oxidase activity in absence of added diferric transferrin can be attributed to a small amount of ferric transferrin bound to the receptors on the membrane.

NADPH can also serve as an electron donor for reduction of diferric transferrin by rat liver plasma membranes.[10] Since the effect of apotransferrin on this activity has not been tested, it is not clear how much of this activity can be based on NADPH ferric chelate reductase activity, or if NADPH can serve as a donor for ferric transferrin reduction.

XVIII. METHEMOGLOBIN REDUCTASE AND FATTY ACYL DESATURASE

NADH cytochrome b_5 reductase and cytochrome b_5 have been detected in small amounts in many plasma membrane preparations.[1,3,24] In some cells the reductase and cytochrome b_5 are linked to the fatty acyl coenzyme A desaturase which desaturates fatty acids through a mixed function oxidase reaction.[4,5] The desaturase is an iron protein which reacts with oxygen and is cyanide sensitive. The NADH oxidase activity of this enzyme combination depends on the presence of fatty acyl coenzyme A.

The combined NADH cytochrome b_5 reductase and cytochrome b_5 act as an NADH cytochrome c reductase. Low levels of this activity are found in most plasma membranes, and the presence of the proteins has been verified by antibodies to the reductase or cytochrome b_5.[1,139,140] In erythrocytes, the reductase-cytochrome b_5 combination acts as a methemoglobin reductase,[139] and a lack of the enzyme in some individuals results in methemoglobinemia. A large portion of the NADH ferricyanide reductase activity associated with the plasma membrane has been shown by antibody inhibition to be based on the NADH cytochrome b_5 reductase flavoprotein, which can use ferricyanide as an alternative electron acceptor.[1,78] In cells other than erythrocytes, the high level of NADH cytochrome b_5 reductase plus cytochrome b_5 produces a very high level of NADH cytochrome c reductase activity (antimycin or rotenone insensitive) in the endoplasmic reticulum, which is ten times greater than the activity found in plasma membranes (Table 14). Until the advent of highly purified plasma membranes, the NADH cytochrome c reductase in the plasma membrane was often attributed to contamination by endoplasmic reticulum.[3,141] The presence of this activity in mammalian erythrocyte membranes clearly shows that endomembrane contamination is not the basis for the activity.

Both NADH cytochrome b_5 reductase and cytochrome b_5 attach to the cytosolic side of membranes through carboxyterminal hydrophobic helices, and the proteins are not transmembrane.[142] The observation of NADH cytochrome c reductase only on the inside of erythrocyte membranes is consistent with the known structure of these enzymes and their lack of a transmembrane domain.[143-145]

The presence of NADH cytochrome b_5 reductase in rat liver plasma membrane has been confirmed by use of a radioimmunoblotting assay.[146] The cytochrome b_5 reductase activity in plasma membrane was 0.7 times that of total homogenate (231 nmol min^{-1} per mg $protein^{-1}$), and the immunoblotting gave a value 0.97 times total homogenate.

XIX. MIXED FUNCTION OXIDASES

Plasma membranes in general have a very low level of cytochrome P450 and its associated NADPH-P450 reductase. The reductase part of the activity is assayed as NADPH cytochrome

TABLE 14
Cytochrome c Reductase Activity in Endomembranes

	Cyt c reduction (nmol/min/mg)	
	NADH→cyt c + rotenone	NADPH→cyt c
Rat liver membranes[a]		
Nuclear membrane	263	12
RER	1180	111
RER (phenobarb rate)	1130	167
SER (phenobarb rate)	1010	167
Golgi	467	20
PM	173	15
PM (phenobarb rate)	124	20
Mouse liver membranes		
Endoplasmic reticulum	1000	240
Outer mitochrondrial	1000	
Plasma membrane	70	6.4

[a] Rat liver membranes salt washed 1 h to remove extrinsic protein. Data from References 24 and 140. Phenobarb rate indicates activity in membranes from phenobarbital-treated rats. SER amd RER, smooth and rough endoplasmic reticulum, respectively.

c reductase (Table 14). The activity in well-purified rat liver plasma membrane is 9% of the activity in endoplasmic reticulum from the same cells.[140] NADPH cytochrome c reductase is also low in rat liver plasma membranes isolated by curtin electrophoresis (6.7 nmol min^{-1} per mg $protein^{-1}$). Both P450 and P420 cytochromes are present in small amounts in isolated plasma membranes.[147] Human erythrocyte membranes contain P420 equivalent to other plasma membranes but have no NADPH dehydrogenase or oxidase activity.[22,71,147]

In endoplasmic reticulum the NADPH dehydrogenase and cytochrome P450 are responsible for NADPH-dependent oxidation of xenobiotics to facilitate detoxification.[142] This function may operate in plasma membranes.[148]

P450 is exposed on the cytosolic surface of endoplasmic reticulum membranes as is the NADPH cytochrome P450 reductase,[137,149] so they would not be expected to catalyze transmembrane electron transport.

XX. SEMIDEHYDROASCORBATE REDUCTASE AND ASCORBATE OXIDASE

Plasma membranes have NADH semidehydroascorbate reductase activity.[6,150-152] In endoplasmic reticulum this activity is attributed to the NADH cytochrome b_5 reductase acting with cytochrome b_5 to reduce semidehydroascorbate.[153,154] In chromaffin granules the membrane contains a cytochrome b_{561} which can reduce semidehydroascorbate inside the granule through the use of NADH in the cytoplasm.[155,156]

Part of the semidehydroascorbate reductase activity in plasma membrane may be based on the NADH cytochrome b_5 reductase system.[153,157] There are properties of the activity that indicate that a different enzyme may also be involved.[154] These properties include hormone control[150] and inhibition by specific lectins.[152] The lectins react with glyoproteins exposed on the outer surface, whereas the cytochrome b_5 reductase is a nonglycosylated protein located on the cytosolic surface of the plasma membrane.

The relation between semidehydroascorbate reductase activity in the plasma membrane and the stimulation of external ferricyanide reduction by dehydroascorbate remains to be elucidated. Dehydroascorbate stimulates ferricyandie reduction by intact erythrocytes and liver cells.[39,64]

Orringer and Roer[39] proposed that this stimulation is based on a dehydroascorbate-ascorbate shuttle across the plasma membrane, with dehydroascorbate reduction on the inside of the cell producing ascorbate which diffuses across the membrane to reduce the external ferricyanide. The kinetics of the reaction showed essential saturation at 0.1 m*M* dehydroascorbate, which would be more consistent with an enzymatic or binding reaction at the surface rather than simple diffusion. With liver cells the ferricyanide reduction was stimulated by dehydroascorbate, but incubation of cells with dehydroascorbate did not lead to ascorbate accumulation in the external medium.[64] The mechanism of dehydroascorbate stimulation of transmembrane electron flow and the nature of the semidehydroascorbate reductase and its role in transmembrane electron transport remain to be clarified.[158]

XXI. ASCORBATE OXIDASE

Plasma membranes from mouse liver are able to catalyze the oxidation of ascorbic acid. This ascorbate oxidase activity is also present in mouse liver Golgi, and smooth endoplasmic reticulum (SER) and rough endoplasmic reticulum (RER) membranes. This activity is more inhibited by thyroxine (T_4) at $2 \times 10^{-6}M$ than by triiodothyronine (T_3). It is also inhibited by diethyldithiocarbamate (0.1 mg/1.5 ml), polylysine (0.01 mg/1.5 ml), but not by 1 m*M* cyanide (Table 3). The relation of this activity to semidehydroascorbate reductase has not been investigated. The NADH semidehydroascorbate reductase together with the ascorbate oxidase would make a complete system to catalyze NADH oxygen reductase activity in the presence of catalytic amounts of ascorbate.[158] The identity of the enzyme is unknown.

XXII. PROSTHETIC GROUPS FOR OXIDATION-REDUCTION ENZYMES IN THE PLASMA MEMBRANE

The nature of the enzymes which carry out oxidoreductase activity in the plasma membrane are not well known. There is evidence for the presence of flavin, quinone, iron and possibly copper in the membrane (Table 15). A small amount of different types of cytochrome b have been noted. These may include cytochrome b_5 and cytochromes P450/P420. The amount of flavin in erythrocyte membranes from different species is roughly correlated with NADH ferricyanide reductase activity in the membranes[159] (Table 16) (except for dog erythrocytes). The lack of cytochrome c reductase activity in some membranes may be based on lack of functional cytochrome b_5.[161] Iron has been reported in some plasma membrane preparations. Early studies on plasma membranes showed no ESR signals that would be expected from iron sulfur proteins, but Dreyer and Treichler[164] have reported that detergent treatment of plasma membrane induces the appearance of an iron-sulfur ESR signal. Since iron sulfur centers are destroyed at acid pH, it is clear that the iron sulfur center cannot be on the outer surface of the membrane since the transmembrane enzyme can reduce diferric transferrin iron at pH 5.0.

Coenzyme Q has been found in highly purified rat liver plasma membranes[160,161] and in dog erythrocyte ghosts. No evidence has been found for coenzyme Q function as a redox carrier in the plasma membrane.[165] Tocopherol has also been reported in plasma membranes.[162,163]

XXIII. THIOREDOXIN REDUCTASE AS A TRANSPLASMA MEMBRANE ENZYME

Part of the ferricyanide reduction by intact cells may be attributed to the activity of thioredoxin reductase in the plasma membrane. Histochemical studies show thioredoxin reductase in the plasma membrane. Evidence has also been reported that thioredoxin reductase

TABLE 15
Prosthetic Groups for Oxidation Reduction Reactions which have been Reported in Plasma Membranes

	Content in nmol per mg protein	Ref.
Flavin		
Cat erythrocytes	24.6	159
Human erythrocytes	12.4	159
Sheep erythrocytes	5.6	159
Kidney brush border	0.2	159
Nonheme Iron		
Porcine erythrocytes	8.2	1
Mouse liver	7.4	1
Rat liver	5.7	1
Placenta microvilli	3.5	1
Kidney brush border	4.5	1
Cytochromes		
Cytochrome b_5	6—236	1
Cytochrome P450	0—17	1
Cytochrome P420	24—388	1
Quinines: coenzyme Q		
Dog erythrocytes	0.012	
Rat liver	0.33	160
Rat liver	0.2	161
Rat kidney	0.12	161
Tocopherol		
Erythrocytes	4—7.5	162,163
Human erythrocytes	1.6	162,163
Thiol		
Human erythrocytes	74—90	1

TABLE 16
Relationship between NADH Ferricyanide Reductase and Flavin Content of Erythrocyte Membranes

Species	NADH-$Fe(CN)_6^{3-}$ reductase (nmol $mg^{-1}min^{-1}$)	FAD (nmol/g protein)	NADH Cyt c reductase (nmol mg^{-1} min^{-1})
Cat	489 ± 45	24.6 ± 4	26 ± 1
Human	241 ± 33	12.4 ± 1.2	13 ± 2
Sheep	132 ± 27	5.6 ± 0.8	0
Rabbit	80 ± 12	9.4 ± 1.2	11 ± 2
Guinea pig	76 ± 5	6.3 ± 2.3	0
Dog	15 ± 2	15.0 ± 4	3 ± 1

Note: Assay with 0.05% saponin. Data based on Reference 159.

can carry out transmembrane electron transport through the plasma membrane.[18] Since thioredoxin reductase can act as ferricyanide reductase,[11] part of the ferricyanide reduction by cells may be based on transmembrane thioredoxin reductase. If thioredoxin reductase is acting as a transmembrane enzyme, then NADPH must act as a source of electrons for the thioredoxin reductase contribution to transmembrane electron transport. An important role for thioredoxin reductase in the plasma membrane may be to control reduction of thiol groups in membrane proteins.[166] This is especially apparent for a form of phospholipase c which acts on phosphatidyl inositides and can therefore control activation of protein kinase c. This phospholipase c has extensive homology with thioredoxin and acts as an electron acceptor for thioredoxin reductase.[167] If the reduction of disulfide groups in this phospholipase can control its activity, then the thioredoxin reductase would be a mechanism for achieving control of this phospholipase and ultimately protein kinase c. Thiol rich receptors such as the PDGF receptor may also require thioredoxin to keep thiol groups reduced in the external domain in order to carry out transmembrane signaling.[168]

XXIV. REDUCTION OF ELECTRON ACCEPTORS IN THE LIPID BILAYER

The transplasma membrane electron transport system will have prosthetic groups buried in the lipid bilayer to carry electrons through the membrane. The presence of these sites would allow transfer of electrons to oxidants present in the bilayer. These oxidants can include fatty acid peroxides which may be produced naturally from unsaturated fatty acids in the lipid.[169] Nitroxide radicals inserted into the bilayer as ESR spin probes may also be reduced by the dehydrogenase and reoxidized by external ferricyanide.[170] The reduction in the interior of the lipid is enzymatic as shown by heat and inhibitor sensitivity. With intact TB cells 20% of the 5 doxyl stearate reduction is rotenone, SKF-525, and propyl gallate insensitive, so it could represent activity of an oxidoreductase in the plasma membrane.[171] Reoxidation by external ferricyanide is a good test for reduction in the plasma membrane instead of in internal membranes.

XXV. CONCLUSIONS

There are enzymes in the plasma membrane which can carry out oxidoreduction reactions on both surfaces of the membrane as well as through the membrane. The primary enzyme on the inside of the membrane is NADH cytochrome b_5 reductase. The enzyme on the outside is an NADH ferricyanide reductase or indophenol reductase, but its natural acceptor may be oxygen. DT diaphorase activity (not sensitive to dicoumarol as compared to the soluble enzyme) has also been found associated with plasma membranes. The transplasma membrane activity may involve NADH oxygen oxidoreductase, NAD(P)H diferric transferrin reductase, NADH semidehydroascorbate reductase and NADPH thioredoxin reductase, the acceptors for which, except for oxygen, may be reduced on the outside of the cell. Ferricyanide acts as an impermeable artificial acceptor for all of these activities. The identity of the enzymes involved is unknown although flavin, iron, coenzyme Q, and some cytochrome b types are present in the membrane. The transmembrane electron transport has been shown to be involved in growth stimulation of transformed cells. The best inhibitors of the transmembrane enzyme are the antitumor drugs adriamycin, bleomycin, *cis*-diamminedichloroplatinum II, and actinomycin D. The enzymes are also activated by growth factors which is consistent with a role in control of cell growth. The reduction of iron in ferric transferrin may be an essential function in growth control since transferrin is required for growth by almost all cells. The reduction of iron in endosomes by the transmembrane reductase can also aid iron transfer into the cylosol from these organelles.

REFERENCES

1. **Crane, F. L., Löw, H., and Clark, M. G.,** Plasma membrane redox enzymes, in *The Enzymes of Biological Membranes,* Vol. 4, Martonosi, A. N., Ed., Plenum Press, New York, 1985, 465.
2. **Crane, F. L., Goldenberg, H., Morré, D. S., and Löw, H.,** Dehydrogenases of the plasma membrane, in *Subcellular Biochemistry,* Vol. 6, Roodyn, D. B., Ed., Plenum Press, New York, 1979, 345.

3. **Goldenberg, H.,** Plasma membrane redox activities, *Biochim. Biophys. Acta,* 694, 203, 1982.
4. **Holloway, P. W.,** A requirement for three protein components in microsomal stearyl coenzyme A desaturation, *Biochemistry,* 10, 1556, 1971.
5. **Giacobino, J.-P. and Chmelar, M.,** The role of chain elongation systems in the supplying of fatty acids to the adipocyte membrane lipids, *Biochim. Biophys. Acta,* 487, 269, 1977.
6. **Goldenberg, H., Grebing, C., and Löw, H.,** NADH monodehydroascorbate reductase in human erythrocyte membranes, *Biochem. Int.,* 6, 1, 1983.
7. **Gayda, D. P., Crane, F. L., Morré, D. J., and Löw, H.,** Hormone effects on NADH oxidizing enzymes of plasma membranes of rat liver, *Proc. Indiana Acad. Sci.,* 86, 385, 1977.
8. **Morré, D. J., Brightman, A., Wang, J., Barr, R., and Crane, F. L.,** Roles for plasma membrane redox systems in cell growth, in *Plasma Membrane Oxidoreductases in Control of Animal and Plant Growth,* Crane, F. L., Morré, D. J., and Löw, H., Eds., Plenum Press, New York, 1988, 45.
9. **Wang, C.S. and Alaupovic, P.,** Isolation and partial characterization of human erythrocyte membrane NADH: acceptor oxidoreductase, *J. Supramol. Struct.,* 9, 1, 1978.
10. **Sun, I. L., Navas, P., Crane, F. L., Morré, D. J., and Löw, H.,** NADH diferric transferrin reductase in liver plasma membrane, *J. Biol. Chem.,* 262, 15915, 1987.
11. **Holmgren, A.,** Thioredoxin and glutaredoxin: functions in growth-dependent DNA synthesis and in cellular regulation via thiol redox control, in *Plasma Membrane Oxidoreductases in Control of Animal and Plant Growth,* Crane, F. L., Morré, D. J., and Löw, H., Eds., Plenum Press, New York, 1988, 295.
12. **Mersel, M., Malviya, A. N., Hindelang, C., and Mandel, P.,** Isolation of plasma membrane from astrocytes in primary cultures. Acceptor oxidoreductase property of the isolated membrane, *Biochim. Biophys. Acta,* 778, 144, 1984.
13. **Mersel, M., Malviya, A. N., Hindeland, C., and Mandel, P.,** Plasma membrane isolated from astrocytes in primary cultures: its acceptor oxidoreductase properties, *Biochim. Biophys. Acta,* 778, 144, 1984.
14. **Malviya, A. N., Mandel, P., and Mersel, M.,** The nature of DT diaphorase activity in plasma membrane of astrocytes in primary culture, *Biochim. Biophys. Acta,* 849, 288, 1986.
15. **Ormstad, K., Lastbom, T., and Orrenius, S.,** Evidence for a different location of glutathione oxidase and γ glutamyl transferase activity during extracellular glutathione metabolism in isolated perfused rat kidney, *Biochim. Biophys. Acta,* 700, 148, 1982.
16. **Tauber, A. I., Yoshimoto, S., Tringale, S. M., Myers, J., and Klitzman, D.,** The human neutrophil NADH oxidase: the electron transport chain for killing, in *Redox Functions of the Eukaryotic Plasma Membrane,* Ramirez, J., Ed., Consejo Superior de Investigaciones Cientificas, Madrid, 1987, 215.
17. **Rossi, F. and Patriarca, P., Eds.,** *Biochemistry and Function of Phagocytes,* Plenum Press, New York, 1982.
18. **Schallreuter, K. V. and Wood, J. M.,** Thioredoxin reductase reduces free radicals at the cell surface, *Biochem. Biophys. Res. Commun.,* 136, 630, 1986.
19. **Henderson, L. M., Chappell, J. B., and Jones, O. T. G.,** Internal pH changes associated with the activity of NADPH oxidase of human neutrophils, *Biochem. J.,* 251, 563, 1988.
20. **Löw, H. and Crane, F. L.,** Redox function in plasma membranes, *Biochim. Biophys. Acta,* 515, 141, 1978.
21. **Crane, F. L. and Löw, H.,** NAOH oxidation in liver and fat cell membranes, *FEBS Lett.,* 68, 153, 1976.
22. **Crane, F. L., Crane, H. E., Sun, I. L., MacKellar, W. C., Grebing, C., and Löw, H.,** Insulin control of a transplasma membrane NADH dehydrogenase in erythrocyte membranes, *J. Bioenerg. Biomemb.,* 14, 425, 1982.
23. **Löw, H., Crane, F. L., Grebing, C., Hall, K., and Tally, M.,** Metabolic milieu and insulin action, in *Diabetes 1979,* Waldhäusl, W. K., Ed., Excerpta Medica, Amsterdam, 1979, 209.
24. **Goldenberg, H., Crane, F. L., and Morré, D. J.,** NADH oxidoreductase of mouse liver plasma membranes, *J. Biol. Chem.,* 254, 2491, 1979.
25. **Ramasarma, T., MacKellar, W. M., and Crane, F. L.,** Nature of NADH acceptor oxidoreductase in plasma membranes of mouse liver, *Indian J. Biochem. Biophys.,* 17, 163, 1980.
26. **Löw, H.,** The effects of atebrin and chlorpromazine on some adenosine triphosphate-splitting enzymes, *Exp. Cell. Res.,* 16, 456, 1959.
27. **Duine, J. A. and Frank, J., Jr.,** Quinoproteins, a novel class of dehydrogenase, *Trends Biochem. Sci.,* 6, 278, 1981.
28. **Clark, M. G., Partick, E. J., Patten, G. S., Crane, F. L., Löw, H., and Grebing, C.,** Evidence for the extracellular reduction of ferricyanide by rat liver: a transplasma membrane redox system, *Biochem. J.,* 200, 565, 1981.
29. **Crane, F. L., Sun, I. L., Clark, M. G., Grebing, C., and Löw, H.,** Transplasma membrane redox systems in growth and development, *Biochim. Biophys. Acta,* 811, 233, 1985.
30. **Barron, E. S. G. and Hoffman, L. A.,** The catalytic effect of dyes on the oxygen consumption of liver cells, *J. Gen. Physiol.,* 13, 483, 1929.

31. **Manyai, S. and Szekeley, M.,** Die Wirkung von natrium Fluorid und Monojodessigsäure auf die Glycolyse von menschlichen roten Blutkörperchen, *Acta Physiol. Acad. Sci. Hung.*, 5, 7, 1954.
32. **Mishra, R. K. and Passow, H.,** Induction of intracellular ATP synthesis by extracellular ferricyanide in human red blood cells, *J. Mem. Biol.*, 1, 214, 1969.
33. **Mitchell, P.,** Vectorial chemistry and the molecular mechanics of chemiosmotic coupling, *Biochem. Soc. Trans.*, 4, 399, 1976.
34. **Zamudio, I., Cellino, M., and Canessa-Fischer, M.,** The relation between membrane structure and NADH (acceptor) oxidoreductases activity of erythrocyte ghosts, *Arch. Biochem. Biophys.*, 129, 336, 1969.
35. **Arese, P., Bosia, A., and Pescamora, G. P.,** Effect of external oxidant on redox state of internal glutathione and pyridine nucleotides, in *6th Int. Symp. Structur und Function der Erythrozyten,* Rapport, S. and Jung, F., Eds., Akademie Verlag, Berlin, 1972, 91.
36. **Sun, I. L., Crane, F. L., Löw, H., and Grebing, C.,** Transplasma membrane redox stimulates HeLa cell growth, *Biochem. Biophys. Res. Commun.*, 125, 649, 1984.
37. **Crane, F. L.,** Plasma membrane redox reactions involved in signal transduction, in *Second Messengers in Plant Growth and Development,* Boss, W. and Morré, D. J., Eds., A. R. Liss, New York, 1989, 115.
38. **Crane, F. L., Barr, R., Craig, T. A., and Morré, D. J.,** Transplasma membrane electron transport in relation to cell growth and iron uptake, *J. Plant Nutrition,* 11, 1117, 1988.
39. **Orringer, E. P. and Roer, M. E. S.,** An ascorbate mediated transmembrane reducing system of the human erythrocyte, *J. Clin. Invest.*, 63, 53, 1979.
40. **MacKellar, W. M.,** A Transmembrane NADH Dehydrogenase in Porcine Erythrocyte Membranes, Ph.D. thesis, Purdue University, West Lafayette, IN, 1981, 168.
41. **Fridovich, I.,** The biology of oxygen radicals: general concepts, in *Oxygen Radicals and Tissue Injury,* Halliwell, B., Ed., Federation of American Societies for Cell Biology, Bethesda, MD, 1988, 1.
42. **McLoughlin, P. A., Sun, I. L., and Crane, F. L.,** Adriamycin effects on transplasma membrane redox functions in porcine neutrophils, *Biochim. Biophys. Acta,* 802, 71, 1984.
43. **McLoughlin, P., Sun, I. L., and Crane, F. L.,** Membrane redox system in porcine neutrophils, *Proc. Indiana Acad. Sci.*, 92, 333, 1983.
44. **Navas, P., Sun, I. L., Morré, D. J., and Crane, F. L.,** Decrease of NADH in HeLa cells in presence of transferrin or ferricyanide, *Biochem. Biophys. Res. Commun.*, 135, 110, 1986.
45. **Garcia-Cañero, R.,** Possible role of transplasma membrane ferricyanide reductase in mitogenic activation of rat liver cells, in *Plasma Membrane Oxidoreductases in Control of Animal and Plant Growth,* Crane, F. L., Morré, D. J., and Löw, H., Eds., Plenum Press, New York, 1988, 27.
46. **Navas, P., Sun, I. L., Morré, D. J., and Crane, F. L.,** Relationship between NAD^+/NADH levels and animal cell growth, in *Plasma Membrane Oxidoreductases in Control of Animal and Plant Growth,* Crane, F. L., Morré, D. J., and Löw, H., Eds., Plenum Press, New York, 1988, 339.
47. **Golub, E. S., Diaz de Pagan, Sun, I. L., and Crane, F. L.,** The retinoic acid-induced cascade of events leading to granulocyte differentiation in HL60 through a postulated GTP-regulated inducing molecule, in *Plasma Membrane Oxidoreductases in Control of Animal and Plant Growth,* Crane, F. L., Morré, D. J., and Löw, H., Eds., Plenum Press, New York, 1988, 313.
48. **Trentesaux, C., Gerard, B., Mayeux, P., Jeannessou, P., Jacquot, R., and Jardillier, J.-C.,** Differentiating activity of adriamycin in human erythroleukemic cells: effect on globin and heme synthesis, *Biochim. Biophys. Acta,* 1012, 161, 1989.
49. **Zamudio, I. and Canessa, M.,** Nicotinamide-adenine dinucleotide dehydrogenase activity of human erythrocyte membranes, *Biochim. Biophys. Acta,* 120, 165, 1966.
50. **Guellaen, G., Mahn, J.-L., Mavier, P., Bertholet, P., and Hanoune, J.,** RMI/2330A, An inhibitor of adenylate cyclase in rat liver, *Biochim. Biophys. Acta,* 484, 465, 1977.
51. **Sun, I. L., Crane, F. L., Chou, J. Y., Löw, H., and Grebing, C.,** Transformed liver cells have modified transplasma membrane redox activity which is sensitive to adriamycin, *Biochem. Biophys. Res. Commun.*, 116, 210, 1983.
52. **Sun, I. L., Crane, F. L., and Chou, J. Y.,** Modification of transmembrane electron transport in plasma membranes of simian virus 40 transformed pineal cells, *Biochim. Biophys. Acta,* 886, 327, 1986.
53. **Sun, I. L., Crane, F. L., Grebing, C., and Löw, H.,** Properties of a transplasma membrane electron transport system in HeLa cells, *J. Bioenerg. Biomemb.*, 16, 583, 1984.
54. **Waranimman, P., Sun, I. L., and Crane, F. L.,** Evidence for a transplasma membrane electron transport system on intact Ehrlich ascites tumor cells, *Proc. Indiana Acad. Sci.*, 95, 137, 1986.
55. **Löw, H., Lindgren, A., Crane, F. L., Sun, I. L., Toole-Simms, W., and Morré, D. J.,** Transplasma membrane electron transport functions as a ferric reductase, in *Plasma Membrane Oxidoreductases in Control of Animal and Plant Growth,* Crane, F. L., Morré, D. J., and Löw, H., Eds., Plenum Press, New York, 1988, 153.
56. **Löw, H., Crane, F. L., Partick, E. J., Patten, G. S., and Clark, M. G.,** Properties and regulation of the transsarcolemma redox system of perfused rat heart, *Biochim. Biophys. Acta,* 804, 253, 1984.

57. **Sun, I. L. and Crane, F. L.,** Effects of anthracycline compounds on transmembrane redox function in cultured HeLa cells, *Proc. Indiana Acad. Sci.*, 93, 267, 1984.
58. **Sun, I. L. and Crane, F. L.,** The antitumor durg, *cis* diaminedichloroplatinum II, inhibits transplasmalemma electron transport in HeLa cells, *Biochem. Int.*, 9, 229, 1984.
59. **Sun, I. L. and Crane, F. L.,** Bleomycin control of transplasma membrane redox activity and proton movement in HeLa cells, *Biochem. Pharamacol.*, 34, 617, 1985.
60. **Sun, I. L. and Crane, F. L.,** Transplasma membrane NADH dehydrogenase is inhibited by actinomycin D, *Biochem. Biophys. Res. Commun.*, 101, 68, 1981.
61. **Arese, P., Bosio, A., Pescamona, G. P., and Till, U.,** The role of redox cell membrane in the regulation of glycolysis and the 2,3 bis phosphoglycerate cycle, *Acta Biol. Med. Ger.*, 36, 481, 1977.
62. **Sun, I. L., Putnam, J. E., and Crane, F. L.,** Control of cell growth by transplasmalemma redox: stimulation of HeLa cell growth by impermeable oxidants, *Proc. Indiana Acad. Sci.*, 94, 407, 1985.
63. **Löw, H., Crane, F. L., Partick, E. J., and Clark, M. G.,** α adrenergic stimulation of trans-sarcolemma electron transport in perfused rat heart. Possible regulation of the Ca^{2+} channels by a sarcolemma redox system, *Biochim. Biophys. Acta,* 844, 142, 1985.
64. **Clark, M. G., Partick, E. J., and Crane, F. L.,** Properties and regulation of a transplasma membrane redox system in rat liver, *Biochem. J.*, 204, 795, 1982.
65. **Sun, I. L., Navas, P., Crane, F. L., Chou, J. Y., and Löw, H.,** Transplasma membrane electron transport is changed in similar virus 40 transformed liver cells, *J. Bioenerg. Biomemb.*, 18, 471, 1986.
66. **Crane, F. L., Roberts, H., Linnane, A. W., and Löw, H.,** A transplasma membrane NADH dehydrogenase in yeast, *J. Cell. Biol.*, 91, 248a, 1981.
67. **Barr, R., Craig, T. A., and Crane, F. L.,** Transmembrane ferricyanide reduction by carrot cells, *Biochim. Biophys. Acta,* 812, 49, 1985.
68. **Sun, I. L., Crane, F. L., Grebing, C., and Löw, H.,** Transmembrane redox in control of cell growth: stimulation of HeLa cell growth by ferricyanide and insulin, *Exp. Cell. Res.*, 156, 528, 1985.
69. **Garcia-Cañero, R., Diaz-Gil, J. J., and Guerra, M. A.,** Is transmembrane ferricyanide reductase connected with the early expression of mitogenic signals in hepatocytes?, in *Redox Functions of the Eukaryotic Plasma Membrane,* Ramirez, J., Ed., Consejo Superior de Investigaciones Cientificas, Madrid, 1987, 43.
70. **Thorstensen, K. and Romslo, I.,** Uptake of iron from transferrin by isolated rat hepatocytes, *J. Biol. Chem.*, 263, 8844, 1988.
71. **Grebing, C., Crane, F. L., Löw, H., and Hall, K.,** A transmembrane NADH dehdyrogenase in human erythrocyte membranes, *J. Bioenerg. Biomemb.*, 16, 517, 1984.
72. **Masuda, Y., Kuchii, M., Yamamoto, A., Murano, T.,** Characterization of NADH cytochrome c reductase of liver cell membranes in normal and CCl_4 treated rats, *Jpn. J. Pharmacol.*, 23, 653, 1973.
73. **Sun, I. L., Crane, F. L., Löw, H., and Grebing, C.,** Inhibition of plasma membrane NADH dehydrogenase by adriamycin and related anthracycline antibiotics, *J. Bioenerg. Biomemb.*, 16, 209, 1984.
74. **Kilberg, M. S., and Christensen, H. N.,** Electron transferrin enzymes in the plasma membrane of the Ehrlich ascites tumor cell, *Biochemistry,* 18, 1525, 1979.
75. **Crane, F. L., MacKellar, W. C., Morré, D. J., Ramasarma, T., Goldenberg, H., Grebing, C., and Löw, H.,** Adriamycin affects plasma membrane redox function, *Biochem. Biophys. Res. Commun.*, 93, 746, 1980.
76. **Frantz, C. E.,** NADH: Ferricyanide Oxidoreductase in Rat Liver Plasma Membranes, M.S. thesis, Purdue University, West Lafayette, IN, 1973.
77. **Beneditti, E. L. and Emmelot, P.,** Structure and function of plasma membranes isolated from liver, in *The Membranes,* Vol. 4, Dalton, A. J. and Haguenau, F. H., Eds., Academic Press, New York, 1968, 33.
78. **Kitajima, S., Yasukochi, Y., and Minakami, S.,** Purification and properties of human erythrocyte membrane NADH cytochrome b_5 reductase, *Arch. Biochem. Biophys.*, 210, 330, 1981.
79. **Sun, I. L. and Crane, F. L.,** Control of transplasma membrane diferric transferrin reductase by antitumor drugs, in *Plasma Membrane Oxidoreductases in Control of Animal and Plant Growth,* Crane, F. L., Morré, D. J., and Löw, H., Eds., Plenum Press, New York, 1988, 181.
80. **Toole, W.,** Regulation of Proton Release from HeLa Cells by Ferric Reductase, Ph.D. thesis, Purdue University, West Lafayette, IN, 1988, 160.
81. **Huang, C. M., Goldenberg, H., Frantz, C., Morré, D. J., Keenan, T. W., and Crane, F. L.,** Comparison of NADH linked cytochrome c reductases of endoplasmic reticulum, Golgi apparatus and plasma membrane, *Int. J. Biochem.*, 10, 723, 1979.
82. **Morré, D. J., Vigil, E. L., Frantz, C., Goldenberg, H., and Crane, F. L.,** Cytochemical demonstration of glutaraldehyde resistant NADH ferricyanide oxidoreductase activities in rat liver plasma membranes and Golgi apparatus, *Cytobiologie,* 18, 213, 1978.
83. **Goldenberg, H., Crane, F. L., and Morré, D. J.,** Influence of hormones on NADH-dehydrogenases in mouse liver plasma membranes, *Biochem. Biophys. Res. Commun.*, 83, 234, 1978.
84. **Löw, H., Crane, F. L., Grebing, C., Tally, M., and Hall, K.,** Hormone control of plasma membrane oxidation-reduction reactions: effects of insulin, *FEBS Lett.*, 91, 166, 1978.

85. **Ernster, L., Danielson, L., and Ljunggren, M.,** DT diaphorase I purification from the soluble fraction of rat liver cytoplasm and properties, *Biochim. Biophys. Acta,* 58, 171, 1962.
86. **Erdman, E., Krawietz, W., Philipp, G., Hackbarth, I., Schultz, W., Scholz, H., and Crane, F. L.,** Purified caridac cell membranes with high Na^+/K^+-ATPase activity contain significant NADH vanadate reductase activity, *Nature,* 282, 335, 1979.
87. **Blomberg, F. and Berzius, K.,** Epinephrine binding plasma membrane antigens in rat liver, *Eur. J. Biochem.,* 56, 319, 1975.
88. **Mersel, M., Vitkovic, L., Vincendon, G., and Malviya, A. N.,** Modification of plasma membrane DT diaphorase activity upon transformation. A comparison between astrocytes in primary culture and C6 glioblastoma cells, in *Plasma Membrane Oxidoreductases in Control of Animal and Plant Growth,* Crane, F. L., Morré, D. J., and Löw, H., Eds., Plenum Press, New York, 1988, 402.
89. **Gracobino, J. P. and Chmelar, M.,** Comparison of plasma membrane and endoplasmic reticulum fractions obtained from whole white adipocyte tissue and isolated adipocytes, *Biochim. Biophys. Acta,* 406, 68, 1975.
90. **Erdmann, E., Krawietz, W., Phillip, G., Hackbarth, I., Schmitz, W., Scholz, H., and Crane, F. L.,** Purified cardiac cell membranes with high Na,K-ATPase activity contain significant NADH vanadate reductase activity, *Nature,* 282, 335, 1979.
91. **Ramasarma, T., MacKellar, W. C., and Crane, F. L.,** Vanadate stimulated NADH oxidase in plasma membrane, *Biochim. Biophys. Acta,* 646, 88, 1981.
92. **Vijaya, S., Crane, F. L., and Ramasarma, T.,** A vanadate stimulated NADH oxidation in erythrocyte membranes generates hydrogen peroxide, *Mol. Cell. Biochem.,* 62, 175, 1984.
93. **Vyskocil, F., Dlouha, H., and Teisinger, J.,** Importance of cardiac cell membranes in vanadate induced NADH oxidation and reply, *Nature,* 294, 288, 1981.
94. **Ramasarma, T. and Crane, F. L.,** Does vanadium play a role in cellular regulation, in *Current Topics in Cellular Regulation,* Vol. 20, Horecker, B. L. and Stadtman, E. R., Eds., Academic Press, New York, 1981, 247.
95. **Ramasarma, T., Swuroop, A., MacKeller, W. C., and Crane, F. L.,** Generation of hydrogen peroxide on oxidation of NADH by hepatic plasma membranes, *J. Bioenerg. Biomemb.,* 13, 241, 1981.
96. **Vijaya, S., Crane, F. L., and Ramasarma, T.,** A vandate-stimulated NADH oxidase in erythrocyte membrane generates hydrogen peroxide, *Mol. Cell. Biochem.,* 62, 175, 1984.
97. **Bernier, M., Laird, D. M., and Lane, M. D.,** Effect of vanadate on the cellular accumulation of pp15, an apparent product of the insulin receptor tyrosine kinase action, *J. Biol. Chem.,* 263, 13626, 1988.
98. **Tamura, S., Brown, T. A., Whipple, J. H., Fujita-Yamaguchi, Y., Dubler, R. E., Cheng, K., and Larner, J.,** Basis for the insulin-like action of vanadate. Vanadate stimulates phosphorylation of the insulin receptor in a manner similar to insulin, *J. Biol. Chem.,* 259, 6650, 1984.
99. **Ramasarma, T.,** Generation of H_2O_2 in biomembranes, *Biochim. Biophys. Acta,* 694, 69, 1982.
100. **Smith, J. B.,** Vanadium ions stimulate DNA synthesis in Swiss mouse 3T3 and 3T6 cells, *Proc. Natl. Acad. Sci., U.S.A.,* 80, 6162, 1983.
101. **Muchmore, D. B. and de Haën, C.,** Counterregulatory control of intracellular H_2O_2 production by insulin and lipolytic hormone in adipocytes, *Fed. Proc.,* 39, 1926, 1980.
102. **Cassel, D., Zhuang, Y.-X., and Glazer, L.,** Vanadate stimulates Na^+/H^+ exchange activity in A431 cells, *Biochem. Biophys. Res. Commun.,* 118, 675, 1984.
103. **Harris, D. C., Rinhart, A. L., Hereld, D., Schwartz, R. W., Burke, F. P., and Salvador, A. P.,** Reduction potential of iron in transferrin, *Biochim. Biophys. Acta,* 838, 295, 1985.
104. **Löw, H., Sun, I. L., Navas, P., Grebing, C., Crane, F. L., and Morré, D. J.,** Transplasma membrane electron transport is part of a diferric transferrin reductase system, *Biochem. Biophys. Res. Commun.,* 137, 1117, 1986.
105. **Sun, I. L., Toole-Simms, W., Crane, F. L., Morré, D. J., Löw, H., and Chou, J. Y.,** Reduction of diferric transferrin by SV40 transformed pineal cells stimulate Na^+/H^+ antiport activity, *Biochim. Biophys. Acta,* 938, 17, 1988.
106. **Golub, E. S., Diaz de Pagan, T., Sun, I. L., Hall, A. K., Crane, F. L., and Isom, G.,** Can opioids regulate hemopoietic differentiation, *Ann. N.Y. Acad. Sci.,* 521, 123, 1988.
107. **Sun, I. L., Toole-Simms, W., Crane, F. L., Morré, D. J., Löw, H., and Chou, J. Y.,** Transformation with SV40 virus prevents retinoic acid inhibition of plasma membrane diferric transferrin reductase in rat liver cells, *J. Bioenerg. Biomemb.,* 20, 383, 1988.
108. **Fry, M.,** Diferric transferrin reduction by malaria (*Plasmodium falciparum*) infected erythrocytes, *Biochem. Biophys. Res. Commun.,* 158, 469, 1989.
109. **Crane, F. L., Löw, H., Sun, I. L., Morré, D. J., and Faulk, W. P.,** Interaction between oxidoreductase, transferrin receptor and channels in the plasma membrane, in *Growth Factors: From Genes to Clinical Application,* Sara, V., Hall, K., and Löw, H., Eds., Nobel Conference Series, Raven Press, New York, 1990, 129.

110. **Sun, I. L., Garcia-Cañero, R., Liu, W., Toole-Simms, W., Crane, F. L., Morré, D. J., and Löw, H.,** Diferric transferrin reduction stimulates the Na^+/H^+ antiport of HeLa cells, *Biochem. Biophys. Res. Commun.,* 145, 467, 1987.
111. **Sun, I. L., Navas, P., Crane, F. L., Morré, D. J., and Löw, H.,** Diferric transferrin reductase in the plasma membrane is inhibited by adriamycin, *Biochem. Int.,* 14, 119, 1987.
112. **Sun, I. L. and Navas, P.,** Transplasma membrane diferric transferrin reductase. Control by antitumor drugs, in *Redox Functions of the Eukaryotic Plasma Membrane,* Ramirez, J., Ed., Consejo Superior de Investigaciones Cientificas, Madrid, 1987, 67.
113. **Testa, U.,** Transferrin receptors: structure and function, *Curr. Top. Hematol.,* 5, 127, 1985.
114. **Morgan, E. H., Smith, G. D., and Peters, T. J.,** Uptake and subcellular processing of 59Fe-12S I labeled transferrin by rat liver, *Biochem. J.,* 237, 163, 1986.
115. **Löw, H., Grebing, C., Lindgren, A., Tally, M., Sun, I. L., and Crane, F. L.,** Involvement of transferrin in the reduction of iron by the transplasma membrane electron transport system, *J. Bioenerg. Biomemb.,* 19, 535, 1987.
116. **Basset, P.,** Modulation of transferrin receptor gene expression by peptide growth factors and role of the receptor in human fibroblasts, in *Plasma Membrane Oxidoreductases in Control of Animal and Plant Growth,* Crane, F. L., Morré, D. J., and Löw, H., Eds., Plenum Press, New York, 1988, 163.
117. **Seligman, P. A. and Chitambar, C. R.,** Effect of different transferrin metal forms on both cellular iron metabolism and proliferation, in *Proteins of Iron Storage and Transport,* Spik, G., Montreuil, I., Chrichton, R. R., and Mazurier, J., Eds., Elsevier, Amsterdam, 1985, 123.
118. **Hanover, J. A. and Dickson, R. B.,** Transferrin: receptor mediated endocytosis and iron delivery, in *Endocytosis,* Pastan, I. and Willingham, M. L., Eds., Plenum Press, New York, 1985, 131.
119. **Rothenberger, S., Iacopetta, B. J., and Kühn, L. C.,** Endocytosis of the transferrin receptor requires the cytoplasmic domain but not its phosphorylation site, *Cell,* 49, 423, 1987.
120. **Octave, J.-N., Schneider, Y. J., Trouet, A., and Chrichton, R. R.,** Iron uptake and utilization by mammalian cells. I. Cellular uptake of transferrin and iron, *Trends Biochem. Sci.,* 8, 217, 1983.
121. **Cooper, S. R., McArdle, J. V., and Raymond, K. N.,** Siderophore electrochemistry: Relation to intracellular release mechanism, *Proc. Natl. Acad. Sci., U.S.A.,* 75, 3551, 1978.
122. **Allnutt, F. C. T. and Bonner, W. D., Jr.,** Characterization of iron uptake from ferrioxamine B by *Chlorella vulgaris, Plant Physiol.,* 85, 746, 1987.
123. **Morley, C. G. D. and Bezkorovainy, A.,** The removal of iron from transferrin by isolated membranes of rat hepatocytes, *Int. J. Biochem.,* 17, 553, 1985.
124. **Wu, C. Y. and Wu, C. H.,** Targeted inhibition of transferrin mediated iron uptake in HepG2 hepatoma cells, *J. Biol. Chem.,* 261, 16834, 1986.
125. **Harris, D. C. and Aisen, P.,** Facilitation of FeII autooxidation by FeIII complexing agents, *Biochim. Biophys. Acta,* 329, 156, 1973.
126. **Bates, G. W., Workman, E. F., Jr., and Schlabach, M. R.,** Study of FeIII transferrin formation from FeII plus oxygen, *Biochem. Biophys. Res. Commun.,* 50, 84, 1973.
127. **Thorstensen, K. and Romslo, I.,** Uptake of iron from transferrin by isolated hepatocytes, *Biochim. Biophys. Acta,* 804, 200, 1984.
128. **Thorstensen, K., Thommesen, L., and Romslo, I.,** Uptake of iron from ferric iron compounds by isolated hepatocytes, in *Plasma Membrane Oxidoreductases in Control of Animal and Plant Growth,* Crane, F. L., Morré, D. J., and Löw, H., Eds., Plenum Press, New York, 1988, 127.
129. **Thommesen, L., Thorstensen, K., Romslo, I., Nilsen, T., and Gederaas, O.,** Hepatocyte iron uptake — a plasma membrane process or an endosomal process, in *Plasma Membrane Oxidoreductases in Control of Animal and Plant Growth,* Crane, F. L., Morré, D. J., and Löw, H., Eds., Plenum Press, New York, 1988, 135.
130. **Trinder, D., Morgan, E. H., and Baker, E.,** The effects of an antibody to the rat transferrin receptor and of rat serum albumin on the uptake of diferric transferrin by rat hepatocytes, *Biochim. Biophys. Acta,* 943, 440, 1988.
131. **Sorokin, L. M., Morgan, E. H., and Yeoh, G. C. T.,** Transformation induced changes in transferrin and iron metabolism in myogenic cells, *Cancer Res.,* 49, 1941, 1989.
132. **Morgan, E. H.,** Membrance transport of non-transferrin-bound iron by reticulocytes, *Biochim. Biophys. Acta,* 943, 428, 1988.
133. **Medina, M. A., Sanchez-Jiménez, F., Segura, J. A., and Nuñez de Castro, I.,** Transmembrane ferricyanide reductase activity in Ehrlich ascites tumor cells, *Biochim. Biophys. Acta.,* 946, 1, 1988.
134. **Kruger, S. and Böttger, M.,** NADH or NADPH, in *Plasma Membrane Oxidoreductases in Control of Animal and Plant Growth,* Crane, F. L., Morré, D. J., and Löw, H., Eds., Plenum Press, New York, 1988, 105.
135. **Benedetti, E. L. and Emmelot, P.,** Structure and function of plasma membranes isolated from liver, in *The Membranes,* Dalton, A. J. and Haguenau, F., Eds., Academic Press, New York, 1968, 33.

136. **Aisen, P. and Leibman, A.,** Thermodynamic and accessibility factors in the specific binding of iron to ferric transferrin, in *Transport by Proteins,* Blauer, G. and Sund, H., Eds., Walter de Gruyter, Berlin, 1978, 277.
137. **DePierre, J. W. and Ernster, L.,** Enzyme topology of intracellular membranes, *Annu. Rev. Biochem.,* 46, 201, 1977.
138. **Sun, I. L., Toole-Simms, W., Crane, F. L., Golub, E. S., Diaz de Pagan, T., Morré, D. J. and Löw, H.,** Retinoic acid inhibition of transplasmalemma diferric transferrin reductase, *Biochem. Biophys. Res. Commun.,* 146, 976, 1987.
139. **Choury, D., Leroux, A., and Kaplan, J. C.,** Membrane bound cyctochrome b_5 reductase (methemoglobin reductase) in human erythrocytes: study in normal and methemoglobinemia subjects, *J. Clin. Invest.,* 67, 149, 1981.
140. **Jarasch, E.-D., Kartenbeck, J., Bruder, G., Fink, A., Morré, D. J., and Franke, W. W.,** B type cytochromes in plasma membrane isolated from rat liver in comparison with those of endomembranes, *J. Cell. Biol.,* 80, 37, 1979.
141. **Goldenberg, H.,** Density gradient fractionation of digitonin treated rat liver plasma membranes and subcellular localization of NADH oxidoreductases and B-type cytochromes, *Enzyme,* 27, 227, 1982.
142. **Bendzko, P. and Pfeil, W.,** Thermodynamics of cytochrome b_5 unfolding. II. Detergent solubilized cytochrome b_5 in solution and a reconstituted system with dimyrestoylphosphatidyl choline, *Biochim. Biophys. Acta,* 742, 669, 1983.
143. **Kant, J. A. and Steck, T. L.,** Cation impermeable inside-out and right-side-out vesicles from human erythrocyte membranes, *Nature* 240, 26, 1972.
144. **Steck, T. L. and Kant, J. A.,** Preparation of impermeable ghosts and inside-out vesicles from human erythrocyte membranes, in *Methods in Enzymology,* Vol. 31, Fleischer, S. and Packer, L., Eds., Academic Press, New York, 1974, 172.
145. **Steck, T. L.,** Preparation of impermeable inside-out and right-side-out vesicles from erythrocyte membranes, in *Methods in Membrane Biology,* Vol. 2, Korn, E. D., Ed., Plenum Press, New York, 1974, 245.
146. **Borgese, N. and Pietrini, G.,** Distribution of NADH cytochrome b_5 reductase in rat liver cells studied by a quantitative radioimmunoblotting assay, *Biochem. J.,* 239, 393, 1986.
147. **Bruder, G., Bretscher, H., Franke, W. W., and Jarasch, E.-D.,** Plasma membranes from intestinal microvilli and erythrocyte contain cytochrome b_5 and P420, *Biochim. Biophys. Acta,* 600, 739, 1980.
148. **Delaisse, J.-M., Martin, P., Verheyer-Bonnitz, M.-F., and Nyus, E. J.,** Subcellular distribution of enzymes in the yeast *S. lipolytica* grown on hexadecane with special reference to the w hydroxylase, *Biochim. Biophys. Acta,* 676, 77, 1981.
149. **Lodish, H. P., Braell, W. A., Schwartz, A. L., Strous, G. J. A. M., and Zilberstein, A.,** Synthesis and assembly of membrane and organelle proteins, *Int. Rev. Cytol.,* 12, (Suppl), 248, 1981.
150. **Goldenberg, H.,** Insulin inhibits NADH-semidehydroascorbate reductase in rat liver plasma membrane, *Biochem. Biophys. Res. Commun.,* 94, 721, 1980.
151. **Navas, P., Estevez, A., Vallalba, J. M., and Buron, M. I.,** Role of cell surface NADH-MDHA oxidoreductase: lectin inhibition, in *Plasma Membrane Oxidoreductases in Control of Animal and Plant Growth,* Crane, F. L., Morré, D. J., and Löw, H., Eds., Plenum Press, New York, 1988, 429.
152. **Navas, P., Estevez, A., Buron, M. I., Valalba, J. M., and Crane, F. L.,** Cell surface glycoconjugates control the activity of the NADH ascorbate free radical reductase of rat liver plasma membrane, *Biochem. Biophys. Res. Commun.,* 154, 1029, 1988.
153. **Shultze, H.-V., Gallencamp, H., and Staudinger, H.,** Untersuchungen zum mikrosomalen NADH abhängigen Elektrontransport, *Hoppe-Seyler's Z. Physiol. Chem.,* 354, 391, 1973.
154. **Sun, I. L., Crane, F. L., Morré, D. J.,** Calmodulin-NADH semidehydroascorbate oxidoreductase interactions of clathrin coated vesicles, *Biochem. Biophys. Res. Commun.,* 115, 952, 1983.
155. **Njus, D., Kusnetz, N. J., Pacquing, Y. V., and Kelley, P. M.,** Cytochrome b_{561} and the maintenance of redox poise in secretory vesicles, in *Plasma Membrane Oxidoreductases in Control of Animal and Plant Growth,* Crane, F. L., Morré, D. J., and Löw, H., Eds., Plenum Press, New York, 1988, 303.
156. **Wakefield, L. M., Cass, A. E. G., and Radda, G. K.,** Electron transfer across the chromaffin granule membrane, *J. Biol. Chem.,* 261, 9746, 1986.
157. **Hara, T. and Minakami, M.,** Purification and properties of NADH cytochrome b_5 reductase from microsomes, *J. Biochem.,* 69, 325, 19.
158. **Morré, D. J., Crane, F. L., Sun, I. L., and Navas, P.,** The role of ascorbate in biomembrane energetics, *Ann. N.Y. Acad. Sci.,* 498, 153, 1987.
159. **Miner, C., Lopez-Burillo, S., Garcia-Sancho, J., Hereros, B.,** Plasma membrane NADH dehydrogenase and Ca^{+2} dependent potassium transport in erythrocytes from several animal species, *Biochim. Biophys. Acta,* 727, 266, 1983.
160. **Kalen, A., Norling, B., Appelkvist, E. L., and Dallner, G.,** Ubiquinone synthesis by the microsomal fraction from rat liver, *Biochim. Biophys. Acta,* 926, 70, 1987.

161. **Zambrano, F., Fleischer, S., and Fleischer, B.,** Lipid composition of the Golgi apparatus of rat kidney and liver in comparison with other subcellular organelles, *Biochim. Biophys. Acta,* 380, 357, 1975.
162. **Burton, C. W., Joyce, A., and Ingold, K. V.,** Is vitamin E the only lipid soluble chain breaking antioxidant in human blood plasma and erythrocyte membranes?, *Arch. Biochem. Biophys.,* 221, 281, 1983.
163. **Yamamoto, Y., Niki, E., Eguchi, J., Kamiya, Y., and Shimasaki, H.,** Oxidation of biological membranes and its inhibition, *Biochim. Biophys. Acta,* 819, 29, 1985.
164. **Dreyer, J. L. and Treichler, T.,** NADH dehydrogenase in synaptic plasma membranes, in *Plasma Membrane Oxidoreductases in Control of Animal and Plant Growth,* Crane, F. L., Morré, D. J., and Löw, H., Eds., Plenum Press, New York, 1988, 406.
165. **Crane, F. L. and Morré, D. J.,** Evidence for coenzyme Q function in Golgi membranes, in *Biomedical and Clinical Aspects of Coenzyme Q,* Folkers, K., and Yamamura, Y., Elsevier, Amsterdam, 1977, 3.
166. **Holmgren, A.,** The bovine thioredoxin system, *J. Biol. Chem.,* 252, 4600, 1977.
167. **Bennett, C. F., Balcarek, J. M., Varrichio, A., and Crooke, S. T.,** Molecular cloning and complete amino acid sequence of form phosphoinositide specific phospholipase C, *Nature,* 334, 268, 1988.
168. **Williams, L. T.,** Signal transduction by the platelet derived growth factor, *Science,* 243, 1564, 1989.
169. **Agutter P. S., Sherriff, R. V., and Kadlubowski, M.,** An NADH fatty acid hydroperoxide reductase on erythrocyte membranes, *Biochem. Soc. Trans.,* 7, 710, 1979.
170. **DuPone, F. M. and Windle, J. J.,** Do ESR spin probes detect a plasma membrane electron transport system, in *Redox Functions of the Eukaryotic Plasma Membrane,* Ramirez, J., Ed., Consejo Superior de Investigaciones Cientificas, Madrid, 1987, 251.
171. **Chen, K., Morse, P. D., III, and Swartz, H. M.,** Kinetics of enzyme-mediated reduction of lipid soluble nitoxide spin labels by living cells, *Biochim. Biophys. Acta,* 943, 477, 1988.
172. **Goldenberg, H., Eder, M., Pumm, R., and Dodel, B.,** Reductive release of iron from transferrin and receptor mediated recycling, in *Plasma Membrane Oxidoreductases in Control of Animal and Plant Growth,* Crane, F. L., Morré, D. J., and Löw, H., Eds., Plenum Press, New York, 1988, 139.
173. **Ponka, P., Schulman, H. M., and Wilczynska, A.,** Ferric pyridoxyl isonicotinoyl hydrazone can provide iron for heme synthesis in reticulocytes, *Biochim. Biophys. Acta,* 718, 151, 1982.

Chapter 3

NADH OXIDASE

D. James Morré and F. L. Crane

TABLE OF CONTENTS

I. INTRODUCTION

A redox system of the plasma membrane has been described that is responsive to growth factors and inhibited by antitumor drugs.[1] Electrons are transferred from cytoplasmic NADH to external electron acceptors such as ferric iron, ascorbate free radical, or molecular oxygen.[2] Transfer to oxygen appears, under certain conditions, to be rate-limiting to growth.[3] Protons released from oxidized NADH may serve to link the plasma membrane redox system to some aspect of cell cycle entry. Possibilities include activation of the Na^+/H^+ antiport,[4] and an attendant alkalinization of the cytoplasm.[1]

Information leading to the discovery and documentation of the constituents of a plasma membrane redox system in eukaryotic cells has been reviewed extensively.[1,5-8] The transmembrane dehydrogenases have received the most attention because of evidence of a potential direct involvement both in membrane transport and in growth. These dehydrogenases transfer electrons to external acceptors presumably via an electron chain, although the constituents of the chain have not been defined. The dehydrogenases respond also to growth factors and hormones and may be part of the mechanism whereby plasma membrane electron transport and control of growth are linked.

Functioning of the transmembrane electron-transport system in growth control is considered to depend on the movements of electrons or protons across or within the plane of the membrane. Both oxygen and iron-containing compounds are available as acceptors but the relationship of the transmembrane transfer of electrons to oxygen is less clear than to iron compounds. For example, even how oxygen and the transmembrane redox chain are linked remains to be determined. Yet, both NADH- and NADPH-acceptor oxidoreductase activities of the plasma membrane are distributed widely among a variety of tissue and cell types.[9-17] The NADH oxidase that may participate as part of a plasma membrane electron transport chain as a terminal oxidase is the subject of this chapter.

II. EVIDENCE FOR THE PLASMA MEMBRANE LOCATION OF A UNIQUE NADH OXIDASE

NADH/acceptor oxidoreductase activities using oxygen as electron acceptor have been demonstrated most unambiguously with isolated plasma membranes (Table 1).[6,10,11,13,17-24] Membrane sources include rat liver,[10,21] rat adipocytes,[13] renal brush border membranes,[22] mouse liver,[11] soybean hypocotyl,[23] and yeast,[17] That the NADH oxidase of plasma membrane is a distinct enzymatic activity is shown also by inhibitor studies (Figure 1, Table 2). Because the oxidase activity is insensitive to cyanide (Table 2), it is distinguished readily from that of mitochondria. The endogenous NADH oxidase of liver membranes is inhibited by atebrin at millimolar concentrations (Table 2). The NADH oxidase also is insensitive to rotenone and relatively resistant to the sulfhydryl reagent PCMB. The activity of endoplasmic reticulum, Golgi apparatus and other mitochondrial membranes is inhibited completely at 80 *M* PCMB, whereas approximately one third of the plasma membrane activity remains at 160 *M* PCMB (Figure 1). Similarly, the oxidase activity of endoplasmic reticulum, Golgi apparatus, and outer mitochondrial membrane was unaffected or stimulated by the flavin antagonist atebrin over the range 2 to 8 m*M*, whereas that of the plasma membrane was inhibited (Figure 1). Similar effects were observed on cytochrome c reductase by plasma membrane.[10] These data point to the unique characteristics of the NADH oxidase of the plasma membrane.

The NADH of oxidase activity of highly purified plasma membrane fractions is 25% or more than that of endoplasmic reticulum, so that it cannot be attributed to endoplasmic reticulum contamination of plasma membrane fractions (Table 3). Specific activities of electron transfer from NADH to ferricyanide, cytochrome c, DCIP, and other artificial

TABLE 1
NADH Oxidase Activity in Plasma Membrane Preparations

Membrane source	NADH oxidase (nmol/min × mg protein)	Ref.
Rat liver	19	10
Mouse liver	60	11
Human erythrocyte	0	18
Pig erythrocyte	0	19
Squid retinal nerve	17	20
Rat adipocyte	20	21
Rat brush border	20	22
Corn coleoptile	2	12
Soybean hypocotyl	2	23
Yeast	4 (corrected)[a]	24

[a] Corrected on basis of contamination values given in paper.

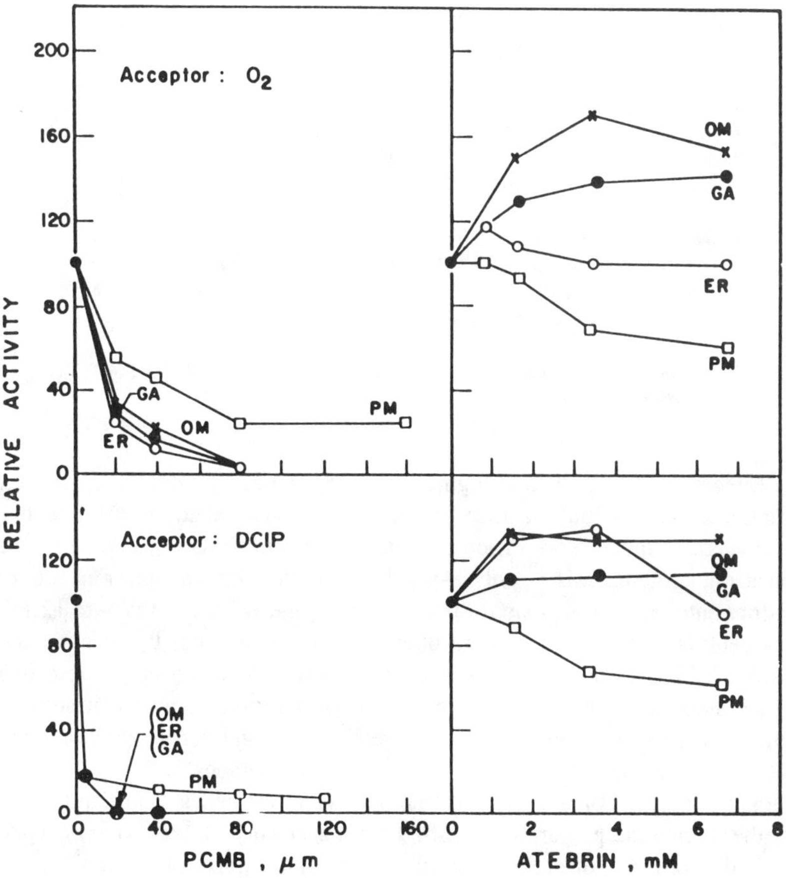

FIGURE 1. Effect of Atebrin and PCMB on NADH-DCIP reductase and NADH oxidase of mouse liver membranes. The compounds were added to the assay cuvette 2 min before starting the reaction with NADH. Plasma membrane (PM); endoplasmic reticulum (ER); Golgi apparatus (GA); outer mitochondrial membrane (OM). (From Goldenberg, H., Crane, F. L., and Morré, D. J., *J. Biol. Chem.*, 254, 2491, 1979. With permission.)

TABLE 2
Inhibition or Stimulation of NADH Oxidase Activity of Rat Liver Plasma Membrane

Agent	Conc	Stimulation (%)	Ref.
Atebrin	3 m*M*	−40	21
Sodium azide	0.1 *M*	+55	10
3,5,3′-Triiodothyronine	1 μ*M*	+240	10
Cyanide	1 m*M*	0	21

TABLE 3
NADH Oxidoreductase Specific Activities of Mouse Liver Endomembranes

	Cell fraction		
Acceptor	Plasma membrane	Endoplasmic reticulum	Outer mitochondrial membrane
Oxygen	60	250 (24)	200 (30)
Cytochrome c	70	1000 (7)	1000 (7)
Ferricyanide	600	3500 (17)	6400 (9)
INT	50	700 (7)	1000 (5)
DCIP	40	400 (10)	500 (8)
Juglone	200	2500 (8)	1800 (11)
Menadione	65	700 (9)	406 (16)

Note: Specific activites are in nm of acceptor reduced per min per mg protein. Figures in parenthesis are the specific activity in plasma membrane as a percentage of the specific activity in endoplasmic reticulum or outer mitochondrial membrane, respectively.

From Goldenberg, H., Crane, F. L., and Morré, D. J., *J. Biol. Chem.*, 254, 2491, 1979. With permission.

electron acceptors are 10- to 20-fold higher for endoplasmic reticulum and outer mitochondrial membranes than for plasma membrane. On the other hand, transfer to the natural acceptor, oxygen, is much more efficient with plasma membrane (Table 3).[11] Transfer to oxygen on a mg protein basis is only 3- to 4-fold higher for endoplasmic reticulum and mitochrondrial outer membrane than for plasma membrane, whereas it is 6- to 12-fold higher for other acceptors. Thus, compared to other electron acceptors, the relative transfer of electrons from NADH to oxygen for plasma membrane is 2 to 3 times more efficient than that for either endoplasmic reticulum or mitochondrial outer membranes. These studies were with plasma membranes from mouse liver, judged by electron microscope morphometry and analysis of marker enzymes to be 95 to 97% plasma membranes.

Additional evidence for a plasma membrane location comes from studies where the activity of the membrane preparations shifted from a density of 1.14 to 1.19 when treated with 1 mg of digitonin/mg of protein according to the procedures of Amar-Costecec et al.[25] In contrast, monoamine oxidase of outer mitochondrial membranes and endoplasmic reticulum markers were unaffected by the digitonin treatment.[11] The digitonin combines with the cholesterol-rich plasma membranes to shift their densities whereas those of cholesterol-poor membranes were unaffected.

TABLE 4
Comparison of the Inhibition by Antitumor Drugs of NAPH Oxidase and NADH-Reductase (Ferricyanide Reductase) Activities of Rat Liver Plasma Membranes

Drug	Conc (*M*)	% Inhibition Reductase	% Inhibition Oxidase
Adriamycin	10^{-7}	25	15
	10^{-6}	56	23
Cis-platinum	10^{-7}	24	11
	10^{-6}	84	12
Trans-platinum	10^{-6}	9	0

III. NADH OXIDASE IS DISTINCT FROM OTHER PLASMA MEMBRANE OXIDOREDUCTASES

The NADH oxidase is not a peroxidase. Hydrogen peroxide can be a product of NADPH oxidation by rat adipocyte plasma membrane,[13,14] but only very limited amounts of hydrogen peroxide were produced during NADH oxidation by rat liver plasma membrane.[26] Decavanadate, but not orthovanadate, induces an NADH oxidase activity in all plasma membranes tested including erythrocyte membranes.[25,29] That hydrogen peroxide formation during NADH oxidation by membranes is increased by polyvanadate ions,[27,28] suggests that the induced activity is a peroxidase different from the endogenous oxidase.

The oxidase activity has many characteristics to distinguish it from the transmembrane reductase. One such characteristic is that of inhibition of the plasma membrane redox system by certain antitumor drugs, including Adriamycin,[30] cis-platinum,[31] and bleomycin.[32] This inhibition may provide an explanation for the ability of the adriamycin, for example, to retain cytotoxic activity when linked to impermeant supports that are unable to enter the cell.[33-35] Inhibition, however, is of the transmembrane dehydrogenase and not of the NADH oxidase. Since these antitumor drugs have less effect on the activity of the oxidase than on the dehydrogenase (Table 4), they provide additional evidence for the distinct response patterns of the NADH reductase and the NADH oxidase activities. However, these impermeant antitumor drugs do inhibit both growth and electron flow from the cytoplasmic NADH across the plasma membrane. Thus, they also provide one line of evidence that electron flow from NADH to oxygen is via an electron transport chain and that the reductase and the oxidase are somehow linked.

Xanthine oxidase can also act as an NADH oxidase. Xanthine oxidase accounts for most of the NADH oxidase activity in the membrane of milk fat globules but not in liver or adipocyte plasma membranes.[6] The activity is distinct from the external oxidase of the mitochondrion and from the superoxide-generating NADH oxidase of neutrophils.

Another aspect of plasma membrane redox activities is the transfer of electrons from NADH to ascorbate free radical, the so-called NADH-ascorbate free radical oxidoreductase. This activity has been related to the activation of coated pits and coated vesicles involved in endocytosis and membrane recycling.[36-38] Oxidation of ascorbate radical has been suggested to energize membrane translocations with internal endomembranes where redox function may depend more strictly upon a pool of regeneratable acceptor such as is provided by the ascorbate free radical once it becomes sequestered within membrane lumens.[37] Here ascorbate oxidase would function as the terminal oxidase. Ascorbate acid oxidase activity was reported to be present in some preparations that contained plasma membrane.[39]

IV. CHARACTERISTICS OF THE OXIDASE FROM RAT LIVER PLASMA MEMBRANES

The NADH oxidase of rat liver plasma membranes has a pH optimum near 7 and, other than NADH, has no obvious cofactor requirements although low levels of Ca^{2+} may be beneficial. The activity is resistant to cyanide and 1 m*M* KCN is normally added to the incubation mixtures as a specificity control and to inhibit activity of mitochondrial origin. The activity shows a low degree of detergent- (structure-)linked latency consistent with an externally oriented active site.

The activity has been partially purified by ion exchange chromatography and HPLC on lipid supports. Plasma membrane vesicles isolated by aqueous two-phase partition[40] were solubilized with the zwitterionic detergent, CHAPS, at a ratio of 2:1 CHAPS:protein (w/w). Solubilized proteins were fractionated first on DEAE-cellulose (DE 52) with stepwise salt elution. Final purification utilized solid phase HPLC supports that mimic biological membranes. These supports were formed by covalently bonding phospholipid monolayers at high densities to rigid, insoluble silica particles.[41] The purified NADH oxidase fraction from rat liver contains two peptides of M_r 32—34 and 70—72 kDa. A similar preparation from plant plasma membranes contained peptides of M_r 36, 52, and 72 kDa.[23] The stoichiometry of NADH oxidized to oxygen consumed by the oxidase of plasma membranes is approximately 2. The purified respiratory burst NAD(P)H oxidase from human neutrophils also showed three prominent bands at M_r of approximately 32, 48(47), and 66(65) kDa. Affinity labeling suggested that M_r 66(65)-kDa band contained the NADPH binding site.[42] The liver activity is distinct from the peroxidase activity of the plasma membrane and is unaffected by superoxide dismutase, catalase or peroxidase inhibitors such as the iron chelator bathophenanthroline phosphosulfonate. The partially purified protein exhibits no Soret band, is not inhibited by cyanide, and appears not to be a heme protein.

V. NADH OXIDASE AS PART OF AN ELECTRON TRANSPORT CHAIN

While oxygen has been suggested to be the natural acceptor of electrons coming from cytoplasmic NADH, the nature of the constituents of the electron transport chain and their organization within the plasma membrane remain to be determined. Presumably, the transmembrane dehydrogenase is oriented with the NADH site at the cytoplasmic membrane surface. This is where deposits of reduced copper ferrocyanide indicative of NADH oxidation are located by electron microscope cytochemistry.[43-45]

If the oxidase is to function as a terminal oxidase to facilitate transfer of electrons to O_2, then the two activities, the oxidase and the reductase, must somehow be coupled. A major limitation in the understanding of plasma membrane electron transport is the need to identify additional electron chain constituents between NADH and molecular oxygen. From theoretical considerations, it is expected that a minimum of three prosthetic groups will be necessary to bridge 35 Å of membrane lipid bilayer because of a 10 Å limitation on electron tunneling.[46]

Flavin, iron, copper, and thiols all have been identified in some plasma membranes.[47] Additionally, cytochromes of the b-type may be present, including both P420 and P450.[48] There are also reports, some conflicting, concerning coenzyme Q[48-50] and a sequence of carriers including flavin, coenzyme Q, and a low-potential cytochrome b in the plasma membrane of neutrophils.[51-52] Other possible candidates as intermediate acceptors would be sulfhydryl (thiol) groups,[53] pterins,[54] and ascorbate free radicals.[36-38] Also not yet adequately analyzed are the nonheme irons and various SH- or organic free radical-generating groups.

There is virtually no information on which constituents participate in the electron transport chain or as to the order of participation of electron transport chain constituents. An involvement of plasma membrane P450 is regarded as unlikely due to lack of carbon monoxide inhibition of the bulk of the transmembrane redox activity. A protein that binds [^{3}H]pteroxylglutamic acid has been described recently for plasma membranes of rat liver,[54] raising the possibility of a pterin involvement. Yet there is no evidence for involvement of nonheme iron, flavin, quinones, or pterins as obligatory chain constituents.

The oxidase activity is inhibited by *N*-ethylmaleimide and is responsive to various sulfhydryl reagents (Hidalgo et al.[95]). These findings suggest some role for sulfhydryl groups in the activity of the oxidase and/or in the linkage between the reductase and the electron transport chain. However, the general lack of information on electron chain constituents does not permit a distinction even, for example, between the oxidase and dehydrogenase of the plasma membrane functioning in parallel or in series.

VI. ORIENTATION WITHIN THE PLASMA MEMBRANE OF THE NADH OXIDASE

Since cells and plasma membranes are impermeable to NADH, measurements with intact cells and right side-out plasma membrane vesicles must involve an external NADH site for the oxidase. Liver plasma membranes are especially favorable for oxidase studies since, due to junctional complexes, formation of sealed vesicles is prevented so that the preparations may contain sheets with both external and cytoplasmic surfaces exposed.

The oxidase is normally assumed to occupy a location in the external leaflet of the plasma membrane bilayer whereas the dehydrogenase would have a transmembrane location (Figure 4). An orientation at one membrane surface or the other is favored by the relative ease by which the oxidase is release from the membrane by detergents and its sensitivity to trypsin and to the membrane-impermeant sulfhydryl reagent *N*-ethylmaleimide. A location in the external membrane leaflet also is favored by the relatively greater sensitivity of NADH oxidase to inhibition by gangliosides which are thought to occupy the outer membrane leaflet of the plasma membrane. An antibody to the purified oxidase and plasma membrane vesicles of known absolute orientation will be required to unambiguously answer this question. The strong inhibition of NADH oxidase by an impermeant inhibitor such as actinomycin D[3,23,55] also suggests an external orientation.

Cytoplasmic side-in and cytoplasmic side-out vesicles may be resolved on the basis of differences in surface charge by the technique of preparative free-flow electrophoresis.[40,56,57] Cytoplasmic side-in vesicles bind concanavalin A and exhibit a latent ATPase activity. In constrast, cytoplasmic side-out vesicles lack externally exposed concanavalin A binding sites and exhibit an ATPase specific activity in the absence of detergent equivalent to that observed with cytoplasmic side-in vesicles in the presence of detergent. Both populations of vesicles exhibit NADH oxidase activity, suggesting that the NADH oxidase may be accessible to NADH from either membrane surface.

VII. LIPID REQUIREMENTS FOR ACTIVITY

The lipid environment of the membrane is increasingly being shown to have profound effects, both direct and indirect, on the signal response cascade of growth control.[58-60] The effect of 1,2-diacylglycerols derived from phosphoinositides on protein kinase C is one example of a direct effect of a lipid modification.[61] The response of protein kinases[62] and the NADH oxidase to positive control by lysophospholipids and fatty acids (Figure 2) and to negative control by gangliosides (Figure 3) provide additional examples.

Gangliosides are concentrated in the external lipid leaflet of the plasma membrane.[64-65] As such, their growth effects also should be most pronounced on membrane proteins as-

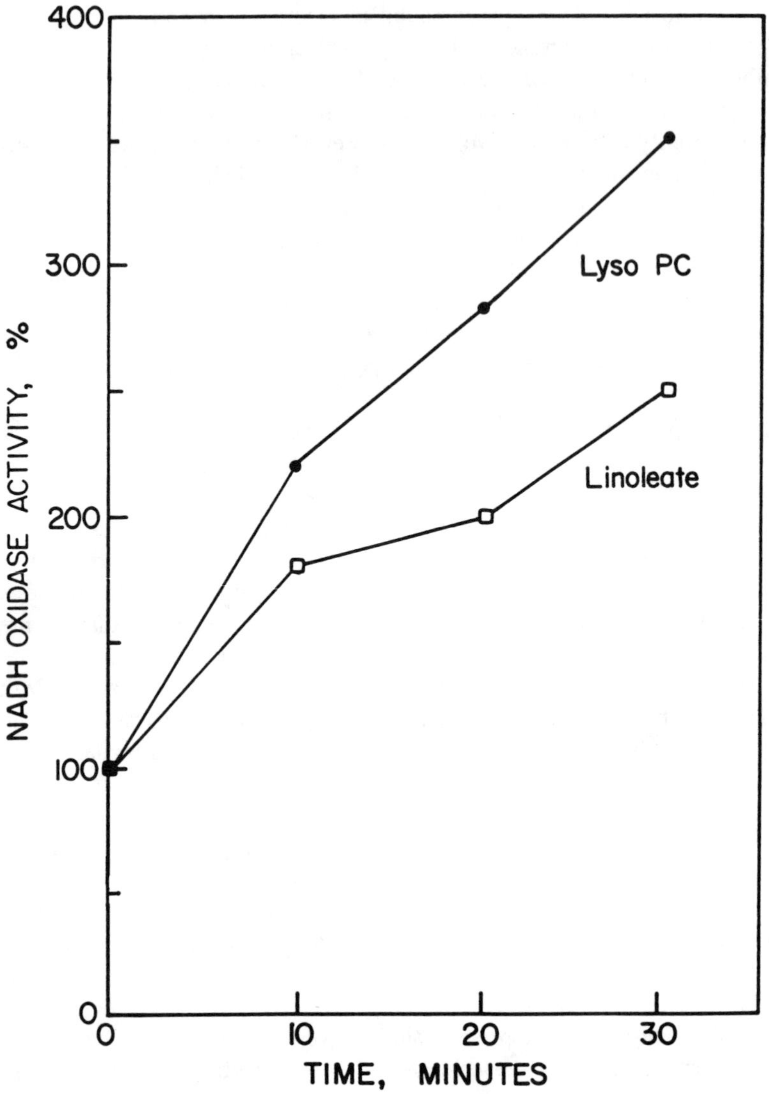

FIGURE 2. Stimulation of NADH oxidase of rat liver plasma membranes by lysophosphatidylcholine (Lyso PC) (20 *M*) and by a free fatty acid (linoleic acid) (200*M*) and the time dependency of stimulation. The stimulation of NADH oxidase by both the lysophospholipid and the fatty acid increase with time of incubation. Such a response would be consistent with an activation mechanism based upon stabilization of an active conformation of the enzyme. Similar results were obtained in the presence of 0.1% Triton X-100 (e.g., Table 5) to eliminate structure-linked latency as the basis for the stimulatory response. (From Brightman, A., Zhu, X.-Z., and Wang, J., Purdue University, unpublished data. With permission.)

sociated with the outer membrane leaflet. This already has been determined for platelet-derived (PDGF)[59] and epidermal (EGF) growth factor receptors.[60] Here growth factor-stimulated membrane phosphorylation or growth factor-stimulated autophosphorylation are inhibited by growth inhibitory gangliosides. A growth-stimulatory ganglioside such as de-*N*-acetyl-G_{M3} acts as a strong promoter for EGF receptor kinase[66] in the outer membrane leaflet of the plasma membrane. NADH oxidase also is inhibited by growth inhibitory gangliosides in a ganglioside-specific and dose-dependent manner (Figure 3) in support of an external

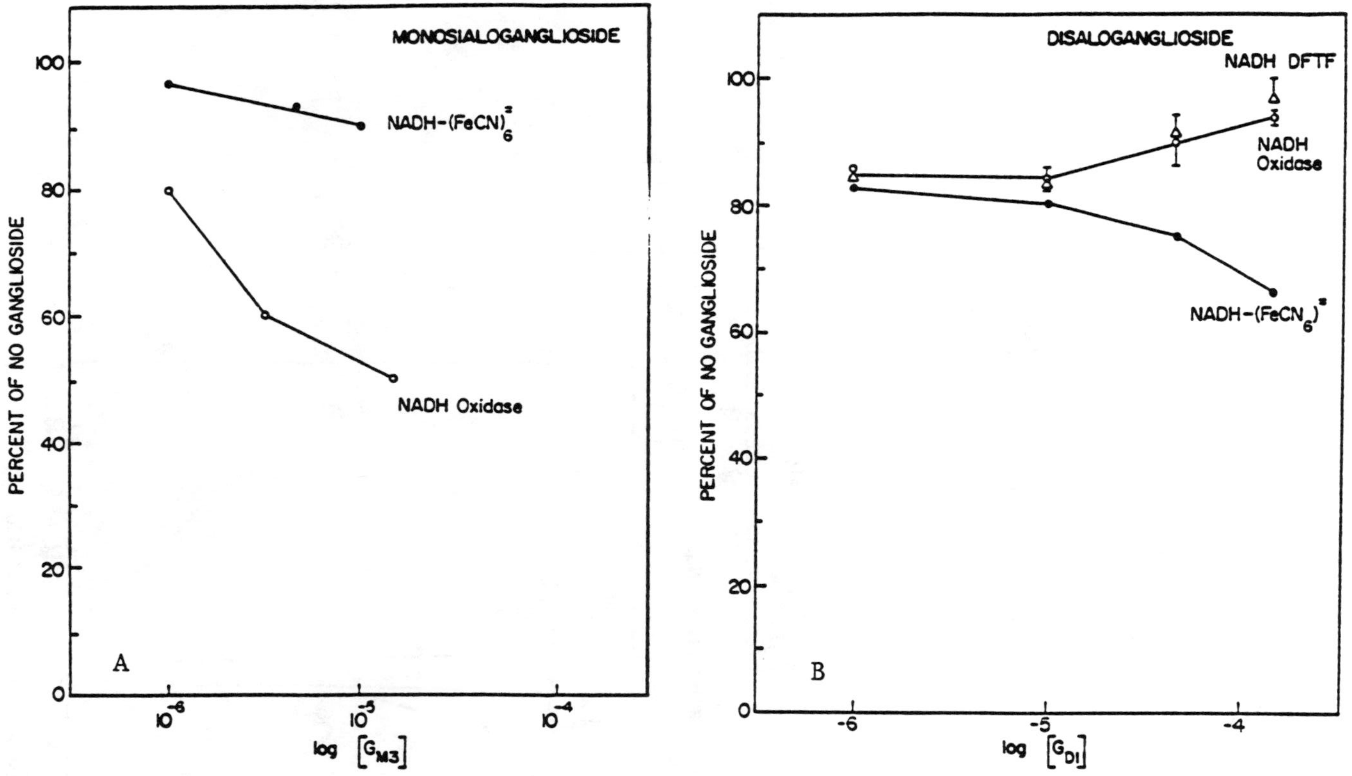

FIGURE 3. Response of redox activities of plasma membranes purified from rat liver to gangliosides prepared from bovine brain. Enzyme activities illustrated include NADH oxidase (inhibited), a transmembrane NADH-ferricyanide oxidoreductase [NADH-$(FeCN)_6^{2-}$] (not inhibited) and NADH-diferric transferrin oxidoreductase (NADH DFTF) not inhibited or stimulated at high concentrations). (A) The monosialoganglioside, G_{M3}. (B) The disialoganglioside, G_{D1a}. (C) The trisialoganglioside, G_{T1b}. (D) Mixed brain gangliosides. Note the preferential inhibition of the NADH oxidase by G_{M3}. Figure continued on next page. (From Morré, D. J., Wilkinson, F. E., Wu, L. Y., and Zhu, X.-Z., Purdue University, unpublished data. With permission.)

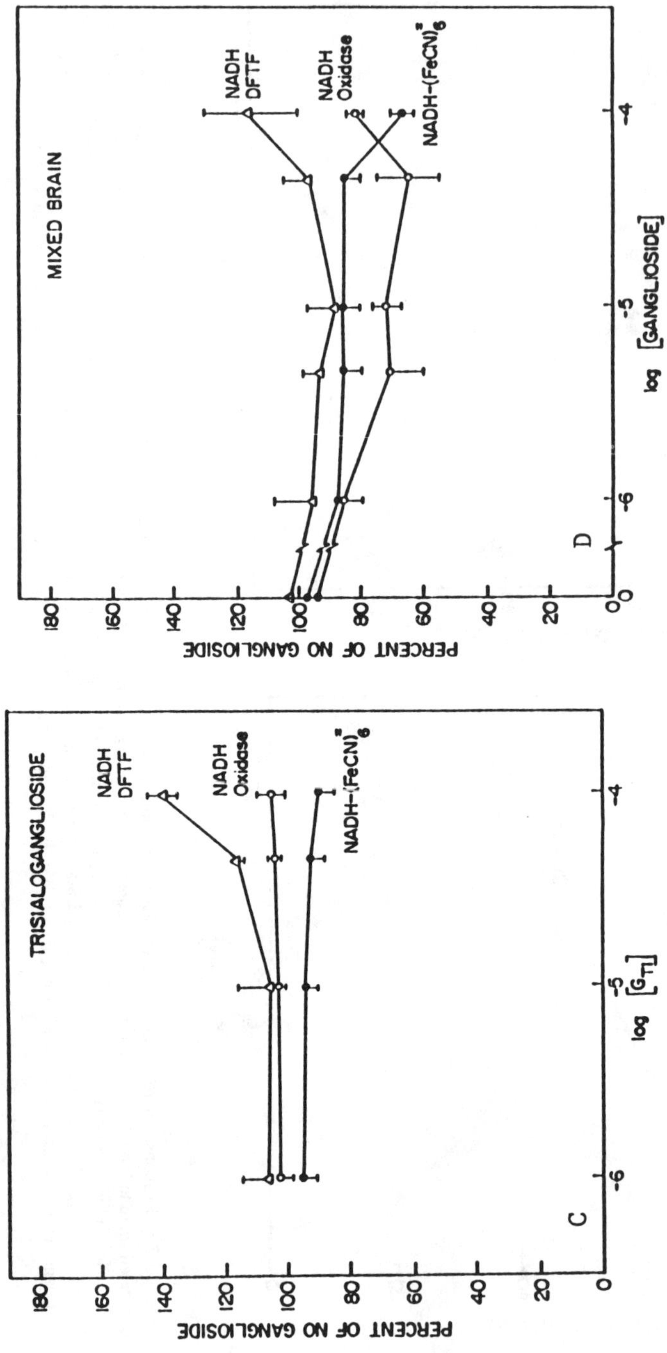

FIGURE 3. Continued.

TABLE 5
Stimulation of NADH Oxidase Activity of Rat Liver Plasma Membrane by 30 μ*M* Lysophosphatidylcholine or Linoleic Acid in the Presence or Absence of 0.1% Triton X-100[97]

Additions	0.1% Triton X-100	nmol/min/mg protein
None	−	1.5
	+	1.6
Lysophosphatidylcholine	−	5.3
	+	5.8
Linoleic acid	−	3.4
	+	5.3

location of the oxidase. As with the effect of G_{M3} or G_{M1} on EGF or PDGF receptor kinase activity, our results show that the inhibitory effects of gangliosides on NADH oxidase are given not only with cells but also with isolated plasma membranes and even with detergent-solubilized or partially purified enzyme. Ganglioside inhibition of the purified EGF receptor already has been demonstrated.[59,60] The inhibition of protein kinases facing the cytoplasm may be more subject to negative control by products of the sphingosine cycle[67] rather than by gangliosides with which, even if inhibitory, they would not be expected to contact. Finally, the NADH oxidase is stimulated by lysophospholipids and fatty acids (Figure 2, Table 5). Both are products of phospholipase A_2 action on the membranes and stimulations of several hundred percent are readily achieved. The NADH oxidase of rat liver is stimulated by both lysophospholipids and free fatty acids under latency-free conditions of 0.1% Triton X-100 (Table 5) to rule out a simple detergent effect. The enhancement is time dependent, (Figure 2) suggestive of an action in which the enzyme activity is stabilized by lysophospholipids and free fatty acids in a more reactive conformation.[68] Thus the hormone-dependent activation of a phospholipase A_2 linked to a GTP-binding protein[69] may be an important component of the signal transduction pathway of oxidase control.

VIII. NADH IS A NATURAL ELECTRON DONOR

NADH has been identified as the natural electron donor for plasma membrane electron transport in HeLa cells.[70] NADH levels have been implicated previously as important to cell proliferation[71] and in SV40-transformed pineal[72] and liver[73] cells.

Numerous studies have shown that pyridine nucleotide pools are significantly altered in proliferating tissues in comparison to those of nonproliferating tissues.[74-80] We have previously observed a decrease in NAD(H) levels of rat liver following administration of a carcinogen (2-acetylaminofluorene).[81] Furthermore, changes in pyridine nucleotide levels as a function of growth in normal and transformed 3T3 cells have been reported.[82] NADH or NADPH may increase[71,83] or decrease[74] as a result of transformation. Although the significance of the relationship between NAD(H) levels and cell proliferation is not clear, their possible involvements in growth control as donors to an electron transport chain involving NADH oxidase emerges as one possiblity.

IX. NADH OXIDASE ACTIVITY AND GROWTH CONTROL

A role of plasma membrane redox activities in growth is supported by two types of findings. First, NADH-ferricyanide reductase activity but not NADH oxidase activity is inhibited by growth-inhibitory drugs in parallel to their inhibition of hormone-induced growth. Second, NADH oxidase activity but not NADH-ferricyanide reductase activity is stimulated strongly by certain plant and animal growth factors. These observations are reconciled diagrammatically by a transmembrane flow of electrons from cytoplasmic NADH outward to the cell surface (Figure 4). In the absence of external electron acceptors, NADH oxidase may serve as a rate limiting terminal oxidase to facilitate transfer of electrons directly to oxygen. What role, if any, the plasma membrane NADH oxidase may play in the oxygen control of cell division[84] is not known. NADH oxidase has been reported to be absent from human and porcine erythrocyte membranes,[19,85] but erythrocytes are classic examples of nongrowing cells.

The transmembrane redox system has been shown directly to stimulate the growth of melanoma cells[86] and HeLa cells[87,88] and to partly replace the essential components of calf serum. It has also been proposed previously that the plasma membrane redox system can regulate cellular function through modification of hormone response.[89] These redox enzymes may also transduce energy for special amino acid transport.[90,91] Tumor or virus transformed cells, which have a defect in growth control, modify the transplasmalemma redox system by changes in the K_m and V_{max} of the enzyme.[72,73]

A. HORMONE EFFECTS

NADPH oxidase of fat cell membranes has been reported to be stimulated by insulin[13] as has NADH oxidase.[92] Diferric transferrin stimulates the NADH oxidase of liver plasma membranes (Table 6) but this stimulation is much less in hepatoma plasma membranes. The NADH oxidase of rat liver plasma membrane is stimulated by both 3,5,3′-triiodothyronine(T_3) and catecholamine.[10] T_3 binding proteins at the plasma membrane have been reported.[93]

With plasma membranes from plants, the NADH oxidase is stimulated by growth hormones.[23,94] Chemically related but growth-inactive analogs of these same compounds are without effect on the oxidase.[97] With the plant oxidase, the hormone responsiveness is retained upon detergent solubilization of the enzyme and with the partially purified enzyme.[23]

B. NADH OXIDASE RESPONSE TO GROWTH FACTORS IS ALTERED BY TRANSFORMATION

As summarized in a preceding section, NADH oxidase activity appears to be growth factor and hormone responsive. On the basis of its growth factor responsiveness, a role for the NADH oxidase in growth has been suggested.[3,55,94] At least with transferrin, the growth factor-responsiveness is lost during hepatocarcinogenesis so that the NADH oxidase of plasma membranes isolated from hyperplastic liver nodules induced by 2-acetylaminofluorene, the response of the oxidase to transferrin is much reduced compared to plasma membrane vesicles of control animals prepared in parallel (Table 6). Additionally, the activity is reduced on confluency in cultured cells (Wilkinson and Zhu.[96])

C. LINKAGE BETWEEN HORMONE OR GROWTH FACTOR RESPONSE AND NADH OXIDASE ACTIVATION

The linkage between hormone or growth factor response and NADH oxidase activation has not been clearly established. However, NADH oxidase activity appears to be responsive to certain lipid-derived second messengers, especially lysophospholipids and free fatty acids (Figure 2, Table 5). The latter would be derived from phospholipase A or B cleavage of the major glycerophospholipids including phosphatidylcholine, phosphatidylethanolamine, and phosphatidylinositol. Hormone responsive A-type phospholipases or phospholipases linked to GTP-binding transducin-type proteins[69] have been described but much more work needs to be done to elucidate the control sequence. A more direct linkage of the oxidase to hormone and growth factor receptors either with or without involvement of a GTP-binding transducin protein also is a possibility.

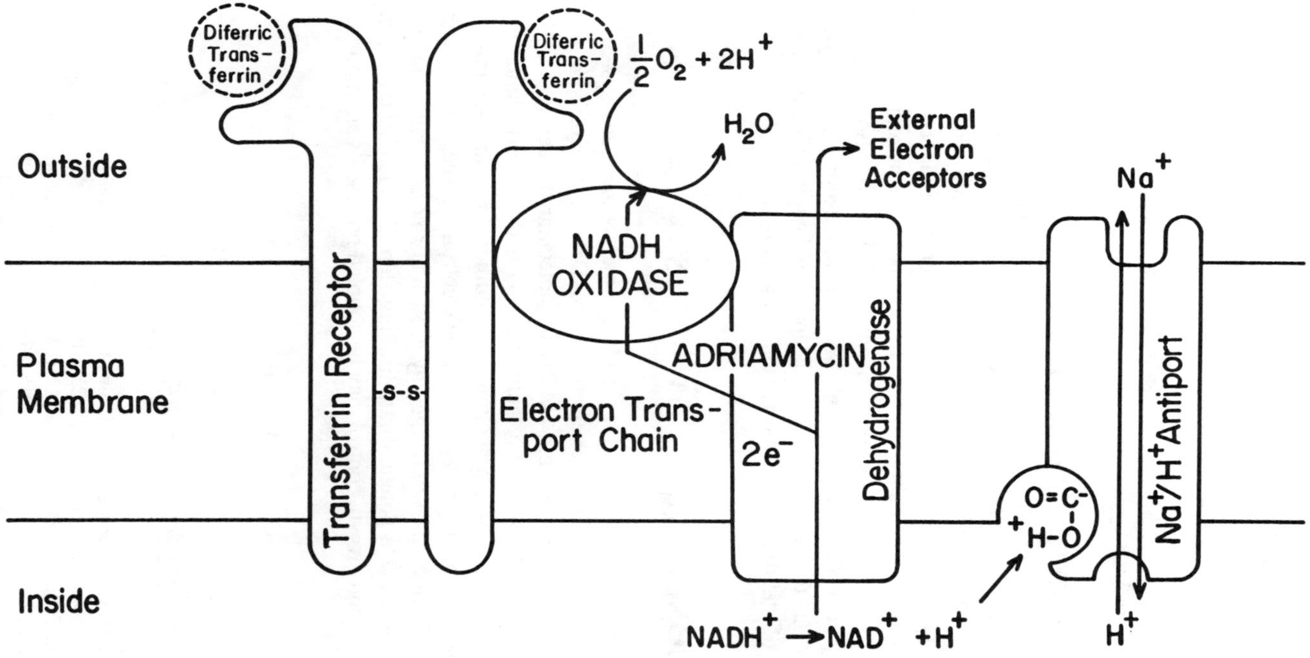

FIGURE 4. Diagram illustrating the hypothetical arrangement of constituents within the mammalian plasma membrane. The flow of electrons from cytoplasmic NADH is dipicted as being coupled to molecular oxygen via the NADH oxidase. The NADH oxidase would be subject both to direct activation via coupling to growth factor (e.g., transferrin) receptors, and to activation via the products of phospholipase A_2 action, lysophospholipids and free fatty acids.

TABLE 6
NADH Oxidase of Plasma Membrane from Rat Liver and Hyperplastic Nodules of the Rat

Source	Addition	nmol/min/mg protein
Liver	None	0.8 ± 0.1
	+ Diferric transferrin	1.3 ± 0.02
Hyperplastic nodules	None	1.5 ± 0.1
	+ Diferric transferrin	1.6 ± 0.1

Note: Assay was in the presence of 1 m*M* KCN, pH 7.4. The diferric transferrin concentration was 3 μ*M*.

D. NADH OXIDASE AND GROWTH CONTROL IN PLANTS

With elongating plant cells, cell expansion is controlled by growth hormones of the auxin type such as indole-3-acetic acid (IAA) or its synthetic analog, 2,4-dichlorophenoxyacetic acid (2,4-D). Auxin-induced growth is blocked by the same antitumor drugs as block growth of mammalian cells,[55] but the oxidase[23,94] and not the reductase is auxin responsive. These findings provide additional evidence for a link between reductase and oxidase in growth control and yet a further example of where NADH oxidase activity of the plasma membrane is subject to hormonal control. As with the mammalian enzyme, activity from plants is enhanced 3- to 4-fold by fatty acids and lysophospholipids in a time-dependent manner suggestive of stabilization of the enzyme in a more active conformation (see Section VII).

X. CONCLUDING COMMENTS

While there is increasing evidence for a role for the plasma membrane redox system in the growth mechanism, the observation of hormone-sensitive NADH oxidases of both plant and animal cells capable of electron transfer from NADH to molecular oxygen now add a new dimension to the phenomenon. Stimulation of animal cell growth by ferricyanide and parallel inhibition of growth and ferricyanide reduction by antitumor drugs in both animals and plants indicate some relationship between transmembrane electron transport and growth. Stimulation of electron transfer from NADH to oxygen by growth hormones, both plant and animal, implicates NADH oxidation as a potential rate-limiting step. To identify the mechanism whereby NADH oxidation and growth control may be coupled represents an important challenge for future research.

REFERENCES

1. **Crane, F. L., Sun, I., Clark, M. G., Grebing, G., and Löw, H.,** Transplasma membrane redox systems in growth and development, *Biochim. Biophys. Acta,* 811, 233, 1985.
2. **Sun, I. L., Navas, P., Crane, F. L., Morré, D. J., and Löw, H.,** NADH diferric transferrin reductase activity in the plasma membrane, *J. Biol. Chem.,* 262, 15915, 1987.
3. **Morré, D. J., Brightman, A., Wang, J., Barr, R., and Crane, F. L.,** Roles for plasma membrane redox systems in cell growth, in *Proc. NATO Advanced Research Workshop, Plasma Membrane Oxidoreductases in Control of Plant and Animal Growth,* Crane, F. L., Löw, H., and Morré, D. J., Eds., Alan R. Liss, New York, 1988, 45.
4. **Sun, I. L., Garcia-Canero, R., Liu, W., Toole-Simms, W., Crane, F. L., and Morré, D. J.,** Diferric transferrin reduction stimulates the Na^+/H^+ antiport of HeLa cells, *Biochem. Biophys. Res. Commun.,* 145, 465, 1987.

5. **Löw, H. and Crane, F. L.,** Redox function in plasma membranes, *Biochim. Biophys. Acta,* 515, 141, 1978.
6. **Crane, F. L., Goldenberg, H., Morré, D. J., and Löw, H.,** Dehydrogenases of the plasma membrane, *Subcell. Biochem.,* 6, 345, 1979.
7. **Goldenberg, H.,** Plasma membrane redox activities, *Biochim. Biophys. Acta,* 694, 203, 1982.
8. **Crane, F. L., Löw, H., and Clark, M. G.,** Plasma membrane redox enzymes, in *The Enzymes of Biological Membranes,* Vol. 4, Martonosi, A., Ed., Plenum Press, New York, 1985, 465.
9. **Crane, F. L.,** Electron transport and cytochromes of subcellular particles from cauliflower buds, *Plant Physiol.,* 32, 619, 1987.
10. **Gayda, D. P., Crane, F. L., Morré, D. J., and Löw, H.,** Hormone effects on NADH-oxidizing enzymes of plasma membranes of rat liver, *Proc. Indiana Acad. Sci.,* 86, 385, 1977.
11. **Goldenberg, H., Crane, F. L., and Morré, D. J.,** NADH oxidoreductase of mouse liver plasma membranes, *J. Biol. Chem.,* 254, 2491, 1979.
12. **Jesaitis, A. J., Heners, P. R., Hertel, R., and Briggs, W. R.,** Characterization of a membrane fraction containing a b type cytochrome, *Plant Physiol.,* 59, 941, 1977.
13. **Mukherjee, S. P. and Lynn, W. S.,** Reduced nicotinamide adenine diculeotide phosphate oxidase in adipocyte plasma membrane and its activation by insulin, *Arch. Biochem. Biophys.,* 184, 69, 1977.
14. **Mukherjee, S. P., Lane, R. H., and Lynn, W. S.,** Endogenous hydrogen peroxide and peroxidative metabolism in adipocytes in response to insulin and sulfhydryl reagents, *Biochem. Pharmacol.,* 27, 2589, 1978.
15. **Lin, W.,** Further characterization on the transport property of plasmalemma NADH oxidation system in isolated corn root protoplasts, *Plant Physiol.,* 74, 219, 1982.
16. **Kochian, L. V. and Lucas, W. J.,** Potassium transport in corn roots: III. Perturbation by exogenous NADH and ferricyanide, *Plant Physiol.,* 77, 429, 1985.
17. **Ramirez, J. M., Gallego, G. G., and Serrano, R.,** Electron transfer constituents in plasma membrane fractions of *Avena sativa* and *Saccharomyces cerivsiae, Plant Sci. Lett.,* 34, 103, 1984.
18. **Zamudio, I. and Canessa, M.,** Nicotinamide-adenine dinucleotide dehydrogenase activity of human erythrocyte membranes, *Biochim. Biophys. Acta,* 120, 165, 1966.
19. **MacKellar, W. C.,** A Transmembrane NADH Dehydrogenase in Porcine Erythrocyte Plasma Membranes, Ph.D. thesis, Purdue University, West Lafayette, IN, 1981, 172.
20. **Fischer, S., Cellino, M., Zambrano, F., Zampighi, G., Nagel, M. T., Marcus, D., and Canessa Fischer, M.,** Isolation characterization of plasma membranes from the retinal axons of the squid: an axolemma rich preparation. *Arch. Biochem. Biophys.,* 138, 1—15, 1970.
21. **Crane, F. L. and Löw, H.,** NADH oxidation in liver and fat cell plasma membranes, *FEBS Lett.,* 68, 153, 1976.
22. **Gimenez-Gallego, G., Benevides, J., Garcia, M. L., and Valdivieso, F.,** Occurence of a reduced nicotinamide adenine nucleotide oxidase activity linked to a cytochrome system in renal brush border membranes, *Biochemistry,* 19, 4839, 1980.
23. **Brightman, A. O., Barr, R., Crane, F. L., and Morré, D. J.,** Auxin-stimulated NADH oxidase purified from plasma membrane, *Plant Physiol.,* 86, 1264, 1988.
24. **Serrano, R.,** Characterization of the plasma membrane ATPase of *Saccharomyces cerevisiae, Mol. Cell Biochem.,* 22, 51, 1978.
25. **Amar-Costesec, A., Wibo, M., Thines-Sempoux, D., Beaufay, H., and Berthet, J.,** Analytical study of microsomes and isolated subcellular membranes from rat liver: IV. Biochemical, physical, and morphological modifications of microsomal components induced by digitonin, EDTA, and pyrophosphate, *J. Cell Biol.,* 62, 717, 1974.
26. **Ramasarma, T., Swaroop, A., MacKellar, W., and Crane, F. L.,** Generation of hydrogen peroxide on oxidation of NADH by hepatic plasma membranes, *J. Bioenerg. Biomemb.,* 13, 241, 1981.
27. **Vijaya, S., Crane, F. L., and Ramasarma, T.,** A vanadate-stimulated NADH oxidase in erythrocyte membrane generates hydrogen peroxide, *Mol. Cell Biochem.,* 62, 175, 1984.
28. **Darr, D. and Fridovich, I.,** Vanadate and molybate stimulate the oxidation of NADH by superoxide radical, *Arch. Biochem. Biophys.,* 232, 562, 1984.
29. **Erdmann, E., Krawietz, W., Philipp, G., Hackbarth, I., Schmitz, W., Scholz, H., and Crane, F. L.,** Purified cardiac cell membranes with high (Na^+ + K^+) ATPase activity contain significant NADH-vanadate reductase activity, *Nature,* 282, 335, 1979.
30: **Sun, I. L., Navas, P., Crane, F. L., Morré, D. J., and Löw, H.,** Diferric transerrin reductase of the plasma membrane is inhibited by adiamycin, *Biochem. Inter.,* 14, 119, 1987.
31. **Sun, I. L. and Crane, F. L.,** The antitumor drug, cis-diamminedichloroplatinum, inhibits trans-plasmalemma electron transport in HeLa cells, *Biochem. Intern.,* 9, 299, 1984.
32. **Sun, I. L. and Crane, F. L.,** Bleomycin control of transplasma membrane redox activity and proton movement in HeLa cells, *Biochem. Pharmacol.,* 34, 617, 1985.

33. **Tökes, S. A., Rogers, K. E., and Rembaum, A.,** Synthesis of adriamycin-coupled polyglutaraldehyde microspheres and evaluation of their cytostatic activity, *Proc. Natl. Acad. Sci. U.S.A.*, 79, 2026, 1982.
34. **Tritton, T. R., Yee, G., and Wingard, L. B., Jr.,** Immobilized adriamycin: a tool for separating cell surface from intracellular mechanisms, *Fed. Proc.*, 42, 284, 1983.
35. **Rogers, K. E., and Tokes, Z. A.,** Novel mode of cytotoxicity obtained by coupling inactive anthracycline to a polymer, *Biochem. Pharmacol.*, 33, 605-608, 1984.
36. **Sun, I. L., Morré, D. J., Crane, F. L., Safranski, K., and Croze, E. M.,** Monodehydroascorbate as an electron acceptor for NADH reduction by coated vesicle and Golgi apparatus fractions of rat liver, *Biochim. Biophys.* vesicle and Golgi apparatus fractions of rat liver, *Biochim. Biophys. Acta*, 797, 266, 1984.
37. **Morré, D. J., Sun, I. L., and Crane, F. L.,** Ascorbate effects on endomembrane electron transport and membrane flux, in *Vitamins and Cancer*, Myeskins, F. L. and Prasad, K. N., Eds., Humana Press, Clifton, NJ, 1985, 83.
38. **Morré, D. J. Crane, F. L., Sun, I. L., and Navas, P.,** The role of ascorbate in biomembrane energetics, *Ann. N.Y. Acad. Sci.*, 498, 153, 1987.
39. **McConnell, D. G.,** The isolation of retinal outer segment fragments, *J. Cell Biol.*, 27, 459-473, 1965.
40. **Navas, P., Nowack, D. D., and Morré, D. J.,** Isolation of highly purified plasma membranes from cultured cells and hepatomas by two-phase partition and preparative free-flow electrophoresis, *Cancer Res.*, 49, 2147, 1989.
41. **Pidgeon, C. and Venkataram, U. V.,** Immobilized artifical membrane chromatography: supports composed of membrane lipids, *Ann. Biochem.*, 176, 36, 1989.
42. **Glass, G. A., DeLisle, D. M., DeTogni, P., Gabig, T. G., Magee, B. H., Markert, M., and Babior, B. M.,** The respiratory burst oxidase of human neutrophils, *J. Biol. Chem.*, 261, 13247, 1986.
43. **Morré, D. J., Vigil, E. L., Frantz, C., Goldenberg, H., and Crane, F. L.,** Cytochemical demonstration of glutaraldehyde-resistant, NADH-ferricyanide oxidoreductase activities in rat-liver plasma membranes and Golgi apparatus, *Eur. J. Cell Biol. (Cytobiologie)*, 18, 213, 1978.
44. **Cherry, J. M., MacKellar, W., Morré, D. J., Crane, F. L., Jacobsen, L. B., and Schirrmacher, V.,** Evidence for a plasma membrane redox system on intact ascites tumor cells with different metastic capacity, *Biochim. Biophys. Acta*, 643, 11, 1981.
45. **Morré, D. J., Auderset, G., Penel, C., and Canut, H.,** Cytochemical localization of NADH-ferranicide oxidoreductase in hypocotyl segments and isolated membrane vesicles of soybean, *Protoplasma*, 140, 133, 1987.
46. **Williams, R. J. P. and Conear, W.,** Long range electron transfer, *Nature*, 322, 213, 1986.
47. **Cunningham, C. C., DeChatelet, L. R., Spach, P. I., Parce, J. W., Thomas, M. J., Lees, C. J., and Shirley, P. S.,** Identification and quantification of electron-transport components in human polymorphonuclear neutrophils, *Biochim. Biophys. Acta*, 682, 430, 1982.
48. **Jarasch, E. D., Kartenbeck, J., Bruder, G., Fink, A., Morré, D. J., and Franke, W. W.,** B-type cytochromes in plasma membranes isolated from rat liver, in comparison with those of endomembranes, *J. Cell Biol.*, 80, 37, 1979.
49. **Crawford, D. R. and Schneider, D. L.,** Identification of ubiquinone 30 in human neutrophils and its role in microbiocidal events, *J. Biol. Chem.*, 257, 6662, 1982.
50. **Lutter, R., van Zwieten, R., Weening, R. S., Hamers, M. H., and Roos, D.,** Cytochrome b, flavins and ubiquinone 50 in untreated human neutrophils, *J. Biol. Chem.*, 259, 9603, 1984., 1984.
51. **Michell, B.,** The lethal oxidase of leucocytes, *Trends Biochem. Sci.*, 8, 117, 1983.
52. **Parkes, C. A., Allen, R. A., Cochrane, C. G., and Jesaitis, A. J.,** Purified cytochrome b_5 from human granulocyte plasma membrane is comprised of two polypeptides with relative molecular weights of 91,000 and 22,000, *J. Clin. Invest.*, 80, 732, 1987.
53. **Holmgren, A.,** Thioredoxin: structure and functions, *TIBS*, 6, 26, 1981.
54. **da Costa, M., and Rothenberg, S. P.,** Characterization of the folate-binding proteins associated with the plasma membrane of rat liver *Biochim. Biophys. Acta*, 939, 533, 1988.
55. **Morré, D. J., Crane, F. L., Barr, R., Penel, C., and Wu, L.-Y.,** Inhibition of plasma membrane redox activities and elongation growth of soybean, *Physiol. Plant.*, 72, 236, 1988.
56. **Morré, D. J.,** Comparison of phase partitioning with alternative methods for isolation of subcellular structures, in *Separations Using Aqueous Phase Systems, Applications in Cell Biology and Biotechnology*, Fisher, D. and Sutherland, I. A., Eds., Plenum Press, New York, 1989, 81.
57. **Morré, D. M., Morré, D. J., Snyder, T., Li, W., Erb, A., and Sammons, D. W.,** Isolation of plasma membrane vesicles from fat cells of epididymal fat pads of the rat by aqueous two-phase partition, in *Separations Using Aqueous Phase Systems. Applications in Cell Biology and Biotechnology*, Fisher, D. and Sutherland, I. A., Eds., Plenum Press, New York, 1989, 113.
58. **Bremer, E. G. and Hakomori, S.,** G_{M3} Ganglioside induces hamster fibroblast growth inhibition in chemically-defined medium: ganglioside may regulate growth factor receptor function, *Biochem. Biophys. Res. Commun.*, 106, 711, 1982.

59. **Bremer, E. G., Hakomori, S., Bowen-Pope, D. F., Raines, E., and Ross, R.,** Ganglioside-mediated modulation of cell growth, growth factor binding, and receptor phosphorylation, *J. Biol. Chem.*, 259, 6818, 1984.
60. **Bremer, E. G., Schlessinger, J., and Hakomori, S.,** Ganglioside-mediated modulation of cell growth: specific effects of G_{M3} on tyrosine phosphorylation of the epidermal growth factor receptor, *J. Biol. Chem.*, 261, 2434, 1986.
61. **Berridge, M. J.,** Inositol triphosphate and diacylglycerol: two interacting second messengers, *Annu. Rev. Biochem.*, 56, 159, 1978.
62. **Tsuchida, T., Saxton, R. E., and Irie, R. F.,** Gangliosides of human melanoma: G_{M2} and tumorigenicity, *J. Nat. Cancer Inst.*, 78, 55, 1987.
63. **Hakomori, S., and Kanfer, J. N.,** Gangliosides, in *Handbook of Lipid Research,* Vol. 3, *Sphingolipid Biochemistry,* Hanahan, D. J. Ed., Plenum Press, New York, 1973, 327.
64. **Op den Kamp, G. A. F.,** Lipid asymmetry in membranes, *Annu. Rev. Biochem.*, 48, 47, 1979.
65. **Wiegandt, H.,** Gangliosides, in *Comprehensive Biochemistry, Glycolipids,* Vol. 10, Nernberger, A. and Van Deenen, L. L. M., Eds., Elsevier, Amsterdam, 1985, 199.
66. **Hanai, N., Nores, G., Torres-Méndez, C.-R., and Hakomori, S.,** Modified ganglioside as a possible modulator of transmembrane signaling mechanism through growth factor receptors, *Biochem. Biophys. Res. Commun.*, 147, 127, 1987.
67. **Hannun, Y. A., Greenberg, C. S., and Bell, R. M.,** Sphingosine inhibition of agonist-dependent secretion and activation of human platelets inplies that protein kinase C is a necessary and common event of the signal transducing pathways, *J. Biol. Chem.*, 262, 13620, 1987.
68. **Scharff, O., Foder, B., and Skibsted, U.,** Hysteretic activation of the Ca^{2+} pump revealed by calcium transients in human red cells, *Biochim. Biophys. Acta,* 730, 295, 1983.
69. **Jelsema, C. L.,** Light activation of phospholipase A_2 in rod outer segments of bovine retina and its modulation by GTP binding proteins, *J. Biol. Chem.*, 282, 163, 1987.
70. **Navas, P., Sun, I. L., Morré, D. J., and Crane, F. L.,** Decrease of NADH in HeLa cells in the presence of transferrin or ferricyanide, *Biochem. Biophys. Res. Commun.*, 135, 110, 1986.
71. **Schwartz, J. P., Passonneau, J. V., Johnson, G. S., and Pastan, I.,** The effect of growth conditions on NAD^+ and NADH concentrations and the NAD^+:NADH ratio in normal and transformed fibroblasts, *J. Biol. Chem.*, 249, 4138, 1974.
72. **Sun, I. L., Crane, F. L., and Chou, J. Y.,** Modification of transmembrane electron transport activity in plasma membranes of Simian virus 40 transformed pineal cells, *Biochim. Biophys. Acta,* 886, 327, 1986.
73. **Sun, I. L., Navas, P., Crane, F. L., Chou, J. Y., and Löw, H.,** Transplasmalemma electron transport is changed in Simian virus 40 transformed liver cells, *J. Bioenerg. Biomemb.*, 18, 471, 1986.
74. **Jedeikin, L. A. and Weinhouse, S.,** Metabolism of neoplastic tissue: assay of oxidized and reduced diphosphopyridine nucleotide in normal and neoplastic cells, *J. Biol. Chem.*, 213, 271, 1955.
75. **Briggs, M. H.,** Vitamin and coenzyme content of hepatomas induced by butter yellow, *Nature (London),* 187, 249, 1960.
76. **Nemeth, A. M. and Dickerman, H.,** Pyridine nucleotides and diphosphopyridine nculeotidase in developing mammalian tissues, *J. Biol. Chem.*, 235, 1761, 1960.
77. **Caiger, P., Morton, R. K., Filsell, O. H., and Jarrett, I. G.,** A comparative study of nicotinamide nucleotide coenzymes during growth of the sheep and rat, *Biochem. J.*, 85, 351, 1962.
78. **Burch, H. B. and Von Dippe, P.,** Pyridine nucleotides in developing rat liver, *J. Biol. Chem.*, 239, 1898, 1964.
79. **Clark, J. B., Greenhaum, A. L., and McLean, P.,** The concentration and biosynthesis of nicotinamide nucleotides in the livers of rats treated with carcinogens, *Biochem. J.*, 98, 546, 1966.
80. **Ferris, G. M. and Clark, J. B.,** The control of nucleic acid and nicotinamide nucleotide synthesis in regenerating rat liver, *Biochem. J.*, 128, 869, 1972.
81. **Sun, I., MacKellar, W. C., Crane, F. L., Barr, R., Elliott, W. L., Lem, N., Varnold, R. L., Heistein, P. F., and Morré, D. J.,** Decreased NADH-oxidoreductase activities as an early response in rat liver to the carcinogen 2-acetylaminofluorene, *Cancer Research,* 45, 157, 1985.
82. **Jacobson, E. L. and Jacobson, M. K.,** Pyridine nucleotide levels as a function of growth in normal and transformed 3T3 cells, *Arch. Biochem. Biophys.*, 175, 627, 1976.
83. **Weber, G.,** Enzymology of cancer cells I and II, *N. Engl. J. Med.*, 296, 486, 541, 1977.
84. **Ghani, Q. P. and Hollenberg, M.,** Poly(adenosine diphosphate ribose) metabolism and regulation of myocardial cell growth by oxygen, *Biochem. J.*, 170, 387, 1978.
85. **MacKellar, W. C. and Crane, F. L.,** Iron and copper in plasma membranes, *J. Bioenerg. Biomembr.*, 14, 241, 1981.
86. **Ellem, K. A. O., and Kay, G. F.,** Ferricyanide can replace pyruvate to stimulate growth and attachment of serum restricted human melanoma cells, *Biochem. Biophys. Res. Commun.*, 112, 183, 1983.
87. **Sun, I. L. and Crane, F. L., Grebing, C., and Löw, H.,** Transmembrane redox in control of cell growth, *Exp. Cell Res.*, 156, 528, 1985.

88. **Sun, I. L., Putnam, J. E., and Crane, F. L.,** Control of cell growth by transplasmalemma redox: Stimulation of HeLa cell growth by impermeable oxidants, *Proc. Indiana Acad. Sci.,* 94, 407, 1985.
89. **Löw, H., Crane, F. L., Grebing, C., Hall, K., and Tally, M.,** Metabolic milieu and insulin action, in *Diabetes,* Waldhäusl, W. K., Ed., Excerpta Medica, Amsterdam, 1979, 209, 1979.
90. **Kilberg, M. S. and Christensen, H. N.,** Electron-transferring enzymes in the plasma membrane of the Ehrlich ascites tumor cell, *Biochemistry,* 18, 1525, 1979.
91. **Ohsawa, M., Kilberg, H. S., Kimmel, G., and Christensen, H. H.,** Energization of amino acid transport in energy-depleted Erlich cells and plasma membrane vesicles, *Biochim. Biophys. Acta,* 599, 175, 1980.
92. **Morré, D. M., Morré, D. J., Reust, T. W., and Sammons, D. W.,** Insulin responsive NADH oxidation by plasma membranes of fat cells from epididymal fat pads of the rat and correlation with surface membrane internatization, *J. Cell Biol.,* 107, (Abstr.), 809a, 1989.
93. **Hasumura, S., Rossi, B., Alderson, R., Pastan, I., and Cheng, S.-Y.,** Antibodies against the plasma membrane 3,3′,5-triiodo-L-thyronine binding protein of rat pituitary GH_3 cells: partial characterization and cross species immunoreactivity, *Biochem. Biophys. Res. Commun.,* 124, 956, 1984.
94. **Morré, D. J., Brightman, A. O., Wu, L.-Y., Leak, B., and Crane, F. L.,** Role of plasma membrane redox activities in elongation growth in plants, *Physiol. Plant.,* 73, 187, 1988.
95. **Hidalgo, A., Brightman, A., Navas, P., and Morré, D. J.,** unpublished results.
96. **Wilkinson, F. E. and Zhu, X.-Z.,** unpublished results, Purdue University.
97. **Brightman, A. and Zhu, X.-Z.,** unpublished data, Purdue University.

Chapter 4

OXIDATION-REDUCTION REACTIONS IN SECRETORY VESICLES

David Njus, Patrick M. Kelley, Gordon J. Harnadek, and Vishram Jalukar

TABLE OF CONTENTS

I. INTRODUCTION

Great progress has been made toward understanding electron transfer reactions in solution,[1] but the application of this insight to systems of biological interest has been limited. Biological systems are fraught with such complexities as membrane-bound electron carriers, reactions that may transfer a number of electrons at a time, and molecular substrates that may undergo disproportionation and autooxidation reactions as well as metabolic transformations. An exception, remarkable for its simplicity, is the electron transfer system in secretory-vesicle membranes. The electron carrier, cytochrome b_{561}, transfers one electron at a time. The single substrate, ascorbic acid, and its free radical function both as electron donor and electron acceptor. These factors make the secretory-vesicle electron transfer system one that may be understood in molecular detail.

The function of the secretory-vesicle electron transfer system is to provide reducing equivalents for monooxygenases contained in the vesicles and catalyzing steps in the biosynthesis of the stored hormones. The immediate electron donor for these monooxygenases seems to be ascorbate stored within the vesicles. Nevertheless, electrons must be imported from the cytosol across the vesicle membrane to maintain intravesicular ascorbate and to provide a constant supply of electrons for the monooxygenases. This is the function of cytochrome b_{561}, an integral membrane protein found exclusively in secretory-vesicle membranes.

Electron transfer across secretory vesicle membranes is a relatively new subject, but it has advanced quickly. Cytochrome b_{561} was discovered in 1971,[2] its electron transfer function was recognized in 1981,[3] and its amino acid sequence was determined in 1988.[4] This is a good time, therefore, to review the mechanism of this electron transfer and its role in supporting redox reactions in secretory vesicles. We will begin with a review of the monooxygenases catalyzing redox reactions in secretory vesicles, follow with a description of the electron transfer system(s) that import reducing equivalents for these enzymes, and conclude with a discussion of cytochrome b_{561} and the mechanism by which it transfers electrons.

Because of the ready availability of bovine adrenal tissue, most is known about chromaffin vesicles, the secretory vesicles that store epinephrine and norepinephrine in the adrenal medulla. Nevertheless, the few studies of other types of secretory vesicles suggest that the bovine chromaffin vesicle is a valid general model. More comprehensive discussions of secretory vesicles are contained in Volume 493 of the *Annals of the New York Academy of Sciences* and in recent reviews.[5-7]

II. SECRETORY-VESICLE MONOOXYGENASES

Secretory vesicles contain two known monooxygenases: dopamine β-monooxygenase and peptidyl glycine α-amidating monooxygenase. Dopamine β-monooxygenase catalyzes the conversion of dopamine to norepinephrine within synaptic vesicles of adrenergic and noradrenergic neurons, and within secretory vesicles of the chromaffin cells in the adrenal medulla. Peptidyl glycine α-amidating monooxygenase catalyzes the cleavage of a C-terminal glycine to amidate peptide hormones in peptide-storing secretory vesicles. The two enzymes seem to have some mechanistic similarities; of particular significance here is a dependence on ascorbate as a reducing agent *in vivo*. Excellent comprehensive reviews of dopamine β-monooxygenase[8] and peptidyl-glycine α-amidating monooxygenase[9] have appeared recently.

A. DOPAMINE β-MONOOXYGENASE

Dopamine β-monooxygenase incorporates one atom of O_2 into the substrate and reduces the other atom to H_2O. The enzyme has a surprisingly low substrate specificity, oxidizing a wide spectrum of aromatic compounds.[10,11] It requires copper as a cofactor and is activated by fumarate.[12]

1. Structure

A cDNA clone for human dopamine β-monooxygenase was recently isolated.[13] The sequence indicates that the protein is synthesized as a single chain with 67,527 mol wt; this is processed to 64,862 by removal of the N-terminal signal sequence. These subunits are linked by disulfide bonds to form dimers. Dimers then associate noncovalently to form the tetramer, which is the predominant oligomeric form of the enzyme. The protein is *N*-glycosylated at an average of 6 sites per tetramer.

Dopamine β-monooxygenase exists in both soluble and membrane-bound forms. It is believed that the soluble enzyme is formed from the membrane-bound species.[14] This idea is supported by mapping analysis, which shows that a single mRNA apparently codes for both soluble and membrane-bound forms of the enzyme.[13] The soluble monomer has a molecular weight between 70,000 and 73,000 and the membrane-bound monomer is somewhat larger (75,000 to 77,000).[14,15] It had been thought that the membrane-bound enzyme could be converted to the soluble form by removal of a small hydrophobic peptide (3 to 5 kDa) which acts as an anchor in the vesicle membrane. The amino acid sequence, however, indicates that the only hydrophobic segment is the N-terminal signal sequence which should be rapidly removed. A second possibility is that the enzyme is anchored via a glycosyl phosphatidyl inositol linkage,[16] but preliminary tests of this hypothesis have proved negative as well.[8] Whether the membrane-bound enzyme contains an additional hydrophobic peptide or a lipid anchor remains to be seen. In any case, the membrane-associated portion of dopamine β-monooxygenase is probably small. This makes it quite unlikely that the membrane-bound enzyme can accept electrons directly from a cytosolic electron donor.

The prosthetic group, copper, seems to be relatively weakly bound to the enzyme and is easily exchanged with free copper in the medium. This "weak" binding gave rise to suspicions of adventitious copper binding. Consequently, the Cu/tetramer stoichiometry was generally accepted as 4 despite the fact that additional copper was needed to maximize activity.[17] It now appears that the true stoichiometry is 8 Cu/tetramer. Villafranca and colleagues[18,19] reported NMR relaxation data indicating 2 Cu/subunit and subsequently demonstrated that the rate of inhibition of dopamine β-monooxygenase using a suicide substrate was maximal at this same stoichiometry. Simultaneously, Klinman et al.[20] reported that maximal hydroxylation of dopamine, the physiological substrate, was obtained at 2 Cu/subunit. The two coppers bind with approximately equal affinity ($K_d \sim 10^{-8}\ M$).

2. Mechanism

There is a wealth of information on the kinetics and mechanism of dopamine β-monooxygenase. This includes structural studies of the Cu sites,[21,22] elegant analyses of kinetic isotope effects,[23,24] and investigations of mechanism-based inhibitors.[11,25,26] For the purposes of this review, however, it is important to summarize only some essential points.

Although ascorbate is the most effective electron donor to dopamine β-monooxygenase, ferrocyanide, hydroquinone and catechols will also work.[27] Ascorbate undergoes a one-electron oxidation producing the semidehydroascorbate free radical and reducing the Cu cofactor.[17,28] This formation of the ascorbate free radical is typical of ascorbate oxidases.[29]

The copper must be in the reduced form (Cu [I]) to bind substrate and O_2. Apparently there is a random association of either dopamine or molecular oxygen with the reduced enzyme.[30] The reaction itself can be divided into two significant steps (Figure 1). First, the C-H bond is broken and a hydrogen is abstracted to yield a radical intermediate. This is then hydroxylated to form the product.[8]

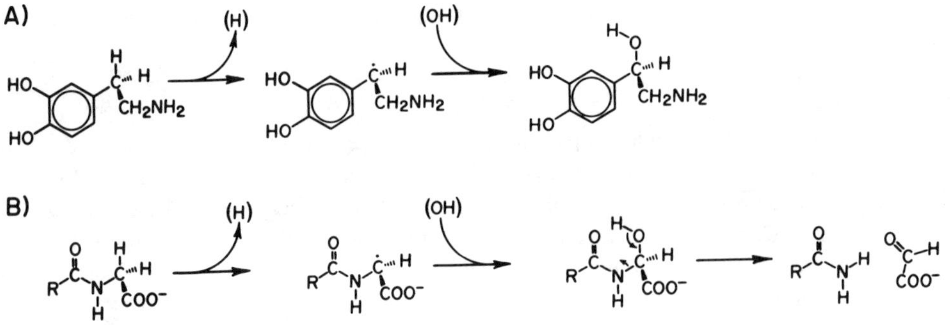

FIGURE 1. Proposed catalytic mechanisms of (A) dopamine β-monooxygenase (after Stewart and Klinman[8]) and (B) peptidyl-glycine α-amidating monooxygenase (after Bateman[33]).

B. PEPTIDYL-GLYCINE α-AMIDATING MONOOXYGENASE

Amidation is a common characteristic of biologically active peptides and the amidating enzyme is accordingly widely distributed. It is most abundant in the anterior, intermediate, and posterior pituitary, the hypothalamus, the atrium of the heart and the submaxillary and parotid glands.[9] Peptidyl-glycine α-amidating monooxygenase (PAM) was discovered only in 1982[31] and is not yet as well characterized as dopamine β-monooxygenase. The two enzymes are similar in their requirement for copper, in their preference for ascorbate as electron donor, and in their localization in the secretory vesicles storing their products.[9] Both enzymes seem to have relatively large, positive primary isotope effects, and both can use ferrocyanide and catechol as electron donors *in vitro*. Unlike dopamine β-monooxygenase, PAM seems to be largely soluble and functions as a monomer with 35,000 to 65,000 mol wt. Furthermore, PAM does not seem to have an anionic activator equivalent to fumarate. The Cu stoichiometry has not yet been reported. PAM may differ from dopamine β-monooxygenase in that it may use other metals as cofactor.[32]

There seems to be a parallel between the mechanisms of the amidation reaction catalyzed by PAM and the hydroxylation reaction catalyzed by dopamine β-monooxygenase. Like dopamine β-monooxygenase, PAM apparently accepts substrate and molecular oxygen in random order. The catalytic mechanism also seems to involve the breaking of the C-H bond to form a radical intermediate followed by hydroxylation (Figure 1). In the case of the peptide, this yields an unstable hydroxylated α-carbon which rearranges to form the amidated peptide and glyoxylate.[33]

It is not yet known whether PAM takes one or two electrons from ascorbate. One would expect PAM to produce semidehydroascorbate following the pattern of dopamine β-monooxygenase and other ascorbate-requiring enzymes. This expectation is further supported by the fact that PAM-containing vesicles have the same cytochrome b_{561}-mediated ascorbate-regenerating system as dopamine β-monooxygenase-containing vesicles. As discussed in the following section, this system will regenerate ascorbate from the free radical but not from dehydroascorbate.

III. ELECTRON TRANSFER ACROSS SECRETORY VESICLE MEMBRANES

A. ASCORBIC ACID REGENERATION

The most likely source of the reducing equivalents needed by secretory vesicle monooxygenases is ascorbate. It is present in these vesicles,[34,35] and it is the most effective electron donor to these enzymes *in vitro*. In fact, serving as the electron donor for redox reactions occurring within secretory vesicles is probably a major biological function of ascorbic acid.

TABLE 1
Properties of Ascorbate and Semidehydroascorbate

	pH 5.5	pH 7.0	Ref.
k_{dis}	$5 \times 10^7\ M^{-1}s^{-1}$	$3 \times 10^6\ M^{-1}s^{-1}$	38
$E°(AH^-/A^{\dot{-}})$	+0.42 V	+0.33 V	39

Note: k_{dis} is the second-order rate constant for disproportionation of semidehydroascorbate. E° is the midpoint reduction potential for the ascorbate/semidehydroascorbate pair.

1. Properties of Ascorbic Acid

Although it can lose two electrons, ascorbate commonly functions as a one-electron donor and is oxidized to its free radical, semidehydroascorbate. That peroxidase and ascorbate oxidase use ascorbate as a one-electron donor was recognized many years ago.[29,36] That dopamine β-monooxygenase also follows this pattern was discovered fairly recently.[17,28] That peptidyl-glycine α-amidating monooxygenase takes only one electron from ascorbate is presumed true but remains an untested hypothesis.

Ascorbic acid is an unusual compound in that its free radical is relatively unreactive[37] and tends to decay by disproportionation (2 semidehydroascorbate + $H^+ \rightarrow$ 1 ascorbate + 1 dehydroascorbate). Disproportionation is a second-order process, and the rate constant is strongly pH-dependent.[37,38] Thus, the free radical has a longer lifetime at alkaline than at acidic pH (Table 1).

Both ascorbate and the free radical are anionic. Their interconversion involves H^+, so the midpoint reduction potential for the ascorbate/semidehydroascorbate pair is pH-dependent.[39] The midpoint reduction potential (E°) is higher at lower pH (Table 1). It should be remembered that, although E° is +330 mV at pH 7, the effective reduction potential is ~60 mV because the semidehydroascorbate concentration is kept low by disproportionation.

2. Mechanism of Ascorbic Acid Regeneration

Secretory vesicles must maintain their supply of ascorbate for use by intravesicular monooxygenases. The mechanism which has emerged over the last several years (Figure 2) specifies (1) that the monooxygenase oxidizes ascorbate to semidehydroascorbate, (2) that this intravesicular semidehydroascorbate is reduced back to ascorbate by cytochrome b_{561}, (3) that cytochrome b_{561} is reduced in turn by cytosolic ascorbic acid, and (4) that cytosolic semidehydroascorbate is reduced back to ascorbate by mitochondrial NADH:semidehydroascorbate oxidoreductase.[3,40-44]

Before focusing on cytochrome b_{561}, the role of NADH:semidehydroascorbate oxidoreductase warrants some comment. In 1982, Diliberto et al.[44] showed that NADH:semidehydroascorbate oxidoreductase activity in the adrenal medulla is localized in the outer mitochondrial membrane. Wakefield et al.[42] observed that the turnover of dopamine β-monooxygenase causes the oxidation of cytochrome b_{561} and that this oxidation is partially reversed by the addition of NADH and mitochondria. Using electron paramagnetic resonance to monitor the ascorbate free radical directly, they also showed that semidehydroascorbate generated by the turnover of dopamine β-monooxygenase is reduced by NADH in the presence of mitochondria.[43] Thus, they were able to demonstrate functional coupling between mitochondrial NADH:semidehydroascorbate oxidoreductase and dopamine β-monooxygenase.

Semidehydroascorbate fills two different coupling functions in this ascorbate-regenerating mechanism. First, in the cytosol, the free radical diffuses from the secretory vesicle to the reductase in the outer mitochondrial membrane (Figure 2, d_1). For this reason, the

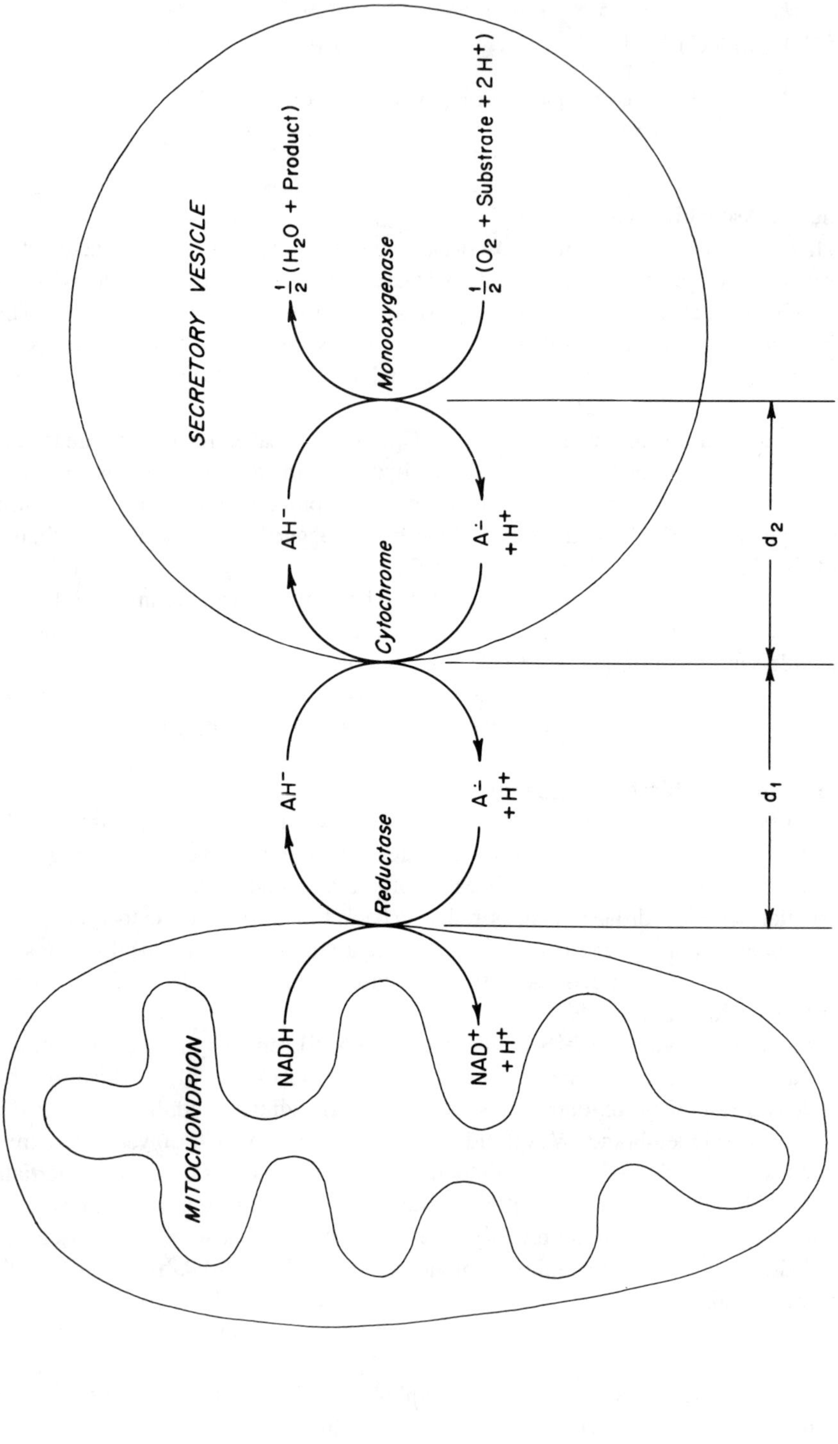

FIGURE 2. Mechanism of ascorbic acid regeneration in secretory vesicles. AH^- is ascorbate; $A^{\dot{-}}$ is semidehydroascorbate.

relatively long lifetime of the free radical is quite important. At the cytosolic pH (7.0) and a concentration of 100 n*M*, semidehydroascorbate has a lifetime of ~3 s. Within the secretory vesicle, the free radical must diffuse from the monooxygenase to cytochrome b_{561} in the vesicle membrane (Figure 2, d_2). Because the intravesicular pH is acidic (5.5), the free radical will disproportionate more quickly. At a concentration of 100 n*M*, semidehydroascorbate will have a lifetime of only 0.2 s, but this is adequate because of the smaller dimensions of the vesicle (diameter < 300 nm).

Secretory vesicles have acidic interiors maintained by an inwardly directed H^+-translocating ATPase.[5,45] The membrane potential and pH gradient established by this ATPase should drive electron transfer from cytosolic ascorbate to intravesicular semidehydroascorbate. The membrane potential (inside positive) should draw electrons into the vesicles. The pH gradient should also contribute to the force drawing electrons into the vesicles because the ascorbate/semidehydroascorbate interconversion involves a proton. Lowering the pH shifts the equilibrium in favor of ascorbate and raises the midpoint reduction potential at pH 5.5 (the intravesicular pH) about 90 mV higher than at pH 7.0 (the cytosolic pH). Consequently, both the pH gradient and membrane potential across the vesicle membrane favor the reduction of intravesicular semidehydroascorbate by cytosolic ascorbate.

B. CYTOCHROME b_{561}-MEDIATED ELECTRON TRANSFER

1. Evidence for Cytochrome b_{561}-Mediated Electron Transfer

Membranes containing cytochrome b_{561} have the capacity to transfer electrons from internal ascorbate to an external acceptor. This outward electron transfer is conveniently assayed by following the reduction of the external acceptor (cytochrome c or ferricyanide) spectrophotometrically.[46] The evidence that cytochrome b_{561} mediates this electron transfer is quite strong. First, rates of electron transfer in different secretory vesicles correlate with amounts of cytochrome b_{561} in the membranes.[47] Second, this electron transfer can be observed in phospholipid vesicles into which cytochrome b_{561} has been reconstituted.[48] Third, cytochrome b_{561} has a midpoint reduction potential appropriate for electron transfer from ascorbate to ferricyanide or cytochrome c.[49,50] Finally, the cytochrome undergoes spectral changes consistent with mediation of this electron transfer.[51]

Electron flow out of secretory vesicles has been well characterized because it is relatively simple to prepare chromaffin-vesicle "ghosts" containing a trapped reductant. Electron flow into secretory vesicles has proved to be more difficult to observe because of technical difficulties associated with loading the vesicles with an oxidant. An observation made by Cidon et al.,[52] however, suggests a possible approach. Phenazine methosulfate (5-methyl phenazinium) concentrates in acidic vesicles, and its fully reduced form is rapidly oxidized by molecular oxygen. When ascorbate is added to a suspension of chromaffin-vesicle ghosts previously acidified by the addition of ATP, it reduces cytochrome b_{561} externally. The cytochrome, in turn, reduces sequestered phenazine methosulfate. Phenazine methosulfate then reduces O_2 to H_2O_2. This consumes H^+, leading to an alkalinization of the intravesicular space. Cidon et al.[52] discovered that this could be monitored spectrophotometrically using acridine orange as an optical indicator of the pH gradient. The gradient created by the H^+-transporting ATPase is quickly collapsed when both phenazine methosulfate and ascorbate are present. This result may also be obtained using the ^{14}C-methylamine distribution technique to measure pH gradients. Furthermore, cytochrome b_{561} spectral changes consistent with oxidation by phenazine methosulfate and reduction by ascorbate may be observed.[76]

2. Evidence for Cytochrome b_{561}-Mediated Ascorbic Acid Regeneration

There is now a substantial body of evidence supporting the ascorbate regeneration mechanism. That cytochrome b_{561} is catalyzing this ascorbate regeneration is suggested by the following: first, the distribution of the cytochrome correlates with the occurrence of dopamine β-monooxygenase and peptide amidating monooxygenase. Second, extravesicular reducing agents generally support intravesicular dopamine hydroxylation only if they also reduce cytochrome b_{561}. Finally, the cytochrome is a transmembrane electron carrier and is kinetically competent to regenerate intravesicular ascorbate.

Cytochrome b_{561} is found not only in chromaffin vesicles but also in dopamine β-monooxygenase-containing adrenergic synaptic vesicles in the central and peripheral nervous systems. It has been identified by its visible absorption spectrum in synaptic vesicles from rat vas dererens[53] and bovine splenic nerve,[54] and immunocytochemical studies have shown that cytochrome b_{561} and dopamine β-monooxygenase colocalize in a variety of tissues.[55] Cytochrome b_{561} is also found in neurohypophyseal secretory vesicles, which contain the peptide amidating monooxygenase.[56] Consequently, the distribution of cytochrome b_{561} seems to coincide with secretory vesicles containing either dopamine β-monooxygenase or PAM or both.

In isolated chromaffin vesicles, dopamine β-monooxygenase will oxidize intravesicular ascorbate if provided with a substrate such as dopamine or tyramine. Depletion of intravesicular ascorbate does not occur, however, if an appropriate electron donor is present externally. For example, in isolated chromaffin vesicles provided with tyramine, net oxidation of intravesicular ascorbate is prevented by extravesicular ascorbate or extravesicular glucoascorbate without apparent entry of either compound into the vesicles.[57-59] Moreover, while external ascorbate and glucoascorbate stimulate norepinephrine synthesis by preventing intravesicular ascorbate depletion, NADH, glutathione, and homocysteine do not.[60] These two groups correlate with the classes of compounds that respectively do and do not reduce cytochrome b_{561}. Ferrocyanide, as discussed in the following section, is a possible exception.

If the ascorbate regeneration mechanism is to work, semidehydroascorbate in the chromaffin-vesicle matrix must be reduced by cytochrome b_{561} more quickly than it disproportionates. The kinetics of electron transfer from internal ascorbate to external ferricyanide can be analyzed quantitatively, and a rate constant $k_0{}^A$ for reduction of cytochrome b_{561} by ascorbate may be determined.[61] The rate constant ($k_{-0}{}^A$) for the reverse reaction, the reduction of internal semidehydroascorbate by cytochrome b_{561}, may be estimated by recognizing that the ratio $k_0{}^A/k_0{}^A$ equals the equilibrium constant which can be calculated from the midpoint reduction potentials of the two redox pairs:

$$k^A_{-0}/k^A_0 = K_{eq} = \exp\{(E^\circ{}_{asc} - E^\circ{}_{b561})(F/RT)\} \qquad (1)$$

The measured value of $k_0{}^A$ and the value of $k_{-0}{}^A$ calculated from it are given in Table 2. Using this value of $k_{-0}{}^A$ and a steady-state semidehydroascorbate concentration of 2×10^{-7} M,[28] we can calculate a semidehydroascorbate reduction rate of 9×10^{-5} M/s. Using the same semidehydroascorbate concentration and a disproportionation rate constant, k_{dis}, of 5×10^7 $M^{-1}s^{-1}$,[38] we obtain an *in vivo* disproportionation rate of 2×10^{-6} M/s. It is apparent that the rate of semidehydroascorbate regeneration is much faster than the rate of disproportionation, indicating that the proposed mechanism for intravesicular ascorbate regeneration is kinetically feasible.[61]

C. OTHER HYPOTHESIZED ELECTRON TRANSFER PATHWAYS

One observation that may not fit this model is that external ferrocyanide stimulates the rate of hydroxylation in chromaffin-vesicle ghosts. In an early experiment, Grouselle and Phillips[62] found that 0.25 mM external ferrocyanide stimulates the rate of intravesicular tyramine hydroxylation by ~3 nmol/min·mg of membrane protein. Since ferrocyanide does not reduce cytochrome b_{561}, they concluded that there must be another pathway for electron flow. Ahn and Klinman[63] have recently reported a similar observation. External ferrocyanide (0.2 to 2 mM) stimulates the hydroxylation of dopamine by about 4 nmol/min·mg. They

TABLE 2
Summary of Rate Constant Measurements

Rate constant	Estimated value	How estimated
Inside	(pH 6.0)	
Outward: k_0^A	34 $M^{-1}s^{-1}$	Measured
Inward: k_{-0}^A	6 × 10^5 $M^{-1}s^{-1}$	From k_0^A (Eq. 1)
Outside	(pH 7.0)	
Outward		
k_1^A	9.6 × 10^5 $M^{-1}s^{-1}$	Measured
k_1^A	1.6 × 10^6 $M^{-1}s^{-1}$	From k_1^F (Eq. 3)
k_1^F	6.7 × 10^5 $M^{-1}s^{-1}$	Measured
Inward		
k_{-1}^A	540 $M^{-1}s^{-1}$	From k_1^A (Eq. 1)
k_{-1}^F	11 $M^{-1}s^{-1}$	From k_1^F (Eq. 1)

Note: To calculate estimated values, the following midpoint reduction potentials were used: $AH^-/A^{\cdot-}$ (pH 6.0) = +0.39 V;[39] $AH^-/A^{\cdot-}$ (pH 7.0) = +0.33 V;[39] cytochrome B_{561} = +0.14 V;[50] ferri/ferrocyanide = +0.42 V.[74] The following self-exchange rates were used: ferricyanide = 5 × 10^3 $M^{-1}s^{-1}$;[74] ascorbate = 10^6 M^{-1} s^{-1}.[75]

hypothesized that membrane bound dopamine β-monooxygenase can accept electrons directly from an external electron donor and that an internal aqueous mediator (ascorbate) is not needed. Using dopamine β-monooxygenase reconstituted into proteoliposomes, Kent and Fleming[64] tested the possiblity that the enzyme can accept electrons from an electron donor on the other side of the membrane in the absence of an aqueous mediator and obtained negative results. Nevertheless, it is possible that the reconstituted system lacked a necessary component, and a second electron transfer system is a valid proposal for resolving the ferrocyanide result.

It is also possible that these observations may be explained within the context of cytochrome b_{561}-mediated electron transfer. While it is true that ferrocyanide does not cause net reduction of cytochrome b_{561}, the cytochrome is capable of mediating electron flow from ferrocyanide. The rate constant for oxidation of cytochrome b_{561} by ferricyanide ($k_1{}^F$) is ~6.7 × 10^5 $M^{-1}s^{-1}$ (Table 2). The rate constant for the reverse reaction, the reduction of cytochrome by ferrocyanide ($k_{-1}{}^F$), may be calculated from Equation 1 (see Table 2). This value (k_{-1}^F = 11 $M^{-1}s^{-1}$) implies that the rate of electron flow from 1 m*M* ferrocyanide to oxidized cytochrome b_{561} should be 1.5 nmol/min · mg, which is comparable to the observed stimulation of dopamine β-monooxygenase activity. Therefore, although ferrocyanide will pass electrons to cytochrome b_{561} only slowly, the rate may be sufficient to support the equally slow rate of hydroxylation.

IV. CYTOCHROME b_{561}

In the chromaffin-vesicle membrane, cytochrome b_{561} is the second-most-abundant protein (after dopamine β-monooxygenase), so it may be obtained in considerable quantity. This should make it a good subject for structural as well as functional studies.

A. SEQUENCE AND STRUCTURE

The amino acid sequence of cytochrome b_{561} was recently determined from a cDNA clone isolated from a bovine brain library.[4] The sequence is consistent with known properties of cytochrome b_{561} although it remains to be seen if translation will yield a functional cytochrome. The cDNA codes for a 273 amino acid polypeptide (30,061 mol wt). Earlier estimates of the molecular weight, determined by polyacrylamide gel electrophoresis, ranged from 20,500 to 30,000.[65-68]

The sequence indicates a high average hydrophobicity, consistent with previous amino acid analysis.[69] A hydrophobicity plot of the primary sequence suggests the existence of six membrane-crossing regions connected by short loops containing clusters of positively charged amino acids. Both the amino and carboxyl termini are on the cytosolic side. This arrangement agrees with earlier labeling studies which showed that cytochrome b_{561} is an asymmetric protein and spans the chromaffin-vesicle membrane.[70,71]

The clusters of cationic amino acids on the two surfaces of the membrane may promote interaction with the negatively charged substrates, ascorbate and semidehydroascorbate. The sequence also contains a large proportion of aromatic amino acids (16%) which may enhance electron transfer rates across the protein.

Duong and Fleming[66] established that the amino acid composition of cytochrome b_{561} is distinct from mitrochondrial cytochrome b. The sequence confirms this and also reveals no similarity to any other sequenced cytochrome.

B. REDOX PROPERTIES

Cytochrome b_{561} is reduced by ascorbate and dithionite but not by NADH, NADPH, glutathione, ferrocyanide or durohydroquinone.[2,49,66,72] It is oxidized by ferricyanide, cytochrome c, and semidehydroascorbate, but not by dehydroascorbate.[2,51] The midpoint reduction potential (+140 mV) is independent of pH.[50]

Cytochrome b_{561} contains only one mol of noncovalently bound heme per mol of protein.[50] Flatmark and Terland[49] found that cytochrome b_{561} exhibits a sigmoidal Nernst plot suggesting that it might exist in two forms. Apps et al.[50] confirmed this and suggested that cytochrome b_{561} might be an oligomer with interactions between the hemes of identical polypeptide subunits. Cross-linking studies done to test this hypothesis, however, were inconclusive.

Whether this heme can be approached from both the cytosolic and intravesicular sides of the membrane or whether additional electron carrying groups are involved is not known. Other proteins are probably not required for electron transfer, since solubilized cytochrome b_{561} seems to interact only weakly if at all with other membrane proteins[67] and since purified cytochrome b_{561} seems to be competent to mediate electron transfer when reconstituted into phospholipid vesicles.[48] The possibility that the protein exists as a multimer cannot be excluded, however.

C. KINETICS OF OXIDATION AND REDUCTION

The simplest model of cytochrome b_{561}-mediated electron transfer presumes two pathways to and from the heme: one from each surface of the membrane (Figure 3). Moreover, it postulates that these electron transfer reactions are bimolecular and exhibit second-order kinetics:

$$\text{(Internal surface)}\quad B_{ox} + X_{red} \xrightarrow{k_0^X} B_{red} + X_{ox}$$

$$B_{red} + X_{ox} \xrightarrow{k^X{}_{-0}} B_{ox} + X_{red}$$

$$\text{(External surface)}\quad B_{red} + Y_{ox} \xrightarrow{k_1{}^Y} B_{ox} + Y_{red}$$

$$B_{ox} + Y_{red} \xrightarrow{k_{-1}{}^Y} B_{red} + Y_{ox}$$

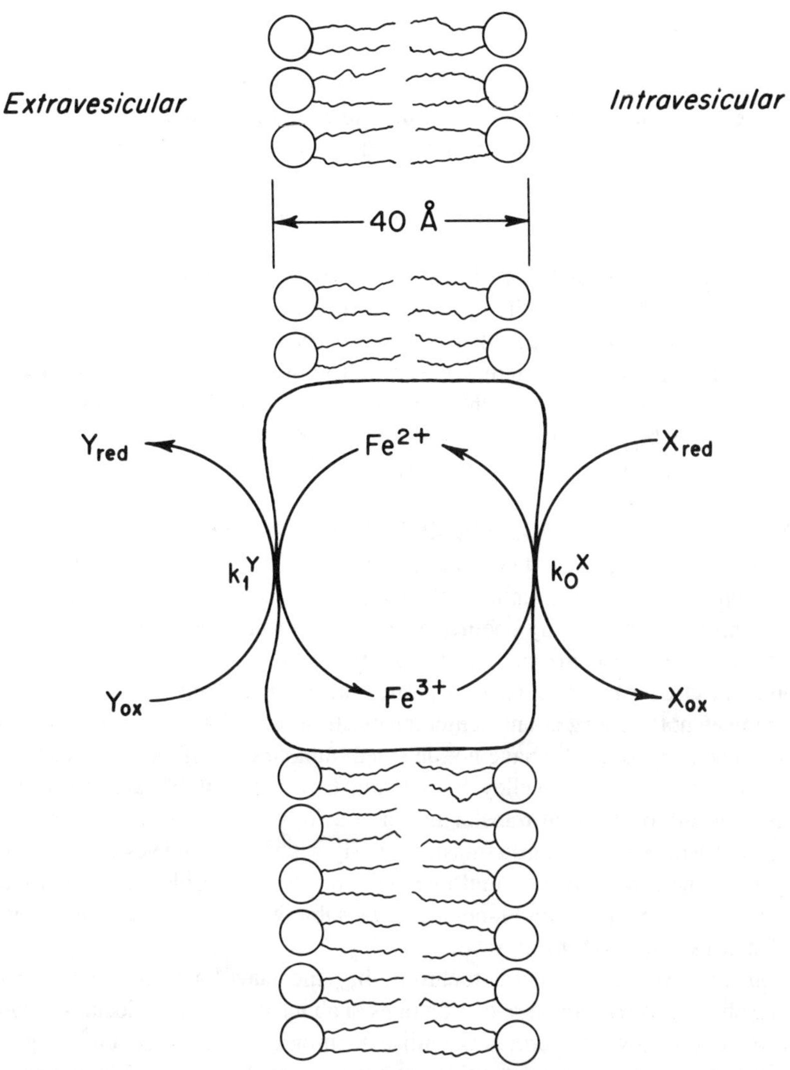

FIGURE 3. Electron transfer through cytochrome b_{561}: k_0^X is the rate constant for reduction of cytochrome b_{561} by intravesicular reductant (X_{red}); k_1^Y is the rate constant for oxidation of cytochrome b_{561} by extravesicular oxidant (Y_{OX}).

Analysis of steady-state electron transfer from internal ascorbate (X = ascorbate) to external ferricyanide (Y = ferricyanide) gives us second-order rate constants k_0^A and k_1^F.[61] We have reevaluated these rate constants at high ionic strength and in the presence of methylamine to minimize surface charge and pH gradient effects. We have also been able to measure k_1^A from an analysis of electron transfer from internal ascorbate to external ascorbate free radical (Y = semidehydroascorbate). These results (Jalukar and Kelley[77]) are summarized in Table 2.

If this simple picture of electron transfer is correct, then measured rate constants should agree with the theory of electron transfer reactions in solution. Accordingly, rate constants should be consistent with the cross-relation, which relates the rate constant for electron transfer from species 1 to species 2 (k_{12}) to the self-exchange rate constants for each species (k_{11} and k_{22}) and the equilibrium constant for the reaction (K_{12}). An approximation to the cross-relation is:[1]

$$k_{12} \approx (k_{11}k_{22}K_{12}f_{12})^{1/2} \quad (2)$$

f_{12} is a collision factor which we will assume is close to unity. If two different electron acceptors (species 2 and species 3) are used to oxidize cytochrome b_{561} (species 1) and both experience the same work functions in approaching the cytochrome, then

$$k_{12}/k_{13} \approx (k_{22}K_{12}/k_{33}K_{13})^{1/2} \quad (3)$$

K_{12} and K_{13} may be calculated from Equation (1) using literature values for the midpoint reduction potentials. Then using literature values for the self-exchange rate constants, one rate constant (e.g., k_1^A) may be calculated from another (e.g., k_1^F). As seen in Table 2, the value of k_1^A calculated from k_1^F agrees fairly well with the measured value of k_1^A. These kinetic studies, therefore, are consistent with the concept that electron transfer between the heme of cytochrome b_{561} and donors/acceptors on either surface of the membrane occurs by ordinary, bimolecular, outer-sphere reactions.

D. MEDIATION OF LONG-RANGE ELECTRON TRANSFER

While cytochrome b_{561} almost certainly provides a pathway for transmembrane electron flow, the nature of this pathway is not yet clear. Since the membrane is about 40 Å thick and the cytochrome contains only one heme, electrons must traverse a distance of about 20 Å to transfer between the heme and either side of the membrane.

No other electron carriers appear to participate in this electron transfer; the purified protein is competent to catalyze transmembrane electron flow.[48] Tests for nonprotein electron carriers have been negative. We have not detected quinones specifically associated with the chromaffin-vesicle membrane (Kelley[78]). Vitamin E, a lipid-soluble electron carrier, does not increase the rate of electron transfer in chromaffin-vesicle ghosts. Moreover, it is not required for reconstitution of electron transfer activity in phospholipid vesicles. The possiblity that the cytochrome functions as a multimer also might be considered. It seems unlikely, however, that two identical membrane-spanning subunits could position heme groups at different distances across the membrane.

Consequently, we believe that cytochrome b_{561} mediates long-range electron transfer. The rates we observe (ferricyanide reduction rates of about 10 s^{-1} per molecule of cytochrome b_{561}[46]) are relatively slow. Experiments with cytochrome c show that long-range (>10Å) electron transfer can occur across peptides at rates faster than this.[73] We suggest that cytochrome b_{561} is specifically adapted to mediate long-range electron transfer. To maximize electron transfer efficiency, the cytochrome does not change conformation upon oxidation or reduction. Interaction with aqueous electron donors and acceptors is relatively nonspecific and does not involve high-affinity binding sites on the surface of the cytochrome. Finally, as suggested by kinetic studies, electrons are transferred between the heme and either surface of the membrane by a simple outer-sphere mechanism. There are two reasons for adopting this hypothesis. First, it is consistent with the data presently available. Second, it is readily testable because rates of electron transfer should agree with the classical theory of electron transfer reactions in solution.[1]

V. CONCLUSION

Compared to other biological electron transfer systems, the secretory vesicle ascorbate-regenerating system is unusually simple. Cytochrome b_{561} seems to mediate long-range electron transfer via two well-defined paths: between the heme and either surface of the membrane. The substrates, ascorbate and its free radical, are aqueous and interact with the cytochrome via bimolecular reactions. Consequently, we are beginning to understand cytochrome b_{561}-mediated electron transfer on the molecular level. With new approaches made possible by the recent cloning and sequencing of the cDNA, this is an excellent model for studying long-range electron transfer in biological systems.

ACKNOWLEDGMENTS

This work was supported by NIH Grant GM30500.

REFERENCES

1. **Marcus, R. A. and Sutin, N.,** Electron transfers in chemistry and biology, *Biochim. Biophys. Acta,* 811, 265, 1985.
2. **Flatmark, T., Terland, O., and Helle, K. B.,** Electron carriers of the bovine adrenal chromaffin granules, *Biochim. Biophys. Acta,* 226, 9, 1971.
3. **Njus, D., Zallakian, M., and Knoth, J.,** The chromaffin granule: proton-cycling in the slow lane, in *Chemiosmotic Proton Circuits in Biological Membranes,* Skulachev, V. P. and Hinkle, P. C., Eds., Addison-Wesley, Reading, MA, 1981, 365.
4. **Perin, M. S., Fried, V. A., Slaughter, C. A., and Sudhof, T. C.,** The structure of cytochrome b_{561}, a secretory vesicle-specific electron transport protein, *EMBO J.,* 7, 2697, 1988.
5. **Njus, D., Kelley, P. M., and Harnadek, G. J.,** Bioenergetics of secretory vesicles, *Biochim. Biophys. Acta,* 853, 237, 1986.
6. **Winkler, H., Apps, D. K., and Fischer-Colbrie, R.,** The molecular function of adrenal chromaffin granules: established facts and unresolved topics, *Neuroscience,* 18, 261, 1986.
7. **Johnson, R. G., Jr.,** Accumulation of biological amines into chromaffin granules: a model for hormone and neurotransmitter transport, *Physiol. Rev.,* 68, 232, 1988.
8. **Stewart, L. C. and Klinman, J. P.,** Dopamine beta-hydroxylase of adrenal chromaffin granules: structure and function, *Annu. Rev. Biochem.,* 57, 551, 1988.
9. **Eipper, B. A. and Mains, R. E.,** Peptide α-amidation, *Annu. Rev. Physiol.,* 50, 333, 1988.
10. **May, S. W., Phillips, R. S., Mueller, P. W., and Herman, H. H.,** Dopamine β-hydroxylase, comparative specificities and mechanism of the oxygenation reactions, *J. Biol. Chem.,* 256, 8470, 1981.
11. **Colombo, G. and Villafranca, J. J.,** An acetylenic mechanism-based inhibitor of dopamine β-hydroxylase, *J. Biol. Chem.,* 259, 15017, 1984.
12. **Friedman, S. and Kaufman, S.,** An electron paramagnetic resonance study of 3,4-dihydroxyphenylethylamine β-hydroxylase, *J. Biol. Chem.,* 241, 2256, 1966.
13. **Lamouroux, A., Vigny, A., Faucon Biguet, N., Darmon, M. C., Franck, R., Henry, J. P., and Mallet, J.,** The primary structure of human dopamine β-hydroxylase: insights into the relationship between the soluble and the membrane-bound forms of the enzyme, *EMBO J.,* 6, 3931, 1987.
14. **Sabban, E. L., Greene, L. A., and Goldstein, M.,** Mechanism of biosynthesis of soluble and membrane-bound forms of dopamine β-hydroxylase in PC12 pheochromocytoma cells, *J. Biol. Chem.,* 258, 7812, 1983.
15. **Saxena, A. and Fleming, P. J.,** Isolation and reconstitution of the membrane-bound form of dopamine β-hydroxylase, *J. Biol. Chem.,* 258, 4147, 1983.
16. **Ferguson, M. A. J. and Williams, A. F.,** Cell-surface anchoring of proteins via glycosyl-phosphatidylinositol structures, *Annu. Rev. Biochem.,* 57, 285, 1988.
17. **Skotland, T. and Ljones, T.,** Direct spectrophotometric detection of ascorbate free radical formed by dopamine β-monooxygenase and by ascorbate oxidase, *Biochim. Biophys. Acta,* 630, 30, 1980.
18. **Villafranca, J. J.,** Dopamine β-hydroxylase, in *Copper Proteins,* Spiro, T. G., Ed., John Wiley & Sons, New York, 1981, 263.
19. **Ash, D. E., Papadopoulos, N. J., Colombo, G., and Villafranca, J. J.,** Kinetic and spectroscopic studies of the interaction of copper with dopamine β-hydroxylase, *J. Biol. Chem.,* 259, 3395, 1984.
20. **Klinman, J. P., Krueger, M., Brenner, M., Edmondson, D. E.,** Evidence for two copper atoms/subunit in dopamine β-monooxygenase catalysis, *J. Biol. Chem.,* 259, 3399, 1984.
21. **Hasnain, S. S., Diakun, G. P., Knowles, P. F., Binsted, N., Garner, C. D., and Blackburn, N. J.,** Direct structural information for the copper site of dopamine β-monooxygenase obtained by using extended X-ray-absorption fine structure, *Biochem. J.,* 221, 545, 1984.
22. **Scott, R. A., Sullivan, R. J., DeWolf, W. E., Jr., Dolle, R. E., and Kruse, L. I.,** The copper sites of dopamine β-hydroxylase: An X-ray absorption spectroscopic study, *Biochemistry,* 27, 5411, 1988.

23. **Ahn, N. G. and Klinman, J. P.,** Mechanism of modulation of dopamine β-monooxygenase by pH and fumarate as deduced from initial rate and primary deuterium isotope effect studies, *Biochemistry,* 22, 3096, 1983.
24. **Miller, S. M. and Klinman, J. P.,** Secondary isotope effects and structure-reactivity correlations in the dopamine β-monooxygenase reaction: Evidence for a chemical mechanism, *Biochemistry,* 24, 2114, 1985.
25. **May, S. W., Mueller, P. W., Padgette, S. R., Herman, H. H., and Phillips, R. S.,** Dopamine-β-hydroxylase: suicide inhibition by the novel olefinic substrate, 1-phenyl-1-aminomethylethene, *Biochem. Biophys. Res. Commun.,* 110, 161, 1983.
26. **Fitzpatrick, P. F., Flory, D. R., Jr., and Villafranca, J. J.,** 3-Phenylpropenes as mechanism-based inhibitors of dopamine β-hydroxylase: evidence for a radical mechanism, *Biochemistry,* 24, 2108, 1985.
27. **Rosenberg, R. C., Gimble, J. M., and Lovenberg, W.,** Inhibition of dopamine-β-hydroxylase by alternative electron donors, *Biochim. Biophys. Acta,* 613, 62, 1980.
28. **Diliberto, E. J., Jr. and Allen, P. L.,** Mechanism of dopamine β-hydroxylation. Semidehydroascorbate as the enzymic oxidation product of ascorbate, *J. Biol. Chem.,* 256, 3385, 1981.
29. **Yamazaki, I. and Piette, L.,** Mechanism of free radical formation and disappearance during the ascorbic acid oxidase and peroxidase reactions, *Biochim. Biophys. Acta,* 50, 62, 1961.
30. **Ljones, T. and Skotland, T.,** Dopamine β-monooxygenase, in *Copper Proteins and Copper Enzymes,* Vol. 2, Lontie, R., Ed., CRC Press, Boca Raton, FL, 1984, 131.
31. **Bradbury, A. F., Finnie, M. D. A., and Smyth, D. G.,** Mechanism of C-terminal amide formation by pituitary enzymes, *Nature,* 298, 686, 1982.
32. **Kizer, J. S., Bateman, L. C., Jr., Miller, C. R., Humm, J., Busby, W. H., Jr., and Youngblood, W. W.,** Purification and characterization of a peptidyl glycine monooxygenase from porcine pituitary, *Endocrinology,* 118, 2262, 1986.
33. **Bateman, R. C., Jr., Youngblood, W. W., Busby, W. H., Jr., and Kizer, J. S.,** Nonenzymatic peptide α-amidation: implications for a novel enzyme mechanism, *J. Biol. Chem.,* 260, 9088, 1985.
34. **Terland, O. and Flatmark, T.,** Ascorbate as a natural constituent of chromaffin granules from the bovine adrenal medulla, *FEBS Lett.,* 59, 52, 1975.
35. **Ingebretsen, O. C., Terland, O., and Flatmark, T.,** Subcellular distribution of ascorbate in bovine adrenal medulla: evidence for accumulation in chromaffin granules against a concentration gradient, *Biochim. Biophys. Acta,* 628, 182, 1980.
36. **Yamazaki, I., Mason, H. S., and Piette, L.,** Identification, by electron paramagnetic resonance spectroscopy, of free radicals generated from substrates by peroxidase, *J. Biol. Chem.,* 235, 2444, 1960.
37. **Bielski, B. H. J.,** Chemistry of ascorbic acid radicals, in *Ascorbic Acid: Chemistry, Metabolism, and Uses,* Seib, P. A. and Tolbert, B. M., Eds., American Chemical Society, Washington, D.C., 1982, 81.
38. **Bielski, B. H. J., Allen, A. O., and Schwarz, H. A.,** Mechanism of disproportionation of ascorbate radicals, *J. Am. Chem. Soc.,* 103, 3516, 1981.
39. **Iyanagi, T., Yamazaki, I., and Anan, K. F.,** One-electron oxidation-reduction properties of ascorbic acid, *Biochim. Biophys. Acta,* 806, 255, 1985.
40. **Njus, D., Knoth, J., Cook, C., and Kelley, P. M.,** Electron transfer across the chromaffin granule membrane, *J. Biol. Chem.,* 258, 27, 1983.
41. **Wakefield, L. M., Cass, A. E., and Radda, G. K.,** Functional coupling between chromaffin granules and mitochondria via the granule cytochrome b, *Fed. Proc.,* 41, 893, 1982.
42. **Wakefield, L. M., Cass, A. E. G., and Radda, G. K.,** Functional coupling between enzymes of the chromaffin granule membrane, *J. Biol. Chem.,* 261, 9739, 1986.
43. **Wakefield, L. M., Cass, A. E. G., and Radda, G. K.,** Electron transfer across the chromaffin granule membrane, *J. Biol. Chem.,* 261, 9746, 1986.
44. **Diliberto, E. J., Jr., Dean, G., Carter, C., and Allen, P. L.,** Tissue, subcellular, and submitochondrial distributions of semidehydroascorbate reductase: possible role of semidehydroascorbate reductase in cofactor regeneration, *J. Neurochem.,* 39, 563, 1982.
45. **Rudnick, G.,** ATP-driven H^+ pumping into intracellular organelles, *Annu. Rev. Physiol.,* 48, 403, 1986.
46. **Harnadek, G. J., Ries, E. A., and Njus, D.,** Rate of transmembrane electron transfer in chromaffin vesicle ghosts, *Biochemistry,* 24, 2640, 1985.
47. **Russell, J. T., Levine, M., and Njus, D.,** Electron transfer across posterior pituitary neurosecretory vesicle membranes, *J. Biol. Chem.,* 260, 226, 1985.
48. **Srivastava, M., Duong, L. T., and Fleming, P. J.,** Cytochrome b_{561} catalyzes transmembrane electron transfer, *J. Biol. Chem.,* 259, 8072, 1984.
49. **Flatmark, T. and Terland, O.,** Cytochrome b_{561} of the bovine adrenal chromaffin granules. A high potential b-type cytochrome, *Biochim. Biophys. Acta,* 253, 487, 1971.
50. **Apps, D. K., Boisclair, M. D., Gavine, F. S., and Pettigrew, G. W.,** Unusual redox behaviour of cytochrome b-561 from bovine chromaffin granule membranes, *Biochim. Biophys. Acta,* 764, 8, 1984.
51. **Kelley, P. M. and Njus, D.,** Cytochrome b_{561} spectral changes associated with electron transfer in chromaffin-vesicle ghosts, *J. Biol. Chem.,* 261, 6429, 1986.

52. **Cidon, S., Ben-David, H., and Nelson, N.,** ATP-driven proton fluxes across membranes of secretory organelles, *J. Biol. Chem.,* 258, 11684, 1983.
53. **Fried, G.,** Cytochrome b_{561} in sympathetic nerve terminal vesicles from rat vas deferens, *Biochim. Biophys. Acta,* 507, 175, 1978.
54. **Flatmark, T., Lagercrantz, H., Terland, O., Helle, K. B., and Stjarne, L.,** Electron carriers of the noradrenaline storage vesicles from bovine splenic nerves, *Biochim. Biophys. Acta,* 245, 249, 1971.
55. **Pruss, R. M. and Shepard, E. A.,** Cytochrome b_{561} can be detected in many neuroendocrine tissues using a specific monoclonal antibody, *Neuroscience,* 22, 149, 1987.
56. **Duong, L. T., Fleming, P. J., and Russell, J. T.,** An identical cytochrome b_{561} is present in bovine adrenal chromaffin vesicles and posterior pituitary neurosecretory vesicles, *J. Biol. Chem.,* 259, 4885, 1984.
57. **Menniti, F. S., Knoth, J., and Diliberto, E. J., Jr.,** Role of ascorbic acid in dopamine β-hydroxylation: The endogenous enzyme cofactor and putative electron donor for cofactor regeneration, *J. Biol. Chem.,* 261, 16901, 1986.
58. **Beers, M. F., Johnson, R. G., and Scarpa, A.,** Evidence for an ascorbate shuttle for the transfer of reducing equivalents across chromaffin granule membranes, *J. Biol. Chem.,* 261, 2529, 1986.
59. **Herman, H. H., Wimalasena, K., Fowler, L. C., Beard, C. A., and May, S. W.,** Demonstration of the ascorbate dependence of membrane-bound dopamine β-monooxygenase in adrenal chromaffin granule ghosts, *J. Biol. Chem.,* 263, 666, 1988.
60. **Levine, M., Morita, K., Heldman, E., and Pollard, H. B.,** Ascorbic acid regulation of norepinephrine biosynthesis in isolated chromaffin granules from bovine adrenal medulla, *J. Biol. Chem.,* 260, 15598, 1985.
61. **Kelley, P. M. and Njus, D.,** A kinetic analysis of electron transport across chromaffin vesicle membranes, *J. Biol. Chem.,* 263, 3799, 1988.
62. **Grouselle, M. and Phillips, J. H.,** Reduction of membrane-bound dopamine β-hydroxylase from the cytoplasmic surface of the chromaffin-granule membrane, *Biochem. J.,* 202, 759, 1982.
63. **Ahn, N. G. and Klinman, J. P.,** Activation of dopamine β-monooxygenase by external and internal electron donors in resealed chromaffin granule ghosts, *J. Biol. Chem.,* 262, 1485, 1987.
64. **Kent, U. M. and Fleming, P. J.,** Purified cytochrome b_{561} catalyzes transmembrane electron transfer for dopamine β-hydroxylase and peptidyl glycine α-amidating monooxygenase activities in reconstituted systems, *J. Biol. Chem.,* 262, 8174, 1987.
65. **Apps, D. K., Pryde, J. G., and Phillips, J. H.,** Cytochrome b_{561} is identical with chromomembrin B, a major polypeptide of chromaffin granule membranes, *Neuroscience,* 5, 2279, 1980.
66. **Duong, L. T. and Fleming, P. J.,** Isolation and properties of cytochrome b_{561} from bovine adrenal chromaffin granules, *J. Biol. Chem.,* 257, 8561, 1982.
67. **Flatmark, T. and Gronberg, M.,** Cytochrome b-561 of the bovine adrenal chromaffin granules. Molecular weight and hydrodynamic properties in micellar solutions of Triton X-100, *Biochem. Biophys. Res. Commun.,* 99, 292, 1981.
68. **Wakefield, L. M., Cass, A. E. G., and Radda, G. K.,** Isolation of a membrane protein by chromatofocusing: cytochrome b-561 of the adrenal chromaffin granule, *Biochem. Biophys. Meth.,* 9, 331, 1984.
69. **Silsand, T. and Flatmark, T.,** Purification of cytochrome b-561. An integral heme protein of the adrenal chromaffin granule membrane, *Biochim. Biophys. Acta,* 395, 257, 1974.
70. **Duong, L. T. and Fleming, P. J.,** The asymmetric orientation of cytochrome b_{561} in bovine chromaffin granule membranes, *Arch. Biochem. Biophys.,* 228, 332, 1984.
71. **Abbs, M. T. and Phillips, J. H.,** Organisation of the proteins of the chromaffin granule membrane, *Biochim. Biophys. Acta,* 595, 200, 1980.
72. **Terland, O. and Flatmark, T.,** Oxidoreductase activities of chromaffin granule ghosts isolated from the bovine adrenal medulla, *Biochim. Biophys. Acta,* 597, 318, 1980.
73. **Isied, S. S.,** Long-range electron transfer in peptides and proteins, *Prog. Inorg. Chem.,* 32, 443, 1984.
74. **Butler, J., Davies, D. M., and Sykes, A. G.,** Kinetic data for redox reactions of cytochrome c with $Fe(CN)_5X$ complexes and the question of association prior to electron transfer, *J. Inorg. Biochem.,* 15, 41, 1981.
75. **Williams, N. H. and Yandell, J. K.,** Outer-sphere electron-transfer reactions of ascorbate anions, *Aust. J. Chem.,* 35, 1133, 1982.
76. **Harnadek, G. J., Ries, E. A., Farhat, A., and Njus, D.,** 5-Methylphenazinium methyl sulfate mediates cyclic electron flow and proton gradient dissipation in chromaffin-vesicle membranes, *J. Biol. Chem.,* in press.
77. **Jalukar, V. and Kelley, P. M.,** unpublished observations.
78. **Kelley, P. M.,** unpublished results.

Chapter 5

HORMONE CONTROL OF PLASMALEMMA ELECTRON TRANSPORT

F. L. Crane, I. L. Sun, and H. Löw

TABLE OF CONTENTS

I. INTRODUCTION

Hormone effects on the transplasma membrane electron transport activity were first observed by Dormandy and Zarday.[1] Insulin at high concentration above normal physiological levels produced a strong stimulation of ferricyanide reduction by human erythrocytes. The ferricyanide-induced proton release was also stimulated. Of course, at the time it was not known that erythrocytes have insulin receptors, and the affinity of the receptors for insulin was not known.[2-6]

Since this early observation, effects of several different hormones on oxidoreductases have been observed in plasma membranes from different tissues and cells. Effects have been observed with glucagon,[7-9,13] ACTH,[9] α- and β-adrenergic agents,[10-12,14] triiodothyronine,[7,10,15] auxins,[16] bombesin, liver growth factor[17] and neurotransmitters,[18] in addition to insulin.[1,19-21] The implications of these observations are that a diverse array of hormone receptors on the plasma membrane can interact with one or more of the plasma membrane oxidoreductases.[21,22] The transmission of the response may be direct or indirect. In other words, the receptor may undergo a conformation change to directly[23] perturb the electron transport enzymes. Indirect effects could be transmitted via changes in membrane lipid orientation or fluidity, through other proteins (e.g., G proteins[24]), or by second messenger activation of other proteins[25-27] (e.g., protein kinase C [PKC] or tyrosine kinase phosphorylation of a redox component). The hormone effects observed on the plasma membrane redox activity have for the most part not been examined beyond the descriptive level. It is clear that much more needs to be done to elucidate the mechanism of action and define the exact oxidoreductase responsible for the response.

II. INSULIN EFFECTS ON PLASMA MEMBRANE OXIDOREDUCTASES

Four different types of insulin effects have been described on the plasma membrane oxidoreductase activity. These effects include the following:

1. Inhibition of electron transport in the membrane by direct addition of insulin to whole cells or isolated membranes: the primary effect has been attributed to the transmembrane NADH dehydrogenase, but dehydrogenases on the outer and inner surfaces may also be affected. The effects are seen at concentrations typical of high affinity insulin receptors.[19,28-30]
2. Activation of an NAD(P)H oxidase in membranes isolated from adipocytes which have been pretreated with insulin: pretreatment with thiol reagents may also activate this oxidase which produces H_2O_2 as a product of the reaction.[31,32] These effects are observed at treatment levels of insulin around 200 μU/ml or 1.4×10^{-6} *M*. NADH oxidase activity on the surface of adipocytes and in isolated membranes has also been reported to be stimulated by insulin.[33]
3. Inhibition of NADH oxidase on the surface of adipocytes.[33]
4. Stimulation of transplasma membrane electron transport detected by external electron acceptors: these effects are observed at high insulin concentrations ($>0.16 \times 10^{-6}$ *M*).[34] An effect on IGF-I receptors is suggested because these high concentrations of insulin stimulate cell growth. High concentrations of insulin bind to the IGF receptors.[35,36]

III. INSULIN EFFECTS ON TRANSMEMBRANE ELECTRON TRANSPORT FROM CELLS

Transplasma membrane electron transport by cells can be measured by the reduction of external ferricyanide or other impermeable oxidants.[37,38] A limited number of experiments

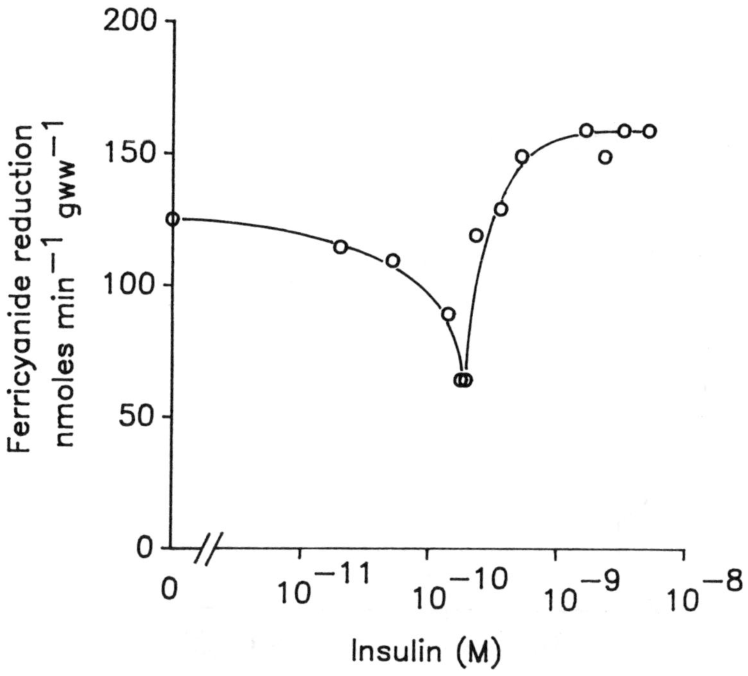

FIGURE 1. Effect of high and low concentrations of insulin on ferricyanide reduction by HeLa cells. Assay in Reference 38.

have tested the effect of insulin on the rate of transplasma membrane electron transport by whole cells. Ferricyanide reduction by rat adipocytes shows a strong inhibition by low concentrations of insulin. Control cells have a rate of ferricyanide reduction of 0.075 μmol $min^{-1}gww^{-1}$, which decreases to 0.033 μmol min $^{-1}gww^{-1}$ after addition of 1.4×10^{-10} *M* insulin.[153] HeLa cells also show a sharp decline in ferricyanide reduction activity at 2×10^{-10} *M* insulin, with an increase in activity to a level higher than the control as the insulin concentration approaches 10^{-9} *M* (Figure 1). The basis for the action of insulin on HeLa cells at low concentrations is not clear since HeLa cells have very little if any high affinity insulin receptors. The low level of insulin I^{125} bound to HeLa cells is only 30% displaced by unlabeled insulin at 15×10^{-10} *M* (Figure 2) (Löw[154]). The stimulation of ferricyanide reduction by HeLa cells by concentrations of insulin over 10^{-9} *M* would not be consistent with insulin binding to the high affinity insulin receptor. High concentrations of insulin do displace IGF-I from its receptor, so the stimulation of transplasma membrane electron transport by concentration of insulin over 10^{-9} *M* may be attributed to effects at the IGF-I receptor.[35] Strong stimulation of ferricyanide reduction by HeLa cells has been observed at concentrations up to 5×10^{-6} *M* insulin[34,39] (Table 1). Since insulin at these high concentrations also stimulates HeLa cell growth,[34] it is logical to propose an effect on the IGF-I receptor to activate ferricyanide reduction by the HeLa cells.

More than one insulin binding site is present on plasma membranes.[40] Insulin is known to bind at a high affinity site on many membranes to rapidly activate functions such as glucose or amino acid transport,[26,41] and activation of certain metabolic enzymes. Insulin also binds to the IGF-I receptor with much lower affinity, and binding at this site can stimulate DNA synthesis and cell growth.[42]

Low concentrations of insulin (1.4×10^{-10} *M*) also partially inhibit ferricyanide reduction by erythrocytes (Figure 3).[20] High concentration of insulin, on the other hand, stimulates ferricyanide reduction by erythrocytes (3.5×10^{-9} *M*).

Transmembrane electron transport from cells shows a consistent pattern of inhibition by insulin at concentrations which should bind to the insulin receptor and stimulation at high

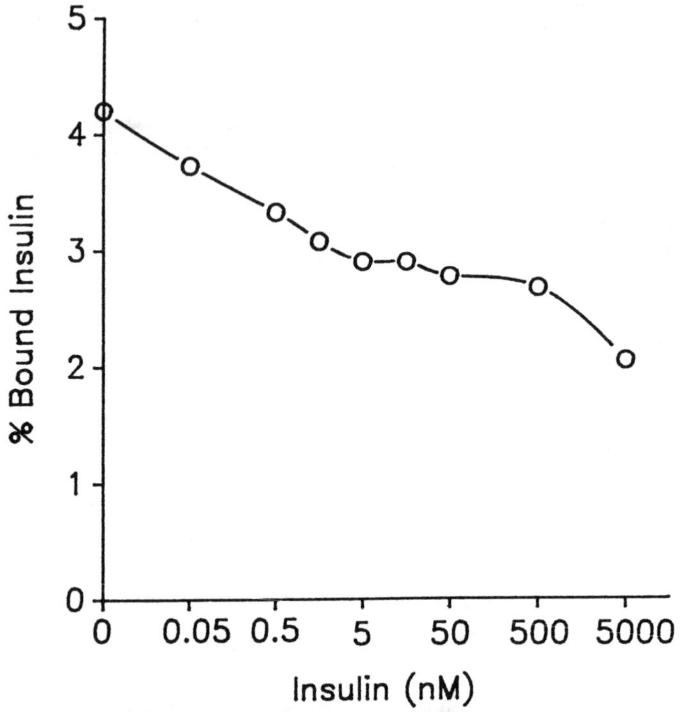

FIGURE 2. Binding and displacement of I^{125} insulin on HeLa cells by unlabeled insulin.

TABLE 1
Effect of Insulin at High Concentration on Ferricyanide Reduction by HeLa Cells

Insulin *M*	Ferricyanide reduction rate (nmol min^{-1} gww^{-1})
	110
4.8×10^{-8}	215
16×10^{-8}	290
48×10^{-8}	310
16×10^{-7}	390
48×10^{-7}	475

Note: Assay with o.1 m*M* ferricyanide, TD-tris buffer (8 g/l NaCl, 0.4 g/l KCl, 0.1 g/l Na_2HPO_4, 3 g/l Trizma base, pH 7.4 with 0.02 g cell w wt/ml. 37°; 3-min incubation with insulin before adding ferricyanide. Reduction determined by absorbance change at 420 minus 500 nm, as described in Reference 38.

concentrations which should bind to the IGF-I receptor. It is not clear if erythrocytes have an IGF-I receptor to account for response to high levels of insulin.

IV. INSULIN AND NAD(P)H DEHYDROGENASE IN ISOLATED MEMBRANES

NAD(P)H dehydrogenase activity in plasma membranes can be assayed using various

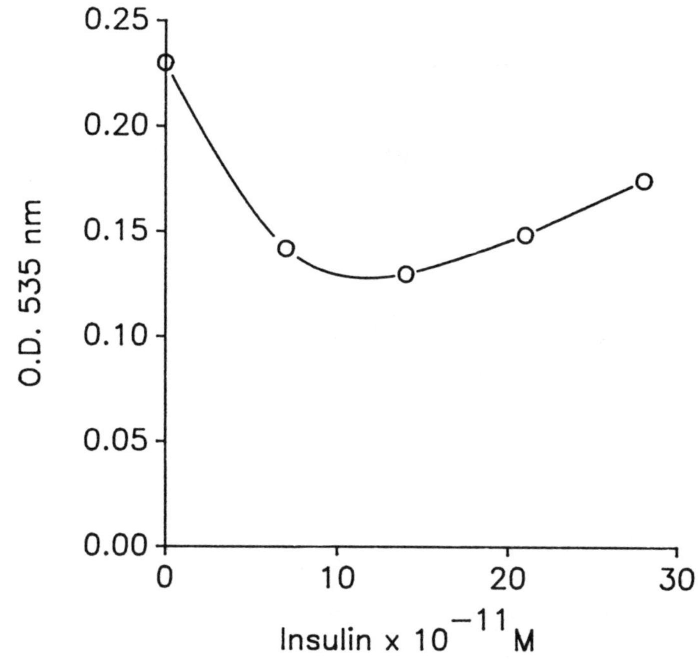

FIGURE 3. Effect of insulin on the rate of ferricyanide reduction by whole pig erythrocytes measured by formation of ferrous bathophenanthroline sulfonate, measuring increase in absorbance at 535 nm.

electron acceptors. The identity of the dehydrogenases that carry out this activity has not been defined, and the activity with a given acceptor may represent the combined activity of more than one dehydrogenase. The degree of vesicularization of the isolated membrane and the orientation of the proteins of the vesicles will also determine what activity is measured, since NAD(P)H and certain acceptors are impermeable to the membrane.[20,21,43]

Liver plasma membrane preparations characteristically have a significant portion of membrane in open sheets with the remainder in vesicles that are mostly inside-out.[44] Dehydrogenase activity in sheets will represent NAD(P)H oxidation sites on both the inner and outer membrane surface, whereas the inside-out vesicles will have NADH dehydrogenase, which transmits electrons to the same side of the membrane when impermeable ferricyanide is the acceptor, but can react with permeable acceptors such as oxygen or dichloroindophenol on both sides of the membrane. NADH cytochrome c reductase in endomembranes and plasma membranes is located on the cytosolic side of the membrane.[45]

Mouse liver plasma membrane preparations show insulin inhibition of NADH dehydrogenase activity when ferricyanide, dichloroindophenol, semidehydroascorbate, and cytochrome c are used as electron acceptors. The concentration of insulin for maximum inhibition is about 2×10^{-10} *M*, which indicates that these membranes have insulin receptors exposed on the surface (Table 2). The effect of insulin concentration on the NADH ferricyanide reductase of plasma membrane is compared with the effect in microsomes in Figure 4. Note the narrow concentration range where inhibition is observed and the stimulation of activity above this concentration. With rat liver plasma membranes, the same biphasic curve is seen with insulin. The maximum inhibition is observed at low concentrations and less inhibition at higher concentrations of insulin (Figure 5).[8,46] The insulin inhibition of NADH ferricyanide reduction by human erythrocyte membranes (open ghosts) and rat adipocyte plasma membranes reaches a maximum at 10^{-10} *M* and increases at higher concentration except in the case of adipocyte membrane (Figure 5). The fat cell plasma membranes used in these assays

TABLE 2
Effect of Insulin on NADH Dehydrogenase Activities of Mouse Liver Plasma Membranes

Enzyme system	Specific activity (nmol/min/mg protein) without insulin	Specific activity (nmol/min/mg protein) with insulin (18×10^{-10} *M*)	% Inhibition
NADH → Fe $(CN)_6$ reductase	983	543	45
NADH → ascorbate free radical reductase	67	33	21
NADH → cytochrome c reductase	41	26	37
NADH → DCPIP[a] reductase	109	51	53

[a] DCPIP, dichlorophenol indophenol.

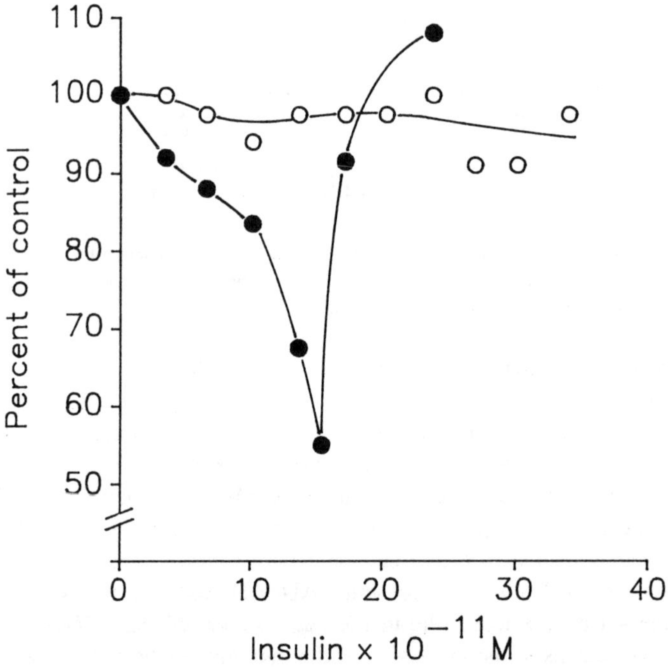

FIGURE 4. Comparison of the effect of insulin on NADH → $Fe(CN)_6$ reductase between plasma membranes and endoplasmic reticulum membranes of mouse liver: (●) plasma membrane, (○) endoplasmic reticulum. Assay as in Reference 20.

would be expected to be mainly in closed vesicles which would not permit analysis of *trans* NADH ferricyanide reductase activity, since ferricyanide and NADH are both impermeable to the membrane. Therefore, the insulin effect on these membranes using the ferricyanide assay is surprising and may indicate an effect on an enzyme which is not transmembrane oriented or the presence of unsealed vesicles. Erythrocyte membranes can be prepared with established surface orientation.[43] The NADH ferricyanide reductase activity of open erythrocyte ghosts, which would include transmembrane electron transport, does show partial insulin inhibition. The NADH ferricyanide reductase of inside-out vesicles is not inhibited by insulin (Figure 6). The latter activity is predominantly based on NADH cytochrome b_5 reductase which is also not inhibited by insulin in microsomes. With fat cell plasma membranes, the insulin inhibition tends to be retained at higher insulin concentrations. The inhibition pattern observed is similar when indophenol is used as an electron acceptor.[7,19,30]

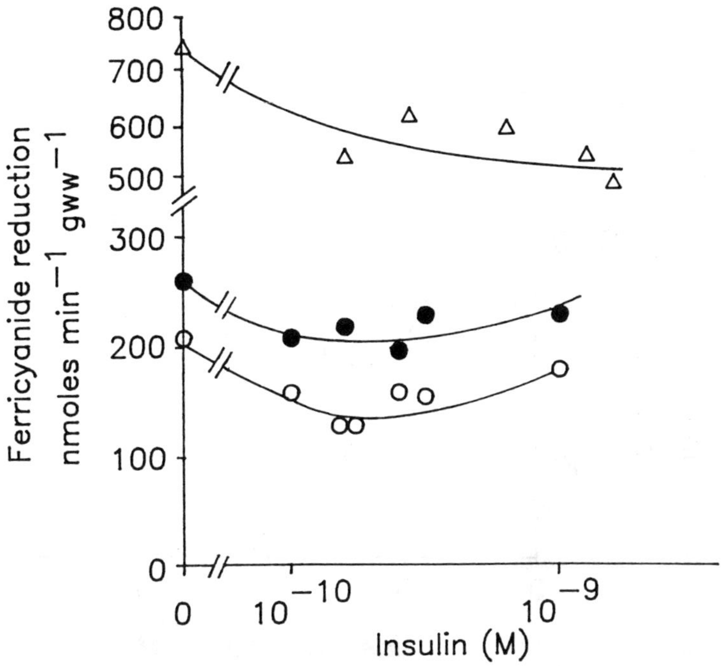

FIGURE 5. Effect of insulin on NADH ferricyanide reductase of (●) liver, (△) adipocyte, and (○) erythrocyte plasma membranes. Assay as in Reference 20

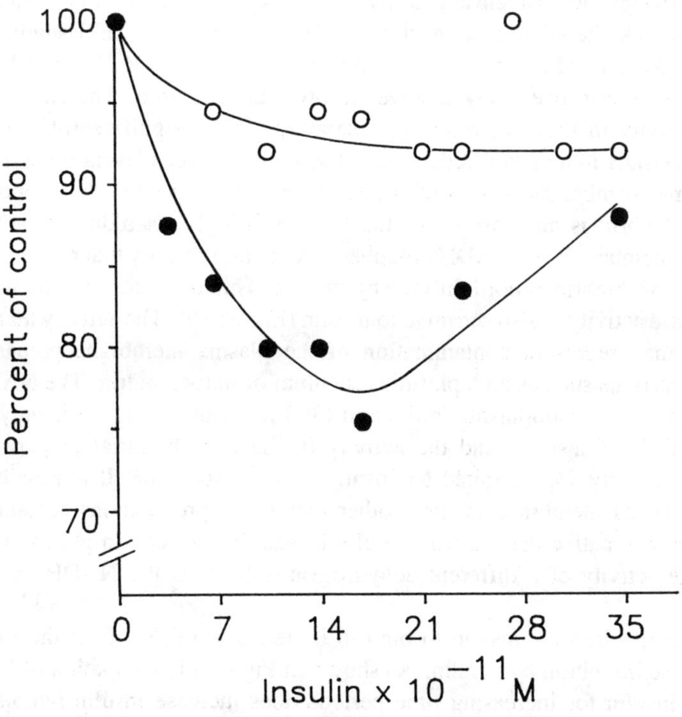

FIGURE 6. Comparison of the effect of insulin on NADH → $Fe(CN)_6$ reductase between (●) open membranes and (○) closed wrong side-out vesicles of pig erythrocytes.

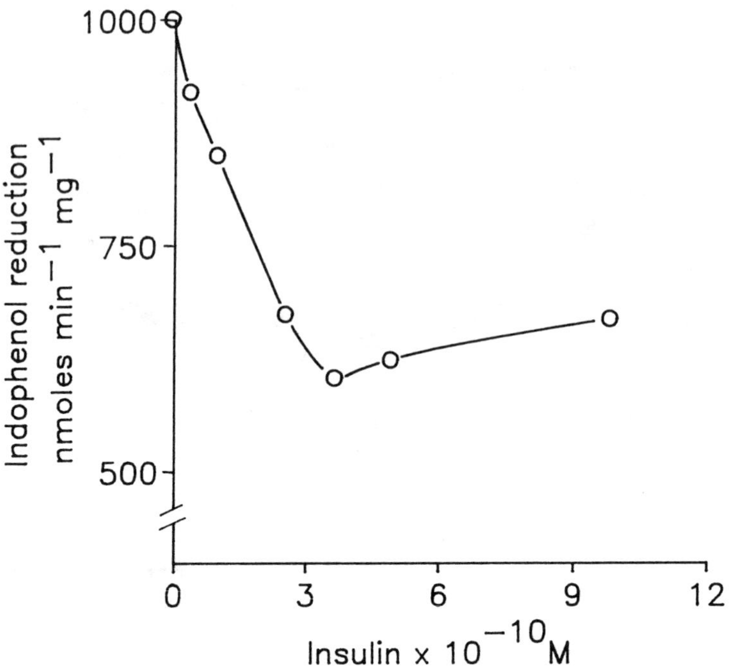

FIGURE 7. Effect of insulin on NADH indophenol reductase of adipocyte plasma membrane.[19]

NADH indophenol reductase activity of adipocyte membranes consistently showed an inhibition of indophenol reduction at 5×10^{-10} *M* (Figure 7). Usually inhibition was observed at lower insulin concentrations down to 7×10^{-11} *M*,[19] but in Figure 8 a slight stimulation is seen in one assay at lower insulin concentration. The NADH cytochrome c reductase activity in these membrane preparations is not significantly affected by insulin (Figure 8). NADH indophenol reductase activity of rat liver plasma membranes and erythrocyte plasma membranes is also inhibited by insulin in the 10^{-10} *M* concentration range (Figure 9). Atebrin is an inhibitor of the NADH indophenol reductase activity. With rat liver plasma membranes the NADH indophenol reductase activity that remains in the presence of 3×10^{-3} *M* Atebrin is not inhibited by insulin. This indicates that the Atebrin-sensitive dehydrogenase activity is also sensitive to insulin (Figure 10). The activity that is not inhibited by Atebrin may represent contamination of the plasma membrane preparation by other membrane fractions such as endoplasmic reticulum or mitochondria. The NADH indophenol reductase activity in endoplasmic reticulum (360 nmol $min^{-1}mg^{-1}$) is only 15% inhibited at 2.5×10^{-10} *M* insulin, and the activity in the mitochondrial preparation (450 nmol $min^{-1}mg^{-1}$) is only 7% inhibited by insulin at 4×10^{-10} *M*. It is possible that a small amount of plasma membrane in these other membrane preparations accounts for the inhibition. As an alternative the Atebrin, insulin insensitive activity in plasma membranes may represent the activity of a different dehydrogenase such as the NADH cytochrome b_5 reductase.

The time required for insulin to bind to its receptor might affect the extent of NADH dehydrogenase inhibition by insulin. As shown in Figure 11, incubation of liver membranes at 37° with insulin for increasing time periods does increase insulin inhibition somewhat. Inhibition at 2 min is 50% of inhibition found after 12 min incubation. The dehydrogenase activity also decreases during the incubation. If a protease inhibitor PCMF is present during the incubation, the decline in dehydrogenase activity during incubation is greatly reduced, so the decrease in activity may depend on an endogenous protease in the membrane.

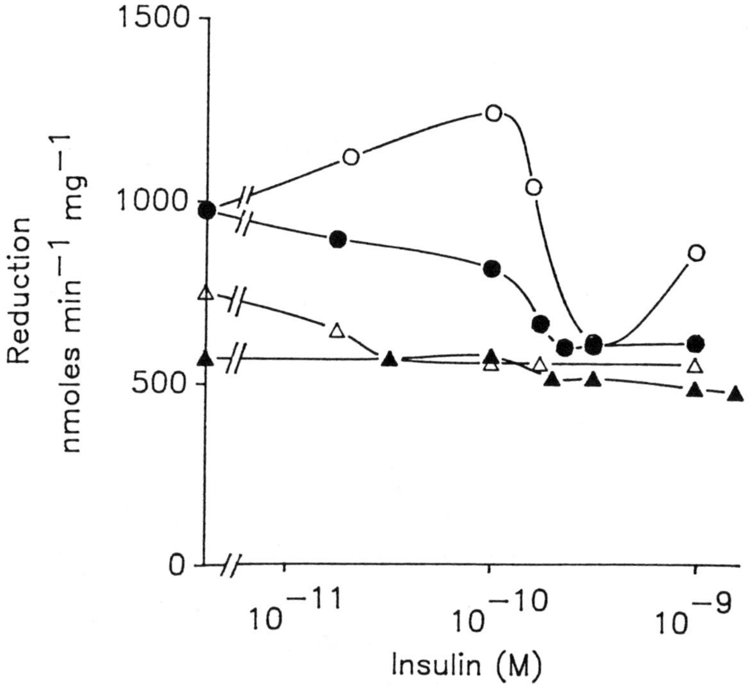

FIGURE 8. Effect of insulin on NADH indophenol reductase at different pH with adipocyte plasma membranes tris pH 6.0. (○); phosphate pH 6.0 (●); tris pH 6.5 (△); and cytochrome c reductase pH 6.6 (▲).

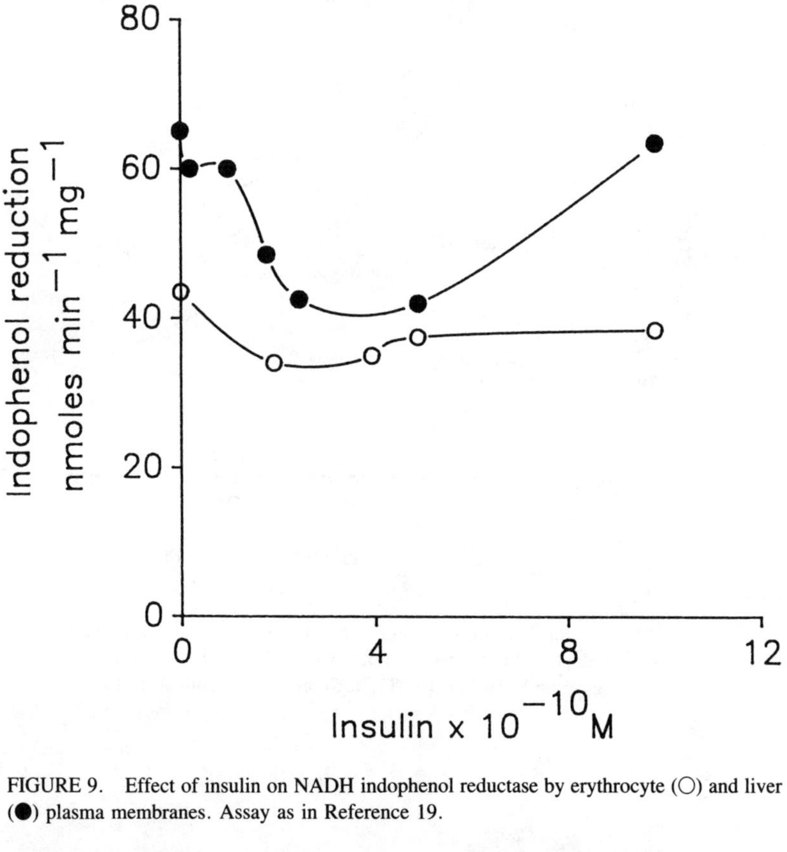

FIGURE 9. Effect of insulin on NADH indophenol reductase by erythrocyte (○) and liver (●) plasma membranes. Assay as in Reference 19.

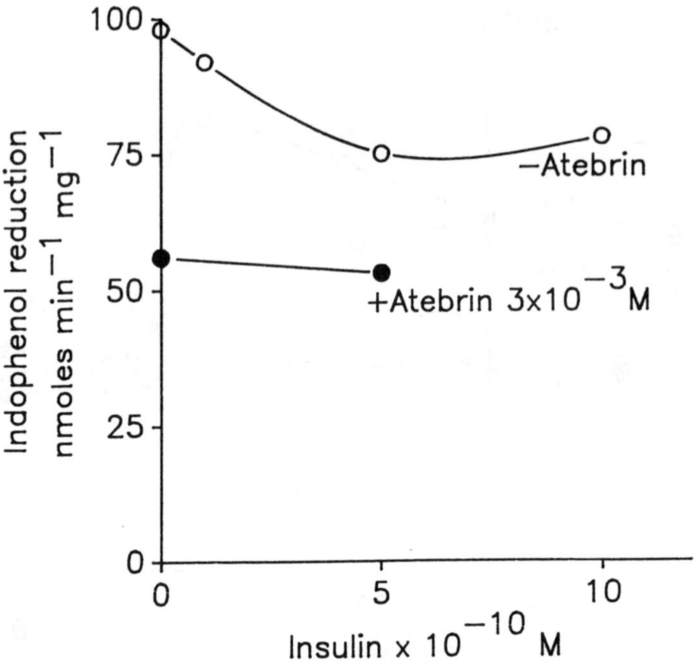

FIGURE 10. Atebrin elimination of insulin inhibition of NADH indophenol reductase of rat liver plasma membrane. Assay as in Reference 19.

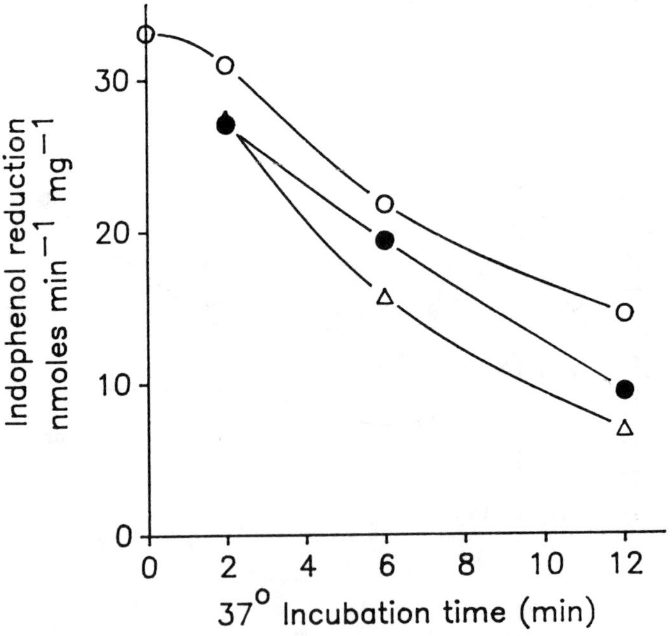

FIGURE 11. Rat liver plasma membrane effect of incubation with insulin on NADH indophenol reductase. (△) 50 μU/ml; (●) 86 μU/ml; and (○) no insulin. Assay as in Reference 19. (From Crane, H. E., unpublished.)

TABLE 3
Effect of Insulin on NADH Semidehydroascorbate Reductase Activity of Rat Liver Plasma Membranes

Insulin conc	NADH semidehydroascorbate reductase % of control
none	100
$2.5 \times 10^{-10}\ M$	100
$3.5 \times 10^{-10}\ M$	35
$5 \times 10^{-10}\ M$	32
$6 \times 10^{-10}\ M$	40
$8.6 \times 10^{-10}\ M$	42

Note: Data based on Reference 46, using rat liver plasma membranes prepared according to Reference 47. Insulin incubated with membrane at pH 8 before assay in tris HCl pH 7.5.

NADH semidehydroascorbate reductase activity of rat liver plasma membrane is also inhibited by insulin at physiological concentrations as shown in Table 3.[46] In contrast to the NADH ferricyanide activity, the inhibition of the NADH semidehydroascorbate reductase does not decrease at higher insulin concentrations.

Lectins such as concanavalin A (con A) compete for insulin binding sites on mouse liver plasma membranes.[48] NADH semidehydroascorbate reductase is also inhibited by con A in rat liver plasma membranes. Wheat germ agglutinin (WGA), however, competes for insulin binding sites but does not inhibit the reductase.[49] Con A and WGA also mimic the effect of insulin on protein phosphorylation in isolated rat liver nuclear membrane.[50] On the other hand, both con A and WGA stimulate lipogenesis and activate glucose transport in rat adipocytes.[51] Therefore, insulin and lectins show similar effects on glucose transport, and since the effects are not additive both ligands may act at the same site on the membrane. The lack of effect of WGA on electron transport indicates that another component may also be involved in transport.

Dehydroascorbate stimulates transplasma membrane electron transport by perfused rat liver[13] and by erythrocytes.[52] This could be accounted for by reduction of the dehydroascorbate to semidehydroascorbate at the cell surface and subsequent reaction with ferricyanide.

V. INSULIN EFFECTS ON REDOX-STIMULATED PROTON RELEASE FROM CELLS

External electron acceptors which can be reduced by the transplasma membrane electron transport stimulate proton release from cells.[38,53] The proton release stimulated by ferricyanide in rat adipocytes is inhibited by insulin starting at $3 \times 10^{-11}\ M$, and this is correlated with the insulin inhibition of the ferricyanide reduction by cells (Table 4 and Figure 12). With HeLa and liver cells the oxidant-stimulated proton release has been demonstrated to be carried out by the Na^+/H^+ antiport.[54,55] The nature of the proton carrier system in adipocytes is not known.

High concentrations of insulin which stimulate ferricyanide reduction by human erythrocytes also stimulate proton release from these cells.[1]

Proton release from neutrophils is observed when the NAD(P)H oxidase responsible for superoxide formation is activated. Proton movement is proposed through an ion exchange channel in the membrane which is activated by the proton production during NADPH oxidation on the inner surface of the membrane.[56]

TABLE 4
Effect of Ferricyanide on Proton Release by Fat Cells

Additions	Proton release (ne H^+/min/mg)	$Fe(CN)_6^{3-}$ Reduction (ne/min/mg cells)
None	0.52	
Insulin 1.4 × $10^{-10}M$	0.48	
Ferricyanide	1.0 ± 0.16	0.12
Ferricyanide + insulin	0.49 ± 0.1	0.03

Note: ΔpH^+ by bromothymol blue, Δ600—500 nm in physiological saline.

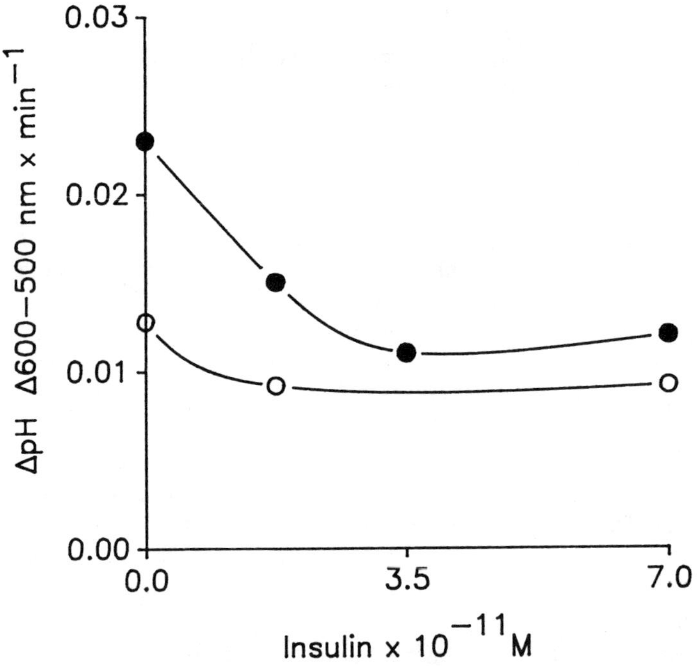

FIGURE 12. Effect of insulin on H^+ release by fat cells in presence (●) and absence (○) of 0.1 m*M* ferricyanide; (Δ) change in 600—500 absorbance is measure of pH change with bromthymol blue.[153]

VI. INSULIN EFFECTS ON NAD(P)H OXIDASE ACTIVITY IN PLASMA MEMBRANES

Plasma membrane preparations from liver and adipocytes, but not from erythrocytes, have NAD(P)H oxidase activity which is greater than can be accounted for by contaminating membranes.[10,57,58]

Only in recent studies has any insulin effect on oxidase activity been observed by direct addition to isolated plasma membranes.[33] On the other hand, isolated membrane from adipocytes that have been pretreated for 15 min with insulin have been reported to have greater NAD(P)H oxidase activity. Pretreatment of cells with thiol reactive reagents such as dithiothreitol, tosyl-1-lysine chloromethyl ketone, or *p*-chloromercuribenzoate (PCMB) also activated the NADPH oxidase. The dehydrogenation site for NADPH is proposed to be on the outer surface of the cell, since oxidation of external NADPH by intact adipocytes was activated by insulin pretreatment.[31] The insulin-activated NADPH oxidase in adipocyte

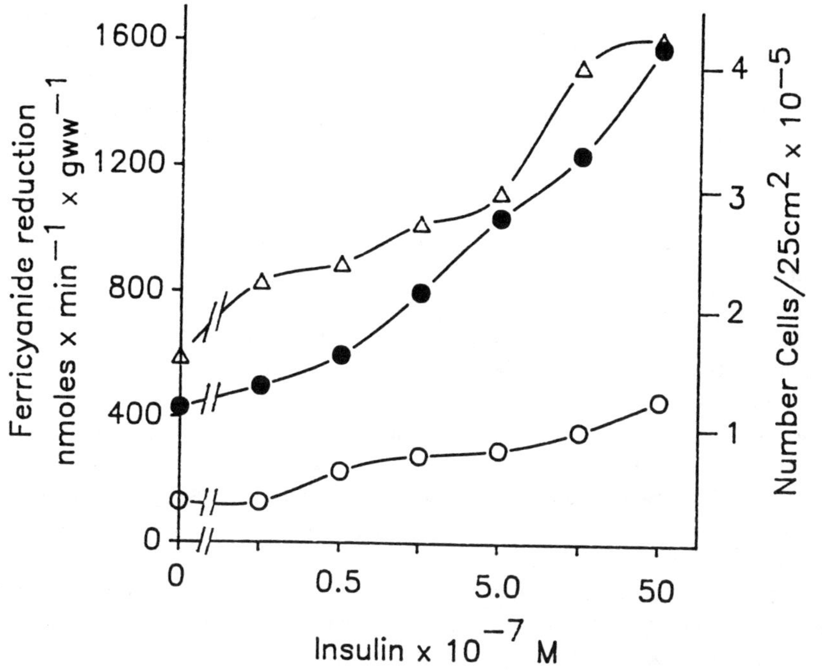

FIGURE 13. Comparison of insulin effects on HeLa cell growth and $Fe(CN)_6$ reduction: △ cell growth; ● fast rate ferricyanide reduction; ○ slow rate. (From Sun, I. L., Crane, F. L., Grebing, C., and Löw, H., *Exp. Cell. Res.*, 156, 528, 1985. With permission.)

plasma membranes was found to produce H_2O_2. The insulin and thiol reagent effects have been rationalized on the basis of common action of H_2O_2 or other thiol reagents at an SH site which would trigger response to insulin.[59-65] Stimulation of a transplasma membrane NADH oxidase by insulin would be more consistent with the evidence that low levels of insulin stimulate the Na^+/H^+ antiport to increase cytosolic pH.[87] The effect would then be consistent with the evidence that transmembrane electron transport stimulates the Na^+/H^+ antiport.

VII. STIMULATION OF TRANSMEMBRANE ELECTRON TRANSPORT BY INSULIN AT HIGH CONCENTRATIONS

Ferricyanide reduction by HeLa cells is significantly stimulated by insulin at over 10 μg/ml (1.6μ*M*). This concentration of insulin is much higher than would saturate the high affinity sites, but is the concentration which would bind to IGF-I binding sites (Figure 12).

High concentrations of insulin have long been associated with mitogenic action on cell growth, and these effects are presumably based on binding to the IGF-I receptor.[42,66,67]

The effect of high concentrations of insulin on ferricyanide reduction and on growth of HeLa cells in absence of serum with ferricyanide is compared in Figure 13. The concentrations of insulin required to stimulate growth and ferricyanide reduction are consistent with binding at the IGF-I site.

VIII. EFFECT OF INSULIN ON MEMBRANE STRUCTURE

Order or viscosity of membrane lipids examined with hexatriene probes is increased in the presence of insulin, and this change in lipid order has been related to enzyme responses in the membrane.[68,69] Insulin has been reported to affect activity of several plasma membrane enzymes such as Na^+/K^+ ATPase, adenylate cyclase, guanylcyclase, acetylcholine esterase, 5′ nucleotidase, glucose, and ion transport, as well as the presence of transferrin receptors.[70-77] The effects of insulin on lipid organization appear to be restricted to specific regions probably associated with protein or cholesterol. Probes such as stearic acid-5-nitroxide failed to report insulin effects on ordering of lipid structure, whereas Ca^{2+} gave strong effects.[78] 8-Anilino-1-naphthalene sulfonate fluorescence is also not changed by insulin.[79]

The nitroxide probes and fluorescence probes used would sample the more fluid portion of membrane lipid,[78,80,81] whereas the diphenylhexatriene may probe regions of lesser fluidity that may be more closely associated with hormone receptors and other proteins. Steroid-based nitroxide probes also show a response to insulin, and these probes are more responsive to change in cholesterol-rich membranes than the aliphatic spin probe.[69]

On the other hand, Luly and Shintizky[68] report less effect of insulin on rat liver membrane viscosity when the membranes have extra cholesterol. They also find no change in microviscosity in human erythrocyte membranes even though these membranes can have insulin receptors and do show dehydrogenase inhibition.[4] Changes in phospholipid composition of membranes in cells in response to insulin (e.g., increase in PS to PC ratio) would be an effect not reproduced in isolated membranes.[82]

Cytochalasin B prevents the increase in membrane lipid order with insulin. It has been proposed that this effect is not based on cytochlasin B binding to the glucose transporter, so the effect may be based on cytochlasin binding to actin associated with the plasma membrane.[69]

Insulin effects on mouse liver plasma membrane fluidity reach saturation at 1 μg/ml (1.7×10^{-10} *M*)[69] with diphenylhexatriene as a probe. For rat liver plasma membranes, saturation of the effect was found at 10^{-9}*M*.[68] These concentrations are close to physiological concentrations and are similar to concentrations where insulin inhibition of the plasma membrane NADH ferricyanide reduction is observed (Figure 14). Maximum inhibition of NADH indophenol reductase in mouse liver membranes has been reported at 10 μU/ml (7×10^{-11} *M*).[7] The maximum insulin inhibition of NADH indophenol reductase activity of rat liver plasma membranes is at 5×10^{-10} *M*, but the effect is complicated by an increase in activity at higher concentrations, so that the true saturation limit on the inhibition curve cannot be measured.[30] Fat cell plasma membranes show maximum inhibition of NADH indophenol reductase at 3×10^{-10} *M* insulin.[19] Erythrocyte ghosts show maximum decline at 3.5×10^{-10}, but in all cases the activity increases at higher concentration, so the concentration for maximum inhibition could be higher if the stimulation effect is at a different site. A comparison of the insulin concentration needed for maximum effect shows changes in membrane viscosity and inhibition of dehydrogenase, ATPase, and other membrane enzymes to occur in a range from 7×10^{-11} *M* to 10^{-9} *M*. Figure 13 shows a comparison of insulin concentration effects on membrane microviscosity and redox activity with rat liver plasma membranes. In contrast, Dutta-Ray et al.[5] describe a decrease in microviscosity in erythrocyte membranes in presence of insulin at 1.6×10^{-9} *M*. A 34% increase of intramembranous particles in plasma membrane of adipocytes exposed to 7×10^{-9} *M* insulin would be consistent with microviscosity change at specific sites.[83] Proinsulin was not as effective as insulin in inhibition of plasma membrane redox activity with both fat and liver membranes.[7,19] Proinsulin also has less effect on microviscosity of rat liver membranes as determined by Luly and Shinitzky.[68]

The inhibitory effects of insulin on oxidoreductase activity of erythrocyte plasma membrane compared with the effects on liver plasma membrane are slightly greater than expected on the basis of insulin receptor concentration on the respective membranes.[6] Rat hepatocytes have 63 and human erythrocytes 14 insulin binding sites per unit of membrane surface. The NADH ferricyanide reductase and NADH indophenol reductase inhibitions show a similar

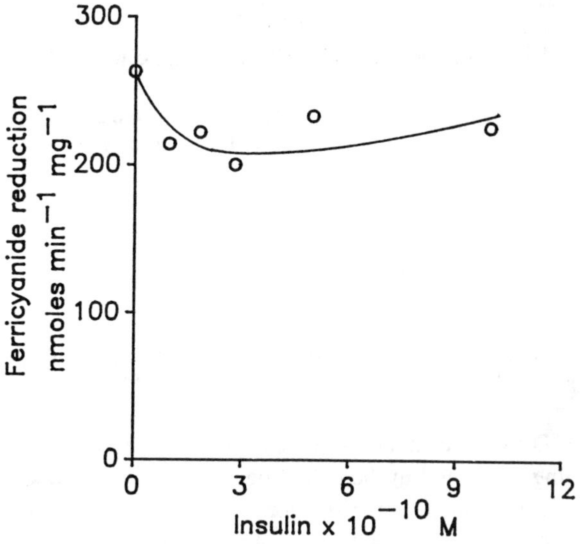

A

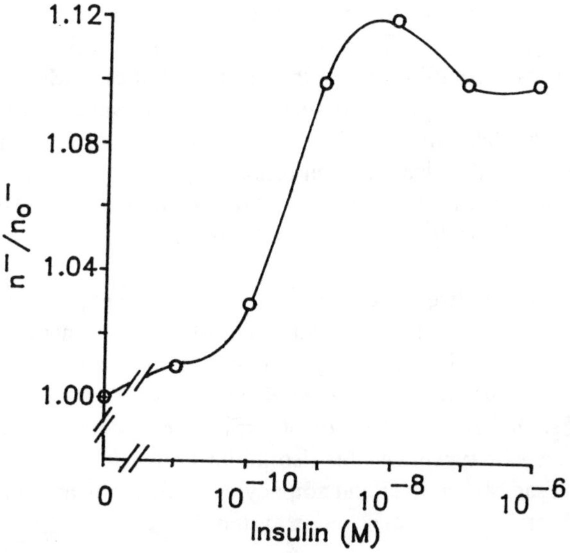

B

FIGURE 14. Comparison of insulin effects on microviscosity and NADH ferricyanide reductase of rat liver plasma membrane. Data for microviscosity from References 68 and 69; ferricyanide reduction from Reference 28.

proportionality with erythrocyte membranes having somewhat more inhibition than expected. Pig erythrocytes have 29 high affinity insulin sites per cell (kDa 1.22 $\times$ 10^{-8} *M*),[3] whereas human erythrocytes have 50 sites/cell (kDa 1.5 $\times$ 10^{-8} *M*)[27] (Table 5).

The greater than expected inhibition of reductase activity in erythrocyte membranes may derive from high levels of NADH cytochrome b_5 reductase in liver plasma membranes. This reductase is less sensitive to insulin inhibition than the transplasma membrane enzyme,[20] as

TABLE 5
Comparison of Insulin Binding Affinity and Indophenol Reductase Inhibition

	Rat hepatocyte	Human erythrocyte	Ratio erythrocyte/hepatocyte
Relative binding/unit surface	63	14	0.22
% Inhibition of NADH-ferricyanide reductase by insulin	45%	24%	0.53
% Inhibition of NADH indophenol reductase by insulin	40%	19%	0.47

From Crane, F. L., Sun, I. L., Clark, M. G., Löw, H., and Grebing, C., in *Hormonal Control of Gluconeogenesis*, Vol. 2, Kraus-Friedman, N., Ed., CRC Press, Boca Raton, FL, 1986, 133. With permission.

shown in Figure 6. Differences may also be caused by different incubation times used with insulin in the binding assay and in the dehydrogenase assay.

Recent advances in membrane purification, such as free flow electrophoresis and two-phase separation, allow much better separation of plasma membranes from other membranes[84] than was possible in the initial study of hormone effects on plasma membrane dehydrogenases. Further purification of plasma membrane fractions from rat adipocytes yields membrane fractions in which insulin stimulates NADH ferricyanide reductase activity as well as fractions which show insulin inhibition of ferricyanide reductase at 10^{-10} *M* insulin.[33] The presence of two membrane types with different response to insulin may account for the biphasic activity changes in response to insulin in the original studies.[19] Moore[87] has proposed that insulin increases intracellular pH to stimulate glycolysis. Since increased transmembrane electron transport stimulates the Na^+/H^+ antiport, insulin stimulation of NADH oxidase would be consistent with the alkalinization of the cytosol by activation of the antiport. The activation of the antiport would account for insulin activation of the Na^+/K^+-ATPase. Insulin inhibition of ferricyanide or dye reduction could be based:

(1) On a tight coupling between electron transport and the antiport when insulin is present. This tight coupling is consistent with amiloride inhibition of the antiport, and the insulin based increase in cytosolic pH, as well as ferricyanide reduction by cells.[17]

(2) As an alternative, insulin may increase electron flow to oxygen at the expense of other acceptors. If so, the insulin inhibition of ferricyanide reduction would only be observed in the presence of oxygen. Insulin also inhibits adenylate cyclase[85] and activates the stearic acid desaturase in liver and adipocytes, which indicates an effect on some part of the cytochrome b_5 dependent redox system.[86]

IX. GLUCAGON EFFECTS ON PLASMA MEMBRANE REDOX FUNCTIONS

Glucagon shows opposing effects on plasma membrane redox activity assayed with intact cells or isolated membranes. With fat cell plasma membranes, glucagon gives a 33% stimulation of NADH trichloroindophenol reductase activity at 3×10^{-8} *M* (Figure 15). With liver cell plasma membranes, increases in activity with glucagon up to 350% have been observed on NADH indophenol reductase activity at 5×10^{-8} *M*. Significant stimulation of NADH cytochrome c reductase activity has also been observed.[7,9,57] The glucagon stimulation is inhibited by 3×10^{-3} *M* Atebrin, which inhibits the plasma membrane redox systems. No stimulation of NADH indophenol reductase or NADH cytochrome c reductase is observed with liver endoplasmic reticulum.

The glucagon stimulation of the plasma membrane NADH indophenol and cytochrome

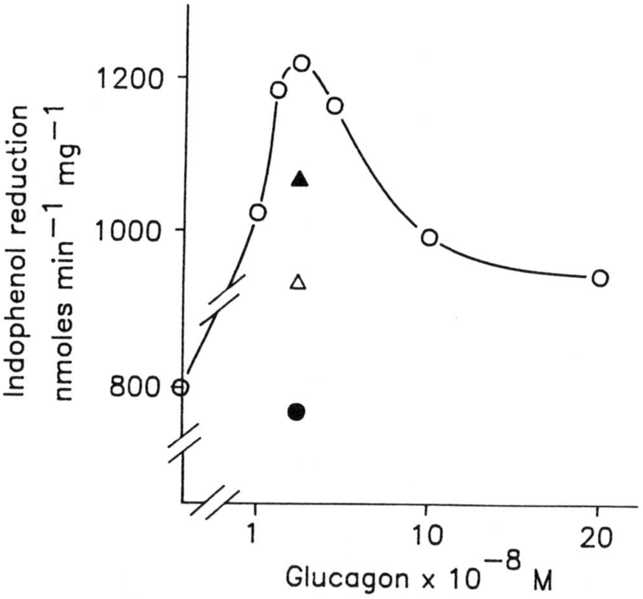

FIGURE 15. Effect of glucagon on NADH indophenol reductase in adipocyte plasma membranes: (○) glucagon in albumin; (●) albumin control 0.5%; (▲) with 2.5 × 10^{-10} insulin; (△) with 5 × 10^{-10} insulin.

TABLE 6
Reversal of Glucagon Stimulation of NADH Indophenol Reductase Activity of Adipocyte and Liver Plasma Membranes by Insulin

Membrane source	Addition	NADH trichloroindophenol reductase activity (nmol $min^{-1}mg^{-1}$)
Adipocyte	None	550
	Glucagon 2 × 10^{-8}	650
	None + insulin 10^{-10} *M*	475
	Glucagon + insulin 10^{-10} *M*	410
Liver	None	62
	Glucagon 2 × 10^{-8} *M*	73
	Insulin 2 × 10^{-10} *M*	43
	Glucagon + insulin	39

Note: Assay in sodium phosphate buffer 0.05 m*M*, pH 6.0, with plasma membranes as in Reference 19.

c reductase activity is restricted to a narrow range of hormone concentrations. Maximum stimulation is observed in the range of 2 to 10 × 10^{-8} *M*, whereas at higher concentrations there is less stimulation. NADH ferricyanide reductase activity of rat liver plasma membranes is stimulated by glucagon.[57] Insulin inhibits glucagon stimulation of redox activity in adipocyte membranes analogous to its inhibition of glucagon stimulation of adenyl cyclase in liver cells[88] (Table 6).

Glucagon does inhibit ferricyanide reduction by perfused liver at 10^{-8} *M*[13] (Table 7). This inhibition of ferricyanide reduction is not observed with added cyclic AMP or dibutyryl cyclic AMP at concentrations which stimulate glucose release, so the inhibitory effect is not mediated through cyclic AMP. The basis for the different effects of glucagon on whole cells and isolated membranes is not clear. With whole cells, the glucagon could be interacting

TABLE 7
Inhibition of Ferricyanide Reduction by Perfused Liver by β-Adrenergic Agonist Ritodrine and Glucagon

	% Inhibition of ferricyanide reduction	
Conc of hormone	**Glucagon**	**Ritodrine**
None[a]	0	0
10^{-12} *M*	3	—
10^{-10} *M*	15	—
10^{-8} *M*	25	3
10^{-7} *M*	28	19
10^{-6} *M*	29	26
10^{-5} *M*	—	29

[a] Initial rate of ferricyanide reduction by the perfused liver was 0.21 μmol min^{-1} gww^{-1}. Data from References 13 and 14.

TABLE 8
Effect of Glucagon on HeLa Cell Ferricyanide Reduction in Presence and Absence of Insulin (1.4 × 10^{-10} *M*)

	Ferricyanide reduction rate (nmol min^{-1} gww^{-1})	
Glucagon μM	**− Insulin**	**+ Insulin**
0	280	190
0.05	330	230
0.1	435	300
0.15	440	405

Note: Insulin at 1.4 × 10^{-10} *M*. Hormones incubated with cells for 3 min before start of assay with 0.1 m*M* potassium ferricyanide. Assay in TD tris buffer pH 7.4 as described in Reference 38.

with ATP-dependent processes inside the cell such as protein phosphorylation, or other redox agents such as glutathione may contribute to the redox function. With isolated membranes, these factors would not be evident. The effect on cytochrome c reductase with membranes also suggests that a different dehydrogenase may be involved than in intact cells.[9] Since glucagon and β-adrenergic agents inhibit transplasma membrane ferricyanide reduction by liver cells,[13,14] it could be expected that activation of the Na^+/H^+ antiport would also be inhibited. Ferricyanide has been shown to activate the Na^+/H^+ antiport in liver and other cells[17,53,55] and increase internal pH of the cells.[89,90] On the other hand, glucagon and adrenergic agonists have been shown not to affect the internal pH of liver cells in the absence of external oxidants.[91] The effect of these hormones on internal pH in presence of ferricyanide or diferric transferrin is not known. In contrast to liver cells, glucagon stimulates the ferricyanide reduction by HeLa cells at 10^{-7} *M*, which is a slightly higher concentration than used with liver cells. Insulin inhibits the activity in HeLa cells at all concentrations of glucagon, but the inhibition is not increased as glucagon increases activity (Table 8). It appears that the glucagon-stimulated activity is not sensitive to insulin using the whole cell assay. The effect of glucagon and NADH dehydrogenase activity of HeLa cell plasma membranes has not been tested.

Glucagon inhibits the Ca^{2+}-ATPase and ATP-dependent Ca^{2+} uptake by liver plasma membrane vesicles at 10^{-6} *M*. These effects parallel the stimulation of the NADH oxidation in concentration of glucagon required.[92]

A role of protein thiol sites in the action of glucagon and isoproterenol has been proposed by Mukherjee and Mukherjee.[93-96] When cells are pretreated with PCMB, $HgCl_2$, or dithiothreitol, the adenylate cyclase in cells or isolated membranes is inhibited. A small amount of H_2O_2 added to the cells will reverse the inhibition. The effect of these treatments on plasma membrane redox activity is not known.

X. EFFECTS OF β-ADRENERGIC AGENTS

β-Adrenergic agents cannot be tested for effects on ferricyanide reduction by cells, since the catechols act as chemical reductants for ferricyanide.[14] Other agents such as ritodrine act as β-agonists and can be tested. Ritodrine, like glucagon, inhibits ferricyanide reduction by perfused liver[14] (Table 7). The effect is blocked by propanolol but not by phenoxybenzamine consistent with a β-adrenergic effect. Effects of ritodrine on NADH dehydrogenase activity with isolated membranes are not known.

The NADH oxidase of isolated membranes shows a strong stimulation (150% at 30 μM) by isoproterenol, but details of the effect have not been studied.[10] Epinephrine stimulation of oxygen uptake by fat cells is not inhibited by oligomycin, whereas resting respiration is inhibited.[97,98] This would be consistent with epinephrine (β-adrenergic) stimulation of a plasma membrane oxidase rather than the mitochondrial oxidase which is oligomycin sensitive.

XI. EFFECT OF TRIIODOTHYRONINE ON PLASMA MEMBRANE REDOX ACTIVITY

It is generally considered that the thyroxines act directly on sites in the cell nucleus and are carried there by cytosolic binding proteins.[99] There have been, however, persistent reports of triiodothyronine (T_3) binding sites on plasma membranes.[100-106]

A 55 kDa protein which specifically binds T_3 has been purified from rat pituitary cell plasma membranes and an antibody has been prepared.[130] The protein is thought to mediate uptake of T_3, but it could also interact with other membrane proteins. Several functions in the plasma membrane respond rapidly to addition of T_3, especially for cells in hypothyroid condition. Effects of T_3 have been shown on amino acid uptake, glucose uptake, Ca^{2+} ATPase, Na^+/K^+-ATPase, and β-adrenergic response (Table 9). Since several of these effects are seen with erythrocytes, which have no nucleus, the T_3 effect cannot be mediated by nuclear proteins. T_3 also affects the Ca^{2+}-ATPase of erythrocytes at 10^{-10} *M*, whereas T_4 only has effects at 10^{-8} *M*.[123] Lack of T_3 in hypothyroid rats decreases Purkinje cell sensitivity to noradrenaline. The effect appears to be on the relation between β-receptor and adenylate cyclase, since the response to butyryl cAMP does not change and cAMP phosphodiesterase activity does not change. The T_3 effect on the redox system could mediate the effect on cyclase.[108] T_3 also rapidly restores glucose transport to astrocytes in thyroid-deficient media in a few minutes, again suggesting a plasma membrane site of action.[107] The concentrations of T_3 used to show these effects are 10^3 higher or 10^{-2} times lower than serum T_3 levels (rat serum 1.3 n*M*). However, the affinity of plasma membrane binding sites for T_3 ranges from 3.5 to 200 n*M*.[100,103,124] Thus, the apparent receptor affinity at the plasma membrane would not indicate a response to levels of T_3 in serum in rats (1.3 n*M*).[122] On the other hand, the increased response of plasma membrane functions to T_3 in the hypothyroid condition suggests that the receptor affinity at the plasma membrane can be modified by lack of hormone.

TABLE 9
Effect of T_3 and T_4 on Plasma Membrane Functions

Cell type	Activity	Effect and specificity	Refs.
Human erythrocyte	Ca^{2+}-ATPase	Stimulation 10^{-10} *M* T_4>T_3	101
Astrocytes	Glucose transporter exposure	50 μ*M* T_3	107
Purkinje neurons	β-Adrenergic response	T_3	108
Thymocytes	Amino acid uptake increase	T_3>>T_4	105
Perfused liver, hypothyroid	O_2 uptake increase	T_3 (1 μ*M* 100 μg/150 ml)	109
Perfused liver, hypothyroid	Amino acid uptake increase	T_3 1μ*M*	109
Perfused liver, hypothyroid	Gluconeogenesis increase	T_3 1 μ*M*	109
Rat liver	Hypothyroid impairs α-adrenergic response		110
Thymocytes	Glucose uptake	T_3 10^{-11} *M*	111,112
Perfused rat liver, hypothyroid	Amino acid uptake	T_3 1μ*M*	113
Rat heart	β-Adrenergic receptor increase	T_3	114
Thymocytes	Amino acid uptake	T_3 0.1 μ*M*	115
Rat erythrocyte	Stimulate Na^+/K^+-ATPase	T_3 10^{-9} *M*	116
Rat heart	β-Receptor increase		118
Rat heart	β-Receptor activation		119
Cultured heart cells	β-Receptor activation		120
Rat heart slices	β-Receptor activation	T_3 15 p*M*	121
Submandibular gland	β-Receptor activation		122

Thyroxine and its analogs can form charge transfer complexes with electron transport prosthetic groups, such as flavins, quinones, and imidazoles, so their inhibitory or stimulatory effects, especially at higher concentrations, may represent an action at a functional site rather than at a traditional protein receptor site.[125]

Respiration stimulated by T_3 is mostly inhibited by ouabain or lack of external sodium ions, which is consistent with inhibition of a plasma membrane oxidase coupled to the Na^+/H^+ antiport. Garcia-Cañero et al.[17] have shown that electron transport from liver cells is inhibited under conditions (low Na, amiloride) which inhibit the antiport, considering this evidence for a coupling between these reactions. Ouabain would prevent release of the accumulated sodium ions to inhibit the oxidase.[126] T_3 does stimulate NADH oxidase in isolated plasma membranes by 330% at 2 μ*M* or 240% at 1 μ*M*[10] (Figure 16). On the other hand, T_3 at 10^{-6} *M* inhibits NADH indophenol reductase activity of both rat liver and adipocyte plasma membranes 60%.[15] Effects on NADH cytochrome c reductase activity of T_3 are not significant at pH 6 or 7 but may reach 40% at pH 8.0, where the inhibition of adipocyte membrane indophenol reductase may reach 90%. Mouse liver plasma membranes also show 70% inhibition of NADH indophenol reductase activity at 10^{-6} *M* (Figure 17). Since T_3 has a growth stimulation effect on liver cells[127,128] and on osteoblasts,[129] it is not surprising that it stimulates NADH oxidase activity in these membranes. The effects on plasma membrane electron transport activity are specific for T_3, and much less effect is seen with T_4.[8] Oxygen uptake by rats is 8 times more stimulated by T_3 than by T_4, which parallels the effect on plasma membrane oxidase,[105] whereas T_4 is more effective than T_3 in stimulation of rat liver mitochondrial NADH oxidase.[117]

XII. EFFECT OF α-ADRENERGIC AGENTS ON FERRICYANIDE REDUCTION IN PERFUSED HEART

Ferricyanide in the perfusion fluid of isolated rat heart is rapidly reduced.[12] The impermeable oxidant is recovered in the reduced form in the affluent. α-Adrenergic agents are

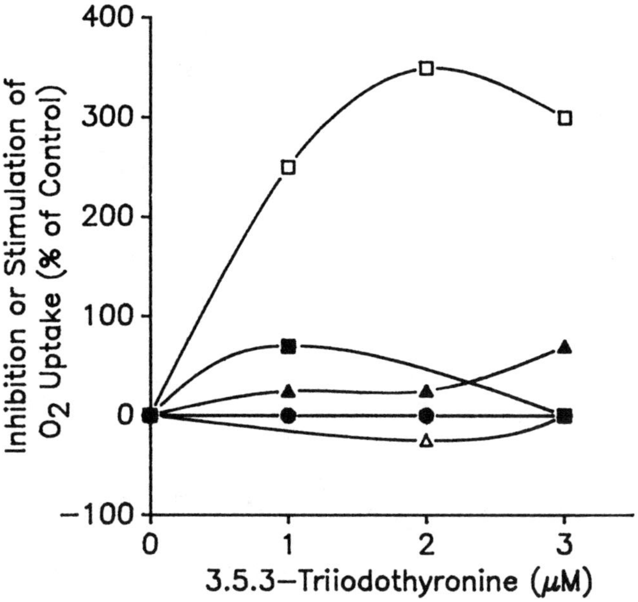

FIGURE 16. T_3 stimulation of plasma membrane NADH oxidase;[10] (□) plasma membrane; (△) endoplasmic reticulum; (●) Golgi; (▲) mitochondria.

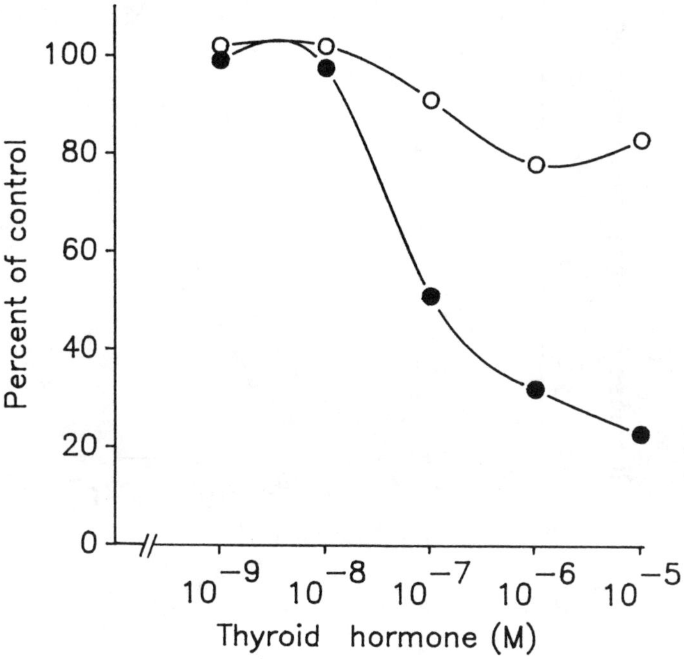

FIGURE 17. Effect of T_3 and T_4 on NADH indophenol reductase of mouse liver plasma membranes. (Modified from Reference 7.) (●) T_4; (○) T_3.

inotropic when perfused into hearts, and this effect is related to an influx of calcium from the outside of the cells. The α-adrenergic agents, 10 μ*M* methoxamine and 10 μ*M* phenylephrine, both increased the rate of ferricyanide reduction in perfused heart (Figure 18). The effect was very rapid, occurring within 6 s of the application of the hormone.[11,12,131]

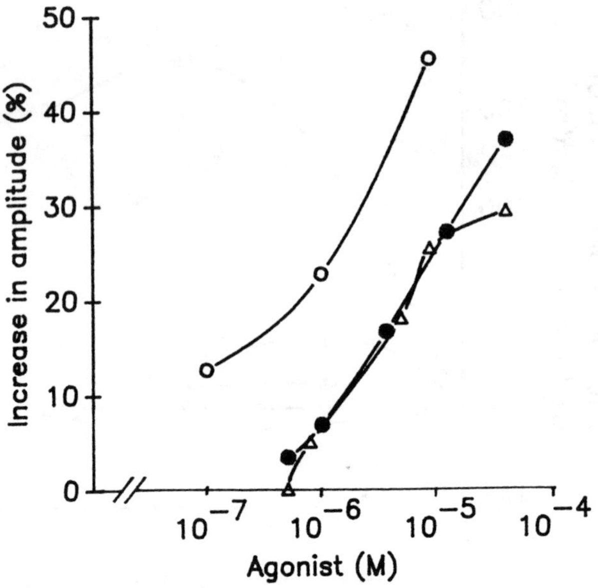

A

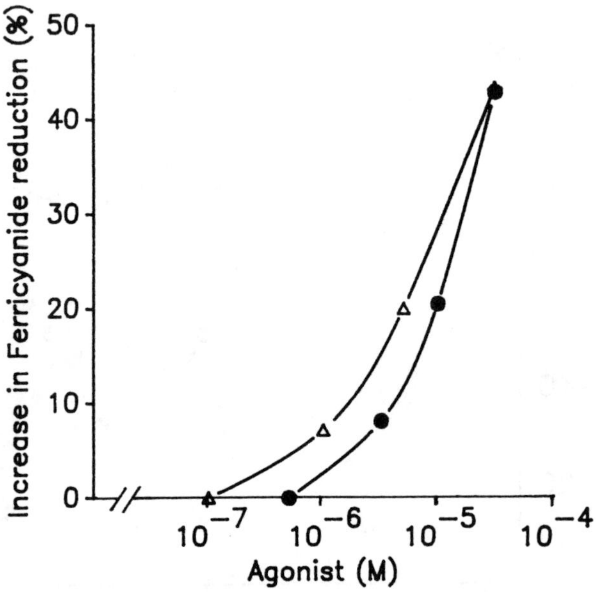

B

FIGURE 18. Methoxamine (●)- or phenylephrine (△)- and epinephrine (□)-stimulated redox in heart. Ferricyanide reduction by perfused rat heart compared to amplitude of heart beat. (From Löw, H., Crane, F. L., Partick, E. J., and Clark, M. G., *Biochim. Biophys. Acta,* 844, 142, 1985. With permission.)

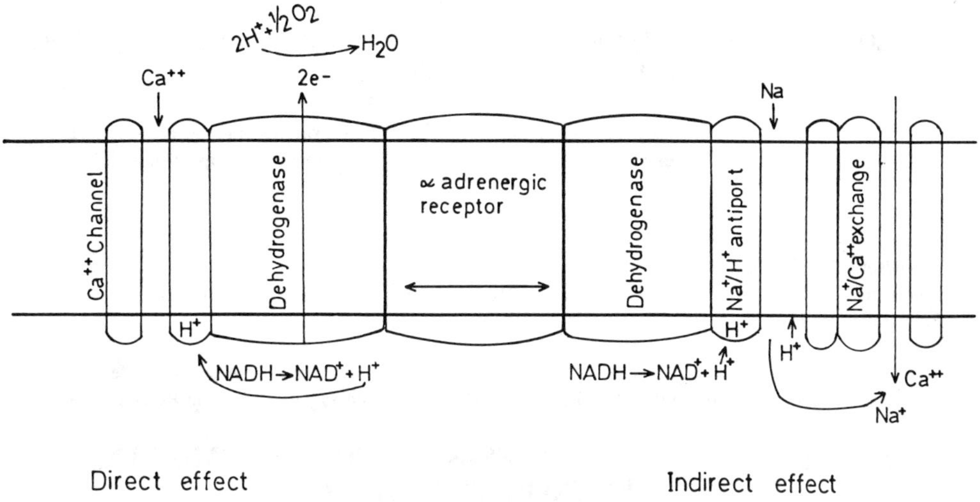

FIGURE 19. Alternatives for redox stimulation of Ca^{2+} channel in heart. Ca^{2+} channel could be activated directly by H^+ release on oxidation of NADH or conformational change of dehydrogenase or kinase activation. Indirect activation could arise by activation of the Na/H antiport.

The effect on ferricyanide reduction is observed even in absence of external calcium ions, so it does not depend on increased cytosolic calcium. Removal of the α-agonist returns the reduction rate to the original value. Nifidipine, which blocks voltage-regulated calcium channels, did not inhibit the methoxamine stimulation of the redox system. It was proposed that the stimulation of electron flow through the redox system to ferricyanide caused an activation of a calcium channel which lets calcium in from outside the cell (Figure 19). The natural electron acceptor for the redox system in the heart cell plasma membrane remains to be determined. Diferric transferrin or oxygen are obvious possibilities. The nature of the internal electron donor is also not known.

The stimulation of calcium uptake may be caused by direct redox action to decrease membrane potential to open a voltage operated channel. Nifidipine inhibition of the channel indicates a voltage-operated channel is involved.[131,136] As an alternative, the stimulation of the Na^+/H^+ antiport by electron transport could increase cytosolic Na^+ to the point that Na^+/Ca^{2+} exchange is activated to bring in more calcium.[137,138] If so, then ouabain should increase the inotropic response of the heart to ferricyanide by inhibition of Na^+ removal through the Na^+/K^+-ATPase. Phorbol myristate acetate, which activates protein kinase C (PKC), did not mimic α-adrenergic agents in stimulation of muscle tension in the perfused heart and did not affect transmembrane electron transport. Activation of PKC is therefore not involved in transsarcolemma electron transport or opening of calcium channels, but the PKC activation may inhibit α-adrenergic stimulation of these functions.[11,132]

XIII. NEUROTRANSMITTER EFFECTS ON REDOX ACTIVITY IN SYNAPTIC MEMBRANES

Glutamate stimulates NADH dichloroindophenol reductase activity in rat brain synaptic membranes.[18,133] Metal chelators azide and *o*-phenanthroline inhibit the binding of glutamate to the receptors. Since the glutamate receptor contains an iron-sulfur complex, the receptor itself may be part of the NADH dehydrogenase. The glutamate stimulation of plasma membrane redox activity may also be related to glutamate-induced Ca^{2+} influx from extracellular calcium as with transmembrane electron transport in heart muscle.[135,139] Neurotransmitter effects are more extensively described in Chapter 15.

TABLE 10
Stimulation of Ferricyanide Reduction and Ferricyanide-Induced Proton Release by Swiss 3T3 Cells with Bombesin

Addition	Ferricyanide reduction rate (nmol min^{-1} gww $cells^{-1}$)	Ferricyanide-induced proton release (neq min^{-1} gww $cells^{-1}$)
Control	175	360
Bombesin 3.1 n*M*	—	500
Bombesin 6.2 n*M*	196	543
Bombesin 12.5 n*M*	330	580
Bombesin 25 n*M*	420	—

Note: Spectrophotometric assay for ferricyanide reduction[38] and assay of proton release when ferricyanide is added to cells in buffer with 0.25 μ*M* tris, pH 7.4. (Method as described as in Reference 38.)

XIV. EFFECTS OF BOMBESIN ON TRANSMEMBRANE ELECTRON TRANSPORT

Neuropeptides such as bombesin are mitogenic in a variety of cells.[140] These agents increase cytosolic alkalinization through activation of the Na^+/H^+ antiport, cause rapid mobilization of calcium and increased phosphotidyl-4,5-bis-phosphate hydrolysis which can activate PKC and release Ca^{2+} from internal stores.[140-142,144,145] In chick embryo cells, however, the bombesin stimulation of growth does not appear to be based on Na^+/H^+ antiport activation.[150] Bombesin also stimulates influx of calcium and 3-*O*-methylglucose across the plasma membrane.[143] The concentrations required for these bombesin effects with Swiss 3T3 cells are from 1 to 10 n*M*.[140,143-145] Bombesin also increases cyclic AMP accumulation in Swiss 3T3 cells,[146] which has been related to the ability of bombesin to stimulate mitogenesis by shifting cells from the G_0 to G_1 stage so that they can respond to other factors, such as epidermal growth factor (EGF) or IGF-I.[147,148] Insulin, IGF-I, and EGF enhance the mitogenic effect of bombesin.[142,143]

Bombesin stimulates the diferric transferrin reduction and ferricyanide reduction by 3T3 cells (Table 10). The concentration required for maximum stimulation is 25 n*M*, which is higher than required in growth studies and in studies of calcium uptake or phosphotidylinositol hydrolysis. The incubation time of cells with bombesin for the electron transport studies was three min; whereas growth studies require a more extensive exposure, so longer incubation may be required for maximum effect in the redox assay. Bombesin also stimulated the ferricyanide-induced proton release from Swiss 3T3 cells, which would be consistent with an activation of the Na^+/H^+ antiport as reported by others[140,142] (Table 10). Activation of a calcium channel[143] would also be consistent with the response of the heart to external oxidants.

XV. OVERALL CONCEPT

Hormones clearly affect plasma membrane electron transport systems, most notably the transmembrane electron transport and the NAD(P)H oxidase. Very little effect is seen with NADH cytochrome b_5 reductase. The external NADH ferricyanide reductase[43,149] has not been investigated. In most cases the hormonal inhibitions are only partial, in the range of 20 to 50%, which suggests that only part of the plasma membrane electron transport is based on a hormone-sensitive redox enzyme or that hormone receptors are not associated with all of the redox enzymes. The differences observed in hormone effects with different electron acceptors or with different membrane orientations can be based on acceptor interaction with different parts of the electron transport system as seen with different sites for ferricyanide, indophenol, and oxygen reduction in mitochondria.[151] The relation between hormone effects on electron transport and signal transduction as indicated by redox effect on growth appears to revolve around the Na^+/H^+ antiport and calcium movement control, either directly or through Na^+/Ca^{2+} exchange. The effects on membrane potential and other redox-related parameters have not been tested extensively.

REFERENCES

1. **Dormandy, T. L. and Zarday, Z.,** The mechanism of insulin action: the immediate electrochemical effects of insulin on red cell systems, *J. Physiol.,* 180, 684, 1965.
2. **Bellomo, G., Bellazzi, R., Nicotera, P. L., Fransesconi, R., and Fratino, P.,** Binding and degradation of 125I-insulin in human erythrocytes, *J. Endocrinol Invest.,* 5, 287, 1982.
3. **Imm, J. H., Zeidler, R. B., Rackley, C. E., and Kim, H. D.,** Insulin receptors on pig red cells, *Arch. Biochem. Biophys.,* 232, 26, 1984.
4. **Dwenger, A. and Zick, R.,** Insulin binding and degradation studies on erythrocytes at different temperatures, *Biochim. Biophys. Acta,* 798, 132, 1984.
5. **Dutta-Ray, A. K., Ray, T. K., and Sinha, A. K.,** Control of erythrocyte membrane microviscosity by insulin, *Biochim. Biophys. Acta,* 816, 187, 1985.
6. **Gambhir, K. K., Archer, J. A., and Bradley, C. J.,** Characteristics of human erythrocyte insulin receptors, *Diabetes,* 27, 701, 1978.
7. **Goldenberg, H., Crane, F. L., and Morré, D. J.,** Influence of hormones on the NADH dehydrogenase in mouse liver plasma membranes, *Biochem. Biophys. Res. Commun.,* 83, 234, 1978.
8. **Ogborne, K. T.,** The Influence of Hormones and Other Agents on NADH Ferricyanide Oxidoreductase of Plasma Membrane Enriched Fractions of the Decidual Endometrium of the Rat, M.S. thesis, State University College, Fredonia, NY, 1980.
9. **Löw, H. and Crane, F. L.,** Hormone regulated redox function in plasma membranes, *FEBS Lett.,* 68, 157, 1976.
10. **Gayda, D. P., Crane, F. L., Morré, D. J., and Löw, H.,** Hormone effects on NADH oxidizing enzymes of plasma membranes of rat liver, *Proc. Indiana Acad. Sci.,* 86, 385, 1977.
11. **Clark, M. G., Rattigan, S., Cleland, P. J. F., Edwards, S. J., and Davison, A. G. M.,** Cardiac Ca^{++} channels and sarcolemma redox, in *Plasma Membrane Oxidoreductases in Control of Animal and Plant Growth,* F. L. Crane, D. J. Morré, and H. Löw, Eds., Plenum Press, New York, 1989, 359.
12. **Löw, H., Crane, F. L., Partick, E. J., and Clark, M. G.,** α adrenergic stimulation of trans-sarcolemma electron efflux in perfused rat heart, *Biochim. Biophys. Acta,* 844, 142, 1985.
13. **Clark, M. G., Partick, E. J., and Crane, F. L.,** Properties and regulation of a transplasma membrane redox system in rat liver, *Biochem. J.,* 204, 795, 1982.
14. **Clark, M. G., Partick, E. J., and Crane, F. L.,** Effects of catecholamines on the transplasma membrane redox system in rat liver, *Biochem. Int.,* 5, 711, 1982.
15. **Crane, F. L. and Löw, H.,** NADH oxidation in liver and fat cell plasma membranes, *FEBS Lett.,* 68, 153, 1976.
16. **Morré, D. J., Brightman, A. O., Wu, L.-Y., Barr, R., Leak, B., and Crane, F. L.,** Role of plasma membrane redox activities in elongation growth in plants, *Physiol. Plant.,* 73, 187, 1988.
17. **Garcia-Cañero, R., Diaz-Gil, J. J., and Guerra, M. A.,** Is transmembrane ferricyanide reductase connected to the early expression of mitogenic signals in hepatocytes, in *Redox Functions of the Eukaryotic Plasma Membrane,* J. M. Ramirez, Ed., Consejo Superior de Investigaciones Cientificas, Madrid, 1987, 41.
18. **Dreyer, J. L. and Treichler, T.,** NADH dehydrogenases in synaptic plasma membranes, in *Plasma Membrane Oxidoreductases in Control of Animal and Plant Growth,* F. L. Crane, D. J. Morré, and H. Löw, Eds., Plenum Press, New York, 1989, 406.
19. **Löw, H., Crane, F. L., Grebing, C., Tally, M., and Hall, K.,** Hormone control of plasma membrane oxidoreduction reactions: effects of insulin, *FEBS Lett.,* 91, 166, 1978.
20. **Crane, F. L., Crane, H. E., Sun, I. L., MacKellar, W. C., Grebing, C., and Löw, H.,** Insulin control of a transplasma membrane NADH dehydrogenase in erythrocyte membranes, *J. Bioenerg. Biomemb.,* 14, 425, 1982.
21. **Crane, F. L., Sun, I. L., Clark, M. G., Grebing, C., and Löw, H.,** Transplasma-membrane redox systems in growth and development, *Biochim. Biophys. Acta,* 811, 233, 1985.

22. **Crane, F. L., Löw, H., and Clark, M. G.,** Plasma membrane redox enzymes, in *The Enzymes of Biological Membranes,* Vol. 4, A. N. Martonosi, Ed., Plenum Press, New York, 1985, 465.
23. **Schenker, E. and Kohanski, R. A.,** Conformational states in the insulin receptor, *Biochem. Biophys. Res. Commun.,* 157, 140, 1988.
24. **Gilman, A. G.,** G-proteins and control of adenylate cyclase, *Cell,* 36, 577, 1984.
25. **Nishizuka, Y.,** The role of protein kinase C in cell surface signal transduction and tumor promotion, *Nature,* 308, 693, 1984.
26. **Larner, J., Cheng, K., Schwartz, C., Kilsuchi, K., Tamura, S., Creasy, S., Dubler, R., Galasko, G., Pullin, C., and Katz, M.,** Insulin mediators and their control of metabolism through protein phosphorylation, *Recent Prog. Horm. Res.,* 38, 511, 1982.
27. **Imm, J. H., Cuppoletti, J., Meezan, E., Rackley, C. E., and Kim, H. D.,** Distribution of insulin receptors in human erythrocyte membranes: insulin binding to right-side-out and inside-out human erythrocyte vesicles, *Biochim. Biophys. Acta,* 775, 260, 1984.
28. **Löw, H., Crane, F. L., Grebing, C., Hall, K., and Tally, M.,** Metabolic milieu and insulin action, in *Diabetes 1979,* K. Waldhäusl, Ed., Elsevier, Amsterdam, 1979, 209.
29. **Goldenberg, H.,** Plasma membrane redox activities, *Biochim. Biophys. Acta,* 694, 203, 1982.
30. **Crane, F. L., Sun, I. L., Clark, M. G., Löw, H., and Grebing, C.,** Hormone sensitive redox reactions in liver plasma membranes, in *Hormonal Control of Gluconeogenesis,* Vol. 2, Kraus-Friedman, N., Ed., CRC Press, Boca Raton, FL, 1986, 133.
31. **Mukherjee, S. P. and Lynn, W. S.,** Reduced nicotinamide adenine dinucleotide phosphate oxidase in adipocyte plasma membrane and its activation by insulin, *Arch. Biochem. Biophys.,* 184, 69, 1977.
32. **Mukherjee, S. P., Lynn, W. S., and Mukherjee, C.,** Direct evidence that H_2O_2 generated by receptor-coupled NADPH oxidase of the plasma membrane is the intracellular messenger of insulin or lectins in adipocytes and thymocytes, Fed. Proc., 41, 1082, 1982.
33. **Morré, D. M., Morré, D. J., Ruest, T. W., and Sammons, D. W.,** Insulin responsive NADH oxidation by plasma membranes of fat cells from epididymal fat pads of the rat and correlation with surface membrane internalization, *J. Cell. Biol.,* 107, 809a, 1988.
34. **Sun, I. L., Crane, F. L., Grebing, C., and Löw, H.,** Transmembrane redox in control of cell growth: stimulation of HeLa cell growth by ferricyanide and insulin, *Exp. Cell. Res.,* 156, 528, 1985.
35. **Takano, K., Hall, K., Fryklund, L., and Sievertsson, H.,** Binding of somatomedins and insulin to plasma membranes prepared from rat and monkey tissue, *Horm. Metab. Res.,* 8, 16, 1976.
36. **Van Wyk, J. J.,** The somatomedins: biological actions and physiologic control mechanisms, in *Hormonal Proteins and Peptides, Growth Factors,* Vol. 12, C. H. Li, Ed., Academic Press, N. Y., 1984, 81.
37. **Clark, M. G., Partick, E. J., Patten, G. S., Crane, F. L., Löw, H., and Grebing, C.,** Evidence for the extracellular reduction of ferricyanide by rat liver: a transplasma membrane redox system, *Biochem. J.,* 200, 565, 1981.
38. **Sun, I. L., Crane, F. L., Grebing, C., and Löw, H.,** Properties of a transplasma membrane electron transport system in HeLa cells, *J. Bioenerg. Biomemb.,* 16, 583, 1984.
39. **Sun, I. L., Putnam, J. E., and Crane, F. L.,** Control of cell growth by transplasmalemma redox: stimulation of HeLa cell growth by impermeable oxidants, *Proc. Indiana Acad. Sci.,* 94, 407, 1985.
40. **King, G. L., Kahn, C. R., Rechler, M. M., and Nissley, S. P.,** Direct demonstration of separate receptors for growth and metabolic activities of insulin and multiplication-stimulating activity, *J. Clin. Invest.,* 66, 130, 1980.
41. **Guma, A., Testar, X., Palacin, M., and Zorzano, A.,** Insulin stimulated α(methyl)amino isobutyric acid uptake in skeletal muscle, *Biochem. J.,* 253, 625, 1988.
42. **Corps, A. N. and Brown, K. D.,** Ligand-receptor interactions involved in the stimulation of Swiss 3T3 fibroblasts by insulin-like growth factors and insulin, *Biochem. J.,* 252, 119, 1988.
43. **Grebing, C., Crane, F. L., Löw, H., and Hall, K.,** A transmembrane NADH dehydrogenase in human erythrocyte membranes, *J. Bioenerg. Biomemb.,* 16, 517, 1984.
44. **Morré, D. J., Vigil, E. L., Frantz, C., Goldenberg, H., and Crane, F. L.,** Cytochemical demonstration of glutaraldehyde-resistant NADH-ferricyanide oxidoreductase activities in rat liver plasma membranes and Golgi apparatus, *Cytobiologie,* 18, 213, 1978.
45. **DePierre and Ernster, L.,** Enzyme topology of intracellular membranes, *Annu. Rev. Biochem.,* 46, 201, 1977.
46. **Goldenberg, H.,** Insulin inhibits NADH semidehydroascorbate reductase in rat liver plasma membrane, *Biochem. Biophys. Res. Commun.,* 94, 721, 1980.
47. **Yunghans, W. N. and Morré, D. J.,** A rapid reproducible homogenization procedure for the isolation of plasma membranes from rat liver, *Prep. Biochem.,* 3, 301, 1973.
48. **Caron, M., Picard, J., and Kern, P.,** Involvement of glycoconjugates in insulin-receptor interactions, *Biochim. Biophys. Acta,* 512, 29, 1978.
49. **Navas, P., Estevez, A., Buron, M. I., Vallalba, J. M., and Crane, F. L.,** Cell surface glycoconjugates control the activity of the NADH ascorbate-free radical reductase of rat liver plasma membrane, *Biochem. Biophys. Res. Commun.,* 154, 1029, 1988.

50. **Purrello, F., Burnham, D. B., and Goldfine, I. D.,** Insulin receptor antiserum and plant lectins mimic the direct effects of insulin on nuclear envelope phosphorylation, *Science,* 221, 462, 1983.
51. **Cherqui, G., Caron, M., Capeau, J., and Picard, J.,** Evidence for surface glycoprotein involvement in the intracellular bioactivity of insulin in rat adipocytes, *Biochem. J.,* 214, 111, 1983.
52. **Orringer, E. P. and Roer, M. E. S.,** An ascorbate mediated transmembrane reducing system of the human erythrocyte, *J. Clin. Invest.,* 63, 53, 1979.
53. **Sun, I. L., Toole-Simms, W., Crane, F. L., Morré, D. J., Löw, H., and Chou, J. Y.,** Reduction of diferric transferrin by SV40 transformed pineal cells stimulates Na^+/H^+ antiport activity, *Biochim. Biophys. Acta,* 938, 17, 1988.
54. **Garcia-Cañero, R.,** Possible role of transplasma membrane ferricyanide reductase activity in mitogenic activation of rat liver cells, in *Plasma Membrane Oxidoreductases in Control of Animal and Plant Growth,* F. L. Crane, D. J. Morré, and H. Löw, Eds., Plenum Press, New York, 1989, 27.
55. **Sun, I. L., Garcia-Cañero, R., Lin, W., Toole-Simms, W., Crane, F. L., Morré, D. J., and Löw, H.,** Diferric transferrin reduction stimulates the Na^+/H^+ antiport of HeLa cells, *Biochem. Biophys. Res. Commun.,* 145, 467, 1987.
56. **Henderson, L. M., Chappell, J. B., and Jones, O. T. G.,** Internal pH changes associated with the activity of NADPH oxidase of human neutrophils, *Biochem. J.,* 251, 563, 1988.
57. **Goldenberg, H., Crane, F. L., and Morré, D. J.,** NADH oxidoreductase of mouse liver plasma membranes, *J. Biol. Chem.,* 254, 2491, 1979.
58. **Ramasarma, T., Swaroop, A., MacKellar, W., and Crane, F. L.,** Generation of hydrogen peroxide on oxidation of NADH by hepatic plasma membranes, *J. Bioenerg. Biomemb.,* 13, 241, 1981.
59. **Mukherjee, S. P. and Mukherjee, C.,** Similar activities of nerve growth factor and its homolog proinsulin in intracellular hydrogen peroxide production and metabolism in adipocytes, *Biochem. Pharm.,* 31, 3163, 1982.
60. **Czech, M. P.,** Differential effects of thiol reagents on activation and deactivation of the fat cell hexose transport system, *J. Biol. Chem.,* 251, 1164, 1976.
61. **Kozka, I. J. and Gould, M. K.,** Inhibitory effects of *N* ethyl maleimide on insulin and oxidant stimulated sugar transport and on ^{125}I-labeled insulin binding by rat soleus muscle, *Biochim. Biophys. Acta,* 797, 212, 1984.
62. **Kwock, L., Wallach, D. F. H., and Hefter, K.,** Involvement of SH groups in the action of insulin and radiation on thymocyte Na^+ dependent amino acid transport, *Biochim. Biophys. Acta,* 419, 93, 1976.
63. **Muchmore, D. B., Little, S. A., and de Haen, C.,** Counter-regulatory control of intracellular H_2O_2 by insulin and lipolytic hormones in isolated rat epididymal fat pads. A role for free fatty acids, *Biochemistry,* 21, 3886, 1982.
64. **Makino, H., Kanatsuka, A., Osegawa, M., and Kumagai, A.,** Effect of dithiothreitol on insulin sensitive phosphodiesterase in rat fat cells, *Biochim. Biophys. Acta,* 704, 31, 1982.
65. **Czech, M. P.,** Molecular basis of insulin action, *Annu. Rev. Biochem.,* 46, 359, 1977.
66. **Petrides, P. E. and Bohlen, P.,** The mitogenic activity of insulin. An intrinsic property of the molecule, *Biochim. Biophys. Acta,* 95, 1138, 1980.
67. **Massague, J. and Czech, M. P.,** The subunit structure of two distinct receptors for insulin-like growth factors I and II and their relationship to the insulin receptor, *J. Biol. Chem.,* 257, 5038, 1982.
68. **Luly, P. and Shinitzky, M.,** Gross structural changes in isolated liver cell plasma membranes upon binding of insulin, *Biochemistry,* 18, 445, 1979.
69. **Hyslop, P. A., York, D. A., and Sauerheber, R. D.,** Effects of insulin on the lipid structure of liver plasma membrane measured with fluorescence and ESR spectroscopic methods, *Biochim. Biophys. Acta,* 776, 267, 1984.
70. **Lelievre, L., Zachowski, A., Aubry, J., Charlemagne, D., Geny, B., and Paraf, A.,** Modulation of some membrane enzymes studied with some variant cells, in *Structural and Kinetic Approaches to Plasma Membrane Functions,* C. Nicolau and A. Paraf, Eds., Springer-Verlag, Berlin, 1977, 152.
71. **Lacko, L., Wittke, B., and Geck, P.,** The temperature dependence of the exchange transport of glucose in human erythrocytes, *J. Cell. Physiol.,* 82, 213, 1973.
72. **Kandutsch, A. A., Chen, H. W., and Heiniger, H.-J.,** Biological activity of some oxygenated sterols, *Science,* 201, 498, 1978.
73. **Davis, R. J., Covera, S., and Czech, M. P.,** Insulin control of transferrin receptor expression and iron uptake in erythrocytes, *J. Biol. Chem.,* 261, 8708, 1986.
74. **Dipple, I. and Houslay, M. D.,** The activity of glucagon-stimulated adenylate cyclase from rat liver plasma membranes is modulated by the fluidity of its lipid environment, *Biochem. J.,* 174, 179, 1978.
75. **Bonne, D., Belhadj, O., and Cohen, P.,** Calcium as modulator of the hormonal receptors — biological response coupling system, *Eur. J. Biochem.,* 86, 261, 1978.
76. **Unates, L. E. and Farias, R. N.,** Biomembrane cooperative enzymes. *In vivo* modulation of rat erythrocyte acetylcholine esterase by insulin under normal and diabetic conditions, *Biochim. Biophys. Acta,* 568, 363, 1979.

77. **de Melian, E. M., Massa, E. M., Morero, R. D., and Farias, R. N.,** Membrane cooperative enzymes interplay of insulin, glucagon and epinephrine on rat erythrocyte acetylcholineesterase, *FEBS Lett.*, 92, 143, 1978.
78. **Sauerheber, R. D., Lewis, U. J., Esgate, J. A., and Gordon, L. M.,** Effect of calcium, insulin and growth hormone on membrane fluidity. A spin label study of rat adipocyte and human erythrocyte ghosts, *Biochim. Biophys. Acta,* 597, 292, 1980.
79. **Avruch, J., Carter, J., and Martin, D. B.,** Insulin stimulated plasma membranes from rat adipocytes: their physiological and physicochemical properties, *Biochim. Biophys. Acta,* 288, 27, 1972.
80. **Butler, K. W., Tattrie, N. H., and Smith, I. C. P.,** The location of spin probes in two phase mixed lipid systems, *Biochim. Biophys. Acta,* 363, 351, 1974.
81. **Bashford, C. L., Morgan, C. G., and Radda, G. K.,** Measurement and interpretation of fluorescence polarizations in phospholipid dispersions, *Biochim. Biophys. Acta,* 426, 157, 1976.
82. **Farese, R. V., Sabir, M. A., Larson, R. E., and Trudeau, J.,** Insulin treatment increases the concentration of phosphotidyl serine in rat adipose tissue, *Biochim. Biophys. Acta,* 750, 200, 1983.
83. **Carpentier, J. L., Perrelet, A., and Orci, L.,** Effects of insulin, glucagon, and epinephrine on the plasma membrane of the white adipose cell: a freeze fracture study, *J. Lipid Res.*, 17, 335, 1976.
84. **Morré, D. J., Navas, P., and Crane, F. L.,** Isolation, purification, evaluation and quantitation of plasma membranes from tissues and cultured cells: methods and applications to cell free systems for analysis of growth processes, in *Redox Functions of the Eukaryotic Plasma Membrane,* Ramirez, J. M., Ed., Consejo Superior de Investigaciones Cientificas, Madrid, 1987, 91.
85. **Runner, R., Hepp, K. D., Humbel, R. E., and Froesch, E. R.,** Mechanism of the antilipolytic action of NSILA-S: inhibition of adenylate cyclase activity in lipocyte ghosts, *Horm. Metab. Res.*, 5, 56, 1973.
86. **Enser, M.,** The role of insulin and regulation of stearic acid desaturase activity in liver and adipose tissue from obese hyperglycaemic ob/ob and lean mice, *Biochem. J.*, 180, 551, 1979.
87. **Moore, R.,** Effects of insulin of ion transport, *Biochim. Biophys. Acta,* 737, 1, 1983.
88. **Kiss, Z.,** Inhibition of the glucagon stimulated adenylate cyclase activity by insulin, *FEBS Lett.*, 92, 29, 1978.
89. **Toole-Simms, W.,** Regulation of Proton Release from HeLa Cells by Ferric Reductase, Ph.D. thesis, Purdue University, West Lafayette, IN, 1988, 160.
90. **Toole-Simms, W., Sun, I. L., Crane, F. L., Golub, E. S., Morré, D. J., and Löw, H.,** Diferric transferrin reductase in the plasma membrane activates the Na^+/H^+ antiport and is inhibited by retinoic acid, *J. Cell. Biol.*, 105, 100a, 1987.
91. **Deutsch, C. J., Kashiwagura, T., Taylor, J., Wilson, D. F., and Erecinska, M.,** The effect of glucagon and adrenergic agonists on intracellular pH of isolated rat hepatocytes, *J. Biol. Chem.*, 260, 6808, 1985.
92. **Lotersztajn, S., Epand, M., Malat, A., and Pecker, F.,** Inhibition by glucagon of the calcium pump in liver plasma membrane, *J. Biol. Chem.*, 259, 8195, 1984.
93. **Mukherjee, S. P. and Mukherjee, C.,** Role of sulfhydryl oxidation in adipocyte plasma membrane in the response of adenylate cyclase to isoproternol and glucagon, *Biochim. Biophys. Acta,* 677, 339, 1981.
94. **Mukherjee, S. P. and Lynn, W.,** Role of cellular redox state and glutathione in adenylate cyclase activity in rat adipocytes, *Biochim. Biophys. Acta,* 568, 224, 1979.
95. **Ziegler, D. M.,** Role of reversible oxidation-reduction of enzyme thiols-disulfides in metabolic regulation, *Annu. Rev. Biochem.*, 54, 305, 1985.
96. **Malbon, C. C., George, S. T., and Moxham, C. P.,** Intramolecular disulfide bridges, answers to receptor activation, *Trends Biochem. Sci.*, 12, 172, 1987.
97. **Challoner, D. R.,** Evidence for uncoupled respiration in thyrotoxic and epinephrine stimulated myocardium, *Am. J. Physiol.*, 214, 365, 1968.
98. **Challoner, D. R.,** A direct effect of triiodothyronine on the oxygen consumption of rat fat cells, *Am. J. Physiol.*, 216, 905, 1969.
99. **Oppenheimer, J. H. and Samuels, H. H., Eds.,** *Molecular Basis of Thyroid Hormone Action,* Academic Press, New York, 1983.
100. **Blondeau, J.,** Saturable binding of thyroid hormone to isolated rat hepatocytes, *FEBS Lett.*, 204, 41, 1986.
101. **Davis, F. B., Cody, V., Davis, P. J., Borzynski, L. J., and Blass, S. D.,** Stimulation by thyroid hormone analogs of red blood cell Ca^{++} ATPase activity *in vitro, J. Biol. Chem.*, 258, 12373, 1983.
102. **Angel, R. C., Botta, J. A., and Farias, R. N.,** Modification of L-T_3 binding sites from rat erythrocyte membranes by heating and by protease treatments, *Biochim. Biophys. Acta,* 897, 488, 1987.
103. **Pliam, N. B. and Goldfine, I. D.,** High affinity thyroid hormone binding sites on purified rat liver plasma membranes, *Biochem. Biophys. Res. Commun.*, 79, 166, 1977.
104. **Gharbi-Chihi, J. and Torresani, J.,** Thyroid hormone binding to plasma membrane preparations: studies in different thyroid states and tissues, *J. Endocrin. Invest.*, 4, 177, 1981.
105. **Goldfine, I. D., Smith, G. J., Simons, C. G., Ingbar, S. H., and Jorgensen, E. C.,** Activities of thyroid hormones and related compounds in an *in vitro* thymocyte assay, *J. Biol. Chem.*, 251, 4233, 1976.

106. **Müller, M. J. and Seitz, H. J.,** Pleiotropic action of thyroid hormones at the target cell level, *Biochem. Pharmacol.,* 33, 1579, 1984.
107. **Roeder, L. M., Hopkins, I. B., Kaiser, J. R., Hanukoglu, L., and Tildon, J. T.,** Thyroid hormone action on the glucose transporter in astrocytes, *Biochem. Biophys. Res. Commun.,* 156, 275, 1988.
108. **Marwaba, J. and Prasad, K. N.,** Hypothyroidism causes electrophysiological nonadrenergic subsensitivity in rat cerebellum, *Science,* 214, 675, 1981.
109. **Müller, M. J. and Seitz, H. J.,** Rapid and direct stimulation of hepatic gluconeogenesis by L-triiodothyronine (T_3) in the isolated-perfused rat liver, *Life Sci.,* 27, 827, 1980.
110. **Sorm, H., van Hardeveld, C., and Kasenar, A. A. H.,** The influence of hypothyroidism on the adrenergic stimulation of glycogenolysis in perfused rat liver, *Biochim. Biophys. Acta,* 798, 350, 1984.
111. **Segal, J. and Ingbar, S. H.,** Studies of the mechanism by which 3,5,3 triiodothyronine stimulates 2-deoxyglucose uptake by rat thymocytes *in vitro, J. Clin. Invest.,* 68, 103, 1981.
112. **Segal, J. and Ingbar, S. H.,** Specific binding sites for triiodothyronine in the plasma membrane of rat thymocytes: correlation with biochemical response, *J. Clin. Invest.,* 70, 919, 1982.
113. **Müller, M. J. and Seitz, H. J.,** Interaction of thyroid hormones and cyclic AMP in the stimulation of hepatic gluconeogenesis, *Biochim. Biophys. Acta,* 756, 360, 1983.
114. **Ciraldi, T. P. and Marinetti, G. V.,** Hormone action at the membrane level. VIII. Adrenergic receptors in rat heart and adipocytes and their modulation by thyroxine, *Biochim. Biophys. Acta,* 541, 334, 1978.
115. **Goldfine, I. D., Simons, C. G., and Ingbar, S. H.,** Stimulation of the uptake of α aminoisobutyric acid in rat thymocytes by L-triiodothyronine: a comparison with insulin and dibutyryl cyclic AMP, Endocrinology, 96, 802, 1975.
116. **De Mendoza, D., Moreno, H., Massa, E. M., Moreno, R. D., and Farias, R. N.,** Thyroid hormone actions and membrane fluidity. Blocking action of T_4 on the T_3 effect, *FEBS Lett.,* 84, 199, 1977.
117. **Dallam, R. D. and Reed, J. M.,** The effect of thyroxine on mitochondrial oxidation and swelling, *J. Biol. Chem.,* 235, 1183, 1960.
118. **Chang, H. Y. and Kunos, G.,** Short term effects of triiodothyronine on rat heart adrenoreceptors, *Biochem. Biophys. Res. Commun.,* 100, 313, 1981.
119. **Malbon, C. C., Moreno, F. J., Cabelli, R. J., and Fain, J. N.,** Fat cell adenylate cyclase and β adrenergic receptors in altered thyroid states, *J. Biol. Chem.,* 253, 671, 1978.
120. **Tsai, J. S. and Chen, H.,** Effect of L-triiodothyronine on $(-)^3$H-dihydroalprenol binding and the cyclic AMP response to (−) adrenaline in cultured heart cells, *Nature,* 275, 138, 1978.
121. **Kempson, S., Marinetti, G. V., and Shaw, A.,** Hormone action at the membrane level. VII. Stimulation of dehydroalprenol binding to β adrenergic receptors of isolated rat heart ventricle slices by triiodothyronine and thyroxine, *Biochim. Biophys. Acta,* 540, 320, 1978.
122. **Medina, J. H., Wolfman, C., Levi de Stein, M., Tumilasci, O., and Houssay, A. B.,** Thyroid hormone regulation of adrenergic receptors and β adrenergic responsiveness of rat submandibular gland, *Life Sci.,* 35, 819, 1984.
123. **Farias, R. N.,** Membrane enzymes and thyroid hormone actions, Proc. 7th Biophys. Congr., Mexico City, 1981, 26.
124. **Samuels, H. H.,** Thyroid hormone receptors, in *Receptors and Hormone Action,* L. Brinbaumer and B. W. O'Malley, Eds., Academic Press, New York, 1978, 35.
125. **Mauchamp, J. and Shinitzky, M.,** Redox properties and charge transfer complexes of thyroxin and analogs, *Biochemistry,* 8, 1554, 1969.
126. **Asano, Y., Liberman, U. A., and Edelman, I. S.,** Thyroid thermogenesis: relation between Na-dependent respiration and Na+K adenosine triphosphatase activity in rat skeletal muscle, *J. Clin. Invest.,* 57, 368, 1976.
127. **Soloman, D. and Mishkin, S.,** Tri-iodothyronine (T_3) stimulates the growth of morris hepatoma cells grown in culture, *Biochem. Biophys. Res. Commun.,* 105, 1611, 1982.
128. **Morré, D. J., Brightman, A. O., Wang, J., Barr, R., and Crane, F. L.,** Roles for plasma membrane redox systems in cell growth, in *Plasma Membrane Oxidoreductases in Control of Animal and Plant Growth,* Crane, F. L., Morré, D. J., and Löw, H., Eds., Plenum Press, New York, 1989, 45.
129. **Ernst, M. and Froesch, E. R.,** Triiodothyronine stimulates the proliferation of osteoblast-like cells in serum free culture, *FEBS Lett.,* 220, 163, 1987.
130. **Hasumura, S., Rossi, B., Alderson, R., Pastan, I., and Cheng, S.-Y.,** Antibodies against the plasma membrane 3,3′,5-triodo-L-thyronine binding protein of rat pituitary GH_3 cells: partial characterization and cross species immunoreactivity, *Biochem. Biophys. Res. Commun.,* 124, 956, 1984.
131. **Löw, H., Crane, F. L., Partick, E. J., Patten, G. S., and Clark, M. J.,** Properties and regulation of a trans-plasma membrane redox system of perfused rat heart, *Biochim. Biophys. Acta,* 804, 253, 1984.
132. **Davison, A. G. M., Cleland, P. J. F., Rattigan, S., and Clark, M. G.,** Effects of PMA on protein kinase c in perfused rat heart, in *Plasma Membrane Oxidoreductases in Control of Animal and Plant Growth,* Crane, F. L., Morré, D. J., and Löw, H., Eds., Plenum Press, New York, 1989, 424.

133. **Treichler, T. and Dreyer, J. L.,** Influence of neurotransmitters on the electron transfer rates of plasma membrane dehydrogenases in rat brain synaptosomes, *Experientia,* 42, 642, 1986.
134. **Michailis, E. K.,** The glutamate receptor-like protein of brain synaptic membranes is a metalloprotein, *Biochem. Biophys. Res. Commun.,* 87, 106, 1979.
135. **Choi, D. W.,** Ionic dependence of glutamate neurotoxicity, *J. Neurosci.,* 7, 369, 1987.
136. **Reynolds, E. E. and Dubyak, G. R.,** Activation of calcium mobilization and calcium influx by α_1 adrenergic receptors in a smooth muscle cell line, *Biochem. Biophys. Res. Commun.,* 130, 627, 1985.
137. **Lougoni, S. and Carafoli, E.,** Sodium calcium exchanges of heart sarcolemma, *Experientia,* 42, 681, 1986.
138. **Grinstein, S., Rotin, D., and Mason, M. J.,** Na^+/H^+ exchange and growth factor induced cytosolic pH changes: role in cellular proliferation, *Biochim. Biophys. Acta,* 988, 73, 1989.
139. **Robinson, M. B. and Coyle, J. T.,** Glutamate and related acidic excitatory neurotransmitters: from basic science to clinical application, *FASEB J.,* 1, 446, 1987.
140. **Zachary, I., Wall, P. J., and Rozengurt, E.,** A role for neuropeptides in control of cell proliferation, *Dev. Biol.,* 124, 295, 1987.
141. **Brown, K. D., Laurie, M. S., Littlewood, C. J., Blakeley, D. M., and Corps, A. N.,** Characterization of the high-affinity receptors on Swiss 3T3 cells which mediate the binding, internalization and degradation of the mitogenic peptide bombesin, *Biochem. J.,* 252, 227, 1988.
142. **Rozengurt, E.,** Early signals in the mitogenic response, *Science,* 234, 161, 1986.
143. **Takawa, N., Takawa, Y., Bollag, W. E., and Rasmussen, H.,** The effects of bombesin on polyphosphoinositide and calcium metabolism in Swiss 3T3 cells, *J. Biol. Chem.,* 262, 182, 1987.
144. **Morris, C., Rice, P., and Rozengurt, E.,** The diacylglycerol kinase inhibitor R59022 potentiates bombesin stimulation of protein kinase c activity and DNA synthesis in Swiss 3T3 cells, *Biochem. Biophys. Res. Commun.,* 155, 561, 1988.
145. **Moody, T. W., Murphy, A., Mahmoud, S., and Fiskum, G.,** Bombesin-like peptides elevate cytosolic Ca^{++} in small cell lung cancer cells, *Biochem. Biophys. Res. Commun.,* 147, 189, 1987.
146. **Millar, J. B. A. and Rozengurt, E.,** Bombesin enhancement of cAMP accumulation in Swiss 3T3 cells: evidence of a dual mechanism of action, *J. Cell. Physiol.,* 137, 214, 1988.
147. **Smets, L. A. and Van Rooy, H.,** Mitogenic and anti-mitogenic effects of cholera toxin mediated cyclic AMP levels in 3T3 cells, *J. Cell. Physiol.,* 133, 395, 1987.
148. **Heikkila, R., Trepel, J. B., Cuthitta, L. M., Neckers, L. M., and Sansville, E.,** Bombesin-related peptides induce Ca^{++} mobilization in a subset of human small cell lung cancer cell lines, *J. Biol. Chem.,* 262, 16456, 1987.
149. **Cherry, J. M., MacKellar, W., Morré, D. J., Crane, F. L., Jacobsen, L., and Schirrmacher, V.,** Evidence for a plasma membrane redox system on intact ascites tumor cells with different metastatic capacity, *Biochim. Biophys. Acta,* 634, 11, 1981.
150. **Gillespie, J. I., Giraldez, F., and Greenwell, J. R.,** Growth stimulation by bombesin does not involve activation of Na^+/H^+ exchange in chick embryo otic vesicles *in vitro, FEBS Lett.,* 243, 17, 1989.
151. **Harmon, H. J. and Crane, F. L.,** Inhibition of mitochondrial electron transport by hydrophilic metal chelators: determination of dehydrogenase topography, *Biochim. Biophys. Acta,* 440, 45, 1976.
152. **Pershadsinge, H. A. and McDonald, J. M.,** Direct addition of insulin inhibits the high affinity Ca^{+2} ATPase in isolated adipocyte plasma membranes, *Nature,* 281, 495, 1979.
153. **Crane, F. L.,** unpublished.
154. **Löw, H.,** unpublished.
155. **Sun, I. L.,** unpublished.
156. **Crane, H. E.,** unpublished.

Chapter 6

THE STIMULATION OF CELL GROWTH BY EXTRACELLULAR OXIDANTS

Charles E. Wenner and Anthony F. Cutry

TABLE OF CONTENTS

I. INTRODUCTION

Several lines of evidence have implicated redox state as a mediator of cell growth. There are several examples of extracellular oxidants which can generate active oxygen or electron acceptors which can induce cell proliferation. Ellem and Kay[1] and Crane and colleagues[2] have used pyruvate or impermeant electron acceptors such as potassium ferricyanide to demonstrate proliferative effects in cells such as HeLa, melanoma, etc. Further, the enzyme system xanthine/xanthine oxidase, which produces H_2O_2 + superoxide, stimulated the growth of cell monolayers of mouse epidermal cells JB6 (clone 41).[3] Similarly, this enzyme system was found to induce DNA replication in quiescent BALB/3T3 monolayer cultures.[4] Further, hydrogen peroxide stimulated DNA synthesis synergistically in the presence of insulin, but alone it did not.[5] An intriguing question concerns the mechanism by which these enzyme systems or extracellular electron acceptors stimulate cell growth when serum is a limiting factor. The structural diversity of extracellular oxidants which affect cell growth suggests that they do not act via specific membrane receptors, as do growth factors. However, these oxidants all share the ability to stimulate transplasmalemma redox activity. Thus, such a system affords the distinct advantage of examining the role of redox systems in signal transduction and cell growth. This system also has the potential to allow one to focus on a single component of the signal transduction machinery. This is in contrast to the study of growth factor-mediated signal transduction, where one must be concerned with a myriad of events triggered upon ligand-receptor interaction.

II. THE ROLE OF ION MOVEMENTS IN OXIDANT-INDUCED CELL GROWTH

A. ROLE OF pH_i INCREASE IN SIGNAL TRANSDUCTION INDUCED BY OXIDANTS

The role of ion movements in signal transduction, particularly Ca^{2+} and Na^+/H^+ movements, has been widely studied. It is of interest to note that the effective concentration of ferricyanide for growth stimulation correlates very closely with the concentration which gives maximum proton release.[6] The contribution of transplasma membrane electron flow to the stimulation of cell growth is supported by the observation that other external electron acceptors such as indigo disulfonate can stimulate cell proliferation in serum-free media.[7] Ferricyanide as well as the aforementioned xanthine/xanthine oxidase system elicit increases in intracellular pH (pH_i) via an amiloride- or superoxide dismutase-sensitive mechanism.[8] These observations suggest a role for the Na^+/H^+ antiporter in the proliferative capabilities of ferricyanide and xanthine/xanthine systems. The antiport system has previously been implicated in cell proliferation,[9,10] presumably via cytosolic alkalizination. Antiport activation has been suggested to be a requirement to enter cell cycle. However, it is not yet established whether it may only be required for correction of pH_i following the imposition of a metabolic load, or whether it is necessary for cell cycle progression.

B. ROLE OF $[Ca^{2+}]$ IN OXIDANT-INDUCED CELL RESPONSES

Larsson and Cerutti[11] recently reported that diamide (which oxidizes cellular SH groups) induced phosphorylation of ribosomal protein S6. This response is commonly observed in growth factor stimulated cells. This study also demonstrated that reagents which generate active oxygen also stimulate S6 phosphorylation. Induction of S6 phosphorylation by diamide occurred in a calcium-dependent fashion. This finding, coupled with observations that both the glucose oxidase system and peroxides increased free intracellular calcium levels $[Ca^{2+}]$, made for an interesting connection. Larsson and Cerutti proposed that oxidants can activate Ca/calmodulin kinases via their ability to increase intracellular Ca^{2+}. This in turn leads to the activation of the S6 kinase. If this hypothesis is correct, it provides a direct link between a plasma membrane redox system and a classical intracellular second messenger system, Ca^{2+}. This is suggestive of a direct role for redox systems in cell growth. One possibility is that the xanthine oxidase system may elicit the production of hydroperoxides which result in the mobilization of calcium from intracellular stores.

The question arises as to how systems such as xanthine-xanthine oxidase or ferricyanide could give rise to increased intracellular calcium and one possibility is that the xanthine oxidase system may give rise to hydroperoxides which result in the mobilization of mitochondrial stores. Ca^{2+} mobilization by hydroperoxides from mitochondria or from endoplasmic reticulum has been demonstrated earlier.[12-14] While it is not expected that ferricyanide can stimulate peroxide formation, its administration could result in the oxidation of pyridine nucleotides. Lehninger and colleagues have proposed that oxidation of intramitochondrial pyridine nucleotides can induce the release of accumulated Ca^{2+} by this organelle (cf. Reference 14). Possibly, externally added ferricyanide acts as an electron acceptor for cytoplasmic pyridine nucleotides, which in turn oxidize intramitochondrial nucleotides, leading to a similar response to that described by Lehninger and colleagues. However, the physiological significance of Ca^{2+} efflux by this mechanism remains to be established because *in vitro* experiments suggest that this type of release mechanism does not operate in the presence of ATP or in the absence of exogenous Ca^{2+} (cf. Reference 14).

C. ROLE OF TYROSINE KINASES IN OXIDANT-MEDIATED PLASMA MEMBRANE RESPONSES

Signal transduction by polypeptide growth factors and oncogene products often involves the activation of a tyrosine kinase. Furthermore, several oncogene products have been shown to possess tyrosine kinase activity. Although precise substrates for these phosphorylations have not as yet been identified, it is obvious that tyrosine kinase activity is necessary for the mitogenic action of the aforementioned agents. For example, epidermal growth factor receptor mutants which lack tyrosine kinase activity cannot stimulate mitogenic activity in any sense, ranging from very early responses to DNA replication.[15-17] Thus, the evidence linking tyrosine kinase activity to cell growth is quite convincing.

It is therefore interesting to note that ferricyanide stimulates tyrosine phosphorylation of the band 3 erythrocyte membrane protein.[18] Tyrosine 8 of band 3 is proposed as the site for reversible association of glyceraldehyde-3-phosphate dehydrogenase.[19] Phosphorylation of this residue apparently affects the stability of the complex between the band 3 protein and the glycolytic enzyme. Low et al.[18] demonstrated that ferricyanide (0.1 m*M* range), as well as hydrogen peroxide was able to induce phosphorylation of a specific site of the anion transporter (tyrosine 8 of band 3) and stimulation of glycolysis by intact human erythrocytes but not by erythrocyte ghosts. The mechanism by which extracellular oxidants may trigger tyrosine kinase activity is not yet known. However, the redox state of membrane thiol groups has been emphasized as a control mechanism for transmembrane signal transduction.[2] Activation of a tyrosine kinase triggered by the oxidation of thiol groups on a transmembrane protein therefore looms as a possible intermediary step in the progression of cell cycle. It should also be noted that tyrosine phosphorylation by oxidants is not limited to erythrocytes, since H_2O_2 has been reported to stimulate phosphorylation of tyrosine in intact adipocytes and crude homogenates.[20]

III. PEROXIDES AND PERIODATES AS STIMULATORS OF DNA SYNTHESIS AND BLASTOGENIC TRANSFORMATION

The prooxidant state has been considered to be involved in modulation of promotion-related genes based on the demonstration that tumor promoters can generate free radicals in mouse epidermis,[21,22] and antioxidants suppress phorbol ester-induced processes such as ornithine decarboxylase (cf. Reference 22). The proposal that free radicals are involved in TPA-induced mitogenesis in hemopoietic systems and in tumor promotion is supported by data based on inhibition of these responses by agents which stimulate superoxide dismutase and H_2O_2 degradation.[21] The use of commercial catalase was reported to prevent TPA induction of the transformed phenotype but this inhibitory effect was not attributable to the ability of catalase to degrade H_2O_2 since a highly purified catalase preparation failed to alter TPAs effects, and it was demonstrated that the catalase contained a hydrolase which inactivated TPA.[23]

Indeed, a number of peroxides which generate reactive oxygen have been reported to stimulate cell growth *in vivo*.[24,25] One of the first peroxides reported to act as a tumor promoter was peroxyacetic acid[24] and other organic peroxides were observed to have tumor-promoting capabilities shortly thereafter.[25] Benzoyl peroxide promotion of transformation of JB6 mouse epidermal cells has been studied by a number of investigators, and it has been reported that it can induce anchorage-independent cell growth.[26] Similarly, an oxidizing agent, sodium metaperiodate, also induced growth of JB6 cells in soft agar.[27] Trisialoganglioside G_T, a major ganglioside of epidermal cells, was able to markedly reduce the number of soft agar colonies induced by benzoyl peroxide[28] or by metaperiodate.[27] Although the molecular targets of these agents remain enigmatic, these findings support a role for reactive oxygen in modulation of tumor promotion and are suggestive that organic peroxides may act primarily at the cell surface with complex oligosaccharides or glycoproteins. Therefore, it is plausible that surface carbohydrates are involved in oxidant-induced signal transduction, and periodate preferentially oxidizes cell surface sialic acid as suggested by these authors.

Treatment of lymphocytes with periodate has also been reported to induce blastogenesis as well as to increase synthesis of nucleic acids and proteins.[29,30] In contrast to lectin-induced blastogenesis, which results in increased hydrolysis and synthesis of phosphoinositides, periodate-induced blastogenesis is independent of phosphoinositide turnover.[31] However, the pattern of protein phosphorylation rates, IL-2R expression, IL-2 production, and DNA synthesis by periodate-oxidized lymphocytes was temporally similar in both lectin and periodate-stimulated lymphocytes. An analogous situation appears to exist between periodate-induced blastogenesis and the aforementioned JB6 transformation studies. Borohydride, which reduces carbonyl groups, abrogated the periodate-induced blastogenic response, as did addition reactions with hydroxylamine or bisulfite (cf. Reference 29). This would suggest that periodate-induced blastogenesis is triggered by the oxidation of cell surface sialic acid residues with subsequent carbonyl induction.

Active oxygen has also been considered to participate in epigenetic mechanisms that result in altered gene expression. Larsson and Cerutti[11] suggested that active oxygen induced ADP-ribosylation of chromosomal proteins may play a role in the modulation of gene expression. Poly ADP-ribosylated proteins from nuclear extracts of active oxygen-treated cells were characterized as histones and also ADP-ribose transferase, topoisomerase I and c-*fos*.[32] The speculation by these authors that ADP ribosylation may be acting to inactivate the nonhistone enzymes is interesting. It is conceivable that modification of the transferase may act as a feedback loop to avoid excessive ADP-ribosylation and consequent energy depletion. These authors also propose that temporary inactivation of topoisomerase may cause transient stoppage of DNA replication allowing time for DNA repair.

IV. ASCORBATE AS A REDOX EFFECTOR

The actions of another redox effector, ascorbic acid and its derivatives, in cellular proliferation are indeed multifarious. Although ascorbate has marked reducing capability, it has been reported that it can stimulate proliferation of fibroblasts in culture.[33] The copresence of *l*-ascorbate and epidermal growth factor gave additive effects on growth and protein synthesis in cultured human skin fibroblasts. However, the effects of *l*-ascorbate and EGF on the extracellular matrix were quite different. The stimulatory effect of ascorbate was cancelled by inhibitors of collagen synthesis, while EGF-stimulated growth was not dependent on increased collagen synthesis. These results led the authors to propose that EGF and ascorbate act independently in the modulation of cell growth.

A possible explanation for the stimulatory response by ascorbate is described in a recent report by Hata and Senoo.[34] These workers reported that ascorbic acid 2-phosphate which provides a source of the normally unstable ascorbate, markedly stimulated collagen accumulation, and reorganization of the monolayer with subsequent formation of a three-dimensional, multilayered cell structure. Although the monolayer system exhibits density-dependent growth, *in vivo* cells are multilayered and exist in an environment of an extracellular matrix which includes collagen and proteoglycans. The authors suggest that ascorbate and its derivatives may function as a cofactor for the hydroxylation of proline and lysine in collagen. This permits the accumulation of collagen which can lead to multilayers in culture plates. Thus, the reorganization of a three-dimensional structure arises by increased collagen synthesis as a consequence of its cofactor activity and/or its ability to increase both the rate of transcription of procollagen genes and stability of procollagen mRNA.

V. REDOX EFFECTORS STIMULATE THE EXPRESSION OF "IMMEDIATE EARLY GENES"

The demonstration of the induction of mRNA expression of "immediate early genes" by means of an impermeant electron acceptor is of interest since it would permit the study of ion movements and membrane processes critical to transmembrane signal transduction independent of receptor-ligand interactions. One of the earliest consequences of mitogenic stimulation of mammalian cells is the activation of a set of immediate early genes, two of which are the proto-oncogenes c-*fos* and c-*myc*.[35] We have examined the ability of ferricyanide to stimulate the expression of these genes in C3H 10T1/2 mouse embryonic fibroblasts.[36] This impermeant extracellular electron acceptor is capable of inducing moderate increases in both c-*fos* and c-*myc* expression. As seen in Figure 1, the kinetics of induction of these mRNAs are similar to those obtained with growth factors such as EGF.[37] It should be pointed out, however, that the induction of these mRNAs is small compared to that observed with serum growth factors. Furthermore, it can also be observed that another electron acceptor, pyruvate, can induce c-*myc* mRNA expression (Figure 2).

The exact mechanism by which ferricyanide stimulates cell growth is not known. The proliferative response is dependent on iron being in the ferric rather than the ferrous state. This is supported by data (Figure 2) which show that ferrocyanide is unable to stimulate the expression of c-*myc* mRNA. Thus, it is conceivable that the transmembrane redox system involved in the reduction of ferricyanide is capable of stimulating "immediate early gene" expression. A model is proposed to explain ferricyanide-stimulated signal transduction (Figure 3). The oxidation of NADH to yield NAD^+ and free protons generates an electron which is extruded to reduce ferricyanide. The concomitant rise in intracellular free protons is detected by the pH sensor of the Na^+/H^+ antiport, which in turn is activated and stimulates a rise in pH_i as described by Pouyssegur.[10] The intracellular signal is then transduced to the nucleus (possibly via a release of calcium from intracellular stores) where expression of c-*fos* and c-*myc* mRNA is stimulated. This model needs to be explored in more detail.

The early reports on the stimulation of cell growth by ferricyanide[2] were followed by studies,[6,38] which showed that the growth activity was both extracellular Na^+ and amiloride-sensitive implicating the involvement of the sodium/proton antiport in the growth response to ferricyanide. Furthermore, this report showed that, while very high concentrations of

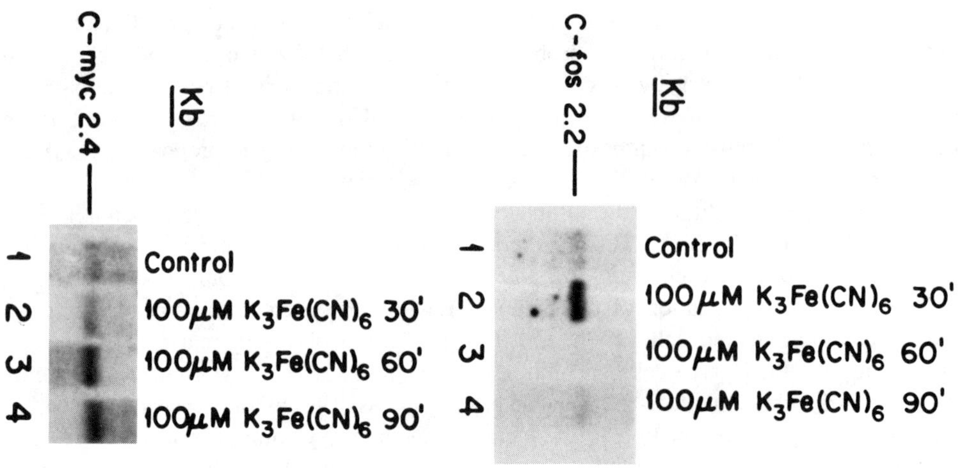

FIGURE 1. Ferricyanide induction of c-*fos* and c-*myc* expression. Quiescent, confluent C3H 10T1/2 fibroblasts were treated with 100 μM ferricyanide for the indicated times. RNA was isolated and analyzed as described in Reference 33.

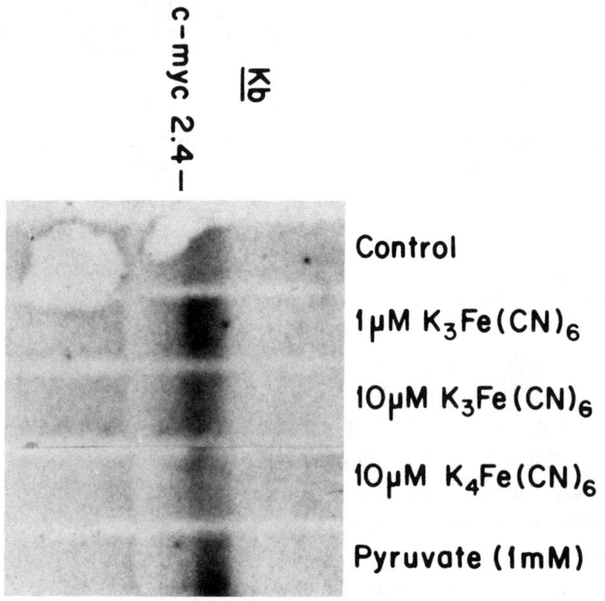

FIGURE 2. Effect of ferricyanide, ferrocyanide, or pyruvate on c-*myc* induction. Quiescent fibroblasts were treated at the designated concentrations for sixty minutes. RNA was isolated and analyzed as described in Reference 33.

ferricyanide could stimulate cell growth, the maximal growth stimulatory effects of ferricyanide correlated quite well with those concentrations of ferricyanide which allotted maximum proton release. This is in line with some of our other observations showing that ferricyanide concentrations, which give the greatest c-*fos* and c-*myc* responses, are also similar to those which have the most pronounced effect on DNA synthesis.[39] Thus, it appears likely that transmembrane redox systems are involved in the stimulation of cell growth. However, we cannot rule out the possibility that FeIII complexes could directly oxidize cell

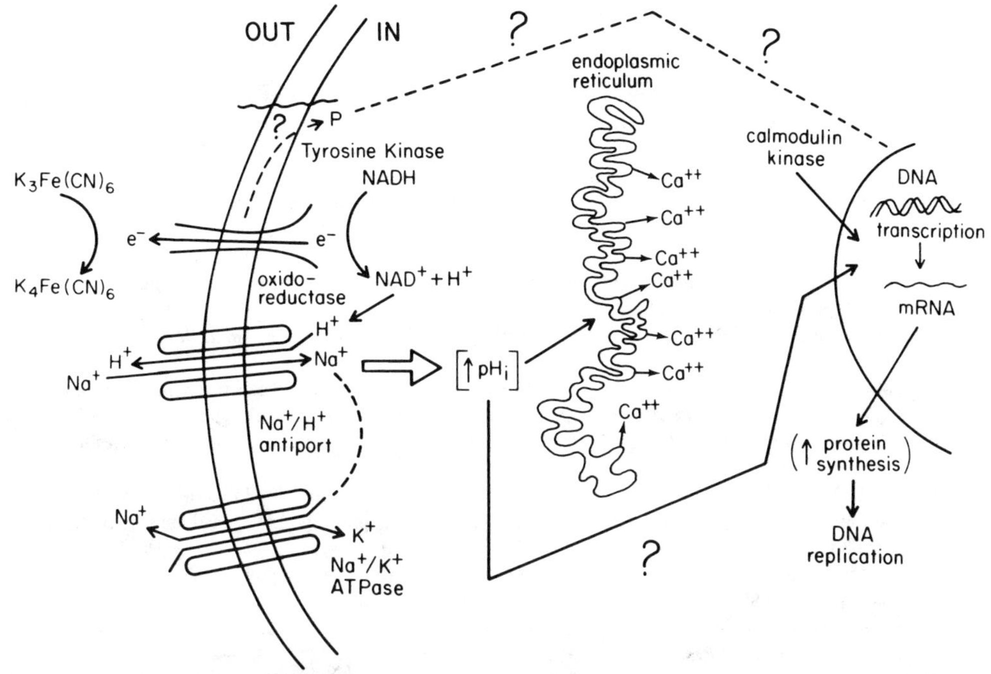

FIGURE 3. Hypothetical model for potential signal transduction pathways induced by oxidants.

surface components such as membrane protein thiols, which could effect anion or cation transport and associated responses, as pointed out by Ellem et al.[40] For example, the presence of ferricyanide may lead to the oxidative inactivation of a plasma membrane Ca^{2+}-ATPase which may alter intracellular Ca^{2+} levels. This, in turn, could cause increases in "immediate early gene" expression. Cerutti et al.[41] suggested that oxidant-induced increase in Ca_i^{2+}, which is required for S6-phosphorylation is the consequence of such inactivation. The calcium ionophore, A23187, which markedly increases the intracellular calcium concentration as measured by aequorin chemiluminescence, has been reported to induce c-*fos* and c-*myc* expression with postconfluent, quiescent C3H 10T1/2 fibroblasts.[42] Preliminary studies indicate that the introduction of ferricyanide leads to a steady but small increase in Ca_i^{2+}, with time in this cell type.[43] This finding would also be in accord with oxidative inactivation of a Ca^{2+}-ATPase. Further studies with ferricyanide and redox involvement in early signaling processes should contribute to the understanding of the mechanisms by which cells are recruited to enter the cell cycle, as well as delineating the contributions of redox systems in the stimulation of gene expression and biosynthetic processes by a variety of mitogens, including serum growth factors.

REFERENCES

1. **Ellem, K. O. and Kay, G. F.,** Ferricyanide can replace pyruvate to stimulate growth and attachment of serum restricted human melanoma cells, *Biochem. Biophys. Res. Commun.*, 112, 183, 1983.
2. **Crane, F. L., Sun, I. L., Clark, M. G., Grebing, C., and Löw, H. L.,** Transplama-membrane redox systems in growth and development, *Biochim. Biophys. Acta,* 811, 233, 1985.
3. **Nakamura, Y., Gindhart, T., Winterstein, D., Tomita, I., Seed, J. L., and Colburn, N. H.,** Early superoxide dismutase-sensitive event promotes neoplastic transformation in mouse epidermal JB6 cells, *Carcinogenesis,* 9, 203, 1988.

4. **Shibanuma, M., Kuroki, T., and Nose, K.,** Induction of DNA replication and expression of proto-oncogene c-myc and c-fos in quiescent Balb/3T3 cells by xanthine/xanthine oxidase, *Oncogene,* 3, 17, 1988.
5. **Nose, K., Kuroki, T., and Shibanuma, M.,** Positive and negative regulation of cell growth by hydrogen peroxide, *J. Cell. Biochem.,* 13B, 37, 1989.
6. **Sun, I. L., Crane, F. L., Grebing, C., and Löw, H. L.,** Properties of a transmembrane electron transport system in HeLa cells, *J. Bioenerg. Biomemb.,* 16, 583, 1984.
7. **Crane, F. L., Löw, H., Sun, I. L., and Isaksson, M.,** The requirement for transmembranal electron flow for cell proliferation, in *Plasma Membrane Oxidoreductase in Control of Animal and Plant Growth,* Crane, F. L., Morré, D. J., and Löw, H. E., Eds., Plenum Press, New York, 1988, 37.
8. **Shibanuma, M., Kuroki, T., and Nose, K.,** Superoxide as a signal for increase in intracellular pH, *J. Cell. Physiol.,* 136, 379, 1988.
9. **Moolenaar, W. H.,** Effects of growth factors on intracellular pH regulation, *Annu. Rev. Physiol.,* 48, 363, 1986.
10. **Pouyssegur, J.,** The growth factor-activatable Na^+/H^+ exchange system: a genetic approach, in *Oncogenes and Growth Factors,* R. Bradshaw and S. Prentis, Eds., Elsevier, New York, 1987, 237.
11. **Larsson, R. and Cerutti, P.,** Oxidants induce phosphorylation of ribosomal protein S6, *J. Biol. Chem.,* 263, 17452, 1988.
12. **Jones, D. P., Thor, M., Smith, M. T., Jewell, S., and Orrenius, S.,** Inhibition of ATP-dependent microsomal Ca^{2+} sequestration during oxidative stress and its prevention by glutathione, *J. Biol. Chem.,* 258, 6390, 1983.
13. **Richter, C. and Frei, B.,** Ca^{2+} release from mitochondria induced by prooxidants, *Free Radical Biol. Med.,* 4, 365, 1988.
14. **Fiskum, G. and Pease, A.,** Hydroperoxide-stimulated release of calcium from rat liver and S-30D hepatoma mitochondria, *Cancer Res.,* 46, 3459, 1986.
15. **Chen, W. S., Lazar, C. S., Poenie, M., Tsien, R. Y., Gill, G. N., and Rosenfeld, M. G.,** Requirement for intrinsic protein tyrosine kinase activity in the immediate and late actions of the EGF receptor, *Nature,* 328, 820, 1987.
16. **Honneger, A. M., Szapary, D., Schmidt, A., Lyall, A., Van Obberghen, E., Dull, T. J., Ullrich, A., and Schlessinger, J.,** A mutant epidermal growth factor receptor with defective protein tyrosine kinase activity is unable to stimulate proto-oncogene expression and DNA synthesis, *Mol. Cell. Biol.,* 7, 4568, 1987.
17. **Moolenaar, W. H., Bierman, A. J., Tilly, B. C., Verlaan, I., Defize, L. H. K., Honneger, A. M., Ullrich, A., and Schlessinger, J.,** A point mutation at the ATP-binding site of the EGF-receptor abolishes signal transduction, *EMBO J.,* 7, 707, 1988.
18. **Low, P. S., Geahlin, R., Mehler, E., and Harrison, M.,** Extracellular control of erythrocyte metabolism mediated by a cytoplasmic tyrosine kinase, *Biomed. Biochim. Acta,* 49, 135, 1990.
19. **Low, P. S.,** Structure and function of the cytoplasmic domain of band 3: center of erythrocyte membrane-peripheral protein interactions, *Biochim. Biophys. Acta,* 864, 145, 1986.
20. **Hayes, G. R. and Lockwood, D. H.,** Role of insulin receptor phosphorylation in the insulinomimetic effects of hydrogen peroxide, *Proc. Natl. Acad. Sci. U.S.A.,* 84, 8115, 1987.
21. **Nakamura, Y., Colburn, N. H., and Gindhart, T.,** Role of reactive oxygen in tumor promotion: implication of superoxide anion in promotion of neoplastic transformation in JB-6 cells by TPA, *Carcinogenesis,* 6, 229, 1985.
22. **Cerutti, P.,** Prooxidant states and tumor promotion, *Science,* 227, 375, 1985.
23. **Dion, L. D., Gindhart, T. D., and Colburn, N. H.,** Four-day duration of tumor promoter exposure required to transform JB6 promotion-sensitive cells to anchorage independence, *Cancer Res.,* 48, 7126, 1988.
24. **Bock, F. G., Myers, H. K., and Fox, H. W.,** Cocarcinogenic activity of peroxy compounds, *J. Nat. Cancer Inst.,* 55, 1359, 1975.
25. **Slaga, T., Klein-Szanto, A., Triplett, L., Yotti, L., and Trosko, J.,** Skin tumor-promoting activity of benzoyl peroxide, a widely used free radical-generating compound, *Science,* 213, 1023, 1981.
26. **Colburn, N., Wendel, E., and Srinivas, L.,** Responses of preneoplastic epidermal cells to tumor promoters and growth factors: use of promoter-resistant variants for mechanism studies, *J. Cell. Biochem.,* 18, 261, 1982.
27. **Srinivasin, L. and Colburn, N. H.,** Preferential oxidation of cell surface by periodate leads to promotion of transformation in JB6 cells, *Carcinogenesis,* 5, 515, 1984.
28. **Gindhart, T., Srinivas, L., and Colburn, N.,** Benzoyl peroxide promotion of JB6 mouse epidermal cells: inhibition by ganglioside G_T but not retinoic acid, *Carcinogenesis,* 2, 309, 1985.
29. **Hornicek, F. J., Malinin, G. I., Gratzner, H., and Malinin, T. I.,** The interaction of bisulfite and its free radical analog-nitroxyldisulfonate with periodic acid-oxidized lymphocytes and their effect on blastogenesis, *Chem.-Biol. Int.,* 55, 259, 1985.

30. **Malinin, G. I., Hornicek, F. J., and Malinin, T. I.,** Enhanced protein phosphorylation, IL-2 receptor expression, and IL-2 production precede cellular proliferation in lymphocytes oxidized by periodic acid, *Biochem. Biophys. Res. Commun.*, 158, 590, 1989.
31. **Hornicek, F. J., Reese, D. H., Lo, H. K., Malinin, T. I., and Malinin, G. I.,** Periodic acid-induced blastogenesis of lymphocytes is independent of phosphoinositide turnover, *Biochem. Biophys. Res. Commun.*, 154, 606, 1988.
32. **Krupitza, G. and Cerutti, P.,** ADP-ribosylation of ADPR-transferase and topoisomerase I in intact mouse epidermal cells JB6, *Biochemistry*, 28, 2034, 1989.
33. **Hata, R., Sunada, H., Arai, K., Sato, T., Ninomiya, Y., Nagai, Y., and Senoo, H.,** Regulation of collagen metabolism and cell growth by epidermal growth factor and ascorbate in cultured human skin fibroblasts, *Eur. J. Biochem.*, 173, 261, 1988.
34. **Hata, R. and Senoo, H.,** L-ascorbic acid 2-phosphate stimulates collagen accumulation, cell proliferation, and formation of a three-dimensional tissuelike substance by skin fibroblasts, *J. Cell. Physiol.*, 138, 8, 1989.
35. **Lau, L. F. and Nathans, D.,** Expression of a set of growth-related immediate early genes in BALB/c 3T3 cells: coordinate regulation with c–*fos* or c-*myc*, *Proc. Natl. Acad. Sci. U.S.A.*, 84, 1182, 1987.
36. **Wenner, C. E., Cutry, A. F., Kinniburgh, A., and Leister, K. J.,** Modulation of redox reactions involved in DNA synthesis by oxygen and artificial electron acceptors, in *Plasma Membrane Oxidoreductase in Control of Animal and Plant Growth*, Crane, F. L., Morré, D. J., and Löw, H. E., Eds., Plenum Press, New York, 1988, 7.
37. **Cutry, A. F., Kinniburgh, A. J., Twardzik, D. R., and Wenner, C.,** Transforming growth factor alpha (TGFα) induction of c-*fos* and c-*myc* expression in C3H 10T1/2 cells, *Biochem. Biophys. Res. Commun.*, 152, 216, 1988.
38. **Sun, I. L., Garcia-Cañero, R., Liu, W., Toole-Sims, W., and Crane, F. L.,** Diferric transferrin reduction stimulates the Na/H antiport of HeLa cells, *Biochem. Biophys. Res. Commun.*, 145, 467, 1987.
39. **Cutry, A. F., Kinniburgh, A., and Wenner, C. E.,** Stimulation of c-*fos* and c-*myc* proto-oncogene expression by ferricyanide, an extracellular electron acceptor, *J. Biol. Chem.*, in press.
40. **Ellem, K. A. O., Kay, G. F., Dunstan, A., Stenzel, D. J.,** Redox agents which modulate the growth of melanoma cells: evaluation of mechanisms, in *Plasma Membrane Oxidoreductase in Control of Animal and Plant Growth*, Crane, F. L., Morré, D. J., and Löw, H. E., Eds., Plenum Press, New York, 1988, 17.
41. **Cerutti, P., Krupitza, G., Larsson, R., Muehlmatter, D., Crawford, D., and Amstad, P.,** Physiological and pathologic effects of oxidants in mouse epidermal cells, *Ann. N.Y. Acad. Sci.*, 551, 75, 1988.
42. **Cutry, A., Kinniburgh, A. J., Krabak, M. J., Hui, S. W., and Wenner, C. E.,** Induction of c-*fos* and c-*myc* proto-oncogene expression by epidermal growth factor and transforming growth factor, *J. Biol. Chem.*, 264, 1970, 1989.
43. **Krabak, M., Cutry, A., Hui, S. W., and Wenner, C. E.,** unpublished results.

Chapter 7

TRANSMEMBRANE ELECTRON TRANSPORT AND GROWTH OF TRANSFORMED CELLS

F. L. Crane, H. Löw, I. L. Sun, and M. Isaksson

TABLE OF CONTENTS

I. INTRODUCTION

The systematic investigation of the effects of iron chelates, including impermeable ferricyanide on the growth of melanoma cells by Ellem and Kay,[1,2] has laid the foundation for investigation of a function of the transplasma membrane electron transport system in cellular growth. The much earlier studies by Brooks[3] could have started this train of investigation on an even broader scale, but these observations appear to have been buried in the literature until Craig rediscovered them in the preparation of his Ph.D. thesis.[4]

The almost universal requirement of transferrin for mammalian cell growth,[5,6] and more especially the evidence that ferric transferrin is required,[7] opened the door for recognition of a general oxidant requirement for growth, even though all attention was focused on iron as an essential growth factor and the facility of iron delivery to the cell by ferric transferrin.[8-13,14] The growth stimulation by iron salts was also interpreted on the basis of requirements for iron in such essential enzymes as ribonucleotide reductase, cytochromes, and other heme and nonheme iron proteins. These iron requirements are clearly a primary part of the stimulation of growth by iron complexes since impermeable ferricyanide, in the absence of iron available for cell uptake, cannot by itself support cell growth.[15,16] On the other hand, all of the iron complexes which have been used to support cell growth, including diferric transferrin and iron chelates of pyridoxyl-isonicotinylhydrazone, can also act as electron acceptors for the transplasma membrane electron transport[17-19] (Table 4, Chapter 2). Serum contains iron in the form of ferritin and transferrin complexes and possibly other forms. Even serum-free media will often contain enough iron to satisfy cellular iron requirements, as evidenced by the inhibition of growth by desferrioxamine which removes ferric iron.[16] The stimulation of growth by ferricyanide in untreated media and lack of stimulation by ferricyanide in the presence of desferroxamine indicate that the ferricyanide cannot supply iron to cells,[16] and that its growth effects are based on other properties. Other aspects of ferricyanide and iron chelators in relation to removal of H_2O_2 and oxidation of external thiol sites are discussed by Ellem in Chapter 8 and Wenner and Cutry in Chapter 6.

In addition to iron, serum supplies other factors which are required for cell growth, including basic nutrients, essential elements, and growth factors. Transformed cells may be cultured through many generations when supplied proper nutrients with a minimum of growth factors, whereas nontransformed cells show density-restricted growth and may require special growth factors or mitogens to initiate proliferation. In general, the less selective growth factor requirements of the transformed cells permit simpler testing for the ability of the transplasma membrane redox system to maintain growth.[20-25]

Growth of cells is determined by direct cell count or by incorporation of thymidine into DNA to measure DNA synthesis. Cell sorter studies can be used to determine the percentage of cells in each stage of the cell cycle in response to growth factors.[26] Most of the studies of growth stimulation by external oxidants have been done with transformed cells in media which does not contain serum or contains a limiting amount of serum.[2,27,28]

The impermeable oxidants which have been used to study growth are compounds which would not be found in the blood serum. These include ferricyanide, indigosulfonates, naphthoquinone sulfonate,[27] nitroprusside,[29] and ferric complexes,[9] or heme proteins such as catalase.[1,2] The natural oxidants for the transmembrane redox system could be any component in the serum or environment that has a redox potential high enough to accept electrons at the cell surface. These compounds are impermeable, which prevent direct interaction with cytosolic reductants such as NADH, but oxygen may be an exception since it would require a catalyst for action on the inside of the plasma membrane.

II. OXIDANTS IN BLOOD SERUM

The most obvious oxidant in blood is oxygen, which is released to cells from the erythrocytes. Since oxygen is freely permeable to membranes, it will be found throughout the cell and it would be able to act as an oxidant on either side of the plasma membrane.[30]

The products of oxygen reduction can be water, hydrogen peroxide, or superoxide. If water is the product, then two protons and four electrons are required during the reaction. For hydrogen peroxide formation only one proton per electron is needed, and for superoxide ($O_2^{\overline{\cdot}}$) formation no protons are involved. A cyanide-insensitive NADH oxidase has been found in plasma membranes, and this enzyme may be stimulated by growth-promoting agents.[31,32] The stoichiometry of the reaction is not well established, but superoxide is not the product.

Peroxide is a possible product of oxygen reduction. Peroxide formation has been related to growth factor effects and regulation by internal Ca^{2+} and protein kinase c activity.[33] Peroxide is also important in cyclooxygenase-dependent prostaglandin production which could affect cAMP formation[34] and can have effects on polyunsaturated fatty acids. Oxidation of a two-electron carrier such as a quinone by a one-electron acceptor such as ferricyanide leads to formation of a semiquinone. Semiquinones in membranes can easily reduce oxygen to form superoxide and, subsequently, peroxide.[35] If the plasma membrane contains a flavin or quinone like coenzyme Q as a two-electron carrier, the oxidation of this carrier by ferricyanide at the cell surface could lead to peroxide formation in the membrane which could then migrate into the cytoplasm. In this way reduction of external ferricyanide would increase internal peroxide. Excess H_2O_2 production by a plasma membrane oxidase might produce a feedback inhibition of growth which would be overcome by catalase. This would allow catalase stimulation of growth in those cells which are sensitive to high levels of peroxide.[2]

Diferric transferrin is another important potential oxidant agent. It is present in human blood at concentrations up to 4.0 mg per ml of blood (equivalent to 50 μM).[8] The protein has a molecular weight of 80,000 and contains two atoms of very tightly bound ferric iron.[36] The high affinity for iron means that the iron in diferric transferrin has a very low redox potential (−400 mV) at neutral pH.[37] Because of the low redox potential, reduction of diferric transferrin at neutral pH is not readily achieved by direct reaction with cellular reductants such as NAD(P)H or glutathione with redox potential in the range from −320 to −280 mV. There is experimental evidence, however, for enzymes in the plasma membrane which can catalyze the reduction[38,39] of ferric transferrin.

The reduction of diferric transferrin at a site on the membrane with a redox potential of over (−)320 mV (e.g., NADH) or higher could be achieved if the diferric transferrin structure is modified by binding to the transferrin receptor[40] or if the reduction site is acidified by some process at the cell surface[39,41,42] to release iron from the transferrin.

Selenite can act as an oxidant for thioredoxin reductase and possibly for other plasma membrane dehydrogenases. When the selenite is reduced to hydrogen selenide, it is rapidly reoxidized by oxygen so that selenite can mediate between the membrane dehydrogenases and oxygen. There is 0.5 μM selenite in blood plasma, so it is present in sufficient supply to act as a natural oxidant. Since selenite is probably easily permeable to the cell membrane, it may act as an oxidant on either side of the membrane.[43-45] Medina and Oborn[45] show a twofold increase in DNA synthesis in mouse mammary epithelial cells with 5×10^{-8} M selenite in addition to insulin, transferrin, epidermal growth factor (EGF), and fibronectin. Higher concentrations of selenium inhibit growth.

Semidehydroascorbate has been shown to be an electron acceptor for plasma membrane electron transport,[46] and detergent activation studies with membrane vesicles have indicated that the enzyme is oriented in a transmembrane position so that internal NADH can reduce external semidehydroascorbate.[47] The permeability of semidehydroascorbate makes it difficult to determine the orientation of the reductase in the membrane, but the reduction inside secretory vesicles (Chapter 4) may provide a model for reduction on the outside of cells. Dehydroascorbate can activate transplasma membrane electron transport as measured by ferricyanide reduction.[48,49] There is also evidence that plasma membrane preparations contain an ascorbate oxidase activity which could reoxidize any ascorbate formed by the semidehydroascorbate reductase.[50]

Ceruloplasmin is a copper-containing protein present in mammalian blood. It has been considered as a possible ferroxidase for aiding iron incorporation from liver cells to transferrin in the plasma membrane. It also can act as a superoxide dismutase which will convert superoxide to H_2O_2 and oxygen.[51]

Disulfides such as glutathione or thioredoxin could also act as electron acceptors for transmembrane reduction by reduced pyridine nucleotides.[52] Some cells have an external glutathione oxidase, but there has been no report of an oxidized glutathione reductase on the cell surface.[53] Thioredoxin reductase has been reported to act as a transmembrane reductase from internal NADPH to external, oxidized thioredoxin.[54] Immune response to thioredoxin and thioredoxin reductase has been demonstrated at the plasma membrane,[55] and thioredoxin reductase can act as a ferricyanide reductase.[56] Thioredoxin present in the serum may serve as an oxidant for a transplasma membrane electron transport system, and thioredoxin reductase could act as an NADPH-specific transmembrane redox enzyme.[52]

III. SIMPLE FERRICYANIDE STIMULATION OF GROWTH

Since the original studies of Ellem and Kay,[1] who used ferricyanide to replace pyruvate to stimulate growth of melanoma cells, several types of cells have shown a positive response to low concentrations of ferricyanide. These include: HeLa, Ehrlich ascites, L1210, HepG2, and SV40-transformed 3T3 cells, as well as rat liver and rat pineal cells transformed with a temperature-sensitive SV40 virus which expresses the SV40 transformation only at low temperature, and cannot express the virus at 40°C.[25] Liver cells in the presence of a mitogen (r-liver growth factor) also show increased DNA synthesis with ferricyanide.[57] The stimulation of growth is not seen with ferrocyanide, so the oxidant form is required,[1,27] and both Na^+ and K^+ ferricyanide are effective.[58]

IV. HeLa CELLS

HeLa cells show growth stimulation by ferricyanide at relatively low concentrations (below 0.1 m*M*) and inhibition of growth above this concentration (Figure 1). These effects are observed in the complete absence of serum or other factors. The cells are grown in αMEM media which contains pyruvate, so the ferricyanide effect is in addition to any pyruvate requirement. Added pyruvate does not increase growth above that obtained with ferricyanide.[27,28] The lack of a pyruvate effect is in contrast to studies with melanoma cells which respond to pyruvate in place of ferricyanide. As discussed by Kay and Ellem,[2] this effect of pyruvate can be attributed to removal of hydrogen peroxide from the RPM1 culture media used for growth of melanoma cells. Since added pyruvate does not stimulate growth of HeLa cells, the inhibition of growth by peroxide and removal of peroxide by ferricyanide are unlikely to be the basis for HeLa cell growth. Ferrocyanide generally gives very little stimulation of growth unless there is some oxidized ferricyanide present which can be expected to give some stimulation at higher concentrations. Both potassium and sodium ferricyanide give similar growth stimulation.[48,58]

With addition of low concentrations of serum which can support only 1/10 of maximum

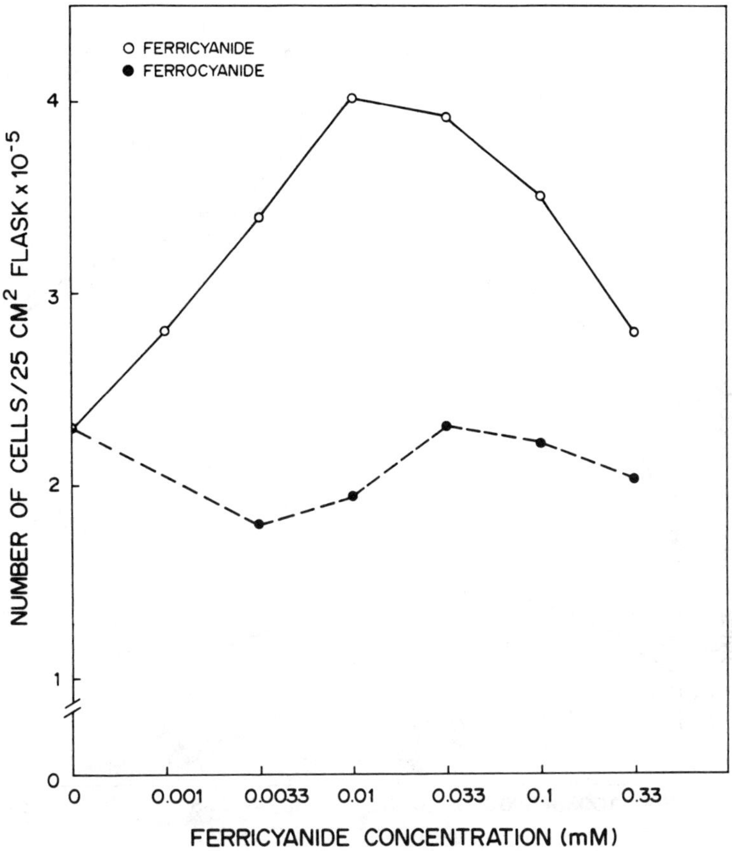

FIGURE 1. Stimulation of HeLa cell growth in serum-free media supplemented with ferricyanide, and lack of stimulation by low concentrations of ferrocyanide. Open circles (○), ferricyanide; closed circles (●), ferrocyanide. (From Sun, I. L., Crane, F. L., Löw, H., and Grebing, C., *Biochem. Biophys. Res. Commun.*, 125, 649, 1984. With permission.)

growth (10% serum), the response of HeLa cells to ferricyanide is greatly enhanced. The growth stimulation is observed even at concentrations up to 0.1 to 0.33 m*M* ferricyanide, which would be inhibitory if the ferricyanide alone was present.[28,59] The additional response in the presence of serum indicates that other growth factors are required in addition to an electron acceptor for transmembrane electron transport (Figures 2, 3), and that some factors can prevent the inhibition at higher ferricyanide concentrations.

Insulin strongly stimulates growth in the presence of ferricyanide, so it may be one of the additional factors present in serum. The concentration of insulin required suggests that the effect is based on the action of insulin as an analog of insulin-like growth factor I since insulin binds to the IGF-I receptor at these high concentrations.[60] Stimulation by higher concentrations of ferricyanide is found when insulin is present compared to ferricyanide alone, similar to the effects of serum with HeLa cells (Figure 4). The same concentrations of insulin (> 1 μg/ml) which increase the growth response to ferricyanide also increase the rate of ferricyanide reduction by HeLa cells (cf. Chapter 5, Figure 13).

Diferric transferrin also stimulates the growth of HeLa cells in serum-free media, and the extent of stimulation can be greater than with ferricyanide (Figure 5). Combination of ferricyanide and diferric transferrin gives only slightly better growth than with diferric transferrin alone, so the diferric transferrin may satisfy the requirement for an external

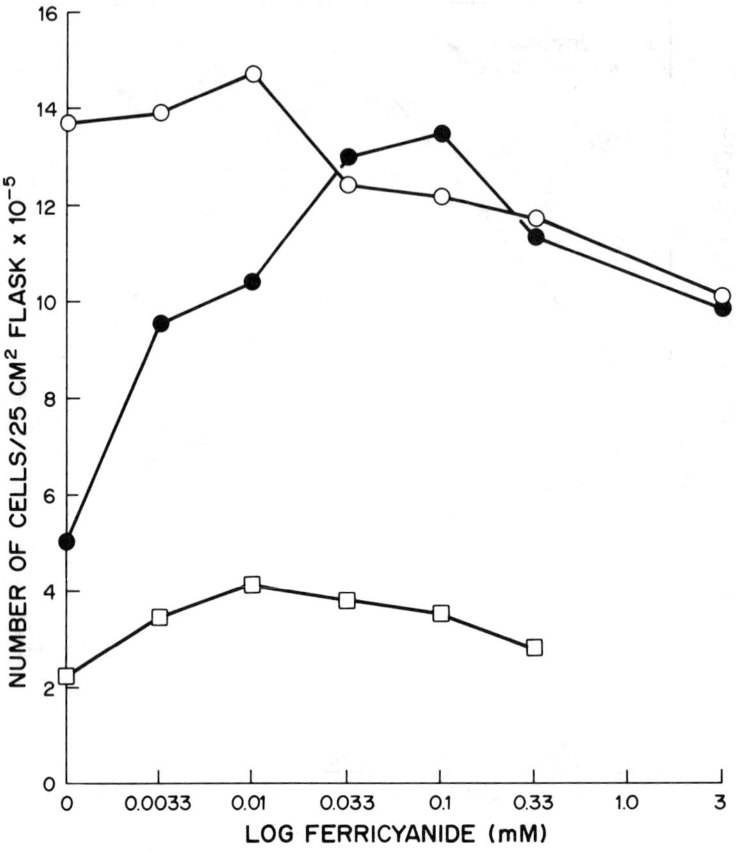

FIGURE 2. Comparison of effects of ferricyanide and serum on growth of HeLa cells. Supplementation of media with 0.75% serum greatly increases the response to ferricyanide.[59] Symbols: (□) $Fe(CN)_6$ only; (●) $FeCN_6$ + 0.75% fetal calf serum; and (○) $Fe(CN)_6$ plus 10% fetal calf serum. (From Crane, F. L., Sun, I. L., Clark, M. G., Grebing, C., and Löw, H., *Biochim. Biophys. Acta,* 811, 233, 1985. With permission.)

oxidant (Figure 6).[61] The diferric transferrin can act as an electron acceptor for the transplasma membrane electron transport, so a role as an external oxidant is possible (Chapter 2).

Apotransferrin, in most experiments, gives a small growth stimulation at low concentrations and has less effect at higher concentrations. The stimulatory effect may depend on the presence of a small amount of iron in the culture media. This iron is often present in culture media as an impurity and can readily combine with the transferrin. Brief exposure to apotransferrin can also inhibit growth of HeLa cells on ferricyanide which may be due to a removal of iron from the media (Figure 7). Ferricyanide cannot act as an iron source for cells.

A strange effect is observed when apotransferrin is added in the presence of diferric transferrin because the growth is inhibited even below the growth observed in the presence of apotransferrin (Figure 8). The basis for this effect may be the presence of excess loosely bound iron in some diferric transferrin preparations. The combination of free iron and diferric transferrin is very effective in stimulating proton release from HeLa cells through the Na/H antiport, and apotransferrin partially inhibits this activity, so the effect on growth may be based on the removal of excess iron from the transferrin.[62]

Gallium transferrin inhibits growth of HL-60 cells and inhibits iron uptake at 2.5 μM.

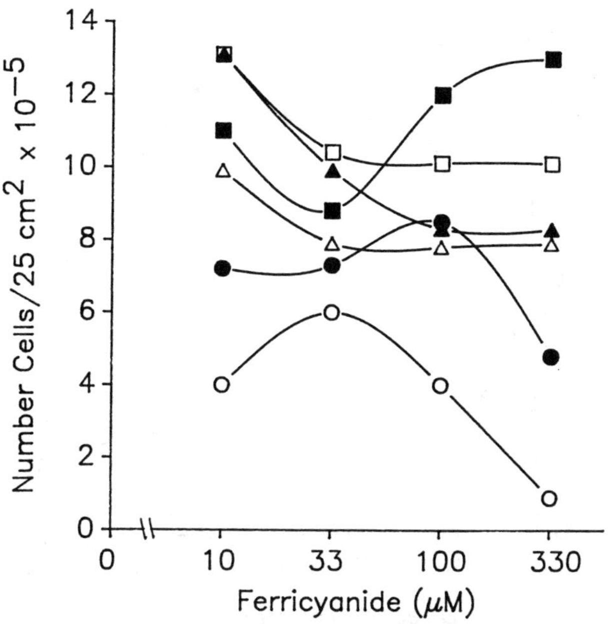

FIGURE 3. Effect of ferricyanide on growth of HeLa cells in presence of increasing concentrations of serum. Low concentrations of serum enhance the effect of increasing ferricyanide, whereas ferricyanide becomes inhibitory with high levels of serum.[58] Symbols: (○) no serum; (●) 2% serum; (■) 4% serum; (□) 6% serum; (▲) 8% serum; and (△) 10% serum. Culture on MEM for 48 h.

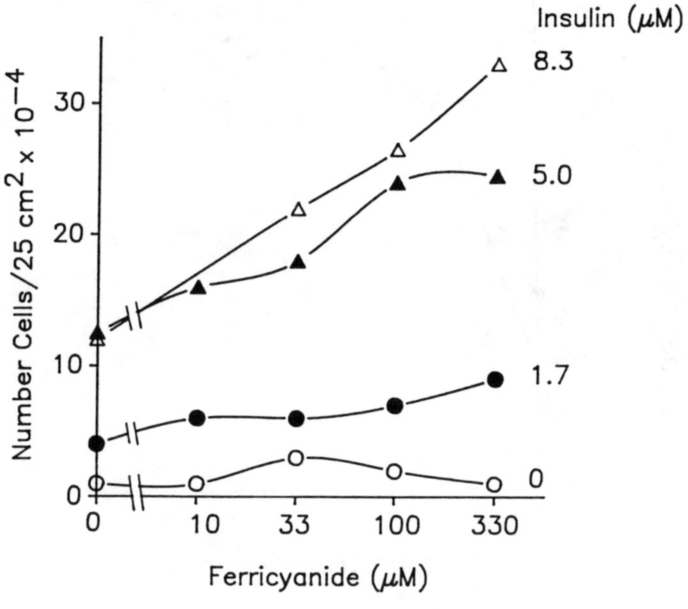

FIGURE 4. Dose-response growth curve of HeLa cells in serum free medium supplemented with potassium ferricyanide and insulin. (From Sun, I. L., Crane, F. L., Grebing, C., and Löw, H., *Exp. Cell Res.*, 156, 528, 1985. With permission.)

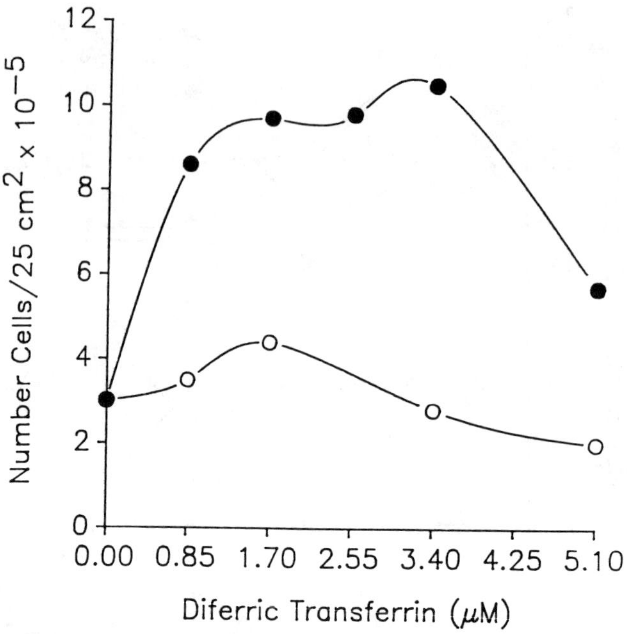

FIGURE 5. Stimulation of HeLa cell growth by ferric transferrin (●), and lack of stimulation by apotransferrin (○).[110] Cells grown in α MEM without serum for 48 h.

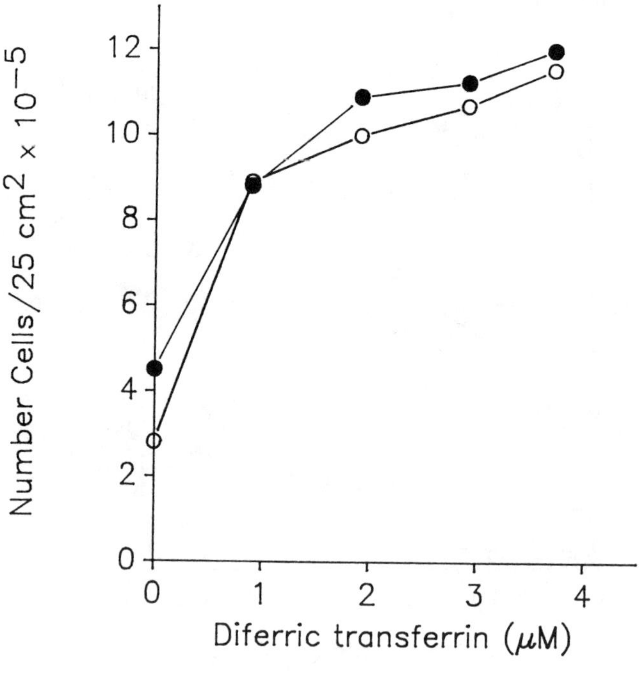

FIGURE 6. Effect of ferricyanide on growth of HeLa cells stimulated by ferric transferrin. Ferricyanide at 0.033 m*M*, (●); no ferricyanide, (○). Cells grown on MEM without serum for 48 h.

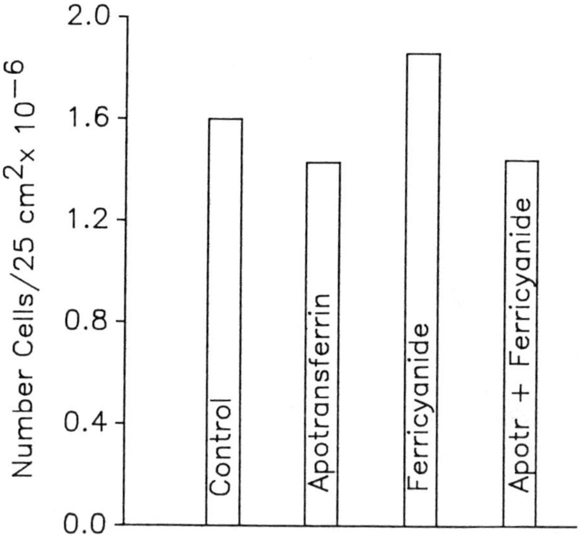

FIGURE 7. Effect of short pulses of ferricyanide and apotransferrin on growth of HeLa cells in a subsequent 20-h period; 10 μM ferricyanide for 10 min, and 10 μM apotransferrin for 5 min treatment as indicated. Cells were separated from the additives and growth was measured without further additions.

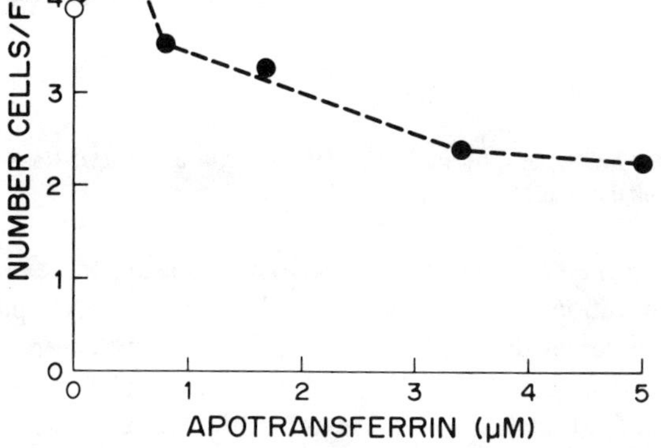

FIGURE 8. Inhibition of diferric transferrin stimulation of HeLa cell growth by apotransferrin. Growth on 3.4 μM diferric transferrin under serum-free conditions with α MEM; (○) apotransferrin only; (●) diferric transferrin plus apotransferrin. (From Crane, F. L., Löw, H., Sun, I. L., Navas, P., and Morré, D. J., in *Redox Functions of the Eukaryotic Plasma Membrane*, Ramirez, J., Ed., CSIC, Madrid. With permission.)

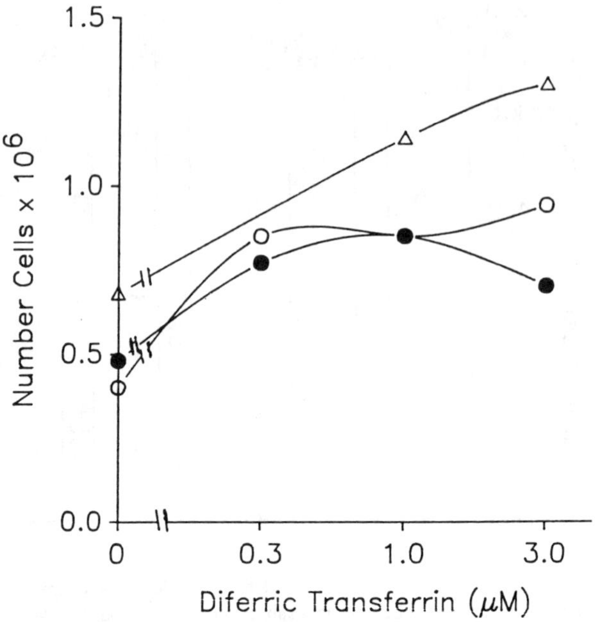

FIGURE 9. Stimulation of L1210 cell growth by diferric transferrin in presence and absence of 1 μg/ml insulin. Based on data in Reference 91; (○) no addition; (●) plus 1 μg/ml insulin; (△) protein increase (μg × 10^{-2}).

TABLE 1
Effect of Insulin on Reduction of Ferricyanide and Diferric Transferrin by HeLa Cells

	Ferricyanide		
	Fast	Slow	Diferric transferrin
Control	145	40	10
30 μg/ml insulin	221	154	14

Note: $Fe(CN)_6$ at 0.01 m*M*, diferric transferrin at 10 μ*M*. Ratio in nmol min^{-1} (g wwt)$^{-1}$.

Exposure to Ga-transferrin also increases transferrin receptor exposure on the inhibited cells. The inhibition of proliferation occurs in the S and G_2 + M cell cycle phases.[63] Addition of 2.5 μ*M* Fe transferrin reverses the inhibition of proliferation by Ga-transferrin. Iron uptake from ferric transferrin is decreased by Ga-transferrin, even though transferrin receptor expression has increased. The Ga-transferrin appears to inhibit iron release from the Fe transferrin and thereby inhibits iron uptake. It is not known if Ga-transferrin inhibits diferric transferrin reduction by cells.

Insulin does not increase the growth response of L1210 cells in the presence of transferrin (Figure 9). Insulin, however, does not increase the reduction rate of transferrin as much as it does with ferricyanide (Table 1), even though insulin (IGF) can rapidly increase expression of transferrin receptors on the surface of cells.[64-67]

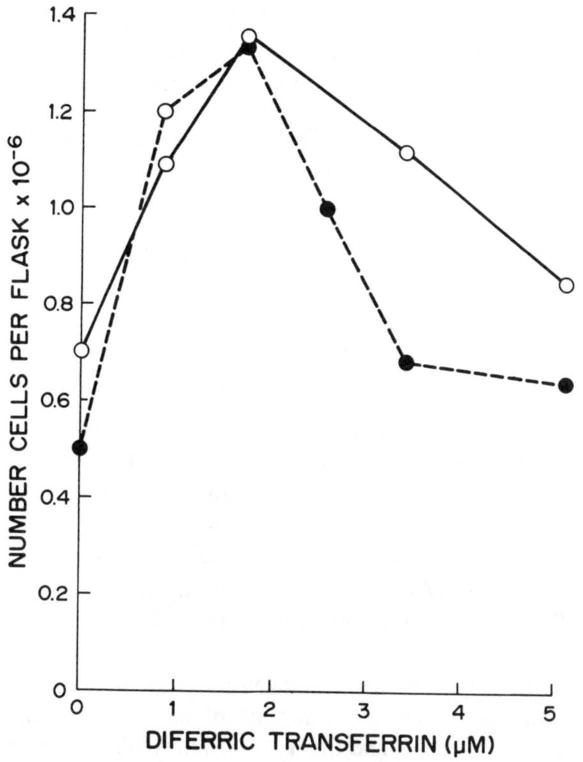

FIGURE 10. Effect of bathophenanthroline disulfonate on growth stimulation by diferric transferrin: (●) only diferric transferrin added; (○) open circles, diferric transferrin plus 10 μ*M* BPS. HeLa cells, 48-h culture in serum-free MEM.

V. EFFECT OF BPS ON GROWTH WITH TRANSFERRIN

Bathophenanthroline disulfonate (BPS) forms a pink complex with ferrous iron produced by reduction of iron in transferrin. This complex remains in the supernatant after centrifugation of cells, so the complex does not enter the cells and the iron is retained outside.[19] Diferric transferrin stimulates HeLa cell growth even in the presence of bathophenanthroline disulfonate, which indicates that reduction of the iron allows growth even if the bulk of the ferrous iron is kept out of the cell by the chelator (Figure 10). Thorstensen and Romslo[39] have shown that BPS inhibits iron uptake by liver cells.

VI. BASIS FOR REDOX EFFECTS ON GROWTH

The basis for the growth stimulation by the external oxidants is not known. In the original studies by Ellem and Kay, the similarities between effects of pyruvate and ferricyanide suggested that an oxidation of pyridine nucleotides could be important. Since pyruvate is present in the HeLa cell media, any effect of ferricyanide or diferric transferrin on NAD levels in the cytoplasm would be in addition to any pyruvate effects. It is known that ferricyanide and diferric transferrin added to HeLa cells in salts-buffer do cause oxidation of the NADH and increase NAD in the cell. It is not known if these oxidation effects can be observed when pyruvate is also present.[68,69]

The long-term (days) effects of the external oxidants on NAD and NADPH levels in HeLa cells are more complex than the short-term effects. These long-term effects would be

TABLE 2
Effect of Low NAD Synthesis on Growth of Chinese Hamster Cells

Cell Type	NAD (pmol mg prot^{-1})	Cell doubling time (h)
V79 control	5854	10
N-3 mutant	90	24

Data from Chatterjee et al.[70]

expected during the 48- to 72-h culture period for the HeLa cells. After 48 h the total pyridine nucleotides in the HeLa cells increases, the NAD/NADH ratio increases, and the total NAD increases over the level seen in the untreated cell.[68] The general effect of the external oxidants is an increase in total pyridine nucleotides and a total increase in oxidized NAD^+ and $NADP^+$ in the cells. The increase in oxidized pyridine nucleotides would favor an increase in glycolysis, ribonucleotide reductase, ADP ribosylation, and guanine nucleotide synthesis. For example, mutant V79 Chinese hamster lung cells with greatly decreased synthesis of NAD show more than a twofold increase in doubling time which can be reversed by incorporation of NAD to permeabilized cells (Table 2).[70] This effect can be based on decrease of poly(ADP ribose).

The growth response of many cells is related to changes in the activity of the Na^+/H^+ antiport in the plasma membrane.[71-74] Inhibition of the antiport with amiloride has been shown to inhibit DNA synthesis,[75] and the use of amiloride as an antitumor agent has been advocated.[76,77] The increased Na^+/H^+ exchange causes expulsion of H^+ from the cell with an associated increase in the pH of the cytoplasm. The increased pH may activate cytoplasmic and nuclear enzymes.[78]

Both ferricyanide and diferric transferrin increase the activity of the Na^+/H^+ antiport in HeLa cells (Figure 11).[62] Ferricyanide increases antiport activity in erythrocytes.[79] The maximum increase in activity is shown with diferric transferrin plus added ferric ions, e.g., ferric ammonium citrate. On the other hand, the internal pH increases with diferric transferrin (Figure 12) but decreases with addition of 0.1 m*M* ferricyanide even though both activate proton release.[80] The basis for this difference can be the difference in reduction rates. The reduction of ferricyanide at 0.1 m*M* by the cells is more than ten times the rate of diferric transferrin reduction. If transmembrane electron transport from NADH only transfers an electron across the membrane, then the proton produced by NADH oxidation remains in the cytoplasm. Thus, 0.1 m*M* ferricyanide would produce internal protons faster than the antiport expels them, whereas diferric transferrin activates much more proton transfer than electron transfer,[81] so the net internal proton level decreases. The stimulation of growth by ferricyanide is maximum at concentrations below 0.1 m*M* where electron transport is slower. The inhibitory effects of higher ferricyanide levels on growth may depend on acidification of the cytoplasm, whereas the stimulation at lower concentrations may be based on a slow rate of electron transport which can activate the antiport fully. The activation of proton release from HeLa cells by ferricyanide has been shown to reach a maximum at a lower concentration of ferricyanide than is required for maximum electron transport.[17] Activation of the antiport even at higher concentrations (0.2 m*M*) of ferricyanide is supported by increased Na^+ uptake by liver cells when ferricyanide is added. A coupling between the oxidoreductase and the antiport may also be indicated by amiloride inhibition of ferricyanide reduction.[82] Other studies have indicated that Na^+/H^+ exchange in the presence of bicarbonate is insufficient to maintain an increased internal pH in physiological conditions,[83] which would argue against growth stimulation through the antiport.

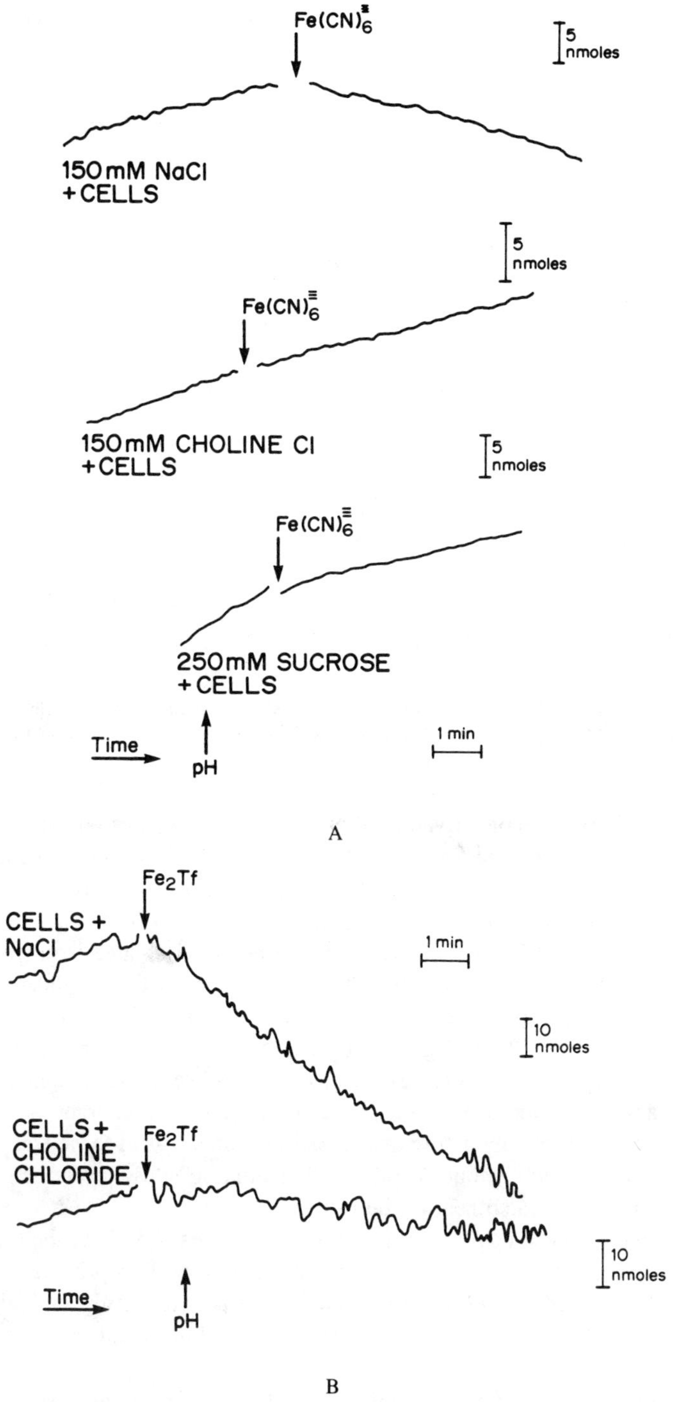

FIGURE 11. Proton release from cells stimulated by external oxidants. (A) Ferricyanide-stimulated change in external pH of pineal cells. 10 mg cells in 1.5 m*M* tris buffer with indicated salts at 150 m*M*. Circulating air used to equilibrate CO_2. pH measured with a glass electrode and proton change titrated with 0.01 *M* HCl at end of experiment. 0.1 m*M* potassium ferricyanide used to start proton release. (B) Diferric transferrin effect on proton release. Cells and procedures same as (A). Proton release started with 17 μM diferric transferrin. (From Sun, I. L., Toole-Simms, W., Crane, F. L., Morré, D. J., Löw, H., and Chou, J. Y., *Biochim. Biophys. Acta,* 938, 17, 1988. With permission.

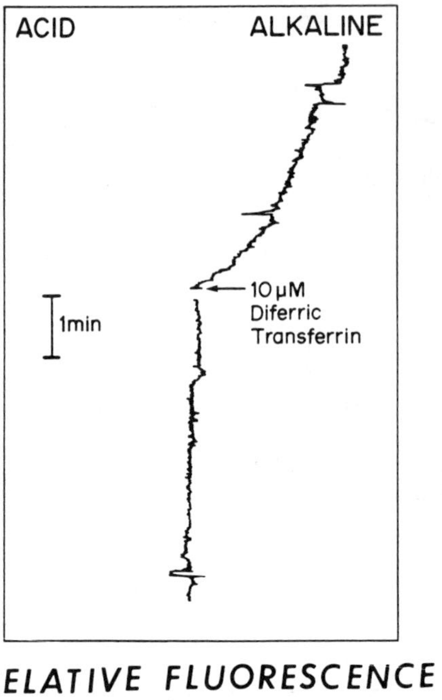

FIGURE 12. Change in internal pH of HeLa cells in response to 10 μ*M* diferric transferrin. Change in internal pH measured with the fluorescent indicator BCECF.[80] The pH change is 0.2 pH units.[111]

A secondary consequence of activation of Na^+/H^+ exchange is an increase in membrane potential because of the increased activity of the Na^+/K^+-ATPase to remove the extra Na^+ taken up by the antiport.[82] Both ferricyanide and diferric transferrin produce an increase in membrane potential, and the increase is ouabain sensitive, so it appears to be based on activation of the Na^+/K^+-ATPase.[80] The membrane potential change is much greater with diferric transferrin than with ferricyanide, which is consistent with the relative activation of the antiport by these oxidants. Likewise, the growth stimulation by diferric transferrin is much greater than the stimulation of growth by ferricyanide (cf. Figures 1 and 5).

The effects of insulin, which increases the growth response to the oxidants, on activation of the Na^+/H^+ antiport, internal pH changes and changes in membrane potential are not known. Since insulin is probably acting as a substitute for IGF-I in these HeLa cells which have very little insulin binding, but do have IGF-I binding sites, the effects of IGF-I on growth and antiport activity also need to be tested.

Transplasma membrane electron transport by erythrocytes stimulates the tyrosine kinase which is associated with the band 3 plasma membrane protein. This protein is an anchoring site for the cytoskeleton and also binds and controls the activity of glyceraldehyde-3-phosphate dehydrogenase. This later effect is involved in control of glycolysis as shown by Löw et al.[84]

Since several oncogenes associated with the plasma membrane are involved in tyrosine kinase activity,[85] the plasma membrane control of tyrosine kinase may be mediated through these oncogenes or the redox system may activate an oncogene tyrosine kinase to initiate stimulation of cellular proliferation.

Other mechanisms have been proposed as a basis for increased proliferation by various cells. These include changes in internal calcium[86] and changes in hydrolysis of phosphatidyl inositol to activate protein phosphorylation and calcium release from internal stores.[87,88] There has been no investigation of the effect of transplasma membrane electron transport

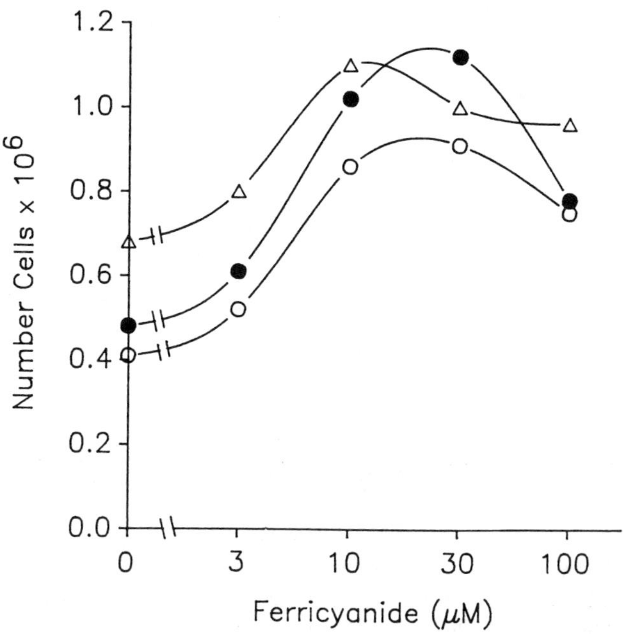

FIGURE 13. Stimulation of L1210 cell growth by ferricyanide in presence and absence of 1 μg/ml insulin. Based on data in Reference 91; (△) protein (÷ 100) μg; (○) $Fe(CN)_6$ only cell count; (●) $Fe(CN_6)$ + insulin 1 μg/ml cell count.

on Ca^{2+} release or hydrolysis of inositol lipids in HeLa cells. In other chapters, the effects of the redox system on oncogene activation and calcium uptake in heart are discussed. Amino acid transport into cells may also be activated by the redox system[89,90] (Chapter 12).

Cell sorter studies to measure the DNA content of cells show that both ferricyanide and diferric transferrin increase the percentage of HeLa cells moving from G_1 through S to G_2 stage. This effect is consistent with an increase in DNA replication as a result of the oxidant treatment. Of course, this observation does not explain how the electron transport at the plasma membrane can activate the replication of DNA.

The activity of the transplasma membrane electron transport during the growth cycle of cells is consistent with high activity during active growth and decreased activity during decreased growth. HeLa cells, which do not go into a truly dormant period, are not the best for studying these effects, but young HeLa cells at the start of cultures show more activity than cells which have matured or have stopped growing because of nutrient exhaustion. An early stimulation is seen in pulse studies in which cells are exposed to ferricyanide for only 10 min at the start of a 20-h growth period (Figure 7).

The lack of a pyruvate effect on growth of HeLa cells indicates that the growth stimulations observed are not based on destruction of H_2O_2, as has been observed with melanoma cells. Ferrocyanide, which would also remove all H_2O_2, does not stimulate growth, which also indicates that removal of H_2O_2 is not the only factor which stimulates HeLa cell growth.

VII. STIMULATION OF L1210 CELL GROWTH

L1210 cells can grow with addition of simple iron salts and do not need diferric transferrin for growth. Both ferricyanide and diferric transferrin stimulate the growth of L1210 cells, as measured by cell count and incorporation of ^3H thymidine (Figures 9, 13, and 14) into DNA.[91] The response to ferricyanide and transferrin is similar in extent, in contrast to the greater growth of HeLa cells on transferrin.

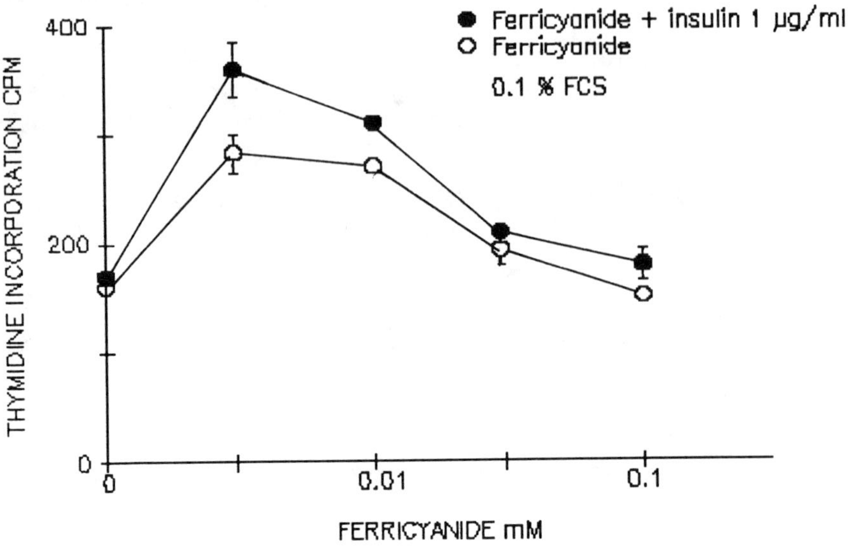

FIGURE 14. Stimulation of thymidine incorporation into L1210 cells by low concentrations of Na ferricyanide. Growth on RPM1 media with additions shown.

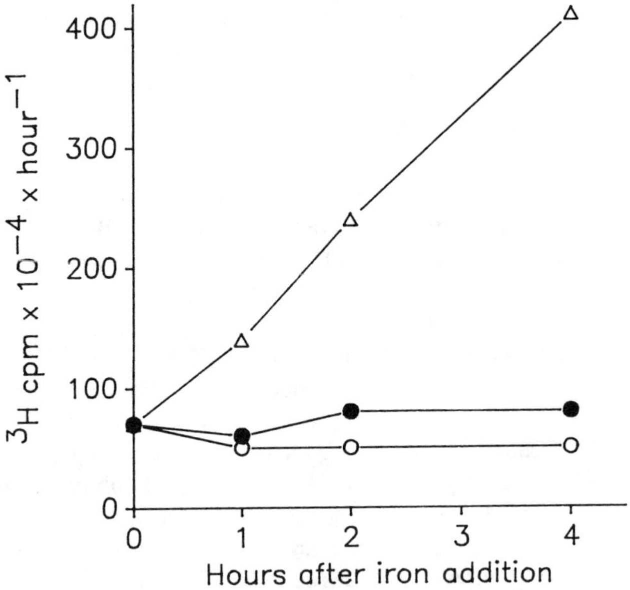

FIGURE 15. DNA synthesis in L1210 cells after iron removal with desferrioxamine using cells adapted to serum free-growth on transferrin. Growth with: (△) ferric citrate; (○) no addition; (●) 0.1 m*M* potassium ferricyanide. The 50% increase in DNA incorporation with ferricyanide contrasts with the 400% increase with ferric citrate.[16] See Chapter 9.

Iron salts have also been used to support the growth of L1210 cells.[15,16] Iron salts have been shown to substitute fully for diferric transferrin in growth of these cells. On the other hand, ferricyanide cannot support growth in the presence of desferrioxamine which removes iron from the media (Figure 15). Since desferrioxamine does not inhibit ferricyanide reduction, the decreased growth with ferricyanide cannot be based on lack of transmembrane

TABLE 3
Reduction of Ferricyanide, Diferric Transferrin and Ferric Ammonium Citrate by L1210 Cells

	Reduction rate (nmol min^{-1} million $cell^{-1}$)		
Fe_2Tf 17 μM		0.035	
FAC 7.5 μM	0.24 (fast)		0.125 (slow)
FAC + FeTf		0.052	
FAC 7.5 μM		0.12	
FAC + Fe_2Tf 7 μM		0.184	
$Fe(CN)_6$ 0.1 mM	1.3 (fast)		0.52 (slow)

Note: Fe_2Tf, diferric transferrin; FAC, ferric ammonium citrate; $Fe(CN)_6$, ferricyanide. The fast rate is initial activity, slow is activity after 3 min. For assay procedure see Reference 18.

electron transport but would be due to lack of available iron, since it has been shown that ferricyanide cannot supply iron to the cells.

Ferricyanide, ferric ammonium citrate and diferric transferrin can all function as electron acceptors for the transplasma membrane electron transport by L1210 cells (Table 3).

Basset has shown that L1210 cells adapted to growth without transferrin take up iron supplied by simple iron salts. It is not known if the transmembrane electron transport is required for iron uptake, but the iron reduction could facilitate uptake. It is clear that the cells cannot grow if iron is removed by desferrioxamine, and the role of iron or diferric transferrin in electron transport cannot be evaluated if the iron is needed for both a redox function and for iron supply. Stimulation of growth by ferricyanide and indigo disulfonate in media which has not been treated to remove iron indicates that a transmembrane redox function is required in addition to the requirement for iron as a nutrient. Indigo disulfonate is an impermeable redox agent that can also accept electrons from the transmembrane electron transport system (Figure 16).

The L1210 cells respond more consistently to ferricyanide if 0.1% fetal calf serum is added, which indicates that other nutrients or factors are limiting maximum growth.

VIII. GROWTH OF EHRLICH ASCITES CELLS

Ehrlich ascites cells can be grown in suspension culture with 10% fetal calf serum. Growth is severely limited on 0.5% fetal calf serum, but can be increased 100% by addition of 0.003 to 0.01 mM potassium ferricyanide (Figure 17). With these cells the rate of ferricyanide reduction is much slower than for HeLa or L1210 cells (Table 4).[92]

IX. GROWTH OF SWISS 3T3 CELLS

It is well known that starved Swiss 3T3 cells require mitogens to initiate cell growth.[93] Starved Swiss 3T3 cells do not initiate growth on ferricyanide alone (Figure 18). The lack of a consistent growth response by starved 3T3 cells to ferricyanide would indicate that ferricyanide cannot substitute for other growth factors needed for the $G_0 \rightarrow G_1$ transition. The effect of additional growth factors EGF and insulin are seen in Figure 19, where ferricyanide gives a 58% stimulation of growth over factors alone.

SV40-transformed 3T3 cells show a much more consistent growth response to ferricyanide (Figure 18).[91] These cells are grown in D-MEM medium which already contains 0.025 μM ferric nitrate. Thus, the growth response to ferricyanide is not just based on a

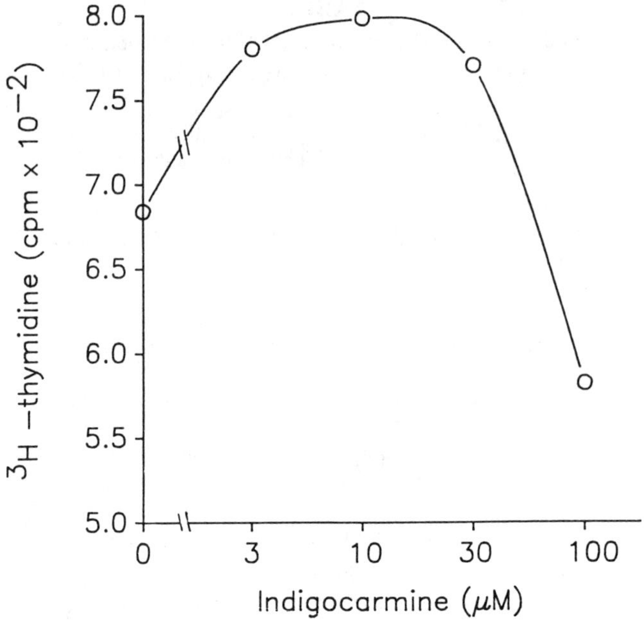

FIGURE 16. Thymidine incorporation by L1210 cells in presence of indigo carmine with 0.1% serum in RPMI 1640 media.

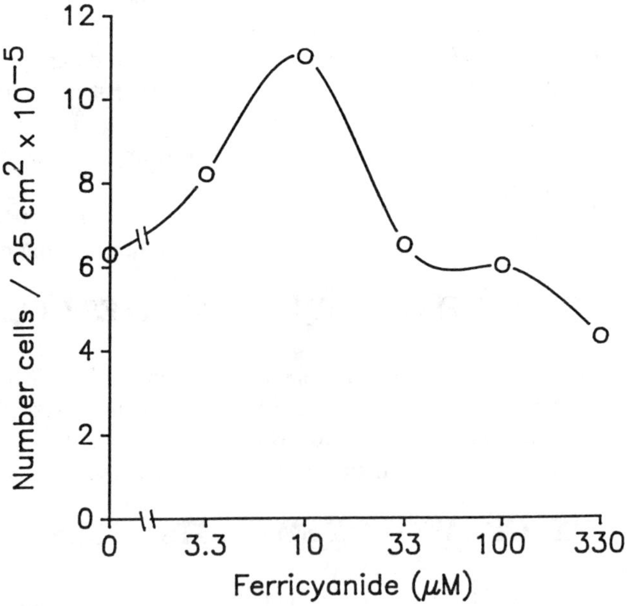

FIGURE 17. Effect of ferricyanide on growth of Ehrlich ascites cells in 0.5% serum.[92]

TABLE 4
Reduction of Ferricyanide by Ehrlich Ascites Cells and Redox-Induced Proton Release from Cells

Rate of ferricyanide reduction (nmol $min^{-1}gww^{-1}$)	Rate of H^+ release (nmol $H^+min^{-1}gww^{-1}$)	Ratio H^+ release/$Fe(CN)_6$ Reduction
15.6	52	3.3

Note: Assay for reduction and H^+ release in sucrose salts solution, 0.1 *M* sucrose, 10 m*M* each of NaCl, KCl, $CaCl_2$ and 0.25 m*M* tris buffer, pH 7.4.

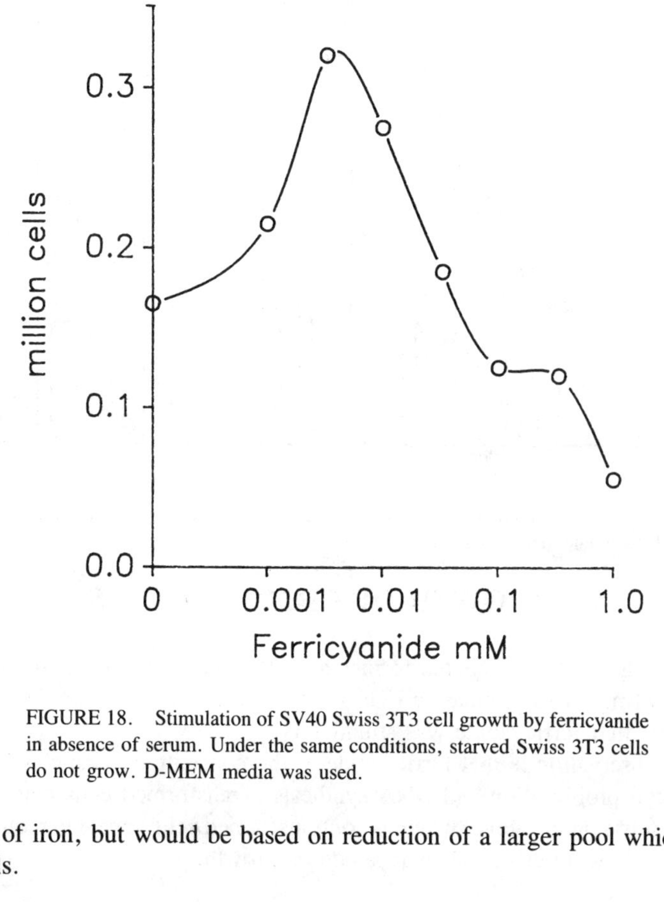

FIGURE 18. Stimulation of SV40 Swiss 3T3 cell growth by ferricyanide in absence of serum. Under the same conditions, starved Swiss 3T3 cells do not grow. D-MEM media was used.

micro supply of iron, but would be based on reduction of a larger pool which is not taken up by the cells.

X. GROWTH OF SV40-TRANSFORMED COMPARED TO UNTRANSFORMED FETAL RAT LIVER CELLS

Temperature-sensitive SV40-transformed fetal rat liver cells do not proliferate in th e presence of ferricyanide alone. The addition of one-tenth the optimum level of fetal ca f serum (0.4% vs. 4%) stimulates growth to 25% of maximal. Addition of low concentratio s of ferricyanide (0.033 m*M*) can then increase growth to 52% of maximal (Figure 20). Th e cells are transformed with a temperature-sensitive SV40, and growth at 40°C gives e nontransformed phenotype.[25] Even with 0.4% serum present, ferricyanide gives much s stimulation of growth at 40°C than at 33°C. The transformed phenotype appears much r e responsive to growth stimulation in the presence of ferricyanide than the untransfo d cells.

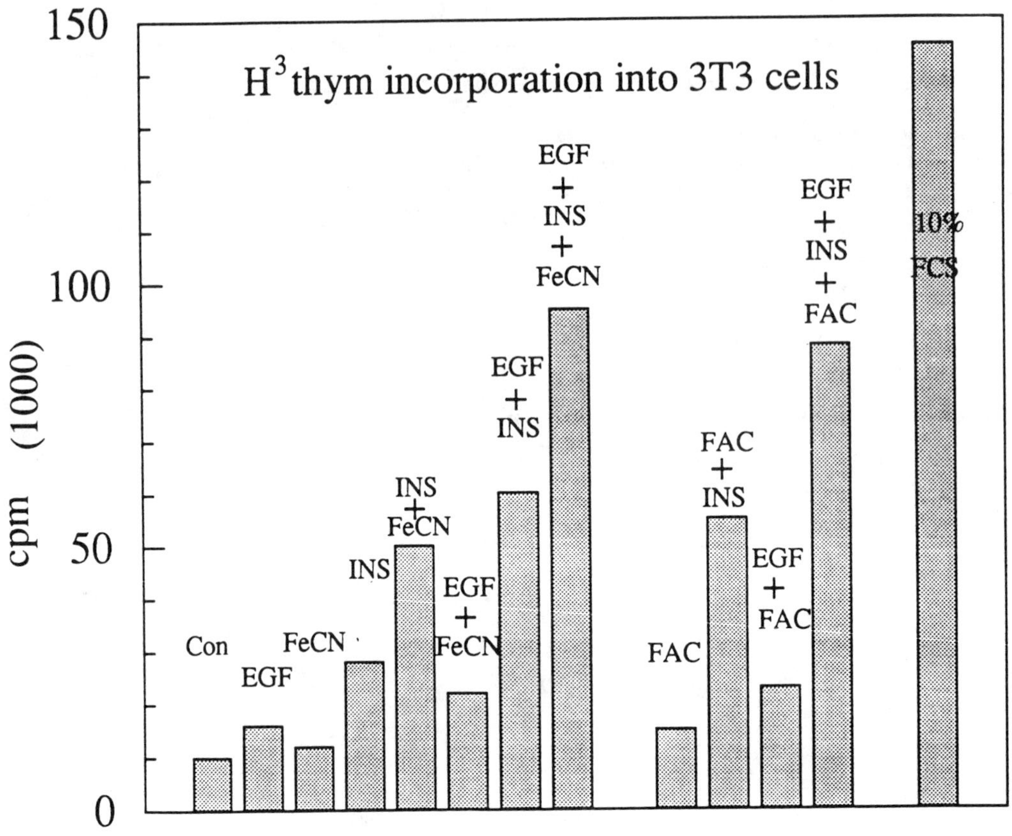

FIGURE 19. Ferric ammonium sulfate or ferricyanide stimulation of growth by 3T3 (Swiss) cells in presence of EGF and insulin. ^{3}H thymidine incorporation in D-MEM media.

XI. GROWTH OF CH310T$^1/_2$ CELLS

CH310T$^1/_2$ cells can be stimulated to increase DNA synthesis in the presence of ferricyanide if phorbol myristate acetate or EGF is also present as a growth factor.[94] In these experiments, 0.01 m*M* ferricyanide was stimulatory.

The overall observation is that ferricyanide in the concentration range of 0.003 to 0.1 m*M* stimulates cell proliferation and DNA synthesis. Transformed cells may show growth stimulation with ferricyanide alone, whereas nontransformed cells appear to require additional growth factors to show an effect of ferricyanide on growth.

XII. GROWTH OF HepG2 CELLS

HepG2 cells show a 50% increase in growth with 0.01 m*M* ferricyanide in the absence of FCS. Concentrations above 0.1 m*M* ferricyanide to 1.0 m*M* do not stimulate but show very little inhibition of growth (Figure 21).[61]

XIII. DNA SYNTHESIS IN HEPATOCYTES

Ferricyanide stimulates ^{3}H thymidine incorporation into DNA in rat hepatocytes in the presence of limited amounts of rLGF liver growth factor. Transferrin plus LGF gives even greater stimulation (Table 5). The addition of rLGF increases oxidation of internal NADH

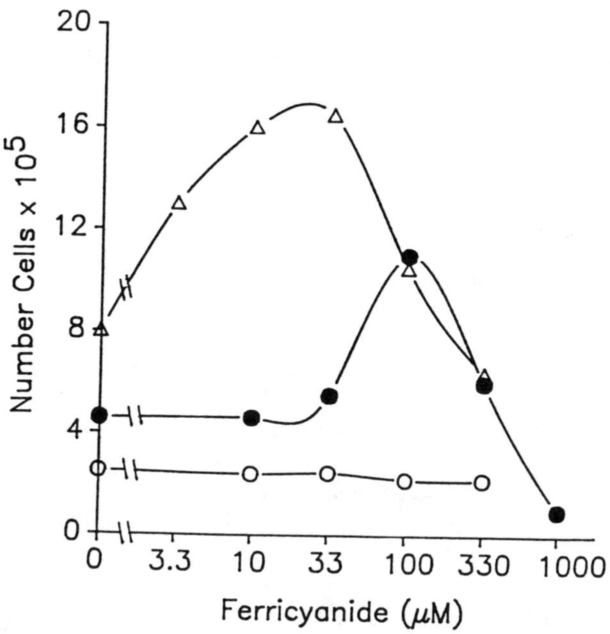

FIGURE 20. Ferricyanide stimulation of growth of SV40-transformed fetal rat liver cells in α-MEM media, with and without low levels of serum; (△) with 0.4% serum SV40 3T3; (○) no serum SV40 3T3; (●) cells grown at 40°C where SV40 cannot be expressed, with 0.4% serum.

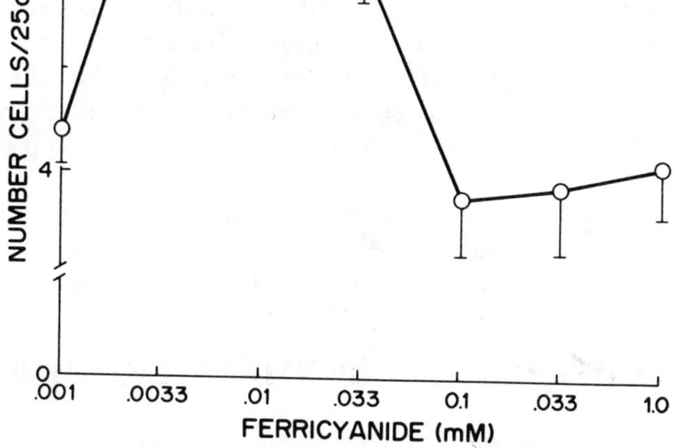

FIGURE 21. Effect of sodium ferricyanide on the growth of HepG2 cells in absence of added serum. (From Crane, F. L., Löw, H., Sun, I. L., Navas, P., and Morré, D. J., in *Redox Functions of the Eukaryotic Plasma Membrane*, Ramirez, J., Ed., CSIC, Madrid, 1988, 3. With permission.)

TABLE 5
Ferricyanide and Diferric Transferrin Stimulation of DNA Synthesis by Rat Hepatocytes

	Stimulation of thymidine incorporation into DNA % over control
rLGF 5 μg/ml	24
rLGF-5 + 0.1 mM ferricyanide	31
rLGF-5 + 5 μg/ml diferric transferrin	48

Note: Data from Garcia-Cañero;[57] rLGF is a growth factor for liver cells.[57]

by ferricyanide in liver cells. The stimulation of DNA synthesis is correlated with stimulation of NADH oxidation. Ferricyanide also stimulates amiloride sensitive proton release from hepatocytes. The rate of proton release with ferricyanide is greatly increased by rLGF whereas rLGF shows only a slight increase when added alone to liver cells. The increase of proton release with ferricyanide is dependent on Na^+ in the external media, which indicates activation of the Na^+/H^+ antiport by ferricyanide as for HeLa cells.[57,82]

XIV. OXIDANTS WHICH CAN STIMULATE GROWTH

In the early studies by Barron and Hoffman[95] it appeared that only impermeable oxidants with a redox potential above −125 mV could be reduced by the transplasma membrane redox system of starfish eggs. Brooks[3] found growth stimulation with compounds that had redox potential as low as −150 mV.

With HeLa cells, Sun et al.[27] have shown good reduction rates with indigodisulfonate (−125 mV) but no activity with anthraquinone sulfonate (−175 mV), so the redox potential will be between those potentials or approximately −150 mV. Compounds with a redox potential high enough to be reduced by the transmembrane redox enzyme also stimulate the growth of HeLa cells (Figure 22). Impermeable compounds with a redox potential −175 mV or lower do not show reduction when added to HeLa cells and do not stimulate HeLa cell growth.

The only exception to this relation between redox potential and growth stimulation is diferric transferrin where the tightly bound ferric iron has an apparent redox potential at −400 mV in solution.[37] Since the iron in the diferric transferrin does get reduced at the plasma membrane, there may be another redox enzyme, or the binding of transferrin to the receptor may change the redox potential to a higher value. As an alternative, Thorstensen and Romslo[39,42] have suggested that the redox system may first acidify the receptor region to release the iron from transferrin. The free iron can then act as a high potential electron acceptor.

Ferric lactoferrin can stimulate growth of macrophages without contributing to iron uptake.[96] Since lactoferrin iron can be reduced by cells,[40,97] the growth stimulation by lactoferrin may provide an example of how ferric transferrin can stimulate growth by acting as an electron acceptor as well as by playing a role in iron supply.

XV. PERMEABLE VS. IMPERMEABLE OXIDANTS

The difference between the effects of 1,4-naphthoquinone and 1,4-naphthoquinone disulfonate indicates the difference in effects of permeable and impermeable redox agents on cellular function or survival. The permeable naphthoquinone (redox potential $E^{\circ\prime}$) 0.036 V stops growth of cells immediately. On the other hand, the nonpermeable naphthoquinone sulfonate $E^{\circ\prime}$ 0.113 V stimulates cell growth until it reaches ten times the concentration of the permeable compound where it shows similar inhibition (Figure 23).

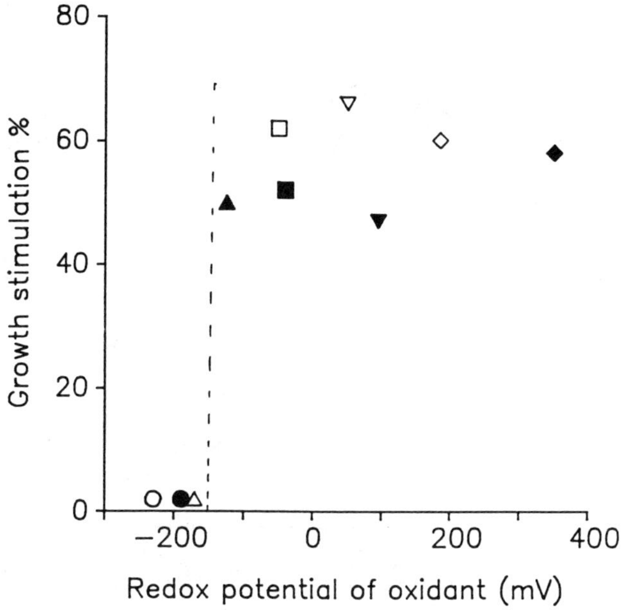

FIGURE 22. Effect of redox potential of external oxidants on growth stimulation of HeLa cells in absence of serum in α-MEM media. Symbols for oxidants: (◆), ferricyanide; (◇), 1,2-naphthoquinone sulfonate; (▼), ferric EDTA; (▽), hexammine ruthenium chloride III; (■), pentammine ruthenium chloride III; (□), indigo tetrasulfonate; (▲), indigo disulfonate; (△), anthraquinone 1,5-disulfonate; (○), anthraquinone 2,6-disulfonate; (●), anthraquinone-2-sulfonate.

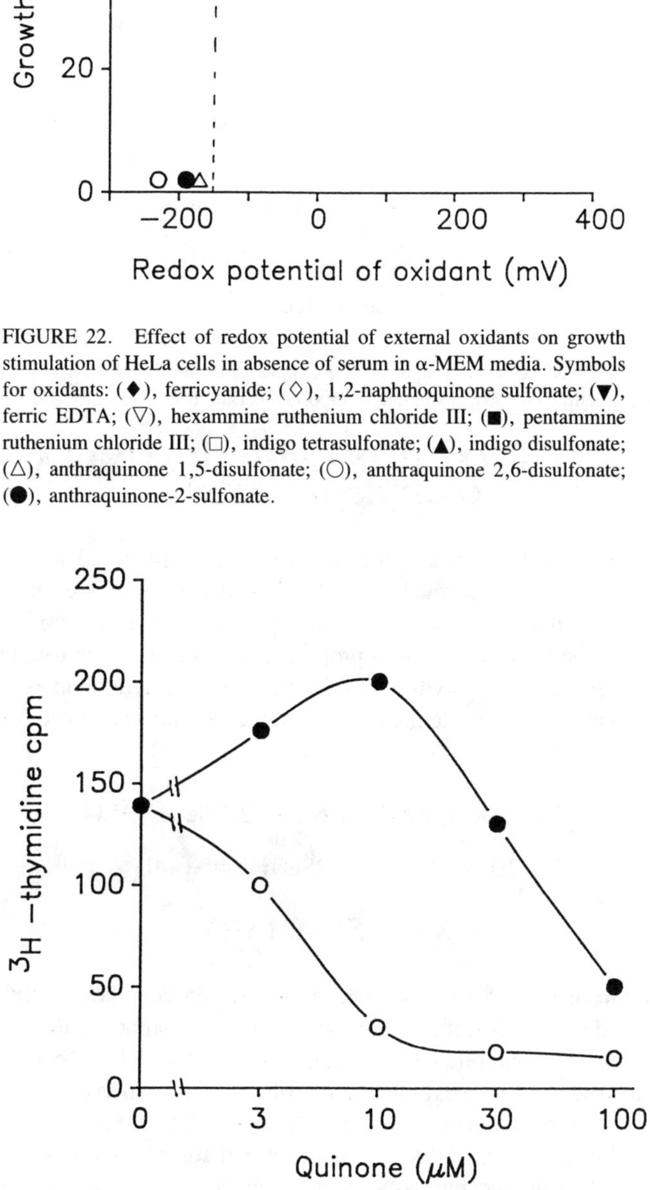

FIGURE 23. Comparison of permeable and impermeable naphthoquinones on growth of L1210 cells in 0.1% serum; growth measured by H^3 thymidine incorporation during a pulse; (●) 1,4-naphthaquinone sulfonate; (○) 1,4-naphthoquinone.

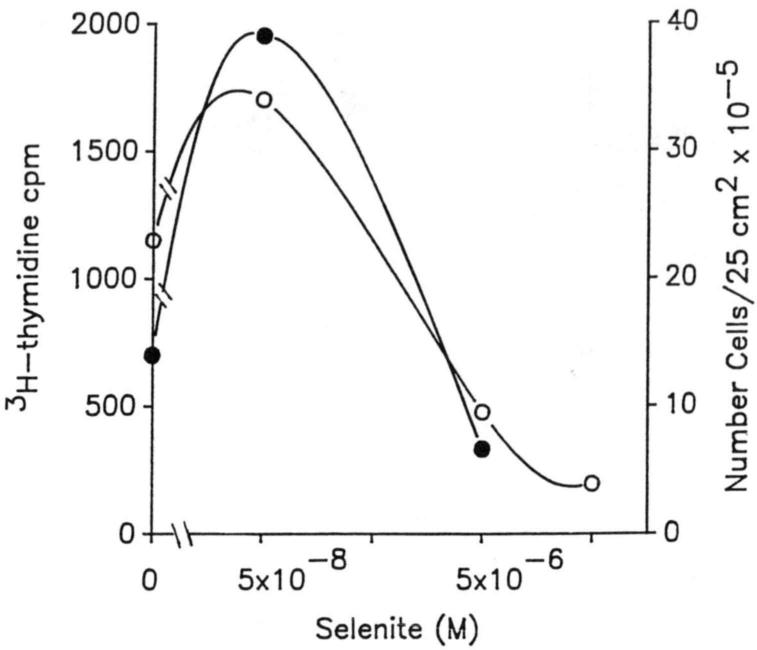

FIGURE 24. Selenite stimulation of growth of mouse mammary YN-4 cells in serum-free media: (○), cell count; (●), thymidine incorporation. Data from Reference 45.

XVI. SELENITE EFFECTS IN GROWTH STIMULATION AND OXIDOREDUCTIONS

Selenite is another natural oxidant found in serum at 0.14 to 0.5 μM which stimulates growth of certain cells.[43-45] Low levels stimulate growth in serum-free media (Figure 24). It is not known if selenite can act as an electron acceptor at the cell surface. If so, it may also activate growth because of its redox properties in addition to its role in enzymes such as glutathione peroxidase. The hydroselenide produced by reduction of selenite can be autooxidized by oxygen, so the selenite may act at the cell surface to catalyze electron flow to oxygen.

$$2H_2SeO_3 + 6XH_2 \rightarrow 6X + 2H_2Se + 6H_2O$$

$$2H_2Se + 3O_2 \rightarrow 2SeO_3^{2-} = + 4H^+$$

XVII. SUMMARY

The stimulation of cell growth by ferricyanide, which does not readily supply iron to cells[16] but is reduced at the cell surface,[98] introduced a new approach to the widely recognized role of diferric transferrin and other iron complexes in cell growth. Clearly, cells need iron for vital redox enzymes,[99] and an external redox role for diferric transferrin was not expected because all evidence pointed to incorporation of the diferric transferrin into endosomes before release and reduction of the iron.[100] It now appears that some of the diferric transferrin can act as an external electron acceptor as well as ferricyanide or other iron complexes. The observation of growth stimulation with impermeable organic oxidants further supports the importance of the transmembrane electron transport. On the other hand, the demonstration by Basset[16] that desferrioxamine treatment can inhibit ferricyanide stimulation of L1210 cell growth shows that available iron is needed in addition to external redox acceptors.

Although oxygen, diferric transferrin, or selenite are available as oxidants for the transplasma membrane electron transport system, it is not clear which one, or the combination of more than one, is the physiological electron acceptor. Thioredoxin can also be associated with the plasma membrane.[101] The natural acceptor may be different with different cells.

The stage in the cell cycle at which the redox agents act is still to be determined. Basset showed sites for iron function both at entry into S and in the G_2 phase of the cell cycle. The requirement for iron in ribonucleotide reductase[102,103] would favor control by internal iron on DNA formation for entry into the S phase of the cell cycle. If so, then the external oxidants may control release from G_2. Gallium transferrin inhibits cell growth in S phase, which suggests that ferric transferrin reduction is necessary for S phase passage.[63] The inability of diferric transferrin or ferricyanide to initiate 3T3 cell growth indicates that they cannot substitute for platelet-derived growth factor (PDGF) or other factors which activate the transition from G_0 (resting) to G_1.[93]

In C3H10T$^1/_2$ cells, ferricyanide alone can stimulate the rapid and transient formation of the protooncogenes c-*myc* and c-*fos* at concentrations even higher than used for growth stimulation (0.1 to 0.5 m*M*). When the C3H10T$^1/_2$ cells are treated with TPA or EGF plus ferricyanide, there is an increase in total DNA synthesis above the effect of the growth factor.[94]

The inhibition of growth observed at high concentrations of external oxidants can be based on excessive oxidation of the internal substrates such as NADH or NADPH, since these artificial oxidants, especially those with a high redox potential such as ferricyanide or naphthoquinone sulfonate, tend to stimulate a high rate of electron transport across the membrane in comparison to diferric transferrin. Other compounds, such as selenite, may also enter the cytoplasm where they would be reduced rapidly by glutathione and various reductase enzymes, but may interfere with cell function through these oxidations.

There is evidence that the transferrin receptor has a controlling effect on the transplasma membrane electron transport and is necessary for electron transport to diferric transferrin (cf. Chapter 10). Evidence includes inhibition by protease treatments or monoclonal antibodies which remove or block the transferrin receptor.[19,38,80] Expression of the transferrin receptor on the cell surface is a feature of proliferating cells,[26,104-109] and growth factors such as $PDGF_1$, insulin, IGF, or EGF can increase the appearance of transferrin receptors on the plasma membrane.[64-66,104] Insulin, at levels which bind to the IGF receptor, activates the transmembrane electron transport in HeLa cells. It is possible that this activation is based on increased transferrin receptors on the cell surface. On the other hand, the SV40-transformed 3T3 cells have more than 2 times as many transferrin receptors as measured by transferrin binding on the cell surface than do 3T3 cells. The reduction of ferricyanide is much slower with SV40 3T3 cells, but diferric transferrin can increase iron reduction by the SV40 cells.[40] Since cyanide inhibits transferrin receptor recycling in adipocytes[65] but does not inhibit reduction of transferrin iron, the reduction does not appear to be based on reduction of iron in internal vesicles followed by reexport to the cell surface by exocytosis.

Overall the evidence is consistent with stimulation of growth by electron transport in the plasma membrane. The mechanism by which the redox activity is translated into growth control remains to be established, but there is evidence that it activates second messenger systems such as changes in pH_i, calcium, kinases, or peroxide to communicate with the nucleus.

REFERENCES

1. **Ellem, K. A. O. and Kay, G. F.,** Ferricyanide can replace pyruvate to stimulate growth and attachment of serum restricted melanoma cells, *Biochem. Biophys. Res. Commun.*, 112, 183, 1983.
2. **Kay, G. F. and Ellem, K. A. O.,** Non-heme complexes of FeIII stimulate cell attachment and growth by a mechanism different from that of serum; 2 oxycarboxylates and hemoproteins, *J. Cell. Physiol.*, 126, 275, 1986.
3. **Brooks, M. M.,** Activation of eggs by oxidoreduction indicators, *Science*, 106, 320, 1947.
4. **Craig, T. A.,** Evidence and Characteristics of a Transmembrane Redox System in the Plasma Membrane of *Daucus carota* L., Ph.D. thesis, Purdue University, West Lafayette, IN, 1985, 96.
5. **Barnes, D. and Sato, G.,** Serum-free cell culture: a unifying approach, *Cell*, 22, 649, 1980.
6. **Wu, R. and Sato, G. L.,** Replacement of serum in cell culture by hormones: a study of hormonal regulation of cell growth and specific gene expression, *J. Toxicol. Environ. Health*, 4, 427, 1978.
7. **Kan, M. and Yamane, I.,** Effect of ferrous iron and transferrin on cell proliferation of human fibroblasts in serum free culture, *In Vitro*, 20, 89, 1984.
8. **Octave, J.-H., Schneider, Y.-J., Trouet, A., and Crichton, R. R.,** Iron uptake and utilization by mammalian cells. Cellular uptake of transferrin and iron, *Trends Biochem. Sci.*, 8, 217, 1983.
9. **Ponka, P., Grady, R. W., Wilcznyska, H., and Schulman, H. M.,** Iron chelates supply iron for cell growth, *Biochim. Biophys. Acta*, 802, 477, 1984.
10. **Landschulz, W., Thesloff, I., and Ekbom, P.,** A lipophilic iron chelator can replace transferrin as a stimulator of cell proliferation and differentiation, *J. Cell. Biol.*, 98, 596, 601.
11. **Laskey, J. D., Ponka, P., and Schulman, H. M.,** Control of heme synthesis during friend cell differentiation: role of iron and transferrin, *J. Cell. Physiol.*, 129, 185, 1986.
12. **Morgan, E. H.,** Transferrin biochemistry, physiology, and clinical significance, *Mol. Aspects Med.*, 4, 1, 1981.
13. **Rudland, P. S., Durbin, H., Chingan, D., and Jimenez de Asna, L.,** Iron salts and transferrin are specifically required for cell division of 3T6 cells, *Biochem. Biophys. Res. Commun.*, 75, 556, 1977.
14. **Brock, J. H., Mainou-Fowler, T., and Wester, L. M.,** Evidence that transferrin may function exclusively as an iron donor in promoting lymphocyte proliferation, *Immunology*, 57, 105, 1986.
15. **Basset, P., Zwiller, J., Revel, M. O., and Vindendon, G.,** Growth promotion of transformed cells by iron in cell free culture, *Carcinogenesis*, 6, 355, 1985.
16. **Basset, P., Quesneau, Y., and Zwiller, J.,** Iron induced L1210 cell growth: evidence for a transferrin-independent iron transport, *Cancer Res.*, 46, 1644, 1986.
17. **Sun, I. L., Crane, F. L., Grebing, C., and Löw, H.,** Properties of a transplasma membrane electron transport system in HeLa cells, *J. Bioenerg. Biomemb.*, 16, 514, 1983.
18. **Löw, H., Grebing, C., Lindgren, A., Tally, M., Sun, I. L., and Crane, F. L.,** Involvement of transferrin in the reduction of iron by the transplasma membrane electron transport system, *J. Bioenerg. Biomemb.*, 19, 535, 1987.
19. **Löw, H., Sun, I. L., Navas, P., Grebing, C., Crane, F. L., and Morré, D. J.,** Transplasmalemma electron transport from cells is part of a diferric transferrin reductase system, *Biochem. Biophys. Res. Commun.*, 139, 1117, 1986.
20. **Milo, G. E., Oldham, J. W., and Zimmerman, R.,** Characterization of human cells transformed by chemical and physical carcinogens, *In Vitro*, 17, 719, 1981.
21. **Steinberg, M. L. and Defendi, V.,** Altered pattern of growth and differentiation in human keratinocytes infected by simian virus 40, *Proc. Natl. Acad. Sci. U.S.A.*, 76, 801, 1979.
22. **Sack, G. H., Jr.,** Human cell transformation by simian virus 40: a review, *In Vitro*, 17, 1, 1981.
23. **Neufeld, D. S., Ripley, S., Henderson, A., and Ozer, H. L.,** Immortalization of human fibroblasts transformed by origin defective simian virus 40, *Mol. Cell. Biol.*, 7, 2794, 1987.
24. **Chang, S. E.,** *In vitro* transformation of human epithelial cells, *Biochim. Biophys. Acta*, 823, 161, 1986.
25. **Chou, J. Y.,** Human placental cells transformed by tsA mutants of simian virus 40: a model for study of placental functions, *Proc. Natl. Acad. Sci. U.S.A.*, 75, 1409, 1978.
26. **Reddell, R. R., Hedley, D. W., and Sutherland, R. L.,** Cell cycle effects of iron depletion on T-47D human breast cancer cells, *Exp. Cell. Res.*, 161, 277, 1985.
27. **Sun, I. L., Crane, F. L., Löw, H., and Grebing, C.,** Transplasma membrane redox stimulates HeLa cell growth, *Biochem. Biophys. Res. Commun.*, 125, 649, 1984.
28. **Sun, I. L., Crane, F. L., Grebing, C., and Löw, H.,** Transmembrane redox in control of cell growth: stimulation of HeLa cell growth by ferricyanide and insulin, *Exp. Cell. Res.*, 156, 528, 1985.
29. **Zwiller, J., Basset, P., Ulrich, G., and Mandel, P.,** Guanylate cyclase activators hemin and sodium nitroprusside stimulate cell growth in serum-free medium, *Exp. Cell. Res.*, 141, 445, 1982.
30. **White, A., Handler, P., Smith, E. L., and Stetten, D.,** in *Principles of Biochemistry*, 2nd ed., McGraw-Hill, New York, 1959, 663.

31. **Gayda, D. P., Crane, F. L., Morré, D. J., and Löw, H.,** Hormone effects on NADH oxidizing enzymes of plasma membranes of rat liver, *Proc. Indiana Acad. Sci. U.S.A.*, 86, 385, 1977.
32. **Morré, D. J., Brightman, A., Wang, J., Barr, R., and Crane, F. L.,** Roles for plasma membrane redox systems in cell growth, in *Plasma Membrane Oxidoreductases in Control of Animal and Plant Growth,* Crane, F. L., Morré, D. J., and Löw, H., Eds., Plenum Press, New York, 1988, 45.
33. **Takasu, N., Komatsu, M., Aizawa, T., and Yamada, T.,** H_2O_2 generation in whole rat pancreatic islets. Synergistic regulation by cytoplasmic free calcium and protein kinase C, *Biochem. Biophys. Res. Commun.*, 155, 569, 1988.
34. **Peterson, D. A., Melita, N., Butterfield, J., Husak, M., Christopher, M. M., Jagarlapundi, S., and Eaton, J. W.,** Polyunsaturated fatty acids stimulate superoxide formation in tumor cells: a model for specific cytotoxicity and a model for tumor necrosis factor, *Biochem. Biophys. Res. Commun.*, 155, 1033, 1988.
35. **Halliwell, B., Ed.,** *Oxygen Radicals and Tissue Injury,* Upjohn Co., Kalamazoo, 1988, 148.
36. **MacGillioray, R. T. A., Mendez, E., Shevale, J. G., Sinha, S. K., Lindback-Zins, J., and Brew, K.,** The primary structure of human transferrin, *J. Biol. Chem.*, 258, 3543, 1983.
37. **Harris, D. C., Rinehart, A. L., Hereld, D., Schwartz, R. W., Burke, F. P., and Salvador, A. P.,** Reduction potential of iron in transferrin, *Biochim. Biophys. Acta,* 838, 295, 1985.
38. **Sun, I. L., Navas, P., Crane, F. L., Morré, D. J., and Löw, H.,** NADH diferric transferrin reductase in liver plasma membranes, *J. Biol. Chem.*, 262, 15915, 1987.
39. **Thorstensen, K. and Romslo, I.,** Uptake of iron from transferrin by isolated rat hepatocytes, *J. Biol. Chem.*, 263, 8844, 1988.
40. **Crane, F. L., Löw, H., Sun, I. L., Morré, D. J., and Faulk, W. P.,** Interaction between oxidoreductase transferrin receptor and channels in the plasma membrane, in *Growth Factors — From Genes to Clinical Applications,* Sara, V., Hall, K., and Löw, H., Eds., Raven Press, New York, 1990, 228.
41. **Shimada, T. and Hoshi, T.,** Na^+ dependent elevation of the acidic surface pH (microclimate pH) of rat *jujunal villus* cells induced by cyclic nucleotide and phorbal ester: possible mediators in the regulation of the Na^+/H^+ antiporter, *Biochim. Biophys. Acta,* 937, 328, 1988.
42. **Thorstensen, K., Thommesen, L., and Romslo, I.,** Uptake of iron from ferric iron compounds by isolated hepatocytes, in *Plasma Membrane Oxidoreductases in Control of Animal and Plant Growth,* Crane, F. L., Morré, D. J., and Löw, H., Eds., Plenum Press, New York, 1988, 127.
43. **Guilberg, L. J. and Iscove, N. N.,** Partial replacement of serum by selenite, transferrin, albumin and lecithin in haemopoietic cell cultures, *Nature,* 263, 594, 1976.
44. **McKeehan, W. L., Hamilton, W. G., and Ham, R. G.,** Selenium is an essential trace element for the growth of WI-38 diploid human fibroblasts, *Proc. Natl. Acad. Sci. U.S.A.*, 73, 2023, 1976.
45. **Medina, D. and Oborn, C. J.,** Selenium inhibition of DNA synthesis in mouse mammary epithelial cell line YN-4, *Cancer Res.*, 44, 4361, 1984.
46. **Morré, D. J., Crane, F. L., Sun, I. L., and Navas, P.,** The role of ascorbate in biomembrane bioenergetics, *Ann. N.Y. Acad. Sci.*, 498, 153, 1987.
47. **Goldenberg, H., Grebing, C., and Löw, H.,** NADH monodehydroascorbate reductase in human erythrocyte membranes, *Biochem. Int.*, 6, 1, 1983.
48. **Orringer, E. P. and Roer, M. E. S.,** An ascorbate mediated transmembrane reducing system of the human erythrocyte, *J. Clin. Invest.*, 63, 53, 1979.
49. **Clark, M. G., Partick, E. J., and Crane, F. L.,** Properties and regulation of a transplasma membrane redox system in rat liver, *Biochem. J.*, 204, 795, 1982.
50. **Crane, F. L., MacKellar, W. C., Morré, D. J., Ramasarma, T., Goldenberg, H., Grebing, C., and Löw, H.,** Adriamycin affects plasma membrane redox functions, *Biochem. Biophys. Res. Commun.*, 93, 746, 1980.
51. **Plonka, A., Motediewa, D., Zgireki, A., Hilewiez, M., and Leyko, W.,** Electron spin resonance evidence of superoxide radical dismutation by human ceruloplasmin, *Biochem. Biophys. Res. Commun.*, 95, 978, 1980.
52. **Holmgren, A.,** Thioredoxin and glutaredoxin systems, *J. Biol. Chem.*, 264, 13963, 1989.
53. **Crane, F. L., Löw, H., and Clark, M. G.,** Plasma membrane redox enzymes, in *The Enzymes of Biological Membranes,* Vol. 4, Martonosi, A., Ed., Plenum Press, New York, 1985, 465.
54. **Schallreuter, K. V. and Wood, J. M.,** The role of thioredoxin reductase in the reduction of free radicals at the surface of the epidermis, *Biochem. Biophys. Res. Commun.*, 136, 630, 1986.
55. **Hansson, H. A., Holmgren, A., Rozell, B., and Täljedal, I.-B.,** Immunohistochemical localization of thioredoxin and thioredoxin reductase in mouse exocrine and endocrine pancreas, *Cell Tissue Res.*, 245, 189, 1986.
56. **Holmgren, A.,** Thioredoxin and gluaredoxin: functions in growth-dependent DNA synthesis and cellular regulation vis thiol redox control, in *Plasma Membrane Oxidoreductases in Control of Animal and Plant Growth,* Crane, F. L., Morré, D. J., and Löw, H., Eds., Plenum Press, New York, 1988, 285.
57. **Garcia-Cañero, R.,** A possible role of transplasma membrane ferricyanide reductase in mitogenic activation of rat liver cells, in *Plasma Membrane Oxidoreductases in Control of Animal and Plant Growth,* Crane, F. L., Morré, D. J., and Löw, H., Eds., Plenum Press, New York, 1988, 27.

58. **Sun, I. L., Putnam, J. E., and Crane, F. L.,** Control of cell growth by transplasmalemma redox: stimulation of HeLa cell growth by impermeable oxidants, *Proc. Indiana Acad. Sci.,* 94, 407, 1985.
59. **Crane, F. L., Sun, I. L., Clark, M. G., Grebing, C., and Löw, H.,** Transplasma-membrane redox in growth and development, *Biochim. Biophys. Acta,* 811, 233, 1985.
60. **Nissley, S. P. and Rechler, M. M.,** Insulin-like growth factors: biosynthesis, receptors and carrier proteins, in *Hormonal Proteins and Peptides, Growth Factors,* Vol. 12, Li, C. H., Ed., Academic Press, San Diego, CA, 1984, 127.
61. **Crane, F. L., Löw, H., Sun, I. L., Navas, P., and Morré, D. J.,** Redox control of cell growth, in *Redox Functions of the Eukaryotic Plasma Membrane,* Ramirez, J., Ed., Consejo Superior Investigaciones Cientificas, Madrid, 1988, 3.
62. **Sun, I. L., Garcia-Cañero, R., Liu, W., Toole-Simms, W., Crane, F. L., Morré, D. J., and Löw, H.,** Diferric transferrin reduction stimulates the Na^+/H^+ antiport of HeLa cells, *Biochem. Biophys. Res. Commun.,* 145, 467, 1987.
63. **Seligman, P. A. and Chitambar, C. R.,** Effect of diferric transferrin-metal forms on both cellular iron metabolism and proliferation, in *Proteins of Iron Storage and Transport,* Spik, G., Montreuil, J., Chrichton, R. R., and Maguier, J., Eds., Elsevier, Amsterdam, 1985, 123.
64. **Davis, R. J., Faucher, M., Racamiello, L. K., Carruthers, A., and Czech, M. P.,** Insulin-like growth factor I and epidermal growth factor regulate the expression of transferrin receptors at the cell surface by different mechanisms, *J. Biol. Chem.,* 262, 13126, 1987.
65. **Davis, R. J., Corvera, S., and Czech, M. P.,** Insulin stimulates cellular iron uptake and causes redistribution of intracellular transferrin receptors to the plasma membrane, *J. Biol. Chem.,* 261, 8708, 1986.
66. **Yu, K. T., Khalaf, M., and Czech, M. P.,** Insulin stimulates the tyrosine phosphorylation of an M_r 160,000 glycoprotein in rat adipocyte plasma membranes, *J. Biol. Chem.,* 262, 7865, 1987.
67. **Tanner, L. and Lienhard, G. E.,** Location of transferrin receptors and insulin-like growth factor II receptors in vesicles from 3T3-L1 adipocytes that contain intracellular glucose transporters, *J. Cell. Biol.,* 108, 1537, 1989.
68. **Navas, P., Sun, I. L., Morré, D. J., and Crane, F. L.,** Relationship between NAD^+/NADH levels and animal cell growth, in *Plasma Membrane Oxidoreductases in Control of Animal and Plant Growth,* Crane, F. L., Morré, D. J., and Löw, H., Eds., Plenum Press, New York, 1988, 339.
69. **Navas, P., Sun, I. L., Morré, D. J., and Crane, F. L.,** Decrease of NADH in HeLa cells in the presence of transferrin or ferricyanide, *Biochim. Biophys. Acta,* 135, 110, 1986.
70. **Chatterjee, S., Hirschler, N. V., Petzald, S. J., Berger, S. J., and Berger, N. A.,** Mutant cells defective in Poly (ADP-ribose) synthesis due to stable alterations in enzyme activity or substrate availability, *Exp. Cell. Res.,* 184, 1, 1989.
71. **Pouyssegur, J., Chambard, J. J. C., Franchi, A., L'Allemain, G., Paris, S., and Obberghen-Schilling, E.,** Growth factor activation of the Na/H antiporter controls growth of fibroblasts by regulating intracellular pH, *Cancer Cells,* 3, 409, 1985.
72. **Moolenaar, W. H.,** Effects of growth factors on cytoplasmic pH regulation, *Annu. Rev. Physiol.,* 48, 363, 1986.
73. **Moolenaar, W. H., Mummery, C. L., van der Saag, P. T., and de Laat, S.,** Rapid ionic events and the initiation of growth in serum stimulationed neuroblastoma cells, *Cell,* 23, 789, 1981.
74. **L'Allemain, G., Franchi, A., Cragoe, E., Jr., and Pouyssegur, J.,** Blockade of the Na^+/H^+ antiport abolishes growth factor-induced DNA synthesis in fibroblasts, *J. Biol. Chem.,* 259, 4313, 1984.
75. **Grinstein, S., Rotin, D., and Mason, M. J.,** Na^+/H^+ exchange and growth-factor induced cytosolic pH changes: role in cellular proliferation, *Biochim. Biophys. Acta,* 988, 73, 1989.
76. **Grunicke, H., Maly, K., Oberhuber, H., Doppler, W., Hoflacher, J., Hochleitner, B. W., Jaggi, R., and Groner, B.,** Role of the Na^+/H^+ antiporter in growth stimulation by Ha-ras, in *Plasma Membrane Oxidoreductases in Control of Animal and Plant Growth,* Crane, F. L., Morré, D. J., and Löw, H., Eds., Plenum Press, New York, 1988, 370.
77. **Harguindey, S., Catalón, J., and Anton Aparicio, L. M.,** Possible exploitation of an H^+ related mechanism in the treatment of breast and colon tumors: a new approach to the adjuvant treatment of cancer, in *Plasma Membrane Oxidoreductases in Control of Animal and Plant Growth,* Crane, F. L., Morré, D. J., and Löw, H., Eds., Plenum Press, New York, 1988, 415.
78. **Perona, R. and Serrano, R.,** Increased pH and tumorigenicity of fibroblasts expressing a yeast proton pump, *Nature,* 334, 438, 1988.
79. **Fuhrmann, G. F., Fehlan, R., Schneider, H., and Knauf, P. A.,** The effect of ferricyanide in calcium free solution on passive cation permeability in human red blood cells: comparison with the Gardos effect and with the influence of PCMBS on passive cation permeability, *Biochim. Biophys. Acta,* 983, 179, 1989.
80. **Toole-Simms, W.,** Regulation of Proton Release from HeLa Cells by Ferric Reductase, Ph.D., Purdue University, West Lafayette, IN, 1988, 160.
81. **Sun, I. L., Toole-Simms, W., Crane, F. L., Morré, D. J., Löw, H., and Chou, J. Y.,** Reduction of diferric transferrin by SV40 transformed pineal cells stimulates the Na^+/H^+ antiport, *Biochim. Biophys. Acta,* 938, 17, 1988.

82. **Garcia-Cañero, R., Diaz-Gis, J. J., and Guerra, M. A.,** Is transmembrane ferricyanide reductase connected with the early expression of mitogenic signals in hepatocytes, in *Redox Functions of the Eukaryotic Plasma Membrane,* Ramirez, J., Ed., Consejo Superior de Investigaciones Cientificas, Madrid, 1987, 41.
83. **Ganz, M. B., Boyarsky, G., Sterzel, R. B., and Boron, W. F.,** Arginine vasopressin enhances pH_i regulation in the presence of HCO_3^- by stimulating three acid-base transport systems, *Nature,* 337, 648, 1989.
84. **Low, P. S., Geahlin, R., Mehler, E., and Harrison, M.,** Extracellular control of erythrocyte metabolism mediated by a cytoplasmic tyrosine kinase, *Biomed. Biochim. Acta,* 49, 135, 1990.
85. **Huber, B. E.,** Therapeutic opportunities involving cellular oncogenes; novel approaches fostered by biotechnology, *FASEB J.,* 3, 5, 1989.
86. **Hesketh, T. R., Moore, J. P., Morris, J. D. H., Taylor, M. V., Rogers, J., Smith, G. A., and Metcalf, J. C.,** A common sequence of calcium and pH signals in the mitogenic stimulation of eukaryotic cells, *Nature,* 313, 481, 1985.
87. **Berridge, M. J.,** Inositol trisphosphate and diacylglycerol, two interacting second messengers, *Annu. Rev. Biochem.,* 56, 159, 1987.
88. **Berridge, M. J.,** Inositol trisphosphate and diacylglycerol as second messengers, *Biochem. J.,* 220, 345, 1984.
89. **Christensen, H. N.,** Amino acid transport, in *Redox Functions of the Eukaryotic Plasma Membrane,* Ramirez, F., Ed., Consejo Superior de Investigaciones Cientificas, Madrid, 1987, 117.
90. **Saier, M. H., Daniels, G. A., Boerner, P., and Lin, J.,** Neutral amino acid transport systems in animal cells: potential targets of oncogene action and regulators of cellular growth, *J. Memb. Biol.,* 104, 1, 1988.
91. **Crane, F. L., Löw, H., Sun, I. L., and Isaksson, M.,** The requirement for transmembranal electron flow for cell proliferation, in *Plasma Membrane Oxidoreductases in Control of Animal and Plant Growth,* Crane, F. L., Morré, D. J., and Löw, H., Eds., Plenum Press, 1988, 37.
92. **Warinimman, P., Sun, I. L., and Crane, F. L.,** Evidence for a transplasma membrane electron transport system on intact Ehrlich ascites tumor cells, *Proc. Indiana Acad. Sci.,* 95, 137, 1986.
93. **Rozengurt, E.,** Early signals in the mitogenic response, *Science,* 95, 137, 1986.
94. **Wenner, C. E., Cutry, A., Kinniburgh, A., Tomei, L. D., and Leister, K. J.,** Modulation of redox reactions involved in DNA synthesis by oxygen and artificial electron acceptors, in *Plasma Membrane Oxidoreductases in Control of Animal and Plant Growth,* Crane, F. L., Morré, D. J., and Löw, H., Eds., Plenum Press, New York, 1988, 7.
95. **Barron, E. S. G. and Hoffman, L. A.,** The catalytic effect of dyes on the oxygen consumption of living cells, *J. Gen. Physiol.,* 13, 483, 1929.
96. **Oria, R., Alvarez-Hernandez, X., Liceaga, J., and Brock, J. H.,** Uptake and handling of iron from transferrin, lactoferrin and immune complexes by a macrophage cell line, *Biochem. J.,* 252, 221, 1988.
97. **Jacob, H. S. and Vercellotti, G. M.,** Granulocyte mediated endothelial injury: oxidant damage amplified by lactoferrin and platelet activating factor, in *Oxygen Radicals and Tissue Injury,* Halliwell, B., Ed., Upjohn Co., Kalamazoo, 1988, 57.
98. **Clark, M. G., Partick, E. J., Patten, G. S., Crane, F. L., Löw, H., and Grebing, C.,** Evidence for the extracellular reduction of ferricyanide by rat liver: a transplasma membrane redox system, *Biochem. J.,* 200, 565, 1981.
99. **Bergeron, R. J.,** Iron: a controlling nutrient in proliferative processes, *Trends Biochem. Sci.,* 11, 133, 1986.
100. **Flatmark, T. and Kahn, M. R.,** The release of iron from a subfraction of rat liver highly enriched in endosomal organelles requires both a functional ATPase and NADH, in *Plasma Membrane Oxidoreductases in Control of Animal and Plant Growth,* Crane, F. L., Morré, D. J., and Löw, H., Eds., Plenum Press, New York, 1988, 117.
101. **Holmgren, A.,** Thioredoxin, *Annu. Rev. Biochem.,* 54, 237, 1985.
102. **Thelander, L., Graslund, A., and Thelander, M.,** Continual presence of oxygen and iron required for mammalian ribonucleotide reduction: possible regulation mechanism, *Biochem. Biophys. Res. Commun.,* 110, 859, 1983.
103. **Reichard, P.,** Regulation of deoxyribonucleotide synthesis, *Biochemistry,* 26, 3245, 1987.
104. **Davis, R. J. and Czech, M. P.,** Regulation of transferrin receptor expression at the cell surface by insulin-like growth factors, epidermal growth factor and platelet-derived growth factor, *EMBO J.,* 5, 653, 1986.
105. **Basset, P.,** Modulation of transferrin receptor in human fibroblasts, in *Plasma Membrane Oxidoreductases in Control of Animal and Plant Growth,* Crane, F. L., Morré, D. J., and Löw, H., Eds., Plenum Press, New York, 1988, 163.
106. **Trowbridge, I. S. and Schakelford, D. A.,** Structure and function of transferrin receptors and their relation to cell growth, *Biochem. Soc. Symp.,* 51, 117, 1986.
107. **Larrick, J. W. and Cresswell, P.,** Modulation of cell surface iron transferrin receptors by cell density and state of activation, *J. Supramol. Struct.,* 11, 579, 1979.

108. **Galbraith, G. M. P., Galbraith, R. M., and Faulk, W. P.,** Transferrin binding by human lymphoblastoid cell lines and other transformed cells, *Cell. Immunol.*, 49, 215, 1980.
109. **Hamilton, T. A., Wada, H. G., and Sussman, H. H.,** Identification of transferrin receptors on the surface of human cultured cells, *Proc. Natl. Acad. Sci. U.S.A.*, 76, 6406, 1979.
110. **Sun, I. L., Navas, P., Morré, D. J., and Crane, F. L.,** Modification of pyridine nucleotide pools in HeLa cells grown under growth promoting factors, *Proc. Indiana Acad. Sci.*, 86, 377, 1987.
111. **Toole-Simms, W. and Crane, F. L.,** unpublished studies.

Chapter 8

CONFOUNDING ACTIVITIES OF PYRUVATE AND FERRIC COORDINATION COMPLEXES IN THE ANALYSIS OF THEIR CONTRIBUTIONS TO CELL GROWTH: HYDROPEROXIDE RESISTANCE OR DIRECT MODULATION OF CELL ACTIVITIES?

K. A. O. Ellem and G. F. Kay

TABLE OF CONTENTS

I. INTRODUCTION

The control or regulation of cell replication is now seen to be effected by peptide growth factors interacting with surface receptors situated in the plasma membrane.[1] The outcome of this active regulation is the algebraic sum of the positive influences such as mitogenic peptide concentrations, e.g., transforming growth factor-α (TGFα),[2] and the density of the specific receptors for the mitogen, e.g., epidermal growth factor receptor (EGFR)[2] with the negative ones, e.g., one of the TGFβ family[3] of peptides and its receptor. The quantitative relationships in the balance of these factors[4] remain to be determined, but will vary with cell phenotype since different differentiation lineages express preferred receptors.[5] The expression of positive and negative factors can be induced in the cells themselves to produce a temporally structured series of responses which can exert autocrine growth effects.[6] For example, the replication of melanocytes in skin following ultraviolet exposure has been recently shown to be explicable by the induction of the growth factor TGFα in normal melanocytes by UV radiation.[7] Aberrant function of this system of growth factors and receptors appears to be responsible for neoplastic growth in many instances, TGFα being commonly found to be constitutively expressed as a candidate autocrine stimulant of autonomous growth in melanoma and other cancers of epithelial cells.[7,8] The interactions which occur between some growth factors and the receptors of others (e.g., TGFβ, causing depression of EGFR density[9]) provides an additional component of complexity.

The vast majority of experiments analyzing cell growth phenomena have been performed in cell culture and the use of this pristine but artificial environment, however compelling it may be technically for observing the consequences of environmental manipulation, carries with it inherent biases and artifactual phenomena which need to be recognized and evaluated. While the growth factor/receptor systems may be categorized as the primary determinants of cell multiplication,[10] many other substances (from protein through to inorganic) may influence the rate at which cells multiply in culture. Whether the active substances have any role to play in modulating normal or abnormal cell growth *in vivo* is a separate question of even greater difficulty to answer. This essay will attempt a balanced survey of modulating influences on cell growth *in vitro*, particularly of cells of the melanocyte lineage, which appear relevant to the redox orientation of this volume.

II. THE SYSTEM

It is prudent to examine the way in which the capacity of a compound is tested for its influence on growth of cell cultures, since many of the strategies used introduce phenomena which are unique to the strategy, and may have little relevance to cell growth control, *in vivo*. Until recently the addition of serum to tissue culture medium formulation has been a prerequisite for cell growth. The functions of serum have been inferred from replacement experiments whereby addition of single constituents of serum allows them to be identified for activity in cultures in which growth is limited by reduction of the serum concentration. Cell growth is variously monitored for short term (1 to 5 d) effects by increase in cell number or changes in the rate of ^{3}H-thymidine incorporation into cell DNA. Long-term influences are conveniently measured by cloning efficiency (i.e., the proportion of cells inoculated which form colonies). The latter is more stringent, since the very low cell density eliminates the cross-feeding between cells, which occurs at high cell densities, of nonessential amino acids and other diffusable metabolites. These reach adequate extracellular levels and thereby avoid the leaching of intracellular pools to further limit cell growth in an "artifactual" fashion at the cell densities used in cloning ($<$100 cells/ml).[11] The stringency of low density cultures allows a wider spectrum of factors/nutrients to be recognized, although these may not be relevant to the *in vivo* milieu.

A prime example of the operation of system related effects arose in our own studies of the production of growth-modulating materials by cells themselves, which began as part of an examination of the autocrine hypothesis for autonomy of melanoma cell growth (e.g., References 7, 12). Serum deprivation was used to limit growth of a melanoma cell line, MM96, and medium conditioned by exposure to confluent cultures of a variety of other cell types was used to identify growth-modulating factors produced and released by those cell types. MM96 culture growth proved to be very sensitive to both serum deprivation and to variations in the density at which the cells were seeded for culture,[12] despite the fact that this cell line is a constitutive producer of the growth factor TGFα.[7] Such properties suggested that factors other than the endogenously produced growth peptides were involved as adjuncts for growth, since it is likely that autocrine peptides bind to cell receptors intracellularly, prior to dilution by the extracellular fluids.[13] We were able to select clones of MM96 which were much more dependent on these putative adjuvant substances than the wild type population,[12] and these served as very sensitive test cultures for subsequent developments.

The adjuvant endogenous growth activity was found to be due to a 2-oxocarboxylate — probably just pyruvate,[12] but every soluble 2-oxocarboxylate that was subsequently tested was also found to be active.[14] All were optimally effective within the concentration range 1 to 10 m*M*, increasing both growth rate and cell attachment to the surface of the culture vessel. These data extended the previous findings of the effectiveness of a number of α-ketoacids in promoting the cloning efficiency of human diploid fibroblasts,[15] but differed in that the latter study did not demonstrate activity with 2-oxocarboxylates with a chain length greater than 6. We had observed activity with octanoate which was equivalent to that of the caproate isomers.

III. THE ARGUMENT

The diversity of structure of active α-ketoacids linked only by the requirement of the 2-oxo-function, led us to consider alternatives to the notion that a mechanism based on intracellular depletion of a metabolic intermediate such as pyruvate was involved.[11,16] The strong correlation between the potentiation of cell growth and cell attachment by both 2-oxocarboxylates and serum[14] focused our attention on cell membrane events as mediators of this activity. Also several studies had suggested that shifting the global redox state of a cell to a more oxidizing condition (increasing the NAD^+/NADH ratio) may potentiate the probability of cell multiplication, while others intimated that electron flow from substrate to oxygen rather than the resulting elevated ATP levels was responsible for modulating cell replication (e.g., Reference 17).

This array of ideas was welded into an hypothesis for experimental testing by the stimulating codification of Crane and Low[18] in the late 1970s concerning the recognition of NAD(P) dehydrogenases as integral plasma membrane components, and the proposition that they may be involved in local membrane energetics. We noted that the 2-oxocarboxylates pyruvate and glyoxylate were both excellent substrates for the plasma membrane dehydrogenases.[19] However, the experimental utility of these substrates suffered from their free transport across the cell membrane, which interfered with the determination of the importance of vectorial electron flow across the cell membrane from, e.g., cytoplasmic NADH to an external acceptor by a transmembrane electron conduit. In addition, these 2-oxocarboxylates complicated the interpretation of the whole cell experiments by the changes in intermediary metabolism which manipulation of their supply may have caused.

Another excellent substrate for plasma membrane dehydrogenases was known[18] to be ferricyanide (FeIIICN) which, since the membranes were impermeable thereto, must be an external electron sink, and in addition, could have no metabolic role other than its redox function. When added to the low serum medium RPMI1640 with 2.5% FBS, FeIIICN was found to promote both growth and attachment of MM96 cells over a range of concentrations from 10 to 100 μ*M*,[17] above which it became inhibitory.

The FeIIICN-FeIICN redox pair have previously been found to act as electron sinks for transmembrane electron flow from intra-erythrocytic NADH, to which the synthesis of ATP was coupled.[20] To determine if any constraints were placed on the stimulating effect of FeIIICN on melanoma cell growth by changing the coordination shell geometry, we tested a variety of FeIII complexes. First, all of the simple complexes (ferri-cyanide, -nitrilotri-acetate, -ethylenediaminetetraacetate, -citrate, -diethylenetriaminepentaacetate, as well as nitroprusside and hemin) were effective over a similar concentration range (3 to 30 μ*M*).[21] Thus, all were 3 orders of magnitude more effective than pyruvate, while producing approximately the same degree of stimulation as the 2-oxocarboxylates. Shell geometry or chemistry therefore had little detectable influence on growth stimulation by ferri coordination complexes. The unlikely trivial explanation that the ferri complexes were acting in part by providing iron as a necessary cellular nutrient was eliminated by the finding that $FeCl_3$[17] and FeIII-loaded transferrin[21] were both ineffective stimulants of growth.

Since the ferri-heme nucleus was effective, we tested various proteins with the heme nucleus present and in which the ferric iron was easily reducible. Cytochrome C and hemoglobin from various sources were effective stimulants of both cell growth and attachment. So far, these data were consistent with an extracellular mode of action of complexed ferric iron, acting as an electron sink for transmembrane redox activity. Since the ferri-heme of catalase is remarkably resistant to reduction,[22] its use was thought to be a sound test of the basic hypothesis. Far from being inactive, catalase was found to be the most active single substance in stimulating growth and cell attachment under these conditions.[21] It produced a much larger stimulation of growth than the other compounds, including pyruvate, and was maximally active at 0.05 to 0.1 μ*M*, two orders of magnitude more effective than the other protein-heme or simple ferric coordination complexes. Like all the other compounds used, above its optimal range of activity it was inhibitory (>0.1 μ*M*).

IV. CAVEATS FROM HYDROPEROXIDE AS CONFOUNDER

These findings with catalase raised a different interpretation of the previous data. Clearly, if H_2O_2 were present in the RPMI1640 formulation, then the various adjuvant growth materials could be acting by removing the cellular inhibition caused by H_2O_2 rather than being primary stimulators of cellular activity. Interestingly, RPMI1640 without serum but supplemented with catalase was capable of supporting colonial growth of MM96 (a melanoma cell line which is a constitutive producer of TGFα and thus presumably, independent of an exogenous supply of peptide growth factors). The clonogenicity of the catalase media was improved by the addition of diferric transferrin (0.5 μg/ml) showing that the catalase was not supplying enough nutrient iron for optimal growth.[21]

Direct examination of serum free RPMI1640 medium, used in a routine fashion in laboratories lit by fluorescent light, detected H_2O_2 levels which were dependent on the volume of medium, time of exposure and whether or not the medium was handled in a biohazard hood with regulation fluorescent lighting.[21] For example, medium dissolved and filtered in the dark and handled for assay in an unlit laboratory on the shady side of the building (light flux 0.6 Wm^{-2}) generated 4 μ*M* in 2 h. 10 ml Aliquots in a biohazard hood (light flux 0.8 Wm^{-2}) generated H_2O_2 at a rate of 5 μ*M*/h. Only if the medium was made and used in a dark room illuminated by a red light suitable for handling photographic emulsion was the H_2O_2 concentration below 0.3 μ*M*. MM96 cultures were sensitive, in a concentration-dependent fashion, to levels of H_2O_2 added to 3 μ*M* and above.

The influence of this photodynamic action cannot be overemphasized. Media prepared in the dark, and handled within 1 d from the initial solution through cell seeding in containers covered with foil, in a laboratory and biohazard hood in which no fluorescent lighting (or any other artificial lighting) was used, still showed at least a two-fold stimulation of growth with added catalase or pyruvate.

Photodynamic production of H_2O_2 in culture media has been previously described and shown to be generated by photoexcitation of riboflavin[23] and tryptophan,[24] principally. It is noteworthy that four commonly used tissue culture media have a very high riboflavin content (MEM, 0.27 μM; RPMI1640, 0.53 μM; McCoy's 5a, 0.53 μM; Dulbecco's MEM, 1.1 μM)) and that pyruvate (1 m*M*) has been included in the last of these.[25] It is even more important to note that the media with the highest riboflavin content (2.7 μM) are MB752/1, MAB87/3 and DM145 and that none of these have added pyruvate or other H_2O_2 scavengers.[25]

The evolution of serum-free medium has been gradual in part due to our ignorance of the role of various factors in the growth and survival of cells. The variability of serum composition and its enormous complexity have mitigated against any useful correlations between its growth effects and its content of specific components.[26] A significant early contribution came from the recognition that transformed cell lines were capable of clonogenic growth in serum-free media (containing riboflavin as in MEM) when catalase was added, instead of serum.[27] The growth-potentiating effect of certain reducing agents was also observed, although they tended to be of variable efficiency with different cell lines. The source of the presumed H_2O_2 causing the inhibition was thought to be from endogenous sources such as the cellular flavin enzymes. The utility of pyruvate as a substitute for catalase, which was also able to overcome the toxicity experienced with various thiols like cysteine, was reported but lost from the context of H_2O_2 scavenging. Subsequently, media were mixed with one another as a starting point for a serum-free base and F-12, a formulation with a larger number of components present, has been frequently used as a common constituent. F-12, interestingly, is characterized by a low riboflavin content (10^{-7} *M*) and also has pyruvate added to 1 m*M*,[25] so that a significant part of its contribution may be the scavenging of H_2O_2 photodynamically produced in the other components of the medium mixture. The growing awareness of photodynamically induced peroxide and other toxic products in media with a reduced serum content needs rapid and universal recognition.

The question arises whether the effects of all the adjuvant growth substances could be ascribed to their peroxide scavenging capability. Certainly serum and the heme-containing proteins removed H_2O_2 in a concentration-dependent fashion, and the 2-oxocarboxylates were all very active in its dissolution.[21] It has long been known that pyruvate undergoes extremely rapid oxidative decarboxylation with H_2O_2 and we found that the other 2-oxocarboxylates (but not 4- or 5-oxocarboxylates) reacted similarly, correlating with their bioactivity in melanoma cell stimulation. Recently, it has been proposed that the elaboration and secretion of pyruvate by cells may be a standard means by which cells both *in vitro* and *in vivo* actively protect themselves against oxidative stress mediated through H_2O_2.[28] In that study it was found that a variety of human and mouse cell lines rapidly released pyruvate into medium until the extracellular concentration reached 90 to 120 μM, which is the level found in normal human plasma and serum. As would be anticipated, the rate of accumulation of pyruvate in the medium was dependent on the initial cell density so that the rate of dissolution of H_2O_2 will be expected to be dependent on the seeding density. This may explain the observation that there is a linear correlation between the number of cells initially attaching to the vessel surface and the measured growth rate of the cell culture 4 days later, when cells are seeded at 10^4 cells/ml into RPMI1640 containing different concentrations of pyruvate or other adjuvant growth substances.[14,17,21]

V. INDIRECT ACTION OF FeIII COMPLEXES

The nonheme ferricoordination complexes could be subdivided into two groups according to whether they significantly reduced the content of preformed H_2O_2 in the medium. Thus the FeIII complexes with citrate (FeIIICT), nitrilotriacetate (FeIIINTA), and ethylenediaminetetraacetate (FeIIIEDTA) all halved an initial concentration of 20 μM H_2O_2 within 1 hour, while FeIII-cyanide (FeIIICN), -diethylenetriaminepentaacetate (FeIIIDTPA), and -nitroprusside (FeIIICNO)[21] did not. Based on the duration of the period of reversibility of the inhibitory effects of fresh, low serum RPMI1640 by subsequent addition of pyruvate, the critical toxic effect occurs within the first 2 to 4 h.[29] Thus, agents achieving an adjuvant growth effect by peroxide scavenging, must remove the peroxide to levels below those which are toxic within this time frame. It seems reasonable, therefore, to attribute the major part of the adjuvant growth effect of FeIIICT, FeIIINTA, and FeIIIEDTA to H_2O_2 dissolution. The coordination structure of these three compounds allows access to aqua sites (free coordination sites), which are necessary for FeIII reactivity in catalysis,[30] in the generation of hydroxyl radicals from the H_2O_2 and superoxide,[31] and in a variety of catalytic oxidants[32-34] which consume H_2O_2.

Concerning mechanisms of H_2O_2 eradication, we found that both hemin and FeIIINTA (but not FeIIICN nor FeIIICNO) catalyzed the oxidation of streptomycin sulfate by H_2O_2 to a product with an absorbance peak at 325 nm.[29] Therefore, a peroxidatic function may supplement catalatic reactions in removing H_2O_2 by FeIIICT, FeIIINTA, and FeIIIEDTA. Peroxidatic activity can be demonstrated even with catalase.[35] Modification of other medium constituents could take place by peroxidatic reactions catalyzed by the FeIII complexes, since the medium evidently has many suitable substrates for horseradish peroxidase (HRP) and H_2O_2. Thus, for complete oxidation of one mole of scopoletin by HRP, 4 moles of H_2O_2 were required in complete medium, whereas the reaction was equimolar when conducted in PBS.[21] Again the caveats: if a peroxidatic mechanism operated for some of the FeIII complexes, then it must have quite different specificity to that of HRP and lactoperoxidase, since the latter proved extremely toxic to the MM96 cultures at 0.1 U/ml and 1 U/ml, respectively. Both peroxidases had a small but significant stimulatory effect at 1/10 to 1/100th of the inhibitory concentration.[21] FeIIINTA has been shown to have peroxidatic and catalatic functions as well as, interestingly, pyrocatechase activity.[32]

Now destruction of streptomycin will not contribute to the growth state of the cells in a culture except by H_2O_2 removal, unless the oxidation products are toxic. However, the ferri coordination complexes may also react with critical nutrients in the medium — which may in a particular medium formulation, be above optimal for a particular cell type[25,36] and, by reducing their concentration, achieve an effective increase in growth rate. To demonstrate the variations in composition in these complex media would be a difficult and not necessarily rewarding study, but this mechanism remains a possibility. Of possible relevance is the observation that simply diluting RPMI1640 1:1 with PBS induces a very significant impetus to MM96 culture growth rate in low serum.[12] Such an effect, however, could also be attributed to reduction in the concentration of preformed H_2O_2 and diminution of the rate of its further photodynamic production.

Thiol oxidation by FeIII complexes[37] could also potentiate growth by preventing the production of H_2O_2 which occurs during "spontaneous" (Cu^{2+} catalyzed) cysteine autoxidation in a fashion similar to the action of the red kidney bean peptide which prevents H_2O_2 production by chelating the traces of Cu^{2+} found in media.[38] Modulation of growth by hemin may come about by altering the amount or nature of extracellular collagen in cultures grown in media containing ascorbic acid, since vitamin C is oxidized by H_2O_2 in the presence of hemin.[39] The growth responses of cells can be significantly altered by the chemistry of the extracellular matrix to which they attach.[40]

VI. DIRECT ACTION OF SATURATED FeIII COMPLEXES

The stimulatory effects on growth and attachment of FeIIICN, FeIIICNO, and FeIIIDTPA do not appear to be attributable to H_2O_2 scavenging, since they will not eradicate preformed peroxide. However, in the context of ongoing generation of H_2O_2 by the photodynamic activation of medium by laboratory lighting, it is relevant that both FeIIICN and FeIIICNO markedly inhibited the photogenesis of H_2O_2,[21] although FeIIIDTPA had little influence on this process. This may be due to the fact that both FeIIICN and FeIIICNO are themselves photoactivable,[30] with the generation of aqua sites for interaction with either H_2O_2 or the photodynamically activated molecules of riboflavin involved in the production of H_2O_2. These structures may facilitate interactions with the excited state of the riboflavin which allow peroxide-free reduction of the flavin to take place (e.g., Reference 41).

FeIIIDTPA seems to occupy a unique position in the series of ferri coordination complexes used. It was chosen to be representative of the stable complexes with completely filled coordination shells which have no available free aqua coordination sites for reactivity with H_2O_2.[31] That FeIIIDTPA did not affect the oxidant stress imposed by H_2O_2 on MM96 cultures can be seen in Figure 1. Ubiquitin is induced in response to many forms of cellular stress,[42] presumably by activation of gene transcription, so that this may thus be used as a signal of cell stress. We found that the addition of 30 μM H_2O_2 to RPMI1640 in low serum induced a marked increase in ubiquitin mRNA (lane 4). Both 10% FBS and 1 m*M* pyruvate reduced the response considerably (lanes 2 and 6, respectively). However, no discernible suppression of ubiquitin mRNA accumulation accompanied the action of FeIIICN or FeIIIDTPA, which thus did not protect the cells from H_2O_2 activation of the sensing systems involved in the increased transcription/decreased degradation of ubiquitin mRNA which may be responsible for the increased levels of the message. This preliminary study appears to suggest that these two ferri coordination complexes protect (or stimulate) cells from H_2O_2 mediated oxidative stress by a direct effect on the cell, rather than by preventing H_2O_2 from exciting some initial damage — enough to evoke at least part of the stress response. Mediation of the stress response does not appear to be based on the alarmone messenger system using diadenosine tetraphosphate[43] which operates in bacteria.

VII. FURTHER COMPLICATING PHENOMENA

Other phenomena add to the complexity of interpreting the results of experiments designed to examine the growth-modulating effects of redox agents on cells as a general phenomenon:

1. Cell imprinting as a manifestation of physiological heterogeneity,
2. Lineage-specific responses possibly related to ontogenetic heterogeneity, and
3. Differences between the effects of different H_2O_2 scavengers.

1. Cell imprinting — It was mentioned above that the period during which cells seeded into low serum medium could be resuscitated to enhanced attachment and growth by the subsequent addition of pyruvate was short (2 to 4 h). This was based on experiments in which the cells used were grown in stock culture in low serum (1% fetal bovine serum [FBS]) medium supplemented with 1 m*M* pyruvate. If, however, the stock cultures were maintained for 3 d prior to the experiment in medium containing high serum (10% FBS) and no added pyruvate, the response was altered. Instead of virtual unresponsiveness to serum stimulation 15 h after seeding, "serum imprinted" cells were still significantly responsive to "serum rescue" after 40 to 50 h.[29] This imprinting phenomenon has been observed not only with cell cultures selected for their sensitivity to rescue by pyruvate,[12]

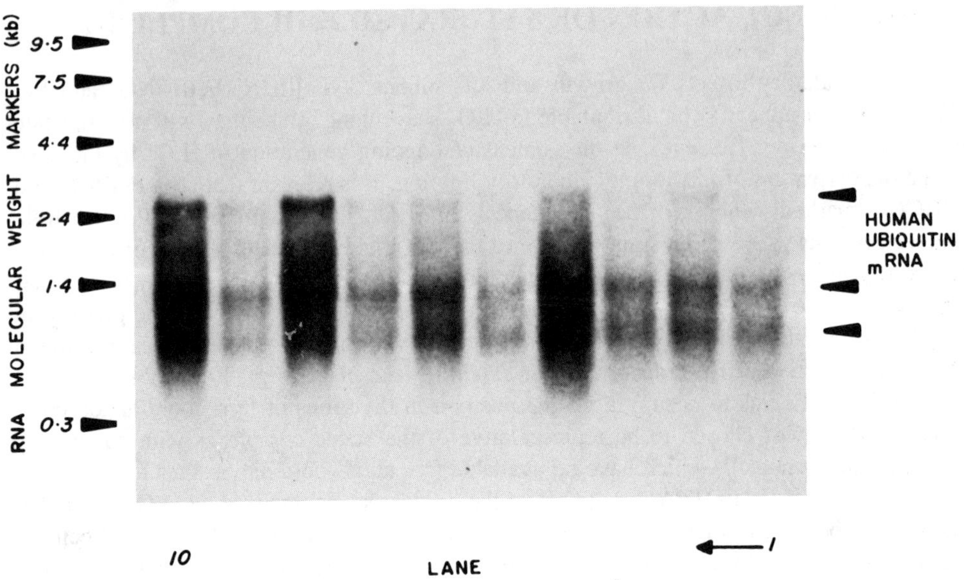

FIGURE 1. Lack of suppression of accumulation of ubiquitin mRNA as a stress signal by FeIIICN and FeIIIDTPA. The medium on replicate cultures of MM96B cells was changed to the formulations listed below (lanes 1 to 10). After 4 h incubation at 37°C, the cells were harvested and the RNA extracted by the quanidine isothiscyanate/CsCl method. In each lane, 20 μg of total cell RNA was electrophoresed in denaturing agarose gels for 20 h, then transferred to Amersham Nylon membranes over at least 18 h. The ubiquitin cDNA probe, cloned into the XhoI site of the plasmid pBR327 and designated pHUB14-38[61] was kindly supplied by Dr. J. Vuust, University of Aarhus, Denmark. The probe was labeled with ^{32}P using an Amersham Nick Translation Kit, and then hybridized to the transfer membranes. These Northern blots were then reprobed with a genomic clone of part of the gene encoding human 28S ribosomal RNA as a check on the equivalence of loading of the RNA in each sample. The lanes, numbered from the right were loaded with RNA from cultures treated as follows:

lane 1	10% FBS/RPMI 1640
lane 2	10% FBS/RPMI 1640
lane 3	0.5% FBS/RPMI 1640 1640
lane 4	0.5% FBS/RPMI 1640 + 30 μ*M* H_2O_2
lane 5	0.5% FBS/RPMI 1640 + 1 m*M* pyruvate
lane 6	0.5% FBS/RPMI 1640 + 1 m*M* pyruvate + 30 μ*M* H_2O_2
lane 7	0.5% FBS/RPMI 1640 + 10 μ*M* FeIII CN
lane 8	0.5% FBS/RPMI 1640 + 10 μ*M* FeIII CN + 30 μ*M* H_2O_2
lane 9	0.5% FBS/RPMI 1640 + 10 μ*M* FeIII DTPA
lane 10	0.5% FBS/RPMI 1640 + 10 μ*M* FeIII DTPA + 30 μ*M* H_2O_2

but also with unselected wild-type MM96 cultures. It is clearly the phenotypic changes brought about by growth in high or low serum which modify the sensitivity of the cells to subsequent oxidative stress. The basis for the phenotypic changes responsible for the imprinting is an open question.

2. Lineage-specific responses — Are the differences between cell lines expressed in their responsiveness to the various adjuvant growth substances? Of seven melanoma cell lines derived from separate human biopsy material, one failed to show a significant growth response to the addition of catalase to low serum medium, in spite of an improvement in the attachment of the seeded cells. Three of the melanoma lines had no significant potentiation of attachment attributable to catalase. Growth of the unrelated ''normal'' rat kidney line, NRK, was doubled by catalase but its attachment was slightly depressed. Without further study there is no obvious explanation for these variations. The important message from these data,[21] however, is that for studies of this nature, where so many variables relevant to cell behavior are concerned, generalization from one cell line to another is hazardous. For example, while hemin proved to be an effective stimulant to both growth and attachment of the human melanoma line MM96[21] as well as rodent glioma and neuroblastoma cell lines,[44] it enhanced the differentiation of mouse 3T3 cells to adipocytes[45] and stimulated neurite outgrowth from mouse neuroblastoma cells.[46] Thus, in both the latter cases hemin favored a differentiation rather than a growth outcome.

3. Difference between adjuvants — A further complexity arises when the stimulatory effects of pyruvate and serum are compared (see Figure 6, Reference 14). Although both serum and pyruvate quickly eliminate H_2O_2, serum has a greater stimulatory capacity which is maximal with 10% FBS. Dialysis of serum removes part of its effectiveness (even from 10% FBS). Pyruvate at 0.3 to 1.0 m*M* can replace this dialysis loss in a concentration-dependent fashion, so that this component of serum action may well be due to its 2-oxocarboxylate content. At concentrations of undialyzed serum below 5%, addition of pyruvate increases both the growth response and the attachment of cells. However, cell attachment is maximized at serum concentrations of 4 to 5% FBS (whether dialyzed or not) but the growth of the cultures can still be significantly increased by pyruvate. Thus the pyruvate effect on growth can be independent of increasing cell attachment.

Pyruvate can maximize attachment at low (0.5%) serum concentrations, but cannot thereby increase the growth parameter to more than 20% of that achievable with serum. Thus, serum and pyruvate have a similar range of stimulatory capacity on cell attachment, but serum has extra capacity to boost growth. This, however, may be due only to its content of other substances essential for growth (e.g., insulin, diferric transferrin, etc.) which are limiting at low serum concentrations, but adequate at about 5% FBS.

VIII. IMPLICATIONS FOR OTHER STUDIES

We have previously identified a number of examples in the recent literature in which growth effects of various transition metal complexes have been described.[21] It is probable that several of these are examples of the elimination of H_2O_2 effects by the metal complexes, rather than being phenomena attributable to some unknown redox reaction at the cell surface or uptake of required trace metals chelated by physiological substances.[47] Similarly, reports of the growth stimulation of various cultured cells by the addition of pyruvate,[48] particularly when using high riboflavin medium such as Waymouth's MAB 87/3,[49] need close examination of the additive for its possible role as a scavenger of H_2O_2, as has been emphasized by others.[28] Again production of H_2O_2 by activated macrophages has been found to be inhibited by media conditioned by many cell types — undoubtedly the result of H_2O_2 scavenging by pyruvate released into the conditioned medium by the inhibitory cells.[50]

Caution in generalizing from one cell system to another has already been justified. There is reason to suspect that melanoma cells may be exceptionally sensitive to oxidative stress. Such sensitivity may mask redox-modulating growth effects mediated by adjuvant stimulants via the plasma membrane. It must be recalled that melanizing cells convert tyrosine to melanin by the polyphenol oxidase tyrosinase. Intermediate compounds include the highly redox-reactive phenol quinoids 3,4-dihydroxyphenylalanine (dopa), 5,6-dihydroxyindole, and their quinones. Exploitation of the unique biochemistry of the melanogenic pathway has been a goal of chemotherapy in melanoma because of the dispensibility of normal epidermal melanocytes (e.g., Reference 51). Extensive studies of the sensitivity of melanoma cell lines to growth inhibitory effects of a series of redox-active catechols found that MM96 cultures, for example, were from 3.6 to 7.7 times more sensitive to noradrenaline, dopa, 3,4-dihydroxybenzylamine (DHBA), dopamine, and 3,4-dihydroxyphenylacetate than HeLa cells. This difference could not be attributed to any deficiencies in intracellular glutathione levels[52] or to any measurable differences in the levels of oxygen-metabolizing enzymes,

DNA repair efficiency, or, in the case of dopa, to catechol uptake.[53] Also, among six different melanoma cell lines there was no correlation between their tyrosinase levels, or their capacity to synthesize melanin, and their sensitivity to these agents. The toxic effects of the catechols were mediated by active oxygen species generated extracellularly, probably by autooxidative reactions, since the cells were protected by catalase, superoxide dismutase, or peroxidase when these were added to the medium. Also the toxicity was potentiated by the inhibitors of these enzymes (3-amino,1,2,4-triazole, diethyldithiocarbamate, and D- or L-penicillamine[53]). Since high cell density was also protective, endogenously produced pyruvate may also be effective by eliminating the H_2O_2 generated by catechol autooxidation.

Curiously when HeLa cells and MM96 cells were tested in a survival/growth assay by seeding into medium containing H_2O_2, they were found to be equally sensitive (90% inhibition by 10 to 12 μM).[53] However, MM96 cultures were 3 times more sensitive than HeLa cells to H_2O_2 generated over a significant length of time by the addition of glucose oxidase to the medium, and were 6 times more sensitive to the combination of superoxide and H_2O_2 produced by the addition of xanthine oxidase.[54] A human retinoblastoma line, of neural crest origin like melanomas, although lacking the melanin-related oxidative hazards, proved to be as sensitive as MM96 to dopa and DHBA. As with melanoma cells and HeLa cells exposed to the higher doses necessary for their inhibition, the effects of the catechols on retinoblastoma cells were shown to be cell cycle specific, causing a block at the G_1-S interface.[54,55]

Further data which highlight the peculiar sensitivity of many melanoma cell lines to agents with direct or indirect redox effects were found during studies involving manipulation of intracellular glutathione levels. Buthionine sulfoximine, an inhibitor of γ-glutamyl cysteine synthetase, the rate-limiting enzyme in glutathione synthesis, although only producing a modest 20% fall in the glutathione pool of both MM96 and HeLa cells, proved to be a very selective inhibitor of melanoma cells (e.g., MM96 were 27 times more sensitive than HeLa cells[52]). Again the cationic dye rhodamine 123 which is concentrated by, and inhibitory to oxidative phosphorylation in, mitochondria, also proved to be 6 to 7 times more toxic to MM96 than HeLa cells.[54] The conclusion seems inescapable that, as a class, melanoma cell lines (and perhaps other neural crest derivatives) show marked sensitivity to diverse agents with known effects on different redox systems. The common relevant property of these agents seems to be the raising of the level of active oxygen species. Demonstration of growth stimulation by putative electron sinks for plasma membrane dehydrogenases appears to be confounded in these cultures by virtue of this unique sensitivity to hydrogen peroxide, etc.

IX. HYDROPEROXIDE AND TUMOR PROGRESSION

The data discussed so far bear on two aspects of the redox/cell growth question: (1) the presence and influence of H_2O_2 in many *in vitro* systems — often as an unsuspected variable with considerable biological effects, and (2) the activity of certain redox compounds (in particular those with a saturated coordination shell like FeIIIDTPA) in stimulating cell growth in a fashion not attributable to H_2O_2 scavenging. These broad conclusions have further biological ramifications.

The occurrence of hydrogen peroxide is universal in oxidative organisms as a byproduct of aerobic metabolism[56] as well as an effector of defensive responses to disturbances of organismal integrity by the respiratory burst of phagocytic cells.[57] Hydroperoxide is thus not only an artifact of high riboflavin media. Pyruvate may be the universal sink for this ubiquitously generated hydroperoxide,[28,58] allowing only very localized build up of peroxide concentration of an evanescent nature. However, hydroperoxide may be an important selective variable in the evolution of tumor metastases, the ultimate step in tumor progression. Several reports have noted an increased resistance to H_2O_2 lysis by cells derived from metastatic deposits or which have undergone extensive selection in tissue culture. The data show a 10- to 100-fold greater sensitivity of various normal cell cultures to H_2O_2 when compared to the sensitivity of various tumor cell lines.[59] Highly metastatic variant cells selected from lung deposits following injection of a poorly metastasizing parental line of spontaneously transformed hamster embryo cells were found to be 10 to 200 times more resistant to H_2O_2 lysis[60] than the parental line. Similarly cell lines derived from tumors growing in subcutaneous injection sites showed enhanced resistance to H_2O_2 lysis. Both results suggest that rare, mutant cells with enhanced resistance survived the H_2O_2-based armaments of the surveillance cells to constitute new H_2O_2-resistant populations highly adapted to *in vivo* growth.[60]

The fascinating finding of relevance to our own studies emerged when the mechanisms of resistance to H_2O_2 cytolysis were sought. Attempts to correlate relative resistance to H_2O_2 of a number of human tumor cell lines with specific activity of their catalase or with the activity of the glutathione redox cycle proved fruitless.[59] Quantitation of these systems in cell lysates predicted neither the rate of H_2O_2 consumption by intact cells, nor their resistance to H_2O_2 lysis. Virtually complete inhibition of the removal of H_2O_2 by the cells with azide did not sensitize the cells to any but millimolar quantities of H_2O_2.[59] A prominent role for the glutathione cycle as a factor in H_2O_2 resistance could be established in only one of the human tumors examined. The authors concluded that, unless certain nonenzymatic antioxidants such as ascorbate or tocopherol were involved, even with catalase and the glutathione redox cycle suppressed, human tumor cells exhibited an inherent resistance to oxidative lysis independent of detectable degradation of H_2O_2. We have not found any of the lipid-soluble antioxidants to be effective.[21] Thus, the parallel with the action of FeIIIDTPA in conferring increased resistance to H_2O_2 in culture medium without scavenging it is obvious. However, whether both phenomena are related through a final common pathway remains to be demonstrated.

REFERENCES

1. **James, R. and Bradshaw, R. A.,** Polypeptide growth factors, *Annu. Rev. Biochem.*, 53, 259, 1984.
2. **Derynck, R.,** Transforming growth factor-alpha: Structure and biological activities, *J. Cell. Biochem.*, 32, 293, 1986.
3. **ten Dijke, P., Hansen, P., Iwata, K. K., Pieler, C., and Foulkes, J. G.,** Identification of another member of the transforming growth factor type β gene family, *Proc. Natl. Acad. Sci. U.S.A.*, 85, 4715, 1988.
4. **Marx, J. L.,** Cell growth control takes balance, *Science*, 239, 975, 1988.
5. **Real, F. X., Retting, W. J., Chesa, P. G., Melamed, M. R., Old, L. J., and Mendelsohn, J.,** Expression of epidermal growth factor receptor in human cultured cells and tissues: relationship to cell lineage and stage of differentiation, *Cancer Res.*, 46, 4726, 1986.
6. **Coffey, R. J., Jr., Derynck, R., Wilcox, J. N., Bringman, T. S., Goustin, A. S., Moses, H. L., and Pittelkow, M. R.,** Production and auto-induction of transforming growth factor-α in human keratinocytes, *Nature*, 328, 817, 1987.
7. **Ellem, K. A. O., Cullinan, M., Baumann, K. C., and Dunstan, A.,** UVR induction of TGFα: a possible autocrine mechanism for the epidermal melanocytic response and for promotion of epidermal carcinogenesis, *Carcinogenesis*, 9, 797, 1988.
8. **Smith, J. J., Derynck, R., and Korc, M.,** Production of transforming growth factor α in human pancreatic cancer cells: evidence for a superagonist autocrine cycle, *Proc. Natl. Acad. Sci. U.S.A.*, 84, 7567, 1987.
9. **Assoian, R. K., Frolik, C. A., Roberts, A. B., Miller, D. M., and Sporn, M. B.,** Transforming growth factor-β controls receptor levels for epidermal growth factor in NRK fibroblasts, *Cell*, 36, 35, 1984.
10. **Pledger, W. J., Stiles, C. D., Antoniades, H. N., and Scher, C. S.,** Induction of DNA synthesis in BALB/c-3T3 cells by serum components: reevaluation of the commitment process, *Proc. Natl. Acad. Sci. U.S.A.*, 74, 4481, 1977.

11. **Eagle, H. and Piez, K.,** The population dependent requirement by cultured mammalian cells for metabolites which they can synthesize, *J. Exp. Med.,* 116, 29, 1962.
12. **Ellem, K. A. O. and Kay, G. F.,** The nature of conditioning nutrients for human malignant melanoma cultures, *J. Cell. Sci.,* 62, 249, 1983.
13. **Keating, M. T. and Williams, L. T.,** Autocrine stimulation of intracellular PDGF receptors in v-sis-transformed cells, *Science,* 239, 914, 1988.
14. **Kay, G. F. and Ellem, K. A. O.,** Use of growth attachment correlation plots to analyze cell dependency on 2-oxocarboxylates and serum, *Cell Tissue Kinet.,* 18, 355, 1985.
15. **McKeehan, W. L. and McKeehan, K. A.,** Oxocarboxylic acids, pyridine nucleotide-linked oxidoreductases and serum factors in regulation of cell proliferation, *J. Cell. Physiol.,* 101, 9, 1979.
16. **Eagle, H.,** Metabolic controls in cultured mammalian cells, *Science,* 148, 42, 1965.
17. **Ellem, K. A. O. and Kay, G. F.,** Ferricyanide can replace pyruvate to stimulate growth and attachment of serum restricted human melanoma cells, *Biochem. Biophys. Res. Commun.,* 112, 183, 1983.
18. **Crane, F. L., Goldenburg, H., Morré, D. J., and Löw, H.,** Dehydrogenases of the plasma membrane, *Subcell. Biochem.,* 6, 345, 1979.
19. **Crane, F. L. and Löw, H.,** NADH oxidation in liver and fat cell plasma membranes, *FEBS Lett.,* 68, 153, 1976.
20. **Mishra, R. K. and Passow, H.,** Induction of intracellular ATP synthesis by extracellular ferricyanide in human red blood cells, *J. Membr. Biol.,* 1, 214, 1969.
21. **Kay, G. F. and Ellem, K. A. O.,** Non haemcomplexes of FeIII stimulate cell attachment and growth by a mechanism different from that of serum, 2-oxocarboxylates and haemproteins, *J. Cell. Physiol.,* 126, 275, 1986.
22. **Deisseroth, A. and Dounce, A. L.,** Catalase: physical and chemical properties, mechanism of catalysis and physiological role, *Physiol. Rev.,* 50, 319, 1970.
23. **Wang, R. J. and Nixon, B. T.,** Identification of hydrogen peroxide as a photoproduct toxic to human cells in tissue-culture medium irradiated with "Daylight" fluorescent light, *In Vitro,* 14, 715, 1978.
24. **McCormick, J. P., Fisher, J. R., Pachlatko, J. P., and Eisenstark, A.,** Characterization of a cell-lethal product from the photooxidation of tryptophan: hydrogen peroxide, *Science,* 191, 468, 1976.
25. **Ham, R. G. and McKeehan, W. L.,** in *Methods in Enzymology,* Vol. 43, Jacoby, W. B. and Pastan, I. H., Eds., Academic Press, New York, 44, 1979.
26. **Price, P. J. and Gregory, E. A.,** Relationship between *in vitro* growth promotion and biophysical and biochemical properties of the serum supplement, *In Vitro,* 18, 576, 1982.
27. **Lieberman, I. and Ove, P.,** Catalase requirement for mammalian cells in culture, *J. Exp. Med.,* 198, 631, 1958.
28. **O'Donnell-Tormy, J., Nathan, G. F., Lanks, K., DeBoer, C. J., and de la Harpe, J.,** Secretion of pyruvate. An antioxidant defense of mammalian cells, *J. Exp. Med.,* 165, 500, 1987.
29. **Ellem, K. A. O., Kay, G. F., Dunstan, A. J., and Stenzel, D. J.,** Redox agents which modulate the growth of melanoma cells: evaluation of mechanisms, in *Plasma Membrane Oxidoreductases in Control of Animal and Plant Growth,* Crane, F. L. and Löw, H., Eds., Plenum Press, 1988, 17.
30. **Baxendale, J. H.,** Decomposition of hydrogen peroxide by catalysts in homogenous aqueous solution, *Adv. Catalysis,* 4, 31, 1952.
31. **Graf, E., Mahoney, J. R., Bryant, R. G., and Eaton, J. W.,** Iron catalyzed hydroxylradical formation: stringent requirement for free iron coordination site, *J. Biol. Chem.,* 259, 3120, 1984.
32. **Weller, M. G.,** Ferric nitrilotriacetate: an active centre analogue of pyrocatechase, in *The Coordination Chemistry of Metalloenzymes,* Bertini, I., Drago, R. S., and Luchinat, C., Eds., D. Reidel Publishing, Dordrecht, Holland, 1983, 273.
33. **Pispisa, B.,** Stereoselective peroxidatic activity of iron (III) complex ions supported on polypeptides, in *The Coordination Chemistry of Metalloenzymes,* Bertini, I., Drago, R. S., and Luchinat, C., Eds., D. Reidel Publishing, Dordrecht, Holland, 1983, 273.
34. **Rush, J. D. and Koppenol, W. H.,** Oxidizing intermediates in the reaction of ferrous EDTA with hydrogen peroxide: reactions with organic molecules and ferricytochrome c, *J. Biol. Chem.,* 261, 6730, 1986.
35. **Oshino, N., Oshino, R., and Chance, B.,** The characteristics of the peroxidatic reaction of catalase in ethanol oxidation, *Biochem. J.,* 131, 555, 1973.
36. **Bottger, W. J. and McKeehan, W. L.,** Mechanisms of cellular nutrition, *Physiol. Rev.,* 66, 1, 1986.
37. **Kidby, D. K.,** Direct spectrophotometric estimation of ferrocyanide and its possible uses in sulfhydril oxidation states, *Anal. Biochem.,* 28, 230, 1969.
38. **Fedorcsak, I., Harms-Ringdahl, M., and Ehrenberg, L.,** Prevention of sulphydryl autoxidation by a polypeptide from red kidney beans, described to be a stimulator of RNA synthesis, *Exp. Cell. Res.,* 108, 331, 1977.
39. **Kremer, M. L.,** Haemin-catalyzed oxidation of ascorbic acid by H_2O_2, *Trans. Faraday Soc.,* 63, 1208, 1967.

40. **Gospodarowicz, D., Greenberg, G., and Birdwell, C. R.,** Determination of cellular shape by the extracellular matrix and its correlation with the control of cellular growth, *Cancer Res.,* 38, 4155, 1978.
41. **Naman, S. A.,** Energy conversion in the decay of triplet lumiflavin in the presence of ferri- and ferrocyanide, *Photochem. Photobiol.,* 47, 43, 1988.
42. **Finley, D. and Varshavsky, A.,** The ubiquitin system: functions and mechanisms, *Trends Biochem. Sci.,* 10, 343, 1985.
43. **Segal, E. and Le Pecq, J.-B.,** Relationship between cellular diadenosine 5,5 - P1,P4 - tetraphosphate level, cell density, cell growth stimulation and toxic stresses, *Exp. Cell. Res.,* 167, 119, 1986.
44. **Zwiller, J., Basset, P., Ulrich, G., and Mandel, P.,** Guanylate cyclase activators hemin and nitroprusside stimulate cell growth in serum-free medium, *Exp. Cell. Res.,* 141, 445, 1982.
45. **Chen, J. J. and London, I. M.,** Hemin enhances the differentiation of mouse 3T3 cells to adipocytes, *Cell,* 26, 117, 1981.
46. **Ishii, D. N. and Maniatis, G. M.,** Haemin promotes rapid neurite outgrowth in cultured mouse neuroblastoma cells, *Nature,* 274, 372, 1978.
47. **Pickart, L. and Thaler, M. M.,** Growth-modulating tripeptide (Glycylhistidyllysine): association with copper and iron in plasma and stimulation of adhesiveness and growth of hepatoma cells in culture by tripeptide-metal ion complexes, *J. Cell. Physiol.,* 102, 129, 1980.
48. **Pickart, L. S. and Thaler, M. M.,** Effect of pyruvate on hepatoma cell contour and macromolecular synthesis, *J. Supramol. Struct.,* 5, 591, 1976.
49. **McGowan, J. A. and Bucher, N. L. R.,** Pyruvate promotion of DNA synthesis in serum-free primary cultures of adult rat hepatocytes, *In Vitro,* 19, 159, 1983.
50. **Nelson, M., Booth, M. L., and Nelson, D. S.,** Effects of tumour cell culture supernatants on some biochemical activities of macrophages, *Aust. J. Exp. Biol. Med. Sci.,* 60, 493, 1982.
51. **Wick, M. M.,** The chemotherapy of malignant melanoma, *J. Invest. Dermatol.,* 80, 61s, 1983.
52. **Kable, E. P. W., Favier, D., and Parsons, P. G.,** Sensitivity of human melanoma cells to L-dopa and DL-buthionine (S,R) sulfoximine, *Cancer Res.,* 49, 2327, 1989.
53. **Parsons, P. G.,** Modification of dopa toxicity in human tumour cells, *Biochem. Pharmacol.,* 34, 1801, 1985.
54. **Kable, E. P. W. and Parsons, P. G.,** Potency, selectivity and cell cycle dependence of catechols in human tumour cells *in vitro, Biochem. Pharmacol.,* 37, 1711, 1988.
55. **Kable, E. P. W. and Parsons, P. G.,** Melanin synthesis and the action of L-dopa and 3,4 dihydroxybenzylamine in human melanoma cells, *Cancer Chemother. Pharmacol.,* 23, 1, 1989.
56. **Chance, B., Sies, H., and Boveris, A.,** Hydroperoxide metabolism in mammalian organs, *Physiol. Rev.,* 59, 527, 1979.
57. **Badwey, J. A. and Karnovsky, M. L.,** Active oxygen species and the functions of phagocytic leucocytes, *Annu. Rev. Biochem.,* 49, 695, 1980.
58. **Andrae, U., Singh, J., and Ziegler-Skylakakis, K.,** Pyruvate and related α-ketoacids protect mammalian cells in culture against hydrogen peroxide-induced cytotoxicity, *Toxicol. Lett.,* 28, 93, 1985.
59. **O'Donnell-Tormey, J., DeBoer, C. J., and Nathan, C. F.,** Resistance of human tumor cells *in vitro* to oxidative cytolysis, *J. Clin. Invest.,* 76, 80, 1985.
60. **Deichman, G. I. and Vendrov, E. L.,** Characteristics of *in vitro* transformed cells essential for their *in vivo* survival, selection and metastasizing activity, *Int. J. Cancer,* 37, 401, 1986.
61. **Wiborg et al.,** The human ubiquitin multigene family: some genes contain multiple directly repeated ubiquitin coding sequences, *EMBO J.,* 4, 755, 1985.

Chapter 9

IRON SUPPLY AND CELL GROWTH CONTROL

Paul Basset

TABLE OF CONTENTS

I. INTRODUCTION

Iron is the fourth most abundant element in the earth's crust,[1] and with the possible exception of the lactic acid bacteria, an anaerobic bacterium which possesses no oxygen-requiring enzymes, all organisms require iron for growth.[1-3] Since in the presence of molecular oxygen iron is oxidized to insoluble ferric hydroxide, specific iron-sequestering molecules have been selected during evolution. In the vertebrate world, these functions are managed by transferrin for iron transport and by ferritin for iron storage.[3] Iron supply to organisms is a controlled process, and lack of iron as well as excess of iron interferes with cell growth and function. In humans, iron deficiency leads to anemia and widespread tissue abnormalities, reflecting the role of both nonheme iron and heme as cofactors of several cellular proteins and enzymes.[4] On the other hand, iron overload can lead to severe, and in some cases fatal, tissue damage. Specific target organs for iron excess include liver, myocardium, and endocrine glands.[4] It has also been suggested that hepatoma and other neoplasms might occur with a higher frequency in patients with idiopathic hemochromatosis,[5] a disease associated with tissue iron overloading.

In this review, we will consider the relationship between iron supply and cell growth, referring mainly, but not only, to cell proliferation. We will first consider how iron is provided to cells and then how iron supply may control cell proliferation, limiting the discussion to what is observed in mammalian cells. We will focus on some aspects of iron metabolism by addressing two main questions. First, is iron always supplied to cells through transferrin-transferrin receptor interaction, or could iron be also delivered to cells by a transferrin-independent process? Second, must iron be considered a vitamin-like nutrient, interfering with cell growth on a relatively long-term basis, or, could iron be also considered as a regulatory molecule, playing a role in the signal transduction of some mitogens?

II. IRON SUPPLY TO MAMMALIAN CELLS

A. TRANSFERRIN-DEPENDENT IRON SUPPLY

Transferrin is a glycoprotein with a single polypeptide chain.[3,6] Transferrin is mainly found in plasma and is synthesized primarily in the liver.[7] Transferrin has two binding sites for ferric iron, but serum transferrin normally is only about 30% saturated with iron. Transferrin is also produced, at lower levels, by nonhepatic tissues.[7] The transferrin gene has been cloned and mapped to chromosome 3 in humans, close to the genes encoding lactoferrin, the melanoma antigen p97 (a transferrin-like molecule of unknown function[8,9]) and the transferrin receptor.[7] Furthermore, in various species of birds, the transferrin gene also codes for ovotransferrin (also known as conalbumin), which differs from serum transferrin by its carbohydrate content.[6]

1. The Transferrin Cycle

Under most normal conditions, iron gains intracellular access by transferrin through its interaction with the transferrin receptor, a transmembrane glycoprotein consisting of 2 disulfide-bonded subunits of 95,000.[10,11] Although only one transferrin receptor gene has been described,[12,13] several forms of transferrin receptors have been observed, which could indicate the existence of different transcriptional or posttranscriptional processing.[14,15] Once the iron-transferrin complex binds to the transferrin receptor, internalization occurs via clathrin-coated pits and iron dissociates from transferrin in endocytic vesicles when they become acidified as a result of the action of a proton pump.[16-18] The mechanism of transfer of the iron from the lumen of the endosome to the cytoplasm is unclear, but reduction of the ferric iron to more permeable ferrous iron by a NADH diferric transferrin reductase has been proposed.[19,20] Under these acidic conditions, apotransferrin remains bound to the receptor and is recycled to the cell surface where apotransferrin is released to the blood stream or the extracellular milieu. Each receptor molecule can undergo many rounds of recycling in its lifetime.

The mechanisms involved in the regulation of transferrin cycling are important to understand since they could be involved in the way iron supply to cells is controlled. While there is still debate as to whether under basal conditions the internalization of transferrin receptor is ligand dependent,[21-25] several studies have been performed to test the hypothesis that transferrin receptor internalization and recycling could be under the control of peptide growth factors known to trigger cell proliferation. In this regard, it has been demonstrated that peptides such as epidermal growth factor (EGF), platelet-derived growth factor (PDGF), or insulin-like growth factor I (IGF-I) can induce the translocation of previously internalized transferrin receptors to the cell surface.[26,27] The exact mechanism by which these growth factors are acting is not understood, but the redistribution of transferrin receptors they induced has been found to be calcium dependent.[24] Furthermore, transferrin receptor cycling is stimulated by phorbol esters.[21,28-30] The effect of phorbol esters is cell-line specific, receptor externalization being stimulated in some cell types and receptor internalization in some others. Since phorbol esters also induce a reversible hyperphosphorylation of transferrin receptor,[21,28] it has been proposed that receptor phosphorylation may be involved in the regulation of transferrin receptor cycling. This latter mechanism could function through the action of protein kinase C on serine 24, a serine residue located within the intracellular domain of the receptor.[31] However, when mutant receptors lacking serine[24] are generated by site-directed mutagenesis of transferrin receptor cDNA, the cycling of phosphorylation-defective receptors is regulated by phorbol diester in a manner similar to that observed for the wild-type receptor.[32,33] Nevertheless, the cytoplasmic domain of the transferrin receptor is essential for internalization of receptors and could play an active role in endocytosis by inducing coat assembly of the receptors.[25,34]

From these studies, it remains possible that peptide growth factors regulate, at least to some extent, the amount of iron taken up by cells through a control occurring at the level of transferrin receptor cycling. However, it has not been proven that it is actually the case under physiological conditions,[35] and iron uptake by cells could be primarily controlled by changes in the receptor number,[36] as shall be discussed in more detail below.

2. Iron Supply by Nonhepatic Transferrin

Although transferrin is primarily synthesized in the liver, transferrin mRNA has also been detected in several extrahepatic sites.[7] In rats, with the exception of the brain, transferrin gene expression in nonhepatic tissues is maximal just before birth and then decreases during postnatal development.[37] In adult animals, significant levels of transferrin mRNA are detected in several tissues, including lactating mammary gland, Sertoli cells of testis, brain, and helper T-lymphocytes.[7,37-42]

It has been proposed that Sertoli cells constitute a blood-testis barrier, explaining why transferrin must be produced by Sertoli cells in order to provide the transferrin needed by germinal cells for their development.[43] In fact, the situation might be more complex, since plasma transferrin, contributed by liver, could also play a role in providing iron to testis cells.[41,44] A comparable situation is observed in the case of the brain. Transferrin mRNA is detected in the brain, with particularly high levels in the choroid plexus of the lateral and third ventricules,[42] but plasma transferrin could also contribute to the brain iron supply since transferrin receptors are found on the endothelium of brain capillaries.[45] Interestingly, the transferrin mRNA content of testis, brain, and other nonhepatic tissues remains unchanged during iron deficiency, in contrast to liver transferrin mRNA level which increases under similar conditions.[40] This further illustrates the different physiological significance of plasma transferrin and nonhepatic transferrins.

In contrast to the situation observed in the mammary gland, testis, or brain, no anatomical blood-tissue barrier can be put forward to explain why helper T-lymphocytes also express the transferrin gene during T-cell activation.[39] It is not unreasonable, however, to postulate that during the progression of inflammation, a compartmentalization of the local milieu is occurring, which must be overcome by a local production of transferrin. Transferrin synthesis by helper T-lymphocytes may be involved in an autocrine loop designed to favor proliferation of the helper subset, since these cells also express transferrin receptor gene during T-cell activation (see below).

3. Lactoferrin

It has also been proposed that transferrin synthesized by T-lymphocytes might act as an antibacterial agent.[42] In fact, such antibacterial activity could be exerted not by transferrin but by lactoferrin. Lactoferrin is found in milk and other biological fluids and shares with transferrin the property of reversibly binding two atoms of iron. Although there is no immunological cross-reactivity between the two proteins,[46] their genes derive from a common ancestor.[7] It has been demonstrated that iron-lactoferrin could supply the iron requirement for growing cells in culture.[47] However, the *in vivo* function of lactoferrin, which is produced by neutrophils, is more likely to be involved with the inflammatory process. Lactoferrin could function by chelating iron, thus inhibiting the growth of microorganisms, while at the same time participating in the reduction of plasma iron associated with acute inflammation.[4] It has also been reported that lactoferrin enhanced phagocytosis by macrophages whether or not iron was bound to it, but it could augment intracellular killing of microorganisms only when it contained iron.[48]

4. Iron Uptake from Transferrin without Endocytosis

Some investigators have reported results that do not support the reversible endocytosis hypothesis for the uptake of iron from transferrin.[49,50] By the nonentrance hypothesis, transferrin attaches itself to the cell membrane and iron is released from transferrin to the receptor directly. In hepatocytes, it has been recently suggested that the process would involve reductive removal of iron from transferrin under the action of a NADH reductase, and then translocation of ferrous iron across the plasma membrane.[51,52] Whether or not mobilization of iron from transferrin requires binding of transferrin to its receptor is uncertain.[51] Furthermore, this putative iron uptake in the absence of transferrin endocytosis was observed only at transferrin concentrations above that needed to saturate the transferrin receptor,[51] suggesting that the observed results could be interpreted in the more general context of transferrin-independent iron uptake, a concept to be developed in the next section.

B. TRANSFERRIN-INDEPENDENT IRON SUPPLY

1. Cell Culture under Transferrin-Free Conditions

It is well known that the iron requirement of some transformed cell lines and of diploid fibroblasts, cultured under serum-free conditions, can be satisfied by supplementing the culture medium with iron salts instead of transferrin.[53-56] Other iron-containing compounds, including heme[57,58] and lipophilic iron chelators[59-61] have also been used for sustaining *in vitro* cell proliferation in the absence of serum and transferrin. Furthermore, some leukemic cells, which are cultured in suspension, can be grown continuously in serum-free medium supplemented solely with iron salts.[58,62-65] In some cases, it appears that cell proliferation in the absence of exogenous transferrin is explained because of the endogenous production of transferrin by the cells in culture.[66,67] However, in other cases transferrin synthesis does not appear to be involved. Some cell lines are capable of taking up iron independently of transferrin and its receptor.[63,64,68,69] Furthermore, inhibition of the proliferation of some cells by antibodies to transferrin receptor can be overcome by the addition of soluble iron salts.[70,71]

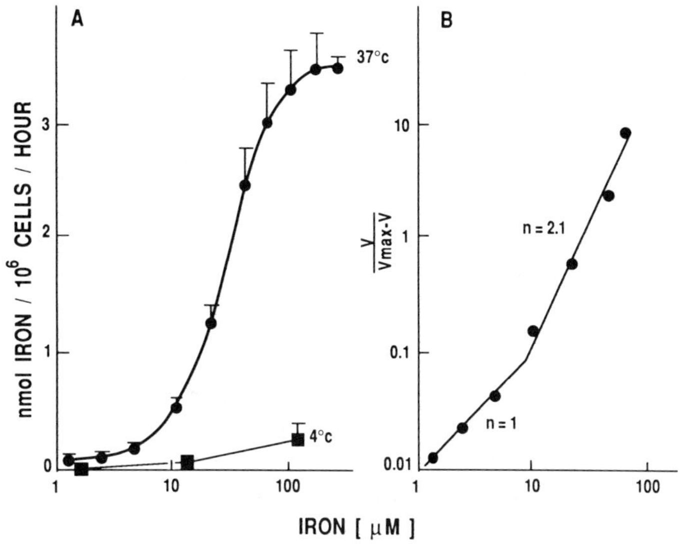

FIGURE 1. Iron uptake by L_{1210} leukemic cells adapted to serum-free transferrin-free culture. (A) Iron uptake as a function of the concentration of ferric citrate in the culture medium. Cells were pulsed with [^{59}Fe] ferric citrate (0.1 to 0.2 μCi/ml) for 30 min at (●) 37°C or (■) 4°C, and the amount of [^{59}Fe] ferric citrate taken up measured. Each result was corrected by subtracting a blank value obtained by running in parallel samples in which [^{59}Fe] ferric citrate was added for 2 min before cell harvesting. The data points represent the means of four measurements. (B) Hill plot of the data from the uptake experiment performed at 37°C. (From Basset, P., Quesneau, Y., and Zwiller, J., *Cancer Res.*, 46, 1644, 1986. With permission.)

The molecular mechanisms underlaying transferrin-independent iron uptake are poorly understood. In L_{1210} mouse leukemic cells adapted to permanent growth in serum-free transferrin-free culture conditions, iron uptake, evaluated as ferric citrate uptake, was found to occur through a saturable transport system.[64] As shown in Figure 1, the uptake exhibited cooperativity, an observation compatible with the involvement of an allosteric transport protein. The observed $K_{0.5}$ for ferric citrate was 30 μM and, above this concentration of ferric citrate in the culture medium, transferrin-independent iron uptake by L1210 cells occurred at a considerably higher rate than the coexisting transferrin-dependent uptake.[64] However, it is not known whether the transport system is a carrier for iron itself or just a citrate carrier, reminiscent of the ferric-citrate receptor known to exist in bacteria.[72]

2. *In vivo* Transferrin-Independent Iron Uptake

In the context of the findings reported above, the question is not to know whether transferrin-independent iron uptake by cells exists, but what significance it could have. *In vivo*, transferrin-independent iron uptake could play a role in some pathological circumstances associated with tissue iron overloading.

a. Atransferrinemia

Congenital atransferrinemia in humans is very rare. However, from the few cases which have been reported in the literature it can be concluded that life with only trace levels of plasma transferrin is possible, at least for some years.[4,6,73] Atransferrinemia is associated with severe hypochromic anemia and possible increased incidence of infection. In one case, death occurred at age 7 and autopsy findings revealed widespread hemosiderosis in most organs, except the bone marrow and the spleen.

b. Hemochromatosis

Idiopathic hemochromatosis in humans is associated with an iron overloading whose exact origin is presently not understood.[74] The excess absorbed iron is deposited at various sites throughout the body, but quantitatively the most important iron storage organ is the liver. Hemochromatosis is characterized by the saturation of plasma transferrin, the presence of nontransferrin iron in plasma and it has been suggested that this form of iron might be responsible for tissue damage associated with iron overloading.[5,75]

Experiments carried out with animal models for hemochromatosis are in agreement with the hypothesis, and indicate that nontransferrin-bound iron can be taken up efficiently by animal tissues *in vivo*.[76,77] Interestingly, transferrin-independent iron uptake by rat liver *in vivo* appears to be much more efficient than transferrin-dependent uptake,[76] an observation consistent with the result obtained *in vitro* with L1210 leukemic cells.[64] Similarly, in both studies the observed K_M values for the transferrin-independent iron transport were of the same order of magnitude and the uptake systems were saturable.[64,76] Recent findings indicate that hepatic uptake of nontransferrin-bound iron could be mediated by a membrane carrier and involve an electrogenic transport mechanism, driven by the transmembrane potential difference.[78]

c. Nature of Nontransferrin Iron in Plasma

It has been proposed that a significant proportion of nontransferrin iron found in sera of patients with thalassemia and idiopathic hemochromatosis is due to the presence of ferritin.[75] The observation is interesting, since it is known that some cells, including hepatocytes, exhibit specific receptors for ferritin.[74,79,80] Besides ferritin, another form of nontransferrin iron found in plasma is heme, which is bound to the carrier protein hemopexin. Receptors have been reported to be present for heme bound to hemopexin on the surface on some cells[81,82] and they could play a role in providing iron to hepatocytes.[36] However, hemopexin is thought to be part of the iron conservation mechanism of the body and, at the present time, there are no experimental data indicating that significant amounts of heme are found in the nontransferrin iron pool of patients with iron overload.

It is apparent that further studies are required to know whether nontransferrin iron found in plasma under some pathological conditions is mainly constituted of ferritin or may be present as low-molecular weight chelates, as also reported.[83] Nevertheless, the data presently available are consistent with the concept that tissue iron overloading observed in patients with atransferrinemia, hemochromatosis, or thalassemia is due to transferrin-independent iron uptake, from nontransferrin iron present in plasma. Thus, while transferrin does not appear to be required to supply cells with iron, the role of transferrin is to deliver iron to cells in a controlled process, through the interaction with transferrin receptor. In contrast, transferrin-independent iron supply is a poorly regulated process (saturation is only observed for very high concentrations of nontransferrin iron in the extracellular milieu) and could lead, for this reason, to major tissue damage. Consistent with this concept, hepatocytes do not express transferrin receptors in patients with advanced cases of hemochromatosis, despite severe hepatic iron overload.[74,84]

III. TRANSFERRIN RECEPTOR EXPRESSION AND CELL GROWTH CONTROL

From the data reported in the previous section, it can be concluded that under physiological conditions iron is mainly supplied to cells through transferrin interacting with transferrin receptor. Since serum transferrin is present at high concentrations, the number and function of transferrin receptors must play a major role in modulating the movement of iron into most of the cells. As already discussed, the rate of transferrin receptor cycling may be under the control of peptide growth factors. A second level of regulation is associated with transferrin receptor gene expression. In the next sections, we are going to see how the number of transferrin receptors expressed by cells is regulated, and how the regulation could be related to cell growth control by iron.

A. BACKGROUND

Specific receptors for transferrin were first identified on immature erythroid cells, and subsequently on placental cells.[11,85,86] However, it was only with the observation that transferrin receptor could be detected on tumor cell lines[87] and on normal proliferating cells,[88] that the study of the transferrin receptor started to be a popular research subject among biologists interested in cell proliferation control. These observations came in a favorable context since it had been demonstrated a few years before that transferrin, the physiological ligand of the transferrin receptor, was an essential component of media allowing mammalian cell culture under serum-free conditions.[89,90]

Thus, when Trowbridge and Bishr Omary[91] demonstrated the identity of a cell surface glycoprotein related to cell proliferation with the transferrin receptor, all the ingredients required for building a new concept were satisfied. Since this time, in 1981, several hundred papers have been published on the transferrin receptor, many of them dealing more or less with the idea that transferrin receptor expression could be part of the process by which cell proliferation is controlled.

B. TRANSFERRIN RECEPTOR AND CELL PROLIFERATION

A major debate for some time has been to know whether the iron-transferrin requirement of cells for proliferation was due only to the iron supply or whether it could also be explained because transferrin would function as a growth factor in a capacity separate from its role in iron transport. This latter hypothesis is now considered to be very unlikely, first, because transferrin cannot exhibit any mitogenic activity in the presence of iron chelators,[62,64,92-94] and second, because the transferrin requirement for cell proliferation in several experimental systems can be completely replaced by iron bound to lipophilic carriers.[59-61]

In this context, a question is whether the level of transferrin receptor expression, by modulating the amount of iron which is delivered to cells, plays role in the regulation of cell proliferation rate. Second, is iron only a vitamin-like nutrient, interfering with cell growth on a relatively long-term basis, or could iron have, in addition, a more specific action in the control of cell proliferation? In other words, is iron a molecule involved in the signal transduction pathway of mitogens? To evaluate these concepts, we are going to examine the available experimental data from two classical systems in which normal cell proliferation can be triggered by extracellular signals, lectin-induced T-lymphocyte proliferation and peptide growth factor-induced fibroblast proliferation, respectively.

1. T-Lymphocyte Proliferation

It is well established that cell surface transferrin receptor expression is a prerequisite for DNA synthesis in mitogen-stimulated T-lymphocytes.[95-97] In these cells, binding of the T-cell growth factor interleukin 2 to its receptor causes the sequential induction of transferrin receptor followed by DNA synthesis, while blocking of the transferrin receptor with specific monoclonal antibodies prevents DNA synthesis. The induction of transferrin receptor expression by interleukin 2 is preceded by an increase of transferrin receptor mRNA level.[98-100] Furthermore, during T-cell activation, transferrin is secreted by helper T-lymphocytes,[39] thus generating an autocrine loop involving transferrin and its receptor in this subset of lymphocytes.

The results indicate that exogenous iron supply is required for the triggering of lymphocyte DNA synthesis by interleukin 2, but they do not demonstrate that iron is a signal molecule in the strict sense of the word. Exogenous iron supply could be required to observe DNA synthesis simply because iron is a cofactor of some proteins of which *de novo* synthesis is required for T-lymphocytes to move from the G_0 to the S phase of the cell cycle. This is the case, for instance, for ribonucleotide reductase,[101] an iron-containing enzyme which catalyzes the first unique step in the biochemical pathway leading to DNA synthesis. Ribonucleotide reductase activity increases when cells are passing from the G_1 to the S phase of the cell cycle and the increase is controlled by *de novo* synthesis of the iron-containing M_2 subunit.[102,103] In this regard, the requirement for extracellular iron would simply indicate that T-lymphocytes cannot use the iron stored in ferritin, as already reported for other hematopoietic cells.[104,105]

Nevertheless, the up-regulation of transferrin receptor gene expression by interleukin 2 is a critical step in the triggering of T-lymphocyte proliferation. It has been proposed that the growth factor could up-regulate transferrin receptor gene expression by depletion of a regulatory intracellular iron pool.[106,107] If the hypothesis is correct, this would indicate that interleukin 2 controls transferrin receptor expression in a way similar to the iron chelator deferoxamine, which has been shown to upregulate transferrin receptor gene expression by decreasing the size of a chelatable intracellular iron pool.[108-110]

2. Fibroblast Proliferation

The studies on T-lymphocytes suggested that transferrin receptor expression was occurring following growth factor interaction with their target cells, and that such upregulation was necessary for cell proliferation to occur. To evaluate whether such concepts are applicable to cells other than lymphocytes, studies have been carried out using fibroblasts stimulated by serum or defined peptide growth factors.

When fibroblasts are stimulated to proliferate, transferrin receptor number,[88,111] and transferrin receptor mRNA levels[68,112,113] increase. In density-arrested human diploid fibroblasts, the up-regulation of transferrin receptor mRNA levels induced by serum was observed to be maximum after 15 h (Figure 2A) and to precede the increase of transferrin receptor number.[68] The effects of serum on transferrin receptor number and mRNA levels could be mimicked by defined growth factors in serum-free conditions.[68,113] PDGF, EGF, or insulin individually each caused mild increases, but the combination of all three growth signals resulted in marked up-regulation of transferrin receptor mRNA levels (Figure 2B). These effects of PDGF, EGF, and insulin are distinct from the control these growth factors exert on transferrin receptor internalization which is a short-term effect, characterized by a transient increase in cell-surface receptor number related to the translocation of preformed intracellular receptors to cell surface,[26,27] as already discussed.

All together, these results demonstrate some similarity between the action of peptide growth factors on diploid fibroblasts and of interleukin 2 on T-lymphocytes. In both conditions, cell proliferation is stimulated and transferrin receptor gene expression is up-regulated. However, and in contrast to what is observed for lymphocytes, the presence of functional transferrin receptors may be not required for the transduction of the mitogenic signals which trigger fibroblast proliferation. As shown in Figure 3, blocking of transferrin binding and of iron uptake by more than 95% with the 42/6 antibody[114] to transferrin receptor did not affect the ability of PDGF, EGF, and insulin to induce DNA synthesis. In contrast, 42/6 antibody prevented T-lymphocyte DNA synthesis induced by a lectin, as already reported by others.[96,97]

It thus appears that exogenous iron supply may not be required for growth signals to trigger and sustain fibroblast proliferation, at least for the one or two rounds of proliferation permitted by the confluent state.[68] However, as we shall discuss in the section E, it is not possible to rule out completely the hypothesis that the fibroblasts are taking up the iron they need from the extracellular milieu by a transferrin-independent mechanism, since both trans-

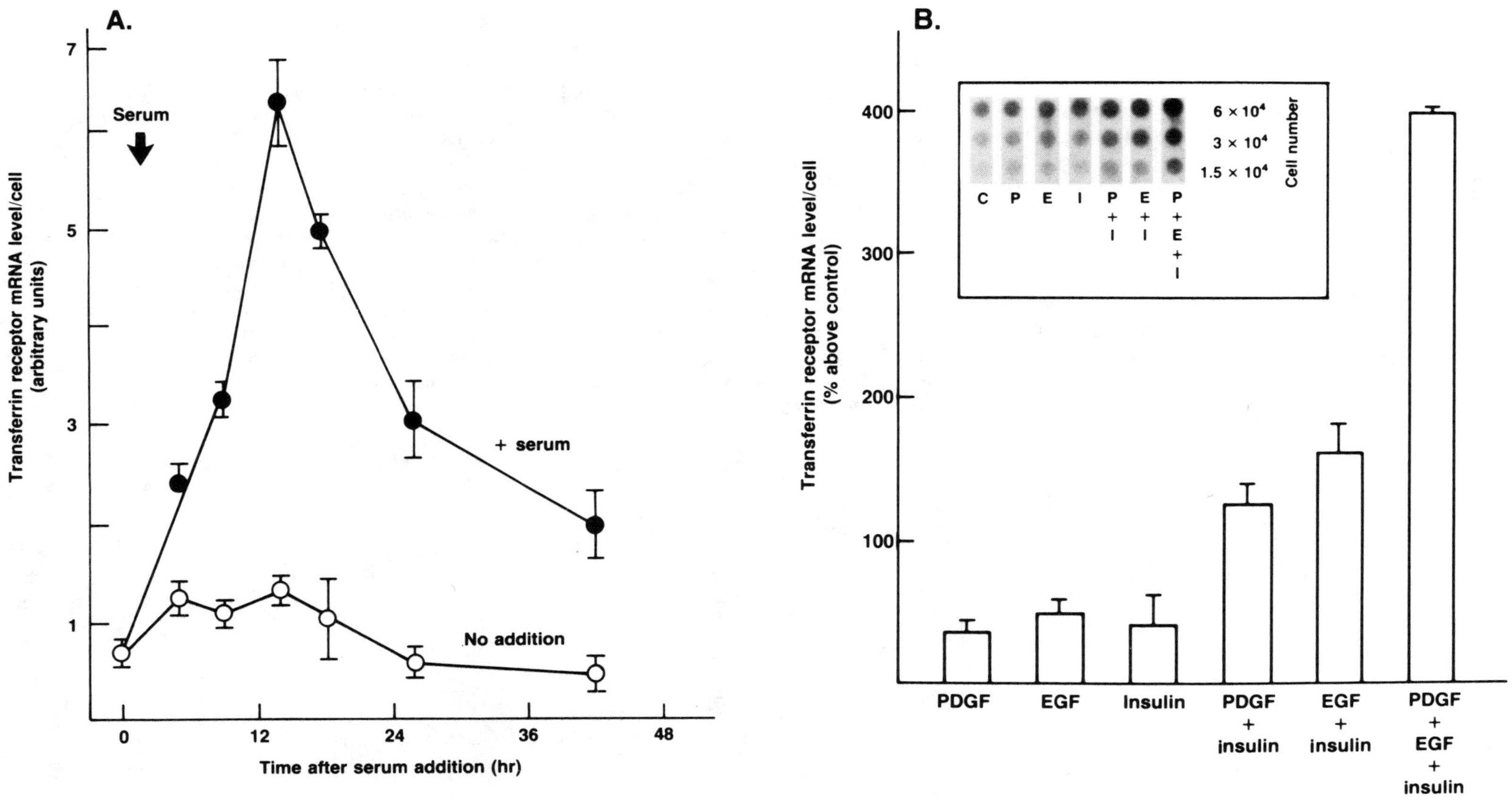

FIGURE 2. Up-regulation of HFL-1 fibroblast transferrin receptor mRNA levels by growth signals. (A) Transferrin receptor mRNA levels following the addition of serum. Density-arrested fibroblasts were cultured in the absence (○) or presence (●) of 10% serum for the times indicated and transferrin receptor mRNA levels were evaluated by cytoplasmic dot hybridization, using a ^{32}P-labeled transferrin receptor cDNA probe. The data points represent the means of triplicate determinations. (B) Transferrin receptor mRNA levels following the addition of defined peptide growth factors. Density-arrested fibroblasts were cultured in serum-free conditions in the absence or presence of PDGF (40 ng/ml), EGF (10 ng/ml) and/or insulin (30 μg/ml) for 15 h. Transferrin receptor mRNA levels were evaluated by cytoplasmic dot hybridization as above. The data represent the means of triplicate determinations. Insert: an example of the cytoplasmic dot hybridization: C = control; P = PDGF; E = EGF; I = insulin. (See Reference 146.)

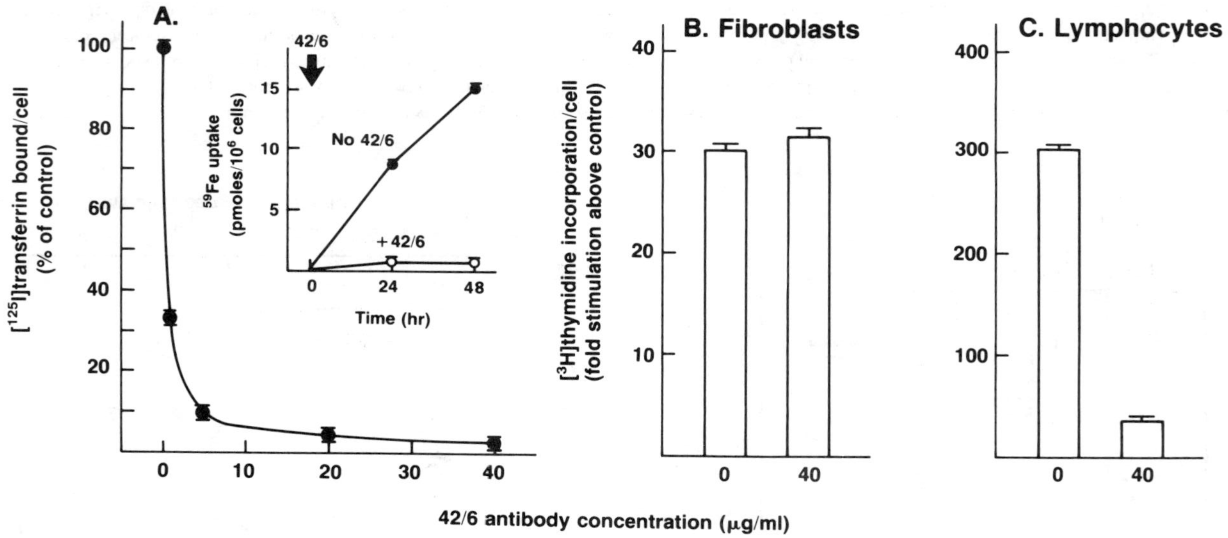

FIGURE 3. Lack of an effect of transferrin receptor blockade on growth factor-induced DNA synthesis in HFL-1 fibroblasts. (A) Prevention of transferrin binding and iron uptake in fibroblasts by the 42/6 antitransferrin receptor antibody. Density-arrested fibroblasts cultured in serum-free conditions were incubated for 1 h alone or with increasing amounts of the 42/6 antibody as indicated. Human diferric [^{125}I] transferrin (2.5 μg/ml) was then added and the specific binding of [^{125}I] transferrin at 37°C was quantified. The data points represent the means of triplicate determinations. Insert: prevention of transferrin-mediated iron uptake by the 42/6 antibody. Density-arrested fibroblasts cultured in serum-free conditions were incubated for 1 h alone (●) or with the 42/6 antibody (○) (20 μg/ml) and supplemented with PDGF (40 ng/ml), EGF (10 ng/ml), insulin (30 μg/ml), transferrin (2.5 μg/ml), and [^{59}Fe] ferrous citrate (1 μCi/ml). At the times indicated, the cells were harvested and transferrin-mediated iron uptake was determined. Blank values, determined by measuring cell-associated [^{59}Fe] radioactivity in the absence of transferrin, have been substracted. The data points represent the means of triplicate determinations. (B) Evaluation of fibroblast DNA synthesis in the presence of the 42/6 antibody. Density-arrested fibroblasts incubated for 1 h alone or with the 42/6 antibody (40 μg/ml) were stimulated by PDGF (40 ng/ml) + EGF (10 ng/ml) + insulin (30 μg/ml). DNA synthesis was determined 16 to 18 h after growth factor addition using a 2-h pulse of [^{3}H] thymidine. (C) Evaluation of lymphocyte DNA synthesis was determined 70 to 72 h after the PHA and serum additions, using a 2-h pulse of [^{3}H] thymidine. In panels B and C, the data represent the means of quadruplicate determinations. (See Reference 146.)

ferrin-dependent and transferrin-independent iron uptake are known to coexist in human diploid fibroblasts.[68]

C. TRANSFERRIN RECEPTOR AND CELL DIFFERENTIATION

During either embryogenesis or during the cell maturation observed in regenerating tissues of adult organisms, several steps of cell proliferation are required before full differentiation is achieved. In this context, it is not surprising to observe that transferrin receptor expression may be required and that transferrin itself has been found in several systems to be an indispensable trophic factor for differentiation to occur. This is the case for erythroid cell differentiation,[115,116] myogenesis,[117-119] or kidney differentiation.[120]

However, it is clear that transferrin receptor and transferrin are not involved in cell differentiation solely because differentiation and proliferation are tightly coupled processes. Differentiated cells which are no longer proliferating are still capable of expressing transferrin receptor at high levels. This is the case for macrophages in which proliferation, if any, is extremely slow. While transferrin receptor is barely detectable on monocytes, once the monocytes become activated macrophages they express high levels of transferrin receptor mRNA and transferrin receptors.[121-124] Similarly, transferrin receptor has been reported to be highly expressed in some adult rat brain neurons, which are not proliferating.[125]

In fact, these observations are not unexpected. There is no reason why iron should be required only for cell proliferation. Several proteins need iron to function and this must be the case in dividing cells as in nondividing cells. For instance, macrophages are involved in killing of microorganisms, a function requiring lactoferrin and iron.[48] Most cells, proliferating or not, produce ATP using the Krebs cycle, whose functioning is dependent on the integrity of iron-containing proteins. Furthermore, several other housekeeping enzymes also require iron to function.[4]

D. TRANSFERRIN RECEPTOR AND CELL TRANSFORMATION

That cancer cells are rapidly proliferating cells is a widespread notion. However, cell transformation is mainly characterized by a blockade of cell maturation and most cancer cells, in primary human tumors, do not grow faster than their normal counterparts.[126] Nevertheless, the fact that transferrin receptor expression was increased in rapidly proliferating cells was used to try to involve transferrin receptor expression in the cancer process. A parallel was made between microorganism multiplication and neoplastic cell dissemination, in which it was proposed that iron availability affected the successful establishment of the disease.[127] When some sequence similarity was observed between the N-terminal region of transferrin and the Blym protooncogene,[128] it was soon reported that ras proteins could function in the regulation of cell growth via interaction with the cell surface receptor for transferrin.[129] However, this finding turned out to be an experimental artifact.[130]

More recently, there have been reported differences in transferrin receptor function between normal and transformed cells.[131,132] In one study,[131] it has been observed that HTLV-1 infected T-lymphocytes expressed higher levels of surface transferrin receptor than their normal counterparts. The overexpression was due to an altered distribution between the cytoplasmic and surface transferrin receptor pools, and the HTLV-1 infected cells internalized 80% less ^{59}Fe-transferrin than the normal T cells. It is assumed that the defect of transferrin receptor internalization may be indicative of a wider defect in membrane trafficking, since interleukin 2 receptor internalization is also defective in HTLV-1 T-lymphocytes.[131]

Thus, while it is clear that cell surface transferrin receptor can be overexpressed in transformed cells, there are no experimental data supporting the concept that the transferrin receptor may be involved in the process by which cells become transformed. Transferrin receptor overexpression at the cell surface can reflect an altered distribution between cell surface and intracellular receptor pools, as illustrated with HTLV-1 infected lymphocytes.

It can also indicate an increased percentage of cycling cells in tumors,[133] since transferrin receptor number is known to be higher in the S and G_2 + M phases than in the G_1 phase of the cell cycle.[134]

E. ANTIBODIES TO TRANSFERRIN RECEPTOR

Several antibodies against human and murine transferrin receptors have been produced and studied by Trowbridge's group.[70,71] Of the antibodies tested so far, only antireceptor antibodies of the IgM or IgA class markedly inhibit transferrin-iron uptake and cell growth, including 42/6 antibody to human transferrin receptor and R17 208 antibody to murine transferrin receptor. As illustrated by the results observed using human T-lymphocytes and human diploid fibroblasts, human cells vary widely in their sensitivity to growth inhibition by 42/6 antibody. Normal myeloid progenitor cells,[135] leukemic myeloid cells,[135] and leukemic lymphoblasts[114] are other examples of cells in which proliferation is prevented by 42/6 antibody, while almost no growth inhibition is observed for melanoma cells and several malignant cell lines of epithelial origin,[136] using the same antibody. Similar variations have been reported for murine cells when treated with R17 208 antibody to transferrin receptor.[70,71]

The reason cells vary widely in their sensitivity to growth inhibition by antitransferrin receptor antibodies is not well understood. One hypothesis is that some cells could use intracellular iron stored in the form of ferritin, whereas other cells would have an obligate requirement for exogenous iron supply. Some cells might recover iron from ferritin through a low molecular weight iron complex[137] and use it for DNA synthesis and proliferation. In contrast, in other cells, this mechanism would not be operational and ferritin would have only a detoxification function.[105] A second hypothesis, which is not contradictory with the first one, is that the level of iron required to trigger cell proliferation may be higher in some cells. For instance, T-lymphocytes are blocked in a deeper G_0 state than density-arrested fibroblasts and, for this reason, may require more iron than fibroblasts for moving toward the S phase of the cell cycle. However, this would not easily explain why the proliferation of some transformed cells is totally prevented by blocking of the transferrin receptor, while the proliferation of some others is scarcely affected.

A third hypothesis to explain why cells vary in their sensitivity to growth inhibition by antitransferrin receptor antibodies is that transferrin-independent iron uptake may be involved in some circumstances. Almost all the experiments studying the effect of transferrin receptor blockage on cell proliferation have been performed under *in vitro* culture conditions. As discussed above, transferrin-independent iron uptake has been demonstrated to occur in these circumstances, and it is not unreasonable to postulate that the aptitude for and the nature of transferrin-independent iron uptake may vary from one cell type to another. Many different proteins involved in iron salt transport have been reported to function in microorganisms,[72] and it may be also the case in animal cells, at least under *in vitro* culture conditions. In this context, it can be assumed that some cell types would be capable of using iron salts, invariably found in culture media,[92] to bypass the blocking of transferrin receptors by antibodies, while other cells would not have the ability. In this regard, what has been observed using lectin-stimulated mouse T-lymphocytes is quite interesting. In these cells, ferric citrate or ferric nitrilotriacetate cannot replace iron-transferrin or reverse the blocking of transferrin receptor by R17 208 antibody to observe proliferation,[60,138] although iron from the ferric chelates is taken up by lymphocytes.[139] In contrast, the addition of pyridoxal isonicotinoyl hydrazone, a lipophilic iron carrier, completely replaces transferrin.[60] Thus, when transferrin-independent iron supply is involved, what may be critical is not only that iron is taken up by cells, but that it can be used in metabolic pathways.

It is now important to know whether cells *in vivo* also vary in their sensitivity to growth inhibition by anti-transferrin receptor antibodies. While it has been suggested that the specificity of transferrin receptor expression for neoplastic cells may be low compared to other tumor markers,[140] Sauvage et al.[141] have tested the effect of monoclonal antibodies which block transferrin receptor function on the *in vivo* growth of a syngeneic murine leukemia. The results showed that the acute toxicity associated with the administration of the antibodies was minimal, while a dose-dependent antitumor effect was observed. Further work is now required to know whether antibodies to transferrin receptor may be used in human diseases as antitumor agents[71] or for controlling the immune response.[142]

IV. CONCLUSIONS

The involvement of iron in cell growth is well documented. What commences to be better understood is that iron supply to cells is a tightly controlled process. During evolution, transferrin has been selected to transport iron in a nontoxic form and to deliver it to cells as a function of their specific need, through interaction with the transferrin receptor. Transferrin receptor function and level of expression are controlled by peptide growth factors including interleukin 2, PDGF, EGF, and IGF-I. The expression of the transferrin receptor is also regulated by the availability of iron and it has been suggested that the regulation of transferrin receptor number by peptide growth factors could be mediated by changes in the size of a regulatory intracellular iron pool. However, at the present time, it is not possible to rule out the existence of two entirely distinct regulatory mechanisms, one involved in the control of receptor expression by iron availability, and one in the control of receptor expression by peptide growth factors. The importance of transferrin-transferrin receptor interaction for iron homeostasis in humans is illustrated by the observations made in patients with atransferrinemia, hemochromatosis, or thalassemia. Under these conditions, tissue damage due to iron overload is observed and can be explained because the physiological iron supply to cells through the transferrin receptor is bypassed by transferrin-independent iron uptake. Consistent with this concept, nontransferrin bound iron is commonly found in the plasma of the patients with iron overload, and transferrin-independent iron uptake by cells has been demonstrated to occur *in vitro* as well as *in vivo*.

The involvement of transferrin-independent iron uptake must be also considered when asking whether or not exogenous iron supply is required for the triggering of cell proliferation. Under *in vitro* conditions, it is clear that lectin-stimulated T-lymphocytes need iron from the extracellular milieu to start proliferation. In contrast, some data suggest it may not be the case with other cell types, including fibroblasts and cells from epithelial origin, in which proliferation is not prevented by specific antibodies to transferrin receptor. These observations suggest the existence of two groups of cells. In the first group, in which T-lymphocytes are the paradigm, exogenous iron supply is required for both the triggering and the maintenance of proliferation. In the second group of cells, in which diploid fibroblasts are the paradigm, exogenous iron supply would be required only for the maintenance of proliferation. However, another hypothesis is that diploid fibroblasts (and not lymphocytes) could use soluble iron usually found in culture media. This would allow the fibroblasts to bypass the blocking of transferrin receptor by antibodies. If the hypothesis is correct, it would indicate that, in fact, all cells require exogenous iron supply to start proliferation because they cannot use intracellular iron associated with ferritin. In this case, ferritin would serve not a storage function, but simply a detoxification function.

These observations indicate that, at least in some cell types, growth factors cannot signal cells to proliferate in the absence of exogenous iron supply. At the present time, there are no precise molecular mechanisms to explain how iron could be a signal molecule, in the strict sense of the word. It has been suggested that ferric iron reduction at the level of the plasma membrane, which is associated with the stimulation of transmembrane electron transport, could play a role in cell cycle control.[143,144] However, the mechanisms by which that regulation might be achieved remain obscure,[145] and it appears reasonable to assume that iron functions mainly as a cofactor, required by several cellular proteins. When the proteins requiring iron have slow turnover rates, exogenous iron must be considered as a vitamin-like nutrient. However, when the proteins have high turnover rates and must be synthesized *de novo* at each G_1/S transition of the cell cycle, exogenous iron, while still a protein cofactor, can be assimilated to a signal molecule. In this regard, a major target for iron action is ribonucleotide reductase, an iron-containing enzyme that may play a role in the control of DNA synthesis. However, the iron requirement by living cells must not be considered as specifically associated with cell proliferation, as illustrated by the high levels of transferrin receptor expression in macrophages and some neurons that do not proliferate. It is obvious that iron-containing proteins are also involved in functions other than proliferation, and that the expression of the transferrin receptor reflects both the iron requirement for proliferation and the metabolic needs for iron unrelated to proliferation.

Finally, it is not clear at the present time whether antibodies to transferrin receptor, either alone or coupled to cytotoxic agents, are potential therapeutic tools for *in vivo* use. Preliminary data have demonstrated some specificity of antibodies to transferrin receptor to prevent the growth of tumor cells in animal models, but further work is required to know whether the approach will be valuable at the clinical level.

REFERENCES

1. **Emery, T. E.,** Iron metabolism in humans and plants, *Am. Scientist,* 70, 626, 1982.
2. **Neilands, J. B.,** Microbial iron compounds, *Annu. Rev. Biochem.,* 50, 715, 1981.
3. **Aisen, P. and Listowsky, I.,** Iron transport and storage proteins, *Annu. Rev. Biochem.,* 49, 357, 1980.
4. **Jacobs, A. and Worwood, M.,** Iron, in *Disorders of Mineral Metabolism,* Vol. 1, Academic Press, New York, 1981, 1.
5. **Propper, R. and Nathan, D.,** Clinical removal of iron, *Annu. Rev. Med.,* 33, 509, 1982.
6. **Putnam, F. W.,** Transferrin, in *The Plasma Proteins,* Vol. 1, 2nd ed., Putnam, F. W., Ed., Academic Press, New York, 1975, 265.
7. **Bowman, B. H., Yang, F., and Adrian, G. S.,** Transferrin: evolution and genetic regulation of expression, *Adv. Genet.,* 25, 1, 1988.
8. **Brown, J. P., Hewick, R. M., Hellström, I., Hellström, K. E., Doolittle, R. F., and Dreyer, W. J.,** Human melanoma-associated antigen p97 is structurally and functionally related to transferrin, *Nature,* 296, 171, 1982.
9. **Rose, T. M., Plowman, G. D., Teplow, D. B., Dreyer, W. J., Hellström, K. E., and Brown, J. P.,** Primary structure of the human melanoma-associated antigen p97 (melanotransferrin) deduced from the mRNA sequence, *Proc. Natl. Acad. Sci. U.S.A.,* 83, 1261, 1986.
10. **Trowbridge, I. S., Newman, R. A., Domingo, D. L., and Sauvage, C.,** Transferrin receptors: structure and function, *Biochem. Pharmacol.,* 33, 925, 1984.
11. **Trowbridge, I. S. and Shackelford, D. A.,** Structure and function of transferrin receptors and their relationship to cell growth, *Biochem. Soc. Symp.,* 51, 117, 1986.
12. **Kühn, L. C., McClelland, A., and Ruddle, F. H.,** Gene transfer, expression, and molecular cloning of the human transferrin receptor gene, *Cell,* 37, 95, 1984.
13. **McClelland, A., Kühn, L. C., and Ruddle, F. H.,** The human transferrin receptor gene: genomic organization, and the complete primary structure of the receptor deduced from a cDNA sequence, *Cell,* 39, 267, 1984.
14. **Poola, I. and Lucas, J. J.,** Purification and characterization of an estrogen-inducible membrane glycoprotein. Evidence that it is a transferrin receptor, *J. Biol. Chem.,* 263, 19137, 1988.
15. **Cotner, T., Das Gupta, A., Papayannopoulou, Th., and Stamatoyannopoulos, G.,** Characterization of a novel form of transferrin receptor preferentially expressed on normal erythroid progenitors and precursors, *Blood,* 73, 214, 1989.
16. **Dautry-Varsat, A., Ciechanover, A., and Lodish, H. F.,** pH and the recycling of transferrin during receptor-mediated endocytosis, *Proc. Natl. Acad. Sci. U.S.A.,* 80, 2258, 1983.
17. **Klausner, R. D., Ashwell, G., Van Renswoude, J., Harford, J. B., and Bridges, K. R.,** Binding of apotransferrin to K562 cells: explanation of the transferrin cycle, *Proc. Natl. Acad. Sci. U.S.A.,* 80, 2263, 1983.

18. **Octave, J.-N., Schneider, Y.-J., Trouet, A., and Crichton, R. R.,** Iron uptake and utilization by mammalian cells. I. Cellular uptake of transferrin and iron, *TIBS*, 8, 217, 1983.
19. **Sun, I. L., Navas, P., Crane, F. L., Morré, D. J., and Löw, H.,** NADH differic transferrin reductase in liver plasma membrane, *J. Biol. Chem.*, 262, 15915, 1987.
20. **Löw, H., Lindgren, A., Crane, F. L., Sun, I. L., Toole-Simms, W., and Morré, D. J.,** Transplasma membrane electron transport functions as a ferric reductase, in *Plasma Membrane Oxidoreductases in Control of Animal and Plant Growth*, Crane, F. L., Morré, D. J., and Löw, H., Eds., Plenum Press, New York, 1988, 153.
21. **Klausner, R. D., Harford, J., and Van Renswoude, J.,** Rapid internalization of the transferrin receptor K562 cells is triggered by ligand binding or treatment with a phorbol ester, *Proc. Natl. Acad. Sci. U.S.A.*, 81, 3005, 1984.
22. **Watts, C.,** Rapid endocytosis of the transferrin receptor in the absence of bound transferrin, *J. Cell Biol.*, 100, 633, 1985.
23. **Ajioka, R. S. and Kaplan, J.,** Intracellular pools of transferrin receptors result from constitutive internalization of unoccupied receptors, *Proc. Natl. Acad. Sci. U.S.A.*, 83, 6445, 1986.
24. **McVey-Ward, D. and Kaplan, J.,** Mitogenic agents induce redistribution of transferrin receptors from internal pools to the cell surface, *Biochem. J.*, 238, 721, 1986.
25. **Iacopetta, B. J., Rothenberger, S., and Kühn, L. C.,** A role for the cytoplasmic domain in transferrin receptor sorting and coated pit formation during endocytosis, *Cell*, 54, 485, 1988.
26. **Wiley, H. S. and Kaplan, J.,** Epidermal growth factor rapidly induces a redistribution of transferrin receptor pools in human fibroblasts, *Proc. Natl. Acad. Sci. U.S.A.*, 81, 7456, 1984.
27. **Davis, R. J. and Czech, M. P.,** Regulation of transferrin receptor expression at the cell surface by insulin-like growth factors, epidermal growth factor and platelet-derived growth factor, *EMBO J.*, 5, 653, 1986.
28. **May, W. S., Jacobs, S., and Cuatrecasas, P.,** Association of phorbol ester-induced hyperphosphorylation and reversible regulation of transferrin membrane receptors in HL60 cells, *Proc. Natl. Acad. Sci. U.S.A.*, 81, 2016, 1984.
29. **Buys, S. S., Keogh, E. A., and Kaplan, J.,** Fusion of intracellular membrane pools with cell surfaces of macrophages stimulated by phorbol esters and calcium ionophores, *Cell*, 38, 569, 1984.
30. **Iacopetta, B., Carpentier, J.-L., Pozzan, T., Lew, D. P., Gorden, P., and Orci, L.,** Role of intracellular calcium and protein kinase C in the endocytosis of transferrin and insulin by HL60 cells, *J. Cell. Biol.*, 103, 851, 1986.
31. **Davis, R. J., Johnson, G. L., Kelleher, D. J., Anderson, J. K., Mole, J. E., and Czech, M. P.,** Identification of serine 24 as the unique site on the transferrin receptor phosphorylated by protein kinase C, *J. Biol. Chem.*, 261, 9034, 1986.
32. **Davis, R. J. and Meisner, H.,** Regulation of transferrin receptor cycling by protein kinase C is independent of receptor phosphorylation at serine 24 in Swiss 3T3 fibroblasts, *J. Biol. Chem.*, 33, 16041, 1987.
33. **McGraw, T. E., Dunn, K. W., and Maxfield, F. R.,** Phorbol ester treatment increases the exocytic rate of the transferrin receptor recycling pathway independent of serine-24 phosphorylation, *J. Cell. Biol.*, 106, 1061, 1988.
34. **Rothenberger, S., Iacopetta, B. J., and Kühn, L. C.,** Endocytosis of the transferrin receptor requires the cytoplasmic domain but not its phosphorylation site, *Cell*, 49, 423, 1987.
35. **Davis, R. J., Faucher, M., Racaniello, L. K., Carruthers, A., and Czech, M. P.,** Insulin-like growth factor I and epidermal growth factor regulate the expression of transferrin receptors at the cell surface by distinct mechanisms, *J. Biol. Chem.*, 262, 13126, 1987.
36. **Bomford, A. B. and Munro, H. N.,** Transferrin and its receptor: their roles in cell function, *Hepatology*, 5, 870, 1985.
37. **Levin, M. J., Tuil, D., Uzan, G., Dreyfus, J.-C., and Kahn, A.,** Expression of the transferrin gene during development of nonhepatic tissues: high level of transferrin mRNA in fetal muscle and adult brain, *Biochem. Biophys. Res. Comm.*, 122, 212, 1984.
38. **Bloch, B., Popovici, T., Levin, M. J., Tuil, D., and Kahn, A.,** Transferrin gene expression visualized in oligodendrocytes of the rat brain by using in situ hybridization and immunohistochemistry, *Proc. Natl. Acad. Sci. U.S.A.*, 82, 6706, 1985.
39. **Lum, J. B., Infante, A. J., Makker, D. M., Yang, F., and Bowman, B. H.,** Transferrin synthesis by inducer T-lymphocytes, *J. Clin. Invest.*, 77, 841, 1986.
40. **Idzerda, R. L., Huebers, H., Finch, C. A., and McKnight, G. S.,** Rat transferrin gene expression: tissue-specific regulation by iron deficiency, *Proc. Natl. Acad. Sci. U.S.A.*, 83, 3723, 1986.
41. **Lee, N. T., Chae, C.-B., and Kierszenbaum, A. L.,** Contrasting levels of transferrin gene activity in cultured rat Sertoli cells and intact seminiferous tubules, *Proc. Natl. Acad. Sci. U.S.A.*, 83, 8177, 1986.
42. **Aldred, A. R., Dickson, P. W., Marley, P. D., and Schreiber, G.,** Distribution of transferrin synthesis in brain and other tissues in the rat, *J. Biol. Chem.*, 262, 5293, 1987.
43. **Skinner, M. K. and Griswold, M. D.,** Sertoli cells synthesize and secrete transferrin-like protein, *J. Biol. Chem.*, 255, 9523, 1980.

44. **Djakiew, D., Hadley, M. A., Byers, S. W., and Dym, M.,** Transferrin-mediated transcellular transport of 59Fe across confluent epithelial sheets of Sertoli cells grown in bicameral cell culture chambers, *J. Androl.*, 7, 355, 1986.
45. **Jefferies, W. A., Brandon, M. R., Hunt, S. V., Williams, A. F., Gatter, K. C., and Mason, D. Y.,** Transferrin receptor on endothelium of brain capillaries, *Nature*, 312, 162, 1984.
46. **Hashizume, S., Kuroda, K., and Murakami, H.,** Cell culture assay of biological activity of lactoferrin and transferrin, *Methods Enzymol.*, 147, 302, 1987.
47. **Hashizume, S., Kuroda, K., and Murakami, H.,** Identification of lactoferrin as an essential growth factor for human lymphocytic cell lines in serum-free medium, *Biochim. Biophys. Acta*, 763, 377, 1983.
48. **Lima, M. F. and Kierszenbaum, F.,** Lactoferrin effects on phagocytic cell function. II. The presence of iron is required for the lactoferrin molecule to stimulate intracellular killing by macrophages but not to enhance the uptake of particles and microorganisms, *J. Immunol.*, 139, 1647, 1987.
49. **Smit, S., Huijskes-Heins, M. I. E., and Leijnse, B.,** Iron uptake by chinese hamster fibroblasts from human transferrin, *Exp. Cell. Res.*, 153, 73, 1984.
50. **Thorstensen, K. and Romslo, I.,** Uptake of iron from transferrin by isolated hepatocytes, *Biochim. Biophys. Acta*, 804, 200, 1984.
51. **Thorstensen, K. and Romslo, I.,** Uptake of iron from transferrin by isolated rat hepatocytes. A redox-mediated plasma membrane process?, *J. Biol. Chem.*, 263, 8844, 1988.
52. **Thorstensen, K.,** Hepatocytes and reticulocytes have different mechanisms for the uptake of iron from transferrin, *J. Biol. Chem.*, 263, 16837, 1988.
53. **Rudland, P. S., Durbin, H., Clingan, D., and de Asua, L. J.,** Iron salts and transferrin are specifically required for cell division of cultured 3T6 cells, *Biochem. Biophys. Res. Commun.*, 75, 556, 1977.
54. **Mather, J. P. and Sato, G. H.,** The growth of mouse melanoma cells in hormone-supplemented, serum-free medium, *Exp. Cell Res.*, 120, 191, 1979.
55. **Young, D. V., Cox, F. W., III, Chipman, S., and Hartman, S. C.,** The growth stimulation of SV3T3 cells by transferrin and its dependence on biotin, *Exp. Cell Res.*, 118, 411, 1979.
56. **Walthall, B. J. and Ham, R. G.,** Multiplication of human diploid fibroblasts in a synthetic medium supplemented with EGF, insulin, and dexamethasone, *Exp. Cell Res.*, 134, 303, 1981.
57. **Zwiller, J., Basset, P., Ulrich, G., and Mandel, P.,** Guanylate cyclase activators hemin and sodium nitroprusside stimulate cell growth in serum-free medium, *Exp. Cell Res.*, 141, 445, 1982.
58. **Titeux, M., Testa, U., Louache, F., Thomopoulos, P., Rochant, H., and Breton-Gorius, J.,** The role of iron in the growth of human leukemia cell lines, *J. Cell Physiol.*, 121, 251, 1984.
59. **Landschulz, W., Thesleff, I., and Ekblom, P.,** A lipophilic iron chelator can replace transferrin as a stimulator of cell proliferation and differentiation, *J. Cell Biol.*, 98, 596, 1984.
60. **Brock, J. H. and Stevenson, J.,** Replacement of transferrin in serum-free cultures of mitogen-stimulated mouse lymphocytes by a lipophilic iron chelator, *Immunol. Lett.*, 15, 23, 1987.
61. **Laskey, J., Webb, I., Schulman, H. M., and Ponka, P.,** Evidence that transferrin supports cell proliferation by supplying iron for DNA synthesis, *Exp. Cell Res.*, 176, 87, 1988.
62. **Basset, P., Zwiller, J., Revel, M. O., and Vincendon, G.,** Growth promotion of transformed cells by iron in serum-free culture, *Carcinogenesis*, 6, 355, 1985.
63. **Taetle, R., Rhyner, K., Castagnola, J., To, D., and Mendelsohn, J.,** Role of transferrin, Fe, and transferrin receptors in myeloid leukemia cell growth, *J. Clin. Invest.*, 75, 1061, 1985.
64. **Basset, P., Quesneau, Y., and Zwiller, J.,** Iron-induced L1210 cell growth: evidence of a transferrin-independent iron transport, *Cancer Res.*, 46, 1644, 1986.
65. **Tsavaler, L., Stein, B. S., and Sussman, H. H.,** Demonstration of the specific binding of bovine transferrin to the human transferrin receptor in K562 cells: evidence for interspecies transferrin internalization, *J. Cell. Physiol.*, 128, 1, 1986.
66. **Kitada, S. and Hays, E. F.,** Transferrin-like activity produced by murine malignant T-lymphoma cell lines, *Cancer Res.*, 45, 3537, 1985.
67. **Vostrejs, M., Moran, P. L., and Seligman, P. A.,** Transferrin synthesis by small cell lung cancer cells acts as an autocrine regulator of cellular proliferation, *J. Clin. Invest.*, 82, 331, 1988.
68. **Basset, P.,** Modulation of transferrin receptor gene expression by peptide growth factors and role of the receptor in human fibroblasts, in *Plasma Membrane Oxidoreductases in Control of Animal and Plant Growth*, Crane, F. L., Morré, D. J., and Löw, H., Eds., Plenum Press, New York, 1988, 163.
69. **Fuchs, O., Borov'a, J., Hradilek, A., Neuwirt, J.,** Non-transferrin donors of iron for heme synthesis in immature erythroid cells, *Biochim. Biophys. Acta*, 969, 158, 1988.
70. **Trowbridge, I. S. and Newman, R. A.,** Monoclonal antibodies to transferrin receptors, in *Monoclonal Antibodies to Receptors: Probes for Receptor Structure and Function, Receptors and Recognition*, Series B, Vol. 17, Greaves, M. F., Ed., Chapman and Hall, London, 1984, 235.
71. **Trowbridge, I. S., Lesley, J. F., Domingo, D., Schulte, R., Sauvage, C., and Rammensee, H.-G.,** Monoclonal antibodies to transferrin receptor and assay of their biological effects, *Methods Enzymol.*, 147, 265, 1987.

72. **Neilands, J. B.,** Microbial envelope proteins related to iron, *Annu. Rev. Microbiol.,* 36, 285, 1982.
73. **Goya, N., Miyazaki, S., Kodate, S., and Ushio, B.,** A family of congenital atransferrinemia, *Blood,* 40, 239, 1972.
74. **Anderson, G. J., Halliday, J. W., and Powell, L. W.,** Transferrin receptors in hemochromatosis, *Hepatology,* 7, 967, 1987.
75. **Pootrakul, P., Josephson, B., Huebers, H. A., and Finch, C. A.,** Quantitation of ferritin iron in plasma, an explanation for non-transferrin iron, *Blood,* 71, 1120, 1988.
76. **Wright, T. L., Brissot, P., Ma, W.-L., and Weisiger, R. A.,** Characterization of non-transferrin-bound iron clearance by rat liver, *J. Biol. Chem.,* 261, 10909, 1986.
77. **Craven, C. M., Alexander, J., Eldridge, M., Kushner, J. P., Bernstein, S., and Kaplan, J.,** Tissue distribution and clearance kinetics of non-transferrin-bound iron in the hypotransferrinemic mouse: a rodent model for hemochromatosis, *Proc. Natl. Acad. Sci. U.S.A.,* 84, 3457, 1987.
78. **Wright, T. L., Fitz, J. G., and Weisiger, R. A.,** Non-transferrin-bound iron uptake by rat liver. Role of membrane potential difference, *J. Biol. Chem.,* 263, 1842, 1988.
79. **Blight, G. D. and Morgan, E. H.,** Ferritin and iron uptake by reticulocytes, *Br. J. Haematol.,* 55, 59, 1983.
80. **Sibille, J.-C., Kondo, H., and Aisen, P.,** Interactions between isolated hepatocytes and Kupffer cells in iron metabolism: a possible role for ferritin as an iron carrier protein, *Hepatology,* 8, 296, 1988.
81. **Galbraith, R. A., Sassa, S., and Kappas, A.,** Heme binding to murine erythroleukemia cells. Evidence for a heme receptor, *J. Biol. Chem.,* 260, 12198, 1985.
82. **Taketani, S., Kohno, H., and Tokunaga, R.,** Cell surface receptor for hemopexin in human leukemia HL60 cells. Specific binding, affinity labeling, and fate of heme, *J. Biol. Chem.,* 262, 4639, 1987.
83. **Morgan, E. H. and Baker, E.,** Iron uptake and metabolism by hepatocytes, *Fed. Proc.,* 45, 2810, 1986.
84. **Sciot, R., Paterson, A. C., Van den Oord, J. J., and Desmet, V. J.,** Lack of hepatic transferrin receptor expression in hemochromatosis, *Hepatology,* 5, 831, 1987.
85. **Garret Wada, H., Hass, P. E., and Sussman, H. H.,** Transferrin receptor in human placental brush border membranes. Studies on the binding of transferrin to placental membrane vesicles and the identification of a placental brush border glycoprotein with high affinity for transferrin, *J. Biol. Chem.,* 254, 12629, 1979.
86. **Seligman, P. A. and Allen, R. H.,** Isolation of transferrin receptor from human placenta, *Methods Enzymol.,* 147, 239, 1987.
87. **Hamilton, T. A., Garret Wada, H., and Sussman, H. H.,** Identification of transferrin receptors on the surface of human cultured cells, *Proc. Natl. Acad. Sci. U.S.A.,* 76, 6406, 1979.
88. **Larrick, J. W. and Cresswell, P.,** Modulation of cell surface iron transferrin receptors by cellular density and state of activation, *J. Supramol. Struct.,* 11, 579, 1979.
89. **Hayashi, I. and Sato, G. H.,** Replacement of serum by hormones permits growth of cells in a defined medium, *Nature,* 259, 132, 1976.
90. **Guilbert, L. J. and Iscove, N. N.,** Partial replacement of serum by selenite, transferrin, albumin and lecithin in haematopoietic cell cultures, *Nature,* 263, 594, 1976.
91. **Trowbridge, I. S. and Bishr Omary, M.,** Human cell surface glycoprotein related to cell proliferation is the receptor for transferrin, *Proc. Natl. Acad. Sci. U.S.A.,* 78, 3309, 1981.
92. **Perez-Infante, V. and Mather, J. P.,** The role of transferrin in the growth of testicular cell lines in serum-free medium, *Exp. Cell Res.,* 142, 325, 1982.
93. **Lederman, H. M., Cohen, A., Lee, J. W. W., Freedman, M. H., and Gelfand, E. W.,** Deferoxamine: a reversible S-phase inhibitor of human lymphocyte proliferation, *Blood,* 64, 748, 1984.
94. **Cavanaugh, P. F., Porter, C. W., Tukalo, D., Frankfurt, O. S., Pavelic, Z. P., and Bergeron, R. J.,** Characterization of L1210 cell growth inhibition by the bacterial iron chelators parabactin and compound II, *Cancer Res.,* 45, 4754, 1985.
95. **Hamilton, T. A.,** Regulation of transferrin receptor expression in concanavalin A stimulated and Gross virus transformed rat lymphoblasts, *J. Cell Physiol.,* 113, 40, 1982.
96. **Mendelsohn, J., Trowbridge, I., and Castagnola, J.,** Inhibition of human lymphocyte proliferation by monoclonal antibody to transferrin receptor, *Blood,* 62, 821, 1983.
97. **Neckers, L. M. and Cossman, J.,** Transferrin receptor induction in mitogen-stimulated human T-lymphocytes is required for DNA synthesis and cell division and is regulated by interleukin 2, *Proc. Natl. Acad. Sci. U.S.A.,* 80, 3494, 1983.
98. **Pauza, C. D., Bleil, J. D., and Lennox, E. S.,** The control of transferrin receptor synthesis in mitogen-stimulated human lymphocytes, *Exp. Cell Res.,* 154, 510, 1984.
99. **Depper, J. M., Leonard, W. J., Drogula, C., Krönke, M., Waldmann, T. A., and Greene, W. C.,** Interleukin 2 (IL-2) augments transcription of the IL-2 receptor gene, *Proc. Natl. Acad. Sci. U.S.A.,* 82, 4230, 1985.
100. **Reed, J. C., Alpers, J. D., Nowell, P. C., and Hoover, R. G.,** Sequential expression of protooncogenes during lectin-stimulated mitogenesis of normal human lymphocytes, *Proc. Natl. Acad. Sci. U.S.A.,* 83, 3982, 1986.

101. **Reichard, P. and Ehrenberg, A.,** Ribonucleotide reductase — A radical enzyme, *Science,* 221, 514, 1983.
102. **Eriksson, S., Gräslund, A., Skog, S., Thelander, L., and Tribukait, B.,** Cell cycle-dependent regulation of mammalian ribonucleotide reductase. The S phase-correlated increase in subunit M2 is regulated by *de novo* protein synthesis, *J. Biol. Chem.,* 259, 11695, 1984.
103. **Engström, Y., Eriksson, S., Jildevik, I., Skog, S., Thelander, L., and Tribukait, B.,** Cell cycle-dependent expression of mammalian ribonucleotide reductase. Differential regulation of the two subunits, *J. Biol. Chem.,* 260, 9114, 1985.
104. **Forsbeck, K. and Nilsson, K.,** Iron metabolism of established human hematopoietic cell lines *in vitro, Exp. Cell Res.,* 144, 323, 1983.
105. **Klausner, R. D.,** From receptors to genes — Insights from molecular iron metabolism, *Clin. Res.,* 36, 494, 1988.
106. **Pelosi, E., Testa, U., Louache, F., Thomopoulos, P., Salvo, G., Samoggia, P., and Peschle, C.,** Expression of transferrin receptors in phytohemagglutinin-stimulated human T-lymphocytes. Evidence for a three-step model, *J. Biol. Chem.,* 261, 3036, 1986.
107. **Pelosi-Testa, E., Samoggia, P., Giannella, G., Montesoro, E., Caravita, T., Salvo, G., Camagna, A., Isacchi, G., Testa, U., and Peschle, C.,** Mechanisms underlying T-lymphocyte activation: mitogen initiates and IL-2 amplifies the expression of transferrin receptors via intracellular iron level, *Immunology,* 64, 273, 1988.
108. **Bridges, K. R. and Cudkowicz, A.,** Effect of iron chelators on the transferrin receptor in K562 cells, *J. Biol. Chem.,* 259, 12970, 1984.
109. **Bottomley, S. S., Wolfe, L. C., and Bridges, K. R.,** Iron metabolism in K562 erythroleukemic cells, *J. Biol. Chem.,* 260, 6811, 1985.
110. **Rouault, T., Rao, K., Harford, J., Mattia, E., and Klausner, R. D.,** Hemin, chelatable iron, and the regulation of transferrin receptor biosynthesis, *J. Biol. Chem.,* 260, 14862, 1985.
111. **Frazier, J. L., Caskey, J. H., Yoffe, M., and Seligman, P. A.,** Studies of the transferrin receptor on both human reticulocytes and nucleated human cells in culture. Comparison of factors regulating receptor density, *J. Clin. Invest.,* 69, 853, 1982.
112. **Miskimins, W. K., McClelland, A., Roberts, M. P., and Ruddle, F. H.,** Cell proliferation and expression of the transferrin receptor gene: promoter sequence homologies and protein interactions, *J. Cell Biol.,* 103, 1781, 1986.
113. **Basset, P., Yamauchi, K., and Crystal, R. G.,** Modulation of transferrin receptor mRNA levels in human fibroblasts by peptide growth factors, *Fed. Proc.,* 46, 1995, 1987.
114. **Trowbridge, I. S. and Lopez, F.,** Monoclonal antibody to transferrin receptor blocks transferrin binding and inhibits human tumor cell growth *in vitro, Proc. Natl. Acad. Sci. U.S.A.,* 79, 1175, 1982.
115. **Schmidt, J. A., Marshall, J., Hayman, M. J., Ponka, P., and Beug, H.,** Control of erythroid differentiation: possible role of the transferrin cycle, *Cell,* 46, 41, 1986.
116. **Sawyer, S. T. and Krantz, S. B.,** Transferrin receptor number, synthesis, and endocytosis during erythropoietin-induced maturation of Friend virus-infected erythroid cells, *J. Biol. Chem.,* 261, 9187, 1986.
117. **Beach, R. L., Popiela, H., and Festoff, B. W.,** The identification of neurotrophic factor as a transferrin, *FEBS Lett.,* 156, 151, 1983.
118. **Shimo-Oka, T., Hagiwara, Y., and Ozawa, E.,** Class specificity of transferrin as a muscle trophic factor, *J. Cell Physiol.,* 126, 341, 1986.
119. **Markelonis, G. J. and Oh, T. H.,** Transferrin: assay of myotrophic effects and method for immunocytochemical localization, *Methods Enzymol.,* 147, 291, 1987.
120. **Ekblom, P., Thesleff, I., Saxen, L., Miettinen, A., and Timpl, R.,** Transferrin as a fetal growth factor: acquisition of responsiveness related to embryonic induction, *Proc. Natl. Acad. Sci. U.S.A.,* 80, 2651, 1983.
121. **Hamilton, T. A., Weiel, J. E., and Adams, D. O.,** Expression of the transferrin receptor in murine peritoneal macrophages is modulated in the different stages of activation, *J. Immunol.,* 132, 2225, 1984.
122. **Andreesen, R., Osterholz, J., Bodemann, H., Bross, K. J., Costabel, U., and Löhr, G. W.,** Expression of transferrin receptors and intracellular ferritin during terminal differentiation of human monocytes, *Blut,* 49, 195, 1984.
123. **Hirata, T., Bitterman, P. B., Mornex, J.-F., and Crystal, R. G.,** Expression of the transferrin receptor gene during the process of mononuclear phagocyte maturation, *J. Immunol.,* 136, 1339, 1986.
124. **Taetle, R. and Honeysett, J. M.,** γ-Interferon modulates human monocyte/macrophage transferrin receptor expression, *Blood,* 71, 1590, 1988.
125. **Hill, J. M., Ruff, M. R., Weber, R. J., and Pert, C. B.,** Transferrin receptors in rat brain: neuropeptide-like pattern and relationship to iron distribution, *Proc. Natl. Acad. Sci. U.S.A.,* 82, 4553, 1985.
126. **Dulbecco, R.,** La nature du cancer, *La Recherche,* 13, 1426, 1982.
127. **Letendre, E. D.,** The importance of iron in the pathogenesis of infection and neoplasia, *TIBS,* 10, 166, 1985.

128. **Diamond, A., Cooper, G. M., Ritz, J., and Lane, M.-A.,** Identification and molecular cloning of the human Blym transforming gene activated in Burkitt's lymphomas, *Nature,* 305, 112, 1983.
129. **Finkel, T. and Cooper, G. M.,** Detection of a molecular complex between ras protein and transferrin receptor, *Cell,* 136, 1115, 1984.
130. **Harford, J.,** An artefact explains the apparent association of the transferrin receptor with a ras gene product, *Nature,* 311, 673, 1984.
131. **Vidal, C., Matsushita, S., Colamonici, O. R., Trepel, J. B., Mitsuya, H., and Neckers, L. M.,** Human T lymphotropic virus I infection deregulates surface expression of the transferrin receptor, *J. Immunol.,* 141, 984, 1988.
132. **Sorokin, L. M., Morgan, E. H., and Yeoh, G. C. T.,** Differences in transferrin receptor function between normal developing and transformed myogenic cells as revealed by differential effects of phorbol ester on receptor distribution and rates of iron uptake, *J. Biol. Chem.,* 263, 14128, 1988.
133. **Trowbridge, I. S.,** Transferrin receptor as a potential therapeutic target, *Prog. Allergy,* 45, 121, 1988.
134. **Musgrove, E., Rugg, C., Taylor, I., and Hedley, D.,** Transferrin receptor expression during exponential and plateau phase growth of human tumour cells in culture, *J. Cell Physiol.,* 118, 6, 1984.
135. **Taetle, R., Honeysett, J. M., and Trowbridge, I.,** Effects of anti-transferrin receptor antibodies on growth of normal and malignant myeloid cells, *Int. J. Cancer,* 32, 343, 1983.
136. **Taetle, R. and Honeysett, J. M.,** Effects of monoclonal anti-transferrin receptor antibodies on *in vitro* growth of human solid tumor cells, *Cancer Res.,* 47, 2040, 1987.
137. **van der Ende, A., du Maine, A., Simmons, C. F., Schwartz, A. L., and Strous, G. J.,** Iron metabolism in BeWo chorion carcinoma cells. Transferrin-mediated uptake and release of iron, *J. Biol. Chem.,* 262, 8910, 1987.
138. **Brock, J. H., Mainou-Fowler, T., and Webster, L. M.,** Evidence that transferrin may function exclusively as an iron donor in promoting lymphocyte proliferation, *Immunology,* 57, 105, 1986.
139. **Brock, J. H.,** The effect of iron and transferrin on the response of serum-free cultures of mouse lymphocytes to Concanavalin A and lipopolysaccharide, *Immunology,* 43, 387, 1981.
140. **Hellström, K. E. and Hellström, I.,** Oncogene-associated tumor antigens as targets for immunotherapy, *FASEB J.,* 3, 1715, 1989.
141. **Sauvage, C. A., Mendelsohn, J. C., Lesley, J. F., and Trowbridge, I. S.,** Effects of monoclonal antibodies that block transferrin receptor function on the *in vivo* growth of a syngeneic murine leukemia, *Cancer Res.,* 47, 747, 1987.
142. **Kemp, J. D., Thorson, J. A., McAlmont, T. H., Horowitz, M., Cowdery, J. S., and Ballas, Z. K.,** Role of the transferrin receptor in lymphocyte growth: a rat IgG monoclonal antibody against the murine transferrin receptor produces highly selective inhibition of T and B cell activation protocols, *J. Immunol.,* 138, 2422, 1987.
143. **Crane, F. L., Sun, I., Clark, M. G., Grebing, G., and Löw, H.,** Transplasma membrane redox systems in growth and development, *Biochim. Biophys. Acta,* 811, 233, 1985.
144. **Crane, F. L., Löw, H., Sun, I. L., and Isaksson, M.,** The requirement for transmembranal electron flow for cell proliferation, in *Plasma Membrane Oxidoreductases in Control of Animal and Plant Growth,* Crane, F. L., Morré, D. J., and Löw, H., Eds., Plenum Press, New York, 1988, 37.
145. **Morré, D. J., Brightman, A., Wang, J., Barr, R., and Crane, F. L.,** Roles for plasma membrane redox system in cell growth, in *Plasma Membrane Oxidoreductases in Control of Animal and Plant Growth,* Crane, F. L., Morré, D. J., and Löw, H., Eds., Plenum Press, New York, 1988, 45.
146. The data presented in Figures 2 and 3 correspond to unpublished results from P. Basset and R. G. Crystal.

Chapter 10

TRANSFERRIN RECEPTOR GROWTH CONTROL IN NORMAL AND TRANSFORMED CELLS

W. Page Faulk, Hava Harats, and Alajos Berczi

TABLE OF CONTENTS

I. INTRODUCTION

Transferrin receptors have been characterized biochemically and found to be transmembrane, homodimeric glycoproteins with a molecular weight of 180 kDa.[1] Each 90-kDa subunit is covalently attached through a single disulfide bond.[2] The gene for human transferrin receptor has been cloned and the primary structure determined.[1,2] Genes for both transferrin and transferrin receptors are located on chromosome 3. The main known role of these receptors is to bind and facilitate entry of holotransferrin into cells via a process known as receptor-mediated endocytosis.[3,4] Red blood cell precursors which require large amounts of iron for hemoglobulin production were the first cells shown to possess specific membrane receptors for transferrin. Other tissues including transformed cells[5,6] and human placentas[7,8] have been found to possess transferrin receptors. Although these cells require a certain amount of iron, it is not known why they manifest permanently upregulated transferrin receptors. In this chapter we explore the possibility that these receptors are involved in the regulation of growth in normal and abnormal cells.

II. STUDIES OF ABNORMAL CELLS

A. CANCER

Insight into the relationship of transferrin receptors with growth has been gained by studying several different medical conditions. For example, cancer cells escape from normal growth control and often metastasize to distant sites where they proliferate and destroy host tissues. Immunohistological studies of biopsies from patients with breast cancer revealed that normal ductal epithelial cells do not manifest transfer receptors, but adenocarcinoma cells in the same biopsies were richly endowed with transferrin receptors.[9] This original observation has been confirmed many times, and a large number of tumors now have been found to react with antibodies to transferrin receptors. These include lymphomas,[10] leukemias,[11] hepatocellular carcinomas,[12] gastrointestinal tumors,[13] bladder cancer,[14] and cancers of kidney, lung, and ovary.[15,16] These observations suggest that the production of transferrin receptors is a characteristic of cancer cells. Indeed, this finding has been used in acute myelogenous leukemia to identify patients in remission who are escaping from therapeutic control of their disease.[17]

Since the presentation of transferrin receptors appears to be a characteristic of cancer cells, a question arises as to whether these are the same receptors that can be identified on normal cells. The intellectual basis of this question comes from reports of receptor polymorphism as defined by biochemical and immunological studies of transferrin receptors and antibodies to transferrin receptors.[18-20] In light of this, normal and tumor cells have been metabolically labeled with ^{35}S-methionine, radiolabeled receptors were isolated, partially digested with *Staphylococcus aureus* protease, and studied for their peptide mapping patterns in SDS polyacrylamide gels. Identical peptide maps were obtained from normal cells, neoplastic cells, and transformed cells from long-term cultures.[21] Another experimental approach used the lactoperoxidase technique to label cells with ^{125}I and obtained a similar result.[16] However, in this study the detergent-solubilized receptors were immunoprecipitated with several different monoclonal antibodies to transferrin receptors. Peptide map analyses of the receptors recognized by each antibody from the same cells were identical, but differences were observed between different types of cells. These observations could reflect differences in amino acid sequence, glycosylation,[5,22] or other types of modifications such as phosphorylation, but they suggest that allotypic variations can exist in transferrin receptors from cells of differing genetic backgrounds.

The appearance of transferrin receptors on cancer cells has been used as an *in vivo* approach to tumor localization and characterization. A patient with nasopharyngeal carcinoma intravenously received ^{131}I-labeled mouse monoclonal IgG1 antibody HuLy-m9 to transferrin receptor, and was scanned with a gamma camera 24 h later. The tumor was accurately localized in the nasopharynx, and a secondary lesion was identified in the brain.[16] Another approach has utilized immunologically compromised nude mice as living test tubes. In these experiments, mice were given known numbers of human tumor cells by either the intravenous route or by injection directly into solid organs. They then were injected intravenously with ^{125}I-labeled human transferrin at various times following inoculation. The animals were scanned for radioactivity at 24, 48, 72, and 96 h. The results showed that small numbers of tumor cells ($<$5000) could be localized, and, depending on the cell type, metastatic tumor cells could be identified in draining lymph nodes.[22a] This model offers a novel method for studying the growth of human cancer cells in an immunologically incompetent laboratory animal. It thus removes complicating microenvironmental variables that could affect the presentation of transferrin receptors, and presents a relatively neutral biological environment in which cell growth can be correlated with receptor presentation.

In light of the presence of transferrin receptors on human cancer cells, an idea was proposed in 1980 that the receptors could be utilized as targets to bind drug conjugates of transferrin.[9] A similar proposal made in 1981 was that transferrin receptors could be used as targets to bind drug conjugates of antibodies to transferrin receptors.[23] The concept of drug targeting with antibody has been actively pursued with the use of antibodies conjugated with doxorubicin,[24] methotrexate,[25] ricin A chain,[26] and other chemotherapeutic agents. This may be a promising approach, but its use of monoclonal antibodies as the delivery system causes certain problems, particularly anaphylactic reactions and antigen shedding.

The problems caused by monoclonal antibodies are diminished if not removed if the delivery system is transferrin, which is the natural ligand for transferrin receptors. Transferrin conjugates of the anticancer drug adriamycin have been shown to selectively target tumor cells *in vivo*, and the targeted cells are killed.[17,27] These conjugates have been used in the treatment of patients with acute leukemia, and the results have shown diminished numbers of leukemic cells in peripheral blood, no increase of leukemic cells in bone marrow biopsies, and no anaphylactic reactions in the patients.[28] The same conjugates have proven to be a useful laboratory tool in studies of the mechanism of death in tumor cells targeted *in vitro* with adriamycin conjugates of human transferrin.[29]

B. INFECTIONS

Pathological conditions of cellular growth in addition to cancer have also been studied for transferrin receptors. Virus infection is an obvious example, because virus-infected cells are often transformed and escape the normal modulating forces of contact inhibition and proliferation control. Indeed, HTLV-1 virus-infected T-lymphocytes have been reported to overexpress interleukin-2 (IL-2) receptors, and it has been suggested that they overexpress transferrin receptors.[30,31] Studies of HTLV-1 virus-infected T cells have shown a fivefold increase in surface expression of transferrin receptors, but the growth rates of the virus-infected cells did not differ from uninfected controls.[32] Although the infected cells hyper-expressed receptor on their plasma membranes, they did not possess a greater total number of receptors than their normal counterparts, and they manifested equivalent transferrin receptor mRNA as uninfected controls. However, the infected cells had a marked defect in ferric-transferrin internalization, receptor phosphorylation, and response to phorbol ester stimulation. These data indicate that HTLV-1 infection of human T cells is accompanied by surface transferrin receptor overexpression secondary to subcellular redistribution and defective internalization.

Another example of infection in which transferrin receptors have been studied is *Plasmodium falciparum*-infected human erythrocytes. This is an instructive model because erythrocytes neither proliferate nor do they manifest transferrin receptors. Plasmodia require iron but they are enzymatically incapable of acquiring iron from the hemoglobin of erythrocytes in which they reside. These observations prompted investigators to search for transferrin receptors on the surface of plasmodia-infected human erythrocytes.[33] Extracts of ^{35}S metabolically labeled *P. falciparum* trophozoites have been passed through solid-phase affinity columns of either antitransferrin or ferric-transferrin. The columns were extensively washed and the affinity beads were removed, boiled in SDS with mercaptoethanol, and the supernatants were studied in SDS polyacrylamide gels. The most prominent protein recovered from both affinity columns was a 93-kDa entity that was shown in subsequent experiments to bind human transferrin. The 93-kDa protein was interpreted as representing a plasmodial transferrin receptor that functioned to bind and transport human ferric transferrin to provide iron for this parasite.

The finding of transferrin receptors on plasmodia-infected erythrocytes has recently been extended to show that the infected cells manifest diferric transferrin reductase.[34] Like other diferric transferrin reductases, the enzyme was driven by ferric transferrin and inhibited by apotransferrin. Interestingly, it was not inhibited by antibody to human transferrin receptor but was inhibited (40%) by antibody to human transferrin. These data taken together support the idea that plasmodia in human erythrocytes assemble a transferrin receptor with specificity for human transferrin. Dissection of the molecular biology of this system holds promise for casting light on the role of transferrin receptors in growth.

C. TRANSPLANTATION

It is not clear why up-regulation of transferrin receptors is associated with cellular growth. Evidence has been put forward that this is related to iron requirements, because treatment of cells with the iron chelator desferrioxamine results in an increase in total cell receptors.[35] The chelator had no effect or receptor half-life, and the increased number of receptors was accounted for by an increase in receptor biosynthesis associated with an increase in mRNA for receptor unaccompanied by an increase in total protein synthesis. However, there are other mechanisms for the up-regulation of other receptors in other cells, and some of these may be relevant to transferrin receptor control, iron concentrations notwithstanding. For example, antigen up-regulates interleukin-2 receptors on activated T-lymphocytes,[36] and cytokines from activated immune cells cause endothelial cells to down-regulate thrombin receptors[37] and up-regulate numerous molecules to the plasma membrane.[38] Potent mediators of receptor mobility in plasma membranes are released *in vitro* during mixed lymphocyte culture reactions,[39] and similar if not identical mediators are generated at the donor-recipient interface of transplanted organs.[40] Thus, the appearance of transferrin receptors on circulating lymphocytes has been studied in human transplant recipients and controls.

Fluorescein-labeled monoclonal antibody OKT-9 was used in flow cytometry to assess the percentage of reactive lymphocytes from 15 healthy nontransplanted controls and 19 cardiac transplant recipients.[41] Results showed 3.5% for controls, 5% for stable recipients, 10.4% for prerejection recipients, and 13.4% for recipients who were rejecting their grafts. After treatment of rejection episodes, reactive cells fell to 5.3%. Inadequate treatment and complicating infections were not associated with a decrease of reactive cells. Thus, the number of circulating lymphocytes with transferrin receptors is a telling index of impending allograft rejection, and the failure of decline in the number of reactive cells suggests inadequate treatment or the presence of a complicating infection.

The observation of increased numbers of transferrin receptor-positive cells in the peripheral blood circulation has been extended to more precisely define the role of rejection versus the contribution of infection in the appearance of reactive cells.[42] This was done by measuring the percentage of lymphocytes with transferrin receptors as well as the T-helper/T-suppressor cytotoxic lymphocyte ratio. Results from heart transplant recipients showed that during stable periods the percentage of receptor-positive cells was 5.9 and the T-helper/

T-suppressor ratio was 1.5. During rejection, the receptor-positive cells were 12% and the T-helper/T-suppressor ratio reached 2.3. Both receptor-positive lymphocytes and T-helper/T-suppressor ratios fell following treatment of the rejection episodes. In contrast, patients with infectious complications had elevated percentages of receptor-positive cells, but their T-helper/T-suppressor ratios were not elevated. These observations suggest that products of the allogeneic recognition and rejection reaction cause peripheral blood lymphocytes to up-regulate transferrin receptors. A similar up-regulation is caused by infection, but infection can be distinguished from rejection by a lack of concomitant changes in the T-helper/T-suppressor ratio of cytotoxic lymphocytes.

The appearance of transferrin receptors on lymphocytes in the blood of transplant recipients has offered a novel approach to the problem of allograft rejection. Since only activated T-lymphocytes express transferrin receptors, it has been reasoned that blocking of cytotoxic antibodies to transferrin receptors might be used as a way of specifically eliminating antigen-activated T cells. The primary *in vitro* cytotoxic T-lymphocyte (CTL) response to alloantigens in the mixed lymphocyte culture (MLC) reaction has been used as a model system for the removal of alloreactive T cells.[43] Two rat mAb to mouse transferrin receptors were studied. These were R17 208, which is a noncomplement-fixing IgM, and REM17.2, which is a complement-fixing IgM. Treatment of MLC cells with the complement-fixing antibody plus complement destroyed all CTL effector activity. This result showed that effector cells expressed the receptor. However, effector CTL were not irreversibly eliminated, because strong effector responses to the alloantigeneic cells were generated in the same culture following restimulation. Since allogeneic organ grafts provide chronic antigeneic stimulation, a method is required to eliminate alloreactive CTL to such a degree that they do not appear after restimulation. This result was obtained when MLC was performed in the presence of noncomplement-fixing antibody to transferrin receptor. These findings offer an exciting new approach to the problem of graft rejection. More important to the aim and goal of this chapter, these results suggest that transferrin receptors are up-regulated as a consequence of soluble mediators that are released from allogeneically activated T cells.

III. MECHANISMS OF RECEPTOR CONTROL

A. STUDIES OF IMMUNE CELLS

The mechanisms of cellular growth have been studied in many different types of cells. Very few general messages have emerged from these studies, but it is generally agreed that proliferating cells manifest transferrin receptors.[42] A marked increase in the number of transferrin receptors occurs when stationary cells are stimulated to proliferate[44-48] and the number of receptors diminishes when cells stop dividing.[49,50] It thus has been concluded that these receptors represent a central mechanism in cellular growth. This is an observational conclusion, and it says nothing about why transferrin receptors are associated with growth. Some investigators have cited evidence to support an idea that the growth function of transferrin receptors is to bind and deliver ferric-transferrin to the intracellular compartment where the iron is released to function in DNA metabolism.[51] This may be true, but it is not proven. Indeed, there are equally compelling data to suggest that transferrin receptors may be associated with an iron-independent system that is necessary for cellular growth.[52] This possibility is represented by the observation that many different cellular activation protocols can be inhibited by antibodies to transferrin receptors that do not inhibit transferrin binding to its receptor.[53]

A useful approach to investigations of the roll of transferrin receptors in cellular growth is to study receptors on immune cells. This in part is because the mechanisms of immune cell activation are fairly well understood, and differentiated function can be measured by studies of antibody synthesis. Generally, T cell activation causes B cell activation, which is followed by antibody production. Although this model has many imponderables, contemporary data indicate that antigens react with T-lymphocytes through the mediation of a heterodimeric receptor that is part of a macromolecular complex involving the three invariant polypeptide chains of the T_3 antigen.[54] Ligand-receptor interaction results in the expression of genes that encode for IL-2 and its high affinity receptor, IL-2R.[55] Nuclear transcription assays performed with isolated nuclei from human T-lymphocytes stimulated by phytohemagglutinin and phorbol myristate acetate have shown that the genes encoding for IL-2 and IL-2R are induced after transcription of c-*myc* and before transcription of the transferrin receptor gene.[56]

The induced IL-2 and IL-2R form a ligand-receptor complex on the plasma membranes of activated T cells. This precedes RNA synthesis, which occurs about 48 h after activation.[57] After IL-2:IL-2R complex formation and the initiation of RNA synthesis, transferrin receptors appear on plasma membranes of the activated cells.[47] The appearance of transferrin receptors is followed by DNA synthesis and cellular proliferation, which occurs about 72 h after activation.[58] Thus, the cell cycle of activated T-lymphocytes is characterized by IL-2R on plasma membranes early in G_1 and by the presence of transferrin receptors late in G_1. These activation receptors are potential restriction points in cellular growth. This has been shown by the ability of antibodies to IL-2R (so-called, anti-Tac) and antibodies to transferrin receptors to inhibit lymphocyte progression through the cell cycle.[59,60]

After the appearance of transferrin receptors on plasma membranes, activated T cells release B cell growth factor (BCGF). This observation is supported by the observation that a monoclonal antibody (NDA9) to transferrin receptor inhibits the release of BCGF.[61] If BCGF is released, it is bound by BCGF receptors on activated B-lymphocytes.[62] After BCGF is bound by its receptor on activated B cells, transferrin receptors appear.[63] Following the appearance of transferrin receptors, the activated B cells proliferate. However, proliferation does not occur in the presence of antibodies to transferrin receptors. Thus, BCGF controls growth in B cells by initiating the appearance of transferrin receptors. The growth promoting function of BCGF is supported by the observation that BCGF from T cell hybridomes causes B cells to proliferate,[64] and the proliferation-associated function of transferrin receptors is supported by observations that antibodies to the receptors inhibit proliferation.

Investigations of B-lymphocytes allow studies of cellular differentiation as well as proliferation, because differentiated B cell synthesize and secrete antibody. Interestingly, blockade of B cell proliferation by antibodies to transferrin receptors does not block differentiation, as measured by antibody synthesis.[63] Indeed, mRNA for the constant region of IgM H-chains is detected 18 to 24 h after stimulation, regardless of the presence of antibody to transferrin receptor. Such antibody-treated B cells thus produce immunoglobulin in the absence of cellular proliferation. Phorbol esters can also induce immunoglobulin mRNA and antibody secretion in the absence of cellular proliferation,[65] but antibody secretion is normally initiated in B cells by T cell replacing factor or B cell differentiating factor from activated T cells.[66,67] Support for the role of T cell replacing factor in B cell differentiation is drawn from the observation that T cell replacing factor causes B cells to secrete immunoglobulin but has no effect on proliferation.[68] In consideration of cellular growth, it is important to distinguish differentiation from proliferation, as exemplified in the above studies on B-lymphocytes. To our knowledge, this has not been done in any other type of cell.

B. STUDIES WITH MONOCLONAL ANTIBODIES

1. Type of Cell Target

In 1980, a mAb (B3/25) was reported that identified a human antigen present on most cultured hematopoietic tumor cell lines and a variety of nonhematopoietic tumor cells.[69] The antigen was shown to be a 200-kDa glycoprotein that reduced to two 100-kDa subunits. Though not present on resting lymphocytes, it was identified on mitogen-stimulated cells.

During the same year, reports appeared of transferrin receptors on many cell lines maintained *in vitro*[70] and on fresh adenocarcinoma tissues obtained from breast biopsies.[9] In 1981, two independent reports appeared which provided evidence that the reducible 200 kDa glycoprotein on proliferating cells was transferrin receptor.[44,6] This conclusion was supported by results from several laboratories published during the same time.[71-73] These observations taken together indicated that tumor cells and mitogen-stimulated normal cells[45,74] manifested transferrin receptors on their plasma membranes.

The B3/25 mAb was shown to inhibit the growth of a human melanoma cell line in nude mice, but it did not interfere with either transferrin binding or cell growth *in vitro*.[23] This report prompted interest in the role of transferrin receptors in growth. In 1982, Trowbridge and Lopez[75] produced another mAb (42/6) to human transferrin receptors and showed that it, in contrast to B3/25, inhibited human tumor cell growth *in vitro*. This mAb blocked transferrin binding to its receptor, giving rise to the possibility that the growth inhibition property of 42/6 was due to its ability to impede the traffic of iron. However, this possibility was discredited by the observation that 42/6-mediated inhibition was not overcome by the addition of iron to the cell cultures. This finding opened the possibility that antibody binding to transferrin receptor might produce a growth inhibitory signal unrelated to the iron transport function of the receptor. However, this possibility was deemed unlikely because results of cross-blocking experiments indicated that the antigenic sites recognized by B3/25 and 42/6 were closely related, though 42/6 blocked growth and B3/25 did not. Another difference between B3/25 and 42/6 was that B3/25 was an IgG and 42/6 was an IgA. The importance of this difference was later to become evident.

Transferrin receptor research during the early 1980s was primarily limited to human cells. A 1981 report had shown striking similarities between murine and human transferrin receptors,[76] though mouse mAb to human receptors did not recognize mouse receptors. The availability of mAb to mouse receptors would allow more flexible experimentation to study the role of transferrin receptors in growth, and in 1982 a rat mAb (R17 208) to murine receptors was reported.[77] This antibody was an IgM that inhibited the uptake of transferrin-bound radiolabeled iron, and inhibited the growth of a murine myeloma cell line *in vitro*. Two puzzling and as yet unexplained observations arose from the growth inhibition studies. First was the finding that mouse myeloma cells grown with R17 208 accumulated in G_2+M rather than S phase, as had been found for human cells grown with 42/6. Second was the finding that R17 208 inhibited transferrin-mediated iron uptake by both mouse myeloma and L cells, but caused growth inhibition only of the myeloma cells.

The second finding suggests mouse L cells may have alternative mechanisms for acquiring iron. In support of this, certain cells do acquire iron by a saturable receptor-mediated process that is inhibited by antibody to transferrin receptor, and by a nonsaturable process that is not inhibited by antibody,[78] though it is not known how these processes are partitioned between different types of cells. Alternatively, the putative growth inhibitory signal unrelated to iron transport function of the receptor could be transmitted subsequent to antibody-receptor binding, and L cells may not manifest such a signaling system. There are other possibilities. Whatever the mechanism, it cannot be accepted as generic that antibodies that block transferrin binding to their receptors are growth inhibitory. This statement is supported by the report of a mouse IgG mAb (GB18) that inhibits the binding of holotransferrin to its receptor, but fails to inhibit the growth of human peripheral blood lymphocytes following mitogenic or alloantigeneic stimulation, and another mouse IgM mAb (GB22) that does not inhibit the binding of holotransferrin to its receptor but inhibits the growth of human peripheral blood lymphocytes following mitogenic or alloantigeneic stimulation.[79]

2. Antibody Isotype and Receptor Number, Affinity and Epitope

A step closer to understanding why some antibodies to transferrin receptors inhibit cellular growth *in vitro* and others do not was made by considering the isotypes of the antibodies. The IgG mAb B3/25 was found not to block either transferrin binding[75] or mitogen-stimulated cell growth,[80] but IgA mAb 42/6 blocked both.[75] This raised the possibility that inhibition of growth was due either to the blockade of transferrin binding or was a property of antibody isotype. The former possibility is unlikely because several different laboratories have reported IgM mAb to either human[79] or mouse[81] transferrin receptors that inhibit cellular proliferation but do not block transferrin binding, although some IgM mAb both inhibit growth and block ligand formation.[77] It thus appears that the ability of antibodies to inhibit growth is primarily a function of their isotype.

The above comments have considered growth inhibition in terms of the properties of mAb to transferrin receptors. The most likely properties of cells that could determine the effects of antibodies are the number of available transferrin receptors on targeted cells, and the affinities of the available receptors. The number of available transferrin receptors, of course, is important, but they do not appear to be critical in normally growing cells. Indeed, no association has been found between antibody-mediated membrane fluorescence of receptors and the ability of the same antibody to inhibit growth of the same cells *in vitro*,[82] though no cell line that was susceptible to antibody-mediated growth inhibition was devoid of membrane immunofluorescence. Some cells were brilliantly fluorescent but were not inhibited, and other lines reacted minimally but were profoundly inhibited. In contrast, when the human promyelocytic leukemia cell line HL60 was differentiated *in vitro* with dimethyl sulfoxide, the display of transferrin receptors on plasma membranes decreased[83] and growth inhibition by mAb 42/6 diminished.[84] It is thus possible to conclude that the presence of large numbers of available transferrin receptors on plasma membranes is not a determining variable of whether growth will be inhibited by antibody to transferrin receptor, but receptor must be present for antibody-mediated inhibition to occur.

The possibility of receptor affinity being a variable in the ability of antibody to inhibit growth is difficult to study. That affinity may play a role is suggested by the observation that more transferrin receptors were identified per cell by antibody to receptor than by the binding of radiolabeled transferrin. The numbers for mitogen-stimulated peripheral blood mononucleated cells were 9.6×10^3 for transferrin and 25×10^3 for mAb 42/6.[60] The numbers for HL60 cells were 2.9×10^4 for transferrin and 1.2×10^5 for mAb 42/6.[84] One possibility for this discrepancy is cells may have low affinity receptors that are recognized by antibody but do not bind measurable amounts of transferrin. In order to test this possibility, cells were grown in serum-free and transferrin-free media to study whether varying transferrin concentrations would alter the affinity of binding. Comparative data for transferrin binding by HL60 cells grown in fetal calf serum and transferrin-supplemented serum-free medium showed no differences in either receptor number or affinity. When the same cells were grown in iron-supplemented, transferrin-free, serum-free medium, the number of receptors were markedly decreased (2.9×10^4 versus 5.0×10^3) but the affinity was not changed ($K_d = 3.8 \times 10^{-9}$ *M* vs. 3.1×10^{-9} *M*). The same experiments showed no change in the binding affinity of 42/6 for receptor ($K_d = 9.6 \times 10^{-9}$ *M* vs. 5.7×10^{-9} *M*). Thus, it is unlikely that variations in ligand affinities account for the effect of antibodies to transferrin receptors on cell growth.

The ability of antibodies to transferrin receptors to inhibit growth does not seem to be dependent on cell type, as long as the cells are grown in complete medium. This suggests the effect is antibody dependent and data have been cited that show IgA and IgM antibodies are more potent inhibitors than IgG. Results of cross-blocking experiments have shown that the epitopes bound by antibodies which inhibit can be partially if not wholly blocked by antibodies that do not inhibit,[75,79,81] indicating the epitope recognized by antibody is not the determining factor in growth inhibition. These observations suggest the effects of IgA and IgM antibodies on transferrin receptors differ from the effects caused by IgG antibodies. Results of several telling experiments have shown that this effect is receptor cross-linking.

The secondary effect of such cross-linking may be iron deprivation to cells, and consequent deficient iron availability for DNA replication.

3. Importance of Receptor Cross-Linking

The cross-linking experiments first done in a human system utilized mouse IgG mAb B3/25, which does not inhibit growth. When anti-mouse immunoglobulin was added to lymphocyte cultures containing B3/25, lymphocyte growth was inhibited.[84] An in-depth study of this phenomenon has been done in the mouse system.[81,85] When AKR/J mouse thymic lymphoma cells (AKR1) were grown in the presence of rat mAb R17 208 (IgM) or REM17 (IgM), the relative cloning frequency for AKR1 was decreased to 5×10^{-5} and 1×10^{-4}, respectively. In contrast, the relative cloning frequency for the same cells grown with R17 217 (IgG2a) was 0.58. However, when the cells were grown in the presence of R17 217 plus anti-rat immunoglobulin, the relative cloning frequency fell to 1×10^{-4}, which is comparable with that obtained when the cells were grown with rat IgM mAb antibodies to mouse transferrin receptors. Two important controls were done with these experiments. First, anti-rat immunoglobulin had no effect on growth of AKR1 cells. Second, two other rat antibodies (anti-T200 and anti-Thy-1) to sites other than transferrin receptors on AKR1 cells had no significant effects on growth when present in cultures with antibodies to rat immunoglobulins. It thus appears that the transmission of a growth inhibiting signal is mediated by the cross-linking of transferrin receptors on plasma membranes.

If IgM anti-transferrin receptor sends a growth inhibition signal by cross-linking receptors, the question was asked how long must antibody be present to transmit the signal? This was approached by pulse exposure of cells to antibody for 45 min at 4°C following by washing and incubation at 37°C for various times. In these pulse experiments, antibody was rapidly cleared from the cell surface.[81] Further insight into the time requirements for antibody was obtained by culture experiments. When cells were cultured with antibody for 24 h, washed and recultured, about half the cells were free of antibody at 24 h, but upon reincubation with antibody to transferrin receptor, it was discovered that new receptor had appeared. These cells grew normally after a 1-d lag. If cells were cultured for 3 d with antibody, washed, and recultured, the growth lag was 2.5 d. If cells were cultured for 7 d with antibody, washed, and recultured, the cells did not grow. Thus, antibody must be present in the growth medium for several days to obtain complete inhibition of growth.

The rapid clearance of IgM antibody from cells, the appearance of new receptors following short incubation times with antibody, and the requirement for prolonged exposure to antibody to impede growth suggest cells respond to receptor cross-linking by up-regulating more receptor. The loss of antibody from plasma membranes could result from a secretion of antibody-receptor complexes, as occurs when IL-2 positive activated T-lymphocytes are incubated with anti-IL-2 (57). However, the appearance of new transferrin receptors would be bound by excess antibody in the medium, accounting for the eventual growth inhibition that occurs with prolonged exposure to antibody. To study if transferrin receptors are made available on the surface of cells grown in the presence of IgM growth inhibiting antibodies, AKR1 mouse cells were grown with rat IgM mAb R17 208 for various times, chilled to 4°C, washed in a sodium azide-containing buffer (to inhibit endocytosis) and then reacted with R17 208 and fluorescein-conjugated anti-rat immunoglobulin and examined in a flow microfluorimeter.[81] The units of mean relative fluorescence were 15.13 at time zero (control), 28.24 at 24 h, and 60 at 48 h, indicating antibody accumulated on plasma membranes, presumably due to the progressive availability of receptor.

The above flow microfluorimeter experiments were repeated with mouse AKR1 cells, and the nongrowth inhibiting rat IgG mAb R17 217, and the results were quite different.[81] After 30 min many of the cells had fewer receptors, after 180 min all but 9% of the cells had fewer receptors, and by 24 h all the cells had fewer receptors. The units of mean relative fluorescence were 38.25 for time zero (control) and 2.17 at 24 h.[81] These values were confirmed by the results of radiolabeled transferrin binding which showed a 3 to 8 times binding decrease at 24 and 48 h as compared with control values. When radiolabeled transferrin binding experiments were done with AKR1 cells grown with the growth inhibiting IgM mAb R17 208, there was not a similar reduction in binding, and some experiments demonstrated increased transferrin binding. It thus appears that IgM cross-linked receptors are secreted or remain on plasma membranes accompanied by new receptors, and IgG-bound receptors may be secreted but do not seem to be accompanied by new receptor. Indeed, it appears likely that IgG treated cells down-regulate receptor.

To investigate the possibility that IgG anti-transferrin receptor treated cells down-regulate and degrade receptors, mouse AKR1 cells were surface iodinated with lactoperoxidase and cultured without (control) or with rat IgG mAb R17 217.[81] Following culture for various times, cells were harvested, solubilized in detergent, immunoprecipitated with R17 217 and anti-rat immunoglobulin, and reduced and electrophoresed in SDS containing polyacrylamide gels from which autoradiograms were produced. The 95-kDa transferrin receptor band disappeared from the control cultures with a half-life of 11 h, and disappeared from the R17 217 treated cells with a half-life of 2 h. In addition, autoradiograms from the antibody-treated cells displayed obvious degradation bands at about 30 kDa. When these experiments were repeated with cells grown in the presence of IgM antibodies to transferrin receptors, the results were the same as those obtained from control cells. It thus appears that IgG antibodies to transferrin receptors initiate accelerated receptor degradation. When this result is considered together with results from the flow microfluorimeter and radiolabeled transferrin binding experiments, it is possible to think that IgG antireceptor causes receptor down-regulation and degradation, and IgM antireceptor does not.

In order to more comprehensively study the effects of antibodies to transferrin receptors on growth, cells have been prepared that are not inhibited by antibody.[85] This was done by growing mouse AKR1 cells in permissive concentrations of rat IgM mAb R17 208 until a cell line arose that was not susceptible to growth inhibition by R17 208. The R17 208-resistant cells have few receptors with the epitope necessary for binding by R17 208, and they have another population of mutant receptors that are recognized by other antibodies to transferrin receptors. This has been interpreted as evidence for a mutation in one copy of the transferrin receptor structural gene.[85] The cells grew in the presence of R17 208, but growth was inhibited in the presence of anti-rat immunoglobulin, indicating the few R17 208 epitope-bearing receptors were cross-linked and inhibited growth of the R17 208-resistant cells. Similarly, rat IgG mAb R17 217 did not inhibit mutants, but inhibition was achieved when anti-rat immunoglobulin was added. Mutant cells grown overnight with IgM R17 208 were found by flow microfluorimeter analysis to have lost most of their surface R17 208 reactivity (units of mean relative fluorescence fell from 5.0 to 1.6), and the reactivity of the same cells with another mAb to transferrin receptors was unchanged. These observations made on cells with low density receptors and thus not cross-linkable receptors by a single IgM antibody, support the idea that cross-linking of receptors inhibits growth, and that non-cross-linked receptors are down-regulated and degraded.

Results thus far suggest that transferrin receptors must maintain their intracellular and plasma membrane traffic of transferrin if cellular growth is to be sustained. Perturbations in trafficking can be accommodated if transferrin internalization is not blocked. This point is supported by observations that

1. Monovalent antibody fragments to transferrin receptors cannot cross-link and do not induce internalization[7]
2. Bivalent IgG antibodies to transferrin receptors do not sufficiently cross-link to impede internalization, though they down-regulate receptors

3. Polyvalent IgM antibodies cross-link transferrin receptors and inhibit growth
4. IgG antibodies to transferrin receptors cross-linked with anti-immunoglobulin exert growth inhibition similar to that caused by IgM antibodies

These findings favor a hypothesis that the primary role of transferrin receptors in growth is their control of transferrin internalization. Receptor cross-linking blocks internalization and stops growth. A reasonable explanation for growth inhibition by receptor cross-linking is that transferrin is not internalized and iron is not delivered to the M2 subunit of ribonucleoside reductase.

4. Role of Iron

It has long been known that iron is necessary for growth. However, an idea arose that iron deprivation was not responsible for growth inhibition by antibody to transferrin receptor, because iron supplementation did not overcome the effect of antibody-mediated inhibition of growth.[75,86] It seems that the iron supplementation experiments are not as clear as they initially appeared, for the antibody-mediated growth inhibiting effect on cells grown in transferrin-supplemented serum-free media is overcome by supplemental iron, and the same cells grown in transferrin-free, iron-supplemented, artificial media are not inhibited by antibody.[84]

Further support for the concept that transferrin receptor-mediated growth control is determined by its ability to deliver iron is drawn from experiments which show that apotransferrin and partially iron-saturated transferrin support growth as a function of their degree of iron saturation.[86] Results from another experiment have shown that cellular growth inhibited by antibody to transferrin receptor could be overcome by the addition of an iron chelate (ferric salicylaldehyde isonicotinoyl hydrazone) that supplies iron to cells by a mechanism which does not utilize transferrin receptors.[51] These experiments have been interpreted as evidence that the normal function of transferrin receptors is to deliver iron necessary for cellular growth.[87] However, supplemental ferric iron also makes more electron acceptors available. This could rescue growth-inhibited cells by accepting electrons from plasma membrane diferric transferrin reductase,[88] and allow growth-inhibiting protons to escape by way of a sodium-hydrogen antiport.

C. MOLECULAR BIOLOGICAL STUDIES

Treatment of K562 cells with either human diferric transferrin or ferric ammonium citrate results in a decreased expression of plasma membrane transferrin receptors, and addition of desferrioxamine, an iron chelator, causes an increase of the receptors on plasma membranes.[89] The molecular biological regulation of receptor presentation can be at three possible levels of control. These are translation, transcription, or replication, and each of these possibilities has been investigated. Incubation of cells with desferrioxamine has no effect on receptor degradation.[35] Incubation of the cells with an iron source or an iron chelator could influence the specific mRNA level. Isolation of cDNA clones for the human transferrin receptor has permitted investigations of the expression of transferrin receptor and the control of cell proliferation.[90] Poly(A)mRNA from K562 cells treated with desferrioxamine for 18 to 24 h was isolated, and it was found by Northern blot analysis that the cells contained 3 to 6 times more specific transferrin receptor mRNA than untreated cells.

The increase in transferrin receptor mRNA began as soon as 2 h after addition of desferrioxamine and was maximal at 18 h. This increase was as much as 6 times more than in untreated cells. *In vitro* translation of mRNA from cells treated with desferrioxamine followed by immunoprecipitation of the translated product with monoclonal antibody to transferrin receptor produced an 80,000-mol wt protein that was threefold more abundant than in untreated cells. Poly(A)mRNA from K562 cells supplemented with diferric transferrin was found by Northern blot analysis to contain an 80% decrease in transferrin receptor mRNA that was maximal after 6 h. Specific mRNA to transferrin receptor in hemin-treated cells decreased even faster than in diferric transferrin-treated cells. Interestingly, the increase of receptor expression as shown by immunoprecipitation of desferrioxamine-treated cells was maximal as soon as 8 h after desferrioxamine, while expression of receptor as measured by ^{125}I-transferrin binding was maximal only after 24 h.

Other evidence suggests the regulation of transferrin receptor is at the transcription level. Transferrin receptor mRNA half-life rapidly changes according to the availability of iron. In desferrioxamine treated cells the half-life is 2.5 h, while in hemin-treated cells the half-life drops to 50 min. In addition, nuclei from desferrioxamine-treated K562 cells incorporate more [α^{32}P] UTP into transferrin receptor mRNA than do hemin-treated cells, as shown by hybridization between the transcript material and transferrin receptor cDNA.[90,91]

By using chimeric constructs consisting of transferrin receptor promotor and CAT gene integrated into murine fibroblasts, it has been shown that the expression of CAT gene is regulated by iron manipulations. However, the regulation of CAT gene expression by iron manipulations was not enough to explain the increase in specific mRNA in desferrioxamine-treated cells. A construct consisting of a heterologous promotor and transferrin receptor cDNA was also iron regulated, which means that other site(s) regulated by iron are within the transferrin receptor gene itself. This regulation was lost by deletion of the DNA corresponding to the 3′ untranslated 3′ transferrin receptor mRNA. Thus, the regulation of expression of transferrin receptor by iron at the DNA level is believed to be at two sites:[91] these are the 3′ untranslated portion of the gene and the 5′ transcription start site.

Similar results obtained by using a construct of human transferrin receptor cDNA inserted into mouse fibroblasts have shown that sequences within the 3′ noncoding region are required for the regulation of the receptor by iron.[92] It has been suggested that the receptor's promotor was not important, for the regulation by iron-transcription also was initiated by using the SV40 early promotor in the presence of an iron chelator.[92] Thus, the only site regulated by iron would be the 3′ noncoding sequence of the gene. Nevertheless, deletion of this portion of the gene does not affect the decrease in transferrin receptor expression in growth arrest. In light of these observations it seems as if the regulation of receptor is under the control of two distinct mechanisms; one that is tightly regulated by iron, and one that is not.

The above findings on the regulation of transferrin receptor expression by iron were obtained by studying the K562-transformed cell line. Studies of the regulation of transferrin receptor expression by normal T cells and HTLV-1 infected T cells have shown that in infected cells there is hyperexpression of surface receptor, while total receptor (cytoplasmic plus surface) was not different from that in normal T cells.[32] mRNA expression also was not different between the two cell populations, but a decrease was shown in the internalization of transferrin receptor in virus-infected cells. This could explain the hyperexpression of receptor independent of the iron concentration in the growth media.

IV. EFFECTS OF TRANSFERRIN-ADRIAMYCIN CONJUGATES

The observation of transferrin receptors on plasma membranes of cancer cells was made a few years after cancer chemotherapeutic agents such as daunomycin and Adriamycin had already been linked to immunoglobulins and used *in vitro* to kill cancer cells.[93] The concept of drug targeting with drug-conjugated antibodies might have been a promising approach but subsequent experience with antibody conjugates indicated that this approach to the clinical problem of cancer was less than satisfactory. Transferrin, which is the natural ligand for transferrin receptors, was recognized as a potent drug carrier to cancer cells,[9] and this observation led to the preparation of transferrin-Adriamycin conjugates that now are being used experimentally to deliver the drug to cancer cells.[17,27-29] It is appropriate to consider

TABLE 1
Parameters of the Transferrin-Adriamycin Conjugation Procedure

	Procedure	
	Original[a]	**Improved**
Step 1: Incubation of transferrin and Adriamycin together	120 min	10 min
Step 2: Incubation of the mixture after glutaraldehyde addition	120 min	10 min
Step 3 Incubation of the mixture after ethanolamine addition	Overnight	10 min
Step 4: Dialysis of the mixture against 200-fold volume of buffer[b]	– – – –	180 min
Step 5: Size exclusion chromatography with	Sepharose CL-6B	Sepharose CL-4B
Ratio of transferrin to Adriamycin, to glutaraldehyde, and to ethanolamine in mole base	1:42:106:4165	1:17:40:62
Yield of compounds in fractions eluted after the void volume of the column ($0.2<K_{av}<0.8$):		
A. Transferrin	n.a.	64%[c]
B. Adriamycin[b]	n.a.	56%[c]

Note: n.a. = not available.

[a] From Reference 27.
[b] There is a considerable loss of Adriamycin during the dialysis step in the form of precipitation, which is removed by centrifugation.
[c] Data for holo-transferrin-Adriamycin conjugates.

these conjugates in some detail, because they appear to be a useful way to study plasma membrane redox systems.[29]

A. PREPARATION OF CONJUGATES

Among three different conjugation procedures, glutaraldehyde coupling has resulted in conjugates that best preserve drug activity.[93] The original procedure used to bind Adriamycin to immunoglobulins was based on a method successfully used to couple enzymes such as peroxidase and glucose-oxidase to antibodies.[94] Transferrin-Adriamycin conjugates were prepared by using the original procedure of Yeh and Faulk.[27] Chromatographic runs on Sepharose CL-6B columns showed that most of the conjugates were eluted with the void volume of the column, suggesting that conjugation resulted in high molecular weight complexes and aggregates. However, reinvestigation of each step in the conjugation procedure has revealed technical shortcomings that now have been modified to produce a more effective method of coupling the ligand to Adriamycin. This was accomplished in part by decreasing the time in each step before chromatography (Table 1). In addition, the ratios of chemical agents used have been changed so that the yields of conjugate are significantly increased and more uniform distribution of Adriamycin on transferrin has been achieved.[95] The modified method works equally well for human albumin, holo- and apo-transferrin, and mouse apotransferrin.

By using this modified conjugation procedure, the average mole ratio for adriamycin-apotransferrin was always lower than that for Adriamycin-holotransferrin, showing the influence of the conformational state of proteins on conjugation number. Glutaraldehyde uses the lysinyl residues of proteins in the coupling procedure,[96] so the difference between the conjugates with apo- and holo-transferrin may reflect the difference in numbers of lysinyl residues exposed on the outer surface of transferrin in its apo- and holo-forms. Since the lysinyl residues at the surface of proteins should be protonated at pH 7, different conjugation numbers may result in different isoelectric points of conjugated proteins. Such changes in molecular charge may impact upon the ability of different populations to bind plasma membranes. It usually is assumed that ligand-conjugates are specifically bound by their receptors, but the binding characteristics also are affected by electrochemical considerations.

B. CONJUGATE BINDING AND KILLING

Binding of transferrin to its receptor is specific, rapid, saturable and reversible, and has a stability constant at pH 7.4 between 10^7 and 10^9 M^{-1}, depending on the type of cell or tissue used as well as on the iron content of transferrin.[97-100] Similar studies have not yet been performed with Adriamycin-transferrin conjugates, probably because of the lack of homogeneity of conjugates in size, charge and conjugation number. The association of conjugates with cells (i.e., the surface binding and uptake by endocytosis) can be monitored in a fluorescence-activated cell sorter (FACS) by using the fluorescence of Adriamycin (Figure 1). Experiments done by FACS analysis with adriamycin and Adriamycin-holotransferrin conjugates have shown that free drug binds faster and is taken up in greater amounts by K562 cells than conjugates (Figure 2). Interestingly, saturation could not be detected for either free drug or conjugate below 10^{-4} *M* Adriamycin. Extrapolation of experimental results obtained by FACS analysis have shown that mean channel shifts are an order of magnitude higher for conjugate than for free drug (2.5×10^{-6} *M* and 3.5×10^{-7} *M*, respectively). Such a big difference cannot be explained by concentration quenching of fluorescence or differences in the quantum yield. Since the two lines on Figure 2 are parallel, the data indicate a central role for Adriamycin in the association of both free drug and conjugate with K562 cells.

C. GROWTH EFFECTS OF CONJUGATES

The biological activity of transferrin-adriamycin conjugates was first evaluated with human tumor cells by measuring the inhibition of ^{3}H-thymidine incorporation as a measure of proliferation, and by the exclusion of trypan blue as a measure of viability.[11,17,27,29] Incubation of conjugates for 3 h at 37°C with Daudi and HL-60 cells followed by two washes and a 16 h culture period in ^{3}H-thymidine-supplemented complete growth medium had a striking effect. This diminished both proliferation and viability, whereas a minimal effect was observed for both ^{3}H-thymidine incorporation and viability when normal cells were tested under the same conditions.[27] The inhibition of ^{3}H-thymidine incorporation by cells from patients diagnosed as having acute myelocytic leukemia was an order of magnitude higher than the inhibition achieved by using normal cells.[29]

Preliminary experiments showed that ^{3}H-thymidine incorporation by K562 cells could be inhibited better with Adriamycin-holotransferrin conjugates than with free Adriamycin, and inhibition in both cases was more than 50% at 10^{-6} *M* Adriamycin concentration. At this concentration, hardly any binding of conjugate could be detected by FACS. This raises the idea that only a few Adriamycin molecules may be able to stop cellular proliferation.

The mechanism of inhibition of ^{3}H-thymidine utilization might be assumed to be that adriamycin was taken up by cells, diffused to the nucleus and intercalated with DNA and inhibited transcription. The uptake of Adriamycin might occur for free drug by facilitated diffusion, and for conjugates by receptor-mediated endocytosis followed by deconjugation within the acidic environment of the receptosome. Such a deconjugation process is supported by the reported observation of Adriamycin in nuclei of tumor cells targeted by transferrin-adriamycin conjugates but not in nuclei of targeted normal cells.[27] It thus appears that a mechanism of tumor cell killing by transferrin-Adriamycin conjugates is a plasma membrane receptor binding of conjugate followed by endocytosis, deconjugation, and diffusion of drug through nuclear membrane pores to intercalate with DNA base pairs and block subsequent cell division. However, results of recent research indicate that Adriamycin inhibits a plasma membrane-located NADH-diferric transferrin oxidoreductase, which is associated with transferrin receptors, both in rat liver,[101] and in different cell lines.[102,103] Since Adriamycin is a potent inhibitor of NADH-diferric transferrin oxidoreductases,[101] this might represent a second mechanism to explain the effects of Adriamycin and its conjugates on proliferation and viability.[104]

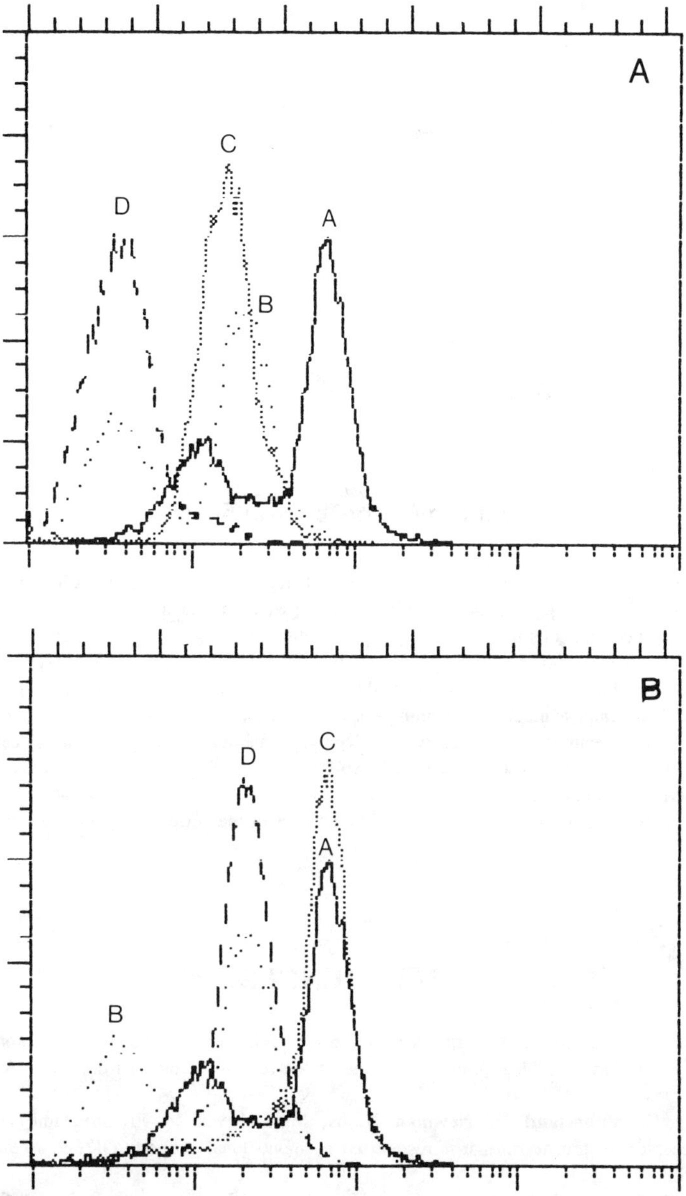

FIGURE 1. FACS analysis of transferrin-Adriamycin conjugate and Adriamycin binding by PBM, K562 cells, and mixtures of PBM with K562 cells. PBM and PBM/K562 mixtures (part A); K562 cells and K562/PBM mixtures (part B). Unmixed cells were incubated with transferrin-Adriamycin conjugate (D) and free Adriamycin (C). Cell mixtures were incubated with transferrin-Adriamycin conjugate (B) and free Adriamycin (A). Note that cell mixtures plot as biphasic curves. PBM are normal control peripheral blood mononuclear cells.

ACKNOWLEDGMENTS

We thank Dr. John McIntyre and Mr. Harold Boldt for assistance with the FACS experiments, and acknowledge collaboration of the Walther Cancer Institute, Indianapolis. Dr. Hava Harats was supported by the Methodist Medical Foundation, and Dr. Alajos Berczi is a Visiting Fellow from the Institute of Biophysics, Biological Research Center, Hungarian Academy of Sciences, Szeged, Hungary.

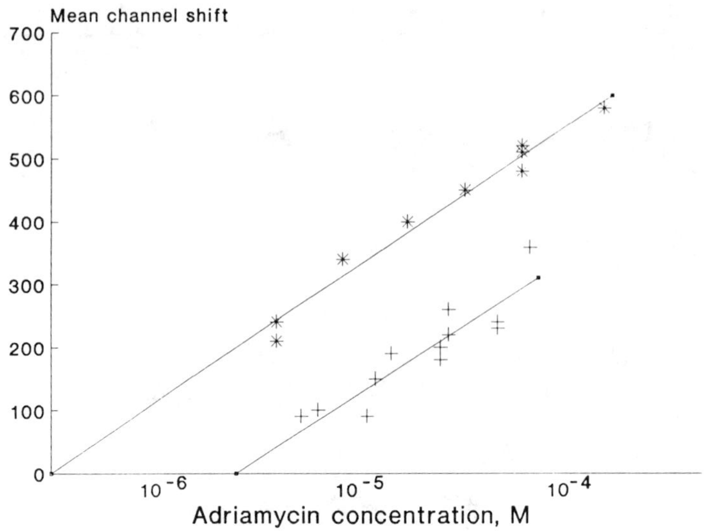

FIGURE 2. Association (binding + uptake) of free Adriamycin (○) and Adriamycin-holotransferrin conjugates (●) with K562 cells measured as mean channel shifts above untreated control cells in a FACS instrument. Cells (10^6/ml) grown in complete growth medium (RPMI 1640 medium with 10 m*M* HEPES buffer, pH 7.4, and 2% (v/v) heat-inactivated fetal calf serum) were washed twice in complete medium, incubated 60 min at 37°C in complete medium containing Adriamycin or transferrin-Adriamycin conjugate at given concentrations, washed twice in complete medium, and the amount of cell-associated Adriamycin was measured by FACS analysis. Straight lines were fitted to experimental points from three independent experiments by least square deviation from means, and the fitting was characterized by squares of the correlation coefficient (r^2).

REFERENCES

1. **McClelland, A., Kuhn, L. C., and Ruddle, F. H.,** The human transferrin receptor gene: genomic organization, and the complete primary structure of the receptor deduced from a cDNA sequence, *Cell,* 39, 276, 1984.
2. **Schneider, C., Sutherland, R., Newman, R. A., and Greaves, M. F.,** Structural features of the cell surface receptor for transferrin that is recognized by monoclonal antibody OKT-9, *J. Biol. Chem.,* 257, 8516, 1982.
3. **Enns, C. A., Larrick, J. W., Suomalainen, H., Schroder, J., and Sussman, H. H.,** Co-migration and internalization of transferrin and its receptor on K562 cells, *J. Cell Biol.,* 97(2), 579, 1983.
4. **Harding, C., Heuser, J., and Stahl, P.,** Receptor mediated endocytosis of transferrin and recycling of the transferrin receptor in rat reticulocytes, *J. Cell Biol.,* 97(2), 329, 1983.
5. **Hemmaplarah, D. and Morgan, E. H.,** Transferrin and iron uptake by human cells in culture, *Exp. Cell. Res.,* 87, 207, 1974.
6. **Trowbridge, I. S. and Omary, M. B.,** Human cell surface glycoprotein related to cell proliferation is the receptor for transferrin, *Proc. Natl. Acad. Sci. U.S.A.,* 78, 3039, 1981.
7. **Seligman, P. A., Schlieder, R. B., and Allen, H. A.,** Isolation and characterization of the transferrin receptor from human placenta, *J. Biol. Chem.,* 254(20), 9943, 1979.
8. **Faulk, W. P. and Galbraith, G. M. P.,** Trophoblast transferrin and transferrin receptors in the host-parasite relationship of human pregnancy, *Proc. Roy. Soc. London Ser. B,* 204, 83, 1979.
9. **Faulk, W. P., Hsi, B.-L., and Stevens, P. J.,** Transferrin and transferrin receptors in carcinoma of the breast, *Lancet,* August, 23, 390, 1980.
10. **Pileri, S., Gordes, J., Rivano, M., Tazari, P. L., and Magnani, M.,** Immunohistochemical determination of growth fractions in human permanent cell lines and lymphoid tumours: a critical comparison of the monoclonal antibodies OKT-9 and KI67, *Br. J. Hematol.,* 65, 271, 1987.

11. **Yeh, C.-J. G., Taylor, C. G., and Faulk, W. P.,** Transferrin binding by peripheral mononuclear cells in human lymphomas, myelomas and leukemias, *Vox. Sang.,* 46, 217, 1984.
12. **Sciot, R., Paterson, A. C., Van-Eyken, P., Callea, F., Kew, M. C., and Desmet, U. J.,** Transferrin receptor expression in human hepatocellular carcinoma: an immunohistochemical study of 34 cases, *Histopathology,* 12, 53, 1988.
13. **Lindholm, M. C., Lindberg, L. A., Vilja, P., Paolakka, V. M., Nording, S., Schroder, T., and Schroder, J.,** Expression of the human transferrin receptor in subrenal capsule assay in the mouse, *J. Surg. Oncol.,* 38, 57, 1988.
14. **Seymour, G. J., Walsh, M. D., Lavin, M. F., Stutton, G., and Gardiner, R. A.,** Transferrin receptor expression by human bladder transitional cell carcinomas, *Urol. Res.,* 15, 341, 1987.
15. **Gatter, K. C., Brown, G., Trowbridge, I. S., Woolstone, R.-E., and Mason, D. Y.,** Transferrin receptors in human tissues: their distribution and possible clinical relevance, *J. Clin. Pathol.,* 36, 539, 1989.
16. **Pannacio, M., Zalcberg, J. R., Thompson, C. M., Leyden, M. J., Sullivan, J. R., Lichtenstein, M., and McKenzie, I. F. C.,** Heterogeneity of the human transferrin receptor and the use of anti-transferrin receptors antibodies to detect tumours *in vivo, Immunol. Cell. Biol.,* 65, 461, 1987.
17. **Yeh, C.-J. G., Taylor, C. G., and Faulk, W. P.,** Targeting of cytotoxic drug by transferrin receptors: selective killing of acute myelogenous leukemia cells, *Protides Biol. Fluids,* 32, 441, 1984.
18. **Wells, M., Yeh, C.-J. G., Hsi, B.-L., and Faulk, W. P.,** Evidence for polymorphism of transferrin receptors in human skin, *J. Clin. Pathol.,* 36, 1414, 1983 (Letter).
19. **Weitkamp, L., Ober, C., and Schacter, B.,** Genetic and immunologic markers of reproductive success, in *Reproductive Immunology,* Clark, D. A. and Croy, B. A., Eds., Elsevier, Amsterdam, 1986, 27.
20. **Plowman, G. D., Brown, J. P., Enns, C. A., Schroder, J., Nikinma, B., Sussman, H., Hellstrom, K. E., and Hellstrom, I.,** Assignment of the gene for human melanoma associated antigen p97 to chromosome 3, *Nature,* 303, 70, 1983.
21. **Stein, B. S. and Sussman, H. H.,** Peptide mapping of the human transferrin receptor in normal and transformed cells, *J. Biol. Chem.,* 255, 2668, 1983.
22. **Van-Driel, I. R. and Goding, J. W.,** Heterogeneous glycosilation of murine transferrin receptor subunit, *Eur. J. Biochem.,* 149, 543, 1985.

22a. **Faulk, W. P.,** Radioodinated apotransferrin for use in tumor diagnosis, imaging, localization or treatment, U.S. Patent No. 4, 895, 714, January 23, 1990.

23. **Trowbridge, I. S. and Domingo, D. L.,** Anti-transferrin receptor monoclonal antibody and toxin antibody conjugates affect growth of human tumours cells, *Nature,* 294, 171, 1981.
24. **Yang, H. M. and Reisfeld, R. A.,** Doxorubicin conjugated with a monoclonal antibody directed to a human melanoma associated proteoglycan supresses the growth of established tumor xenografts in nude mice, *Proc. Natl. Acad. Sci. U.S.A.,* 85, 1189, 1988.
25. **Raso, V., Fehrmann, C., Solan, V. C., and Rosowsk, A.,** Monensin is obligatory for the cytotoxic action of disulfide linked methotrexate-anti-transferrin receptor conjugate, *Biochem. Biophys. Res. Commun.,* 150, 104, 1988.
26. **Griffin, T. W., Richardson, C., Houston, C. C., Le Page, D., Boyden, A., and Raso, V.,** Antitumor activity of intraperitoneal immunotoxins in a nude mouse model of human malignant mesothelioma, *Cancer Res.,* 47, 4266, 1987.
27. **Yeh, C.-J. G. and Faulk, W. P.,** Killing of human tumor cells in culture with adriamycin conjugates of human transferrin, *Clin. Immunol. Immunopathol.,* 32, 1, 1984.
28. **Faulk, W. P., Taylor, C. G., Yeh, C.-J. G., and McIntyre, J. A.,** Preliminary clinical study of transferrin-adriamycin conjugate for drug delivery to acute leukemia patients, *Mol. Biother.,* 2, 57, 1990.
29. **Faulk, W. P., Torry, D. S., Harats, H., McIntyre, J. A., and Taylor, C. G.,** Basic and clinical studies of transferrin-adriamycin conjugates: mechanisms for a new approach of drug targeting, in *Plasma Membrane Oxidoreductases in Control of Animal and Plant Growth,* Plenum Press, New York, 1988, 173.
30. **Greene, W. L. and Robb, R. J.,** Receptors for T-cell growth factor: structure, function and expression on normal and neoplastic cells, *Contemp. Top. Mol. Immunol.,* 10, 1, 1985 (review).
31. **Markham, P. D., Salahuddin, S. Z., and Gallo, R. C.,** *In vitro* cultivation of normal and neoplastic human T-lymphocytes, *Clin. Haematol.,* 13, 423, 1989.
32. **Vidal, C., Matsushita, S., Colamonici, O. R., Trepel, J. B., Mitsuya, M., and Neckers, L. M.,** Human T-lymphotropic virus I infection deregulates surface expression of the transferrin receptor, *J. Immunol.,* 141, 984, 1988.
33. **Rodriguez, M. H. and Junqery, M.,** A protein of *Plasmodium falciparum*-infected erythrocytes functions as a transferrin-receptor, *Nature,* 324, 388, 1983.
34. **Fry, M.,** Diferric transferrin reductase in *Plasmodium falciparum*-infected erythrocytes, *Biochem. Biophys. Res. Commun.,* 158, 469, 1989.
35. **Mattia, E., Rao, K., Shapiro, D. S., Sussman, S. S., and Klausner, R. D.,** Biosynthetic regulation of the human transferrin by desferrioxamine in K562 cells, *J. Biol. Chem.,* 259, 2689, 1984.

36. **Hancock, W. W., Lord, R. H., Colby, A. J., Diamantstein, T., Rickles, F. R., Dijkstra, C., Hogg, N., Tilney, N. L.,** Identification of IL 2R + T-cells and macrophages within rejecting rat cardiac allografts, and comparisons of the effects of treatment with anti-IL 2R monoclonal antibody or cyclosporin, *J. Immunol.*, 138, 164, 1987.
37. **Esmon, C. T.,** The regulation of natural anticoagulant pathways, *Science,* 235, 1348, 1987.
38. **Pober, J. S.,** Cytokine-mediated activation of vascular endothelium, *Am. J. Pathol.*, 133, 426, 1988.
39. **Fajardo, L. F.,** The complexity of endothelial cells, *Am. J. Clin. Pathol.*, 92, 241, 1989.
40. **Faulk, W. P., Gargiulo, P., McIntyre, J. A., and Bang, N. U.,** Hemostasis and fibrinolysis in renal transplantation, *Semin. Thromb. Hemost.*, 15, 88, 1989.
41. **Mohanakumar, T., Hoshinaga, K., Wood, N. L., Szentpetery, S., and Lower, R. R.,** Enumeration of transferrin receptor depressing lymphocytes as a potential marker for rejection in human cardiac transplant recipients, *Transplantation,* 42, 691, 1986.
42. **Hoshinaga, K., Mohanakumar, T., Pascoe, E. A., Szentpetery, S., Lee, H. M., and Lower, R. R.,** Expression of transferrin receptors on lymphocytes: its correlation with T-helper/T-suppressor cytotoxic ratio and rejection in heart transplant recipients, *J. Heart Transp.*, 3, 198, 1988.
43. **Rammensee, H. G., Lesley, J., Trowbridge, I. S., and Bevan, M. J.,** Antibodies against the transferrin receptor block the induction of cytotoxic T-lymphocytes. A new method for antigen-specific negative selection *in vitro, Eur. J. Immunol.*, 15, 687, 1987.
44. **Sutherland, R., Delia, D., Schneider, C., Newman, R., Kemshead, J., and Greaves, M.,** Ubiquitous cell-surface glycoprotein on tumor cells is proliferation-associated receptor for transferrin, *Proc. Natl. Acad. Sci. U.S.A.*, 78, 4516, 1981.
45. **Galbraith, R. M. and Galbraith, G. M. P.,** Expression of transferrin receptors on mitogen-stimulated human peripheral lymphocytes: relation to cellular activation and related metabolic events, *Immunology,* 44, 703, 1981.
46. **Hamilton, T. A.,** Regulation of transferrin receptor expression in concavalin. A stimulated and gross virus transformed rat lymphoblasts, *J. Cell. Physiol.*, 113, 40, 1982.
47. **Neckers, L. M. and Cossman, J.,** Transferrin receptor induction in mitogen stimulated human T-lymphocytes is required for DNA synthesis and cell division, and is regulated by interleukin 2, *Proc. Natl. Acad. Sci. U.S.A.*, 800, 3494, 1983.
48. **Chitambar, C. R., Massey, E. J., and Seligman, P. A.,** Regulation of transferrin receptor expression on human leukemic cells during proliferation and induction of differentiation. Effects of fallium and dimethylsulfoxide, *J. Clin. Invest.*, 72, 1314, 1983.
49. **Tei, I., Makina, Y., Sakakgami, H., Kanamaru, I., and Konno, K.,** Decrease of transferrin receptor during mouse myeloid leukemia (M1) cells differentiation, *Biochem. Biophys. Res. Commun.*, 107, 169, 1982.
50. **Besancon, F., Bourgeade, M.-F., and Testa, U.,** Inhibition of transferrin receptor expression by interferon-α in human lymphoblastoid cells and mitogen-induced lymphocytes, *J. Biol. Chem.*, 260, 13074, 1985.
51. **Laskey, J., Webb, J., Schulman, H. M., Ponka, P.,** Evidence that transferrin supports cell proliferation by supplying iron for DNA synthesis, *Exp. Cell. Res.*, 176, 87, 1988.
52. **May, W. S. and Cuatrecasas, P.,** Transferrin receptor: its biological significance, *J. Membr. Biol.*, 88, 205, 1988.
53. **Kemp, J. P., Thorson, J. A., McAlmont, T. H., Horowitz, M., Cowdery, J. S., and Ballas, Z. K.,** Role of the transferrin receptor in lymphocyte growth: a rat IgG monoclonal antibody against the murine transferrin receptor produces highly selective inhibition of T and B cell activation protocol, *J. Immunol.*, 138, 2422, 1987.
54. **Hood, L., Kronenberg, M., and Hunkapiller, T.,** T-cell antigen receptors and the immunoglobulin supergene family, *Cell,* 40, 225, 1985.
55. **Teshigaward, K., Wang, H.-M., Kato, K., and Smith, K. A.,** Interleukin-2 high-affinity receptor expression requires two distinct binding proteins, *J. Exp. Med.*, 165, 223, 1987.
56. **Krönke, M., Leonard, W. J., Depper, J. M., Greene, W. C.,** Sequential expression of genes involved in human T lymphocyte growth and differentiation, *J. Exp. Med.*, 161, 1593, 1985.
57. **Poulton, T. A., Gallacher, A., Potts, R. C., and Swanson Beck, J.,** Changes in activation markers and cell membrane receptors on human peripheral blood T-lymphocytes during cell cycle progression after PHA progression, *Immunology,* 64, 419, 1988.
58. **Williams, J. M., Loertscher, R., Cotner, T., Reddish, M., Shapiro, H. M., Carpenter, C. B., Stominger, J. L., and Strom, T. B.,** Dual parameter flow cytometric analysis of DNA content, activation antigen expression, and T-cell subset proliferation in the human mixed lymphocyte reaction, *J. Immunol.*, 132, 2336, 1984.
59. **Bettens, F., Kristensen, F., Walker, C., Schwulera, U., Bonnard, G. D., and De Week, A. L.,** Lymphokine regulation of activated (G1) lymphocytes. II. Glucocorticoid and anti-Tac-induced inhibition of human T-lymphocyte proliferation, *J. Immunol.*, 132, 261, 1984.

60. **Taetle, R., Castagnola, J., and Mendelsohn, J.,** Mechanisms of growth inhibition of anti-transferrin receptor monoclonal antibodies, *Cancer Res.,* 46, 1759, 1986.
61. **Suciu-Foca, N., Rosochacki, S. J., Cai, J., Reed, E., Rubinstein, P., and King, D. W.,** Immunological and biochemical characterization of an epitope of the transferrin receptor involved in the production of interferon- and B cell growth factor, *Cell. Immunol.,* 110, 265, 1987.
62. **Yoshizaki, K., Nakagawa, T., Fukunaga, K., Kaieda, T., Maruyama, S., Kishimoto, S., Yamamura, Y., and Kishimoto, T.,** Characterization of human B cell growth factor (BCGF) from cloned T-cells or mitogen-stimulated T-cells, *J. Immunol.,* 130, 1241, 1983.
63. **Neckers, L. M., Yenokida, G., Trepel, J. B., Lipford, E., and James, S.,** Transferrin receptor induction is required for human B-lymphocyte activation but not for immunoglobulin secretion, *J. Cell. Biochem.,* 27, 377, 1985.
64. **Butler, J. L., Muraguchi, A., Lane, H. C., and Fauci, A.-S.,** Development of a human T-T cell hybridoma secreting B cell growth factor, *J. Exp. Med.,* 157, 60, 1983.
65. **Cossman, J., Neckers, L. M., Braziel, R. M., Trepel, J. B., Korsmeyer, S. T., and Bakhshi, A.,** *In vitro* enhancement of immunoglobulin gene expression in chronic lymphocytic leukemia, *J. Clin. Invest.,* 73, 587, 1984.
66. **Howard, M. and Paul, W. E.,** Regulation of B-cell growth and differentiation by soluble factors, *Annu. Rev. Immunol.,* 1, 307, 1983 (review).
67. **Muraguchi, A., Kehrl, J. M., Butler, J. L., and Fauci, A. S.,** Sequential requirements for cell cycle progression of resting human B cell after activation of anti-Ig, *J. Immunol.,* 132, 176, 1984.
68. **Takatsu, K., Tanaka, K., Tominaga, A., Kamahura, Y., and Hamaoka, T.,** Antigen-induced T-cell replacing factor (TRF). III. Establishment of T-cell hybrid clone continuously producing TRF and functional analysis of released TRF, *J. Immunol.,* 125, 2646, 1980.
69. **Omary, M. B., Trowbridge, I. S., and Minowada, J.,** Human cell surface glycoprotein with unusual properties, *Nature,* 286, 888, 1980.
70. **Galbraith, G. M. P., Galbraith, R. M., and Faulk, W. P.,** Transferrin binding by human lymphoblastoid cell lines and other transformed cells, *Cell. Immunol.,* 49, 215, 1980.
71. **Judd, W., Poodry, C. A., and Stominger, J. K.,** Novel surface antigen expressed on dividing cells but absent from nondividing cells, *J. Exp. Med.,* 152, 1430, 1980.
72. **Haynes, B. F., Hemler, M., Cotner, T., Mann, D. L., Eisenbarth, G. S., Stominger, J. L., and Fauci, A. S.,** Characterization of a monoclonal antibody (5E9) that defines a human cell surface antigen of cell activation, *J. Immunol.,* 127, 347, 1981.
73. **Goding, J. W. and Burns, G. F.,** Monoclonal antibody OKT-9 recognizes the receptor for transferrin on human acute lymphocytic leukemia cells, *J. Immunol.,* 12, 1256, 1987.
74. **Galbraith, R. M., Werner, P., Arnaud, P., and Galbraith, G. M. P.,** Transferrin binding to peripheral blood lymphocytes activated by phytohemagglutinin involves a specific receptor-ligand interaction, *J. Clin. Invest.,* 66, 1135, 1980.
75. **Trowbridge, I. S. and Lopez, F.,** Monoclonal antibody to transferrin receptor blocks transferrin binding and inhibits human tumor cell growth *in vitro, Proc. Natl. Acad. Sci. U.S.A.,* 79, 1175, 1982.
76. **Goding, J. W. and Harris, W. A.,** Subunit structure of cell surface proteins: disulfide bonding in antigen receptors, Ly-2/3 antigens and transferrin receptors of murine T and B lymphocytes, *Proc. Natl. Acad. Sci. U.S.A.,* 78, 4530, 1981.
77. **Trowbridge, I. S., Lesley, J., and Schulte, R.,** Murine cell surface transferrin receptor: studies with anti-receptor monoclonal antibody, *J. Cell. Physiol.,* 112, 403, 1982.
78. **Trinder, D., Morgan, E. H., and Baker, E.,** The effects of an antibody to the rat transferrin receptor and of rat serum albumin on the uptake of diferric transferrin by rat hepatocytes, *Biochim. Biophys. Acta,* 943, 440, 1988.
79. **Yeh, C.-J. G., Hsi, B.-L., Samson, M., McIntyre, J. A., Breittmeyer, J.-P., and Fehlman, M.,** Monoclonal antibodies (GB16, GB18, GB19, GB22) raised against human placental microvilli recognize the transferrin receptor, *Placenta,* 8, 627, 1987.
80. **Mendelsohn, J., Trowbridge, I. S., and Castagnola, J.,** Inhibition of human lymphocyte proliferation by monoclonal antibody to transferrin receptor, *Blood,* 62, 821, 1983.
81. **Lesley, J. F. and Schulte, R. J.,** Inhibition of cell growth by monoclonal anti-transferrin receptor antibodies, *Mol. Cell. Biol.,* 5, 1814, 1985.
82. **Vaickus, L. and Levy, R.,** Antiproliferative monoclonal antibodies: detection and initial characterization, *J. Immunol.,* 135, 1987, 1985.
83. **Yeh, C.-J. G., Papamichail, M., and Faulk, W. P.,** Loss of transferrin receptors following induced differentiation of HL-60 promyelocytic leukemia cells, *Exp. Cell. Res.,* 132, 429, 1982.
84. **Taetle, R., Rhyner, K., Castagnola, J., To, P., and Mendelsohn, J.,** Role of transferrin, Fe and transferrin receptors in myeloid leukemia cell growth; studies with an anti-transferrin receptor monoclonal antibody, *J. Clin. Invest.,* 7, 1061, 1985.

85. **Lesley, J., Hyman, R., Schulte, R., and Trotter, J.,** Expression of transferrin receptor on murine hematopoietic progenitors, *Cell. Immunol.*, 83, 14, 1984.
86. **Taettle, R., Monesett, J. M., and Trowbridge, I. S.,** Effects of anti-transferrin receptors antibodies on growth of normal and malignant myeloid cells, *Int. J. Cancer,* 32, 343, 1983.
87. **Brock, J. H., Mainou-Fowler, T., and Webster, L. M.,** Evidence that transferrin may function exclusively as an iron donor in promoting lymphocyte proliferation, *Immunology,* 57, 105, 1986.
88. **Löw, H., Grebing, C., Lindgren, A., Tally, M., Sun, I. L., and Crane, F. L.,** Involvement of transferrin in the reduction of iron by the transplasma membrane electron transport system, *J. Bioenerg. Biomemb.*, 19, 535, 1987.
89. **Rao, K., Marford, J. B., Ronault, T., McClelland, A., Ruddle, F. H., and Klausher, R. D.,** Transcriptional regulation by iron of the gene for the transferrin receptor, *Mol. Cell. Biol.*, 6, 236, 1980.
90. **Scheider, C., Kurkinen, M., and Greaves, M.,** Isolation of cDNA clones for the human transferrin receptor, *EMBO J.*, 2, 2259, 1983.
91. **Casey, J. L., DiJeso, B., Rao, K., Klausner, R. D., and Harford, J. B.,** Two genetic loci participate in the regulation by iron of the gene for the human transferrin receptor, *Proc. Natl. Acad. Sci. U.S.A.*, 85, 1787, 1988.
92. **Owen, D. and Kuhn, L. C.,** Noncoding 3′ sequence of the transferrin receptor gene are required for mRNA regulation by iron, *EMBO J.*, 6, 1287, 1987.
93. **Hurwitz, E., Levy, R., Marou, R., Wilchek, M., Arnou, R., and Sela, M.,** The covalent binding of daunomycin and adriamycin to antibodies with retention of both drug and antibody activities, *Cancer Res.*, 35, 1175, 1975.
94. **Avranmeas, S.,** Coupling of enzymes to proteins with glutaraldehyde. Use of the conjugates for the detection of antigens and antibodies, *Immunochemistry,* 6, 43, 1969.
95. **Berczi, A., Barbabas, K., Sizensky, J. A., and Faulk, W. P.,** Coupling of adriamycin to human transferrin and characterization of the drug-protein conjugate, submitted.
96. **Korn, A. H., Feairhelle, S. H., and Filachione, E. M.,** Glutaraldehyde: nature of the reagent, *J. Mol. Biol.*, 65, 525, 1971.
97. **van Bockxmeer, F. M., Yates, G. K., and Morgan, E. H.,** Interaction of transferrin with solubilized receptors from reticulocytes, *Eur. J. Biochem.*, 92, 147, 1978.
98. **Loh, T. T., Higuchi, D. A., van Bockxmeer, F. M., Smith, C. H., and Brown, E. B.,** Transferrin receptors on the human placental microvillous membrane, *J. Clin. Invest.*, 65, 1182, 1980.
99. **Klausner, R. D., Ashwell, G., van Renswoude, J., Harford, J. B., and Bridges, K. R.,** Binding of apotransferrin to K562 cells: explanation of the transferrin cycle, *Proc. Natl. Acad. Sci. U.S.A.*, 80, 2263, 1983.
100. **McArdle, H. J., Douglas, A. J., and Morgan, E. H.,** Transferrin binding by microvillar vesicles isolated from rat placenta, *Placenta,* 5, 131, 1984.
101. **Sun, I. L., Navas, P., Crane, F. L., Morré, D. J., and Löw, H.,** NADH differic transferrin reductase in liver plasma membrane, *J. Biol. Chem.*, 262, 15915, 1987.
102. **Sun, I. L. and Crane, F. L.,** Control of transplasma membrane diferric transferrin reductase by antitumor drugs, in *Plasma Membrane Oxidoreductases in Control of Animal and Plant Growth,* Crane, F. L., Morré, D. J., and Löw, H., Plenum Press, New York, 1988, 181.
103. **Navas, P., Sun, I. L., Morré, D. J., and Crane, F. L.,** Decrease of NADH in HeLa cells in the presence of transferrin or ferricyanide, *Biochem. Biophys. Res. Commun.*, 135, 110, 1986.
104. **Faulk W. P., Harats, H., McIntyre, J. A., Berczi, A., Sun, I. L., and Crane, F. L.,** Recent advances in cancer research: drug targeting without the use of monoclonal antibodies, *Am. J. Reprod. Immunol.*, 21, 151, 1989.
105. Unpublished data.

Chapter 11

ROLES OF PYRIDINE NUCLEOTIDES IN THE CONTROL OF CELL GROWTH

Plácido Navas and M. Isabel Burón

TABLE OF CONTENTS

I. INTRODUCTION

The pyridine nucleotides play an important role in metabolism since they are the coenzymes in most dehydrogenases, and NAD^+ is consumed in certain metabolic reactions in which it serves as a substrate. Since there are two forms of nicotinamide coenzymes, they must have a significance in the regulation of the metabolism. They have different functions, as was proposed by Kaplan and coworkers in 1956.[1] Their hypothesis points out that NADH is oxidized by mitochondria to yield ATP, whereas NADPH appears to be involved in reductive reactions.

There are several cellular mechanisms by which the concentrations of NAD(H) and NADP(H) can be regulated. One of these mechanisms is due to the pyridine nucleotide transhydrogenase. This reactions catalyzes the direct and reversible transfer of a hydride ion equivalent between NADH and $NADP^+$, first detected by Kaplan's group[2] and recently reviewed by Fisher and Earle.[3]

The existence of specific NAD(H) and NADP(H) dehydrogenases might be other contributing factors. The levels of the various dehydrogenases can play a role in determining the concentrations of NAD(H) and NADP(H). Skeletal muscle, which is rich in NAD-dependent enzymes, contains high levels of the NAD(H) forms and little of the phosphopyridine nucleotide forms. The opposite pattern is observed in liver, where high level of NADP(H) corresponds to a high concentration of NAD-dependent dehydrogenases.[1] In rat livers after administration of a carcinogen, a rapid decline in NADH-dependent redox enzymes has been reported to follow a marked reduction of the NAD(H) pool.[4,5]

Besides the oxidation-reduction reactions, pyridine nucleotides are involved in other nonredox mechanisms. The most intriguing of these reactions in eukaryotic cells is the cleavage of NAD^+ to form a unique polymer, polyadenosine diphosphoribose (poly ADPR) and nicotinamide catalyzed by the enzyme poly ADPR synthetase, although other nonredox reactions are also involved in NAD^+ cleavage.[6] For example, NAD^+ kinase affects the synthesis of $NADP^+$ from NAD^+, and NADPase generates NAD^+ from $NADP^+$.[7]

Further, pyridine nucleotides play an important role in metabolism, which itself changes during growth. From the total pool of NAD(H) synthesized, about 95% replaces the NAD(H) that is catabolized and only 5% maintains the NAD(H) pool size during growth.[8] The nucleus is the organelle in which NAD^+ is actively synthesized and broken down,[9,10] and the cytoplasm includes the classical role in oxidation-reduction reactions of pyridine nucleotides which utilizes the nicotinamide moiety. After the synthesis of NAD^+ in the nucleus there is a rapid equilibrium between the oxidized and reduced forms of the individual pyridine nucleotides,[11] but, in contrast, there is a slow equilibrium between NAD(H) and NADP(H). Although the source of NADP(H) is the NAD(H) pool,[12] the slow equilibrium between these two forms of coenzymes appears to be due to the cytoplasmic localization of NAD^+ kinase.[11] In general, cells do not contain significant amounts of pyridine nucleotide cycle intermediates other than NAD^+,[13] suggesting that recycling occurs very efficiently, representing the majority of NAD^+ formed.[14]

There is, then, evidence to support the concept that evolution of both pyridine coenzymes has functional significance. In this sense, we review here the information regarding the changes of NAD(H) and NADP(H) pools and their oxidation-reduction ratios as a consequence of the stimulation of cell growth, and especially when cells are induced to replicate by the activation of the transplasma membrane redox system.

II. CHANGES OF PYRIDINE NUCLEOTIDES IN GROWING CELLS

Numerous studies have shown that pyridine nucleotide pools are significantly altered in proliferating tissues in comparison to those of nonproliferating ones.[15-20] However, the levels of total nucleotides and the ratios of oxidized to reduced forms seem to be more important to the proliferative response in pyridine nucleotide pools.[21]

Mitogen stimulation of human lymphocytes by phytohemagglutinin (PHA) results in an increase of the $NAD^+/NADH$ ratio, while the $NADP^+/NADPH$ ratio tends to decrease. The possibility that the NAD^+ level increases to provide a substrate for polyadenosine diphosphoribose (ADPR) polymerase, an enzyme whose activity is markedly increased in mitogen stimulated lymphocytes,[22,23] has been indicated. Since poly(ADPR) polymerase seems to be involved in the DNA repair process,[24-26] its increase after mitogen stimulation may be required to maintain the integrity of the newly synthesized DNA. Activities of other enzymes involved in DNA repair have also been shown to increase in mitogen-stimulated lymphocytes,[27] and as a function of the cell cycle in human fibroblasts.[28]

Therefore, increased amounts of reduced NADPH after mitogen stimulation are expected to be required, due to a marked increase in activity of the pentose phosphate pathway.[29] Such an increase may contribute to the reversion of the $NADP^+/NADPH$ ratio.

Addition of fresh serum to starved cell cultures is known to stimulate growth.[30,31] Insulin is a component of serum, which has been shown to be an essential growth factor for some cells.[32] Figure 1 shows the $NAD(P)^+/NAD(P)H$ ratios with respect to different amounts of fetal calf serum added to growing HeLa cells. Both $NAD^+/NADH$ and $NADP^+/NADPH$ ratios increase as cells grow. Similar response is observed when HeLa cells are stimulated to grow by insulin (Figure 2). Addition of fresh serum or insulin has been indicated to result in a decrease of cyclic AMP,[30] while the elevation of cyclic AMP concentration has been reported as a prerequisite to density-dependent inhibition of cell growth.[34,35]

There is considerable evidence that the impermeable electron-acceptor ferricyanide, which can accelerate electron transport through the plasma membrane, stimulates growth.[35] Since insulin increases the ferricyanide reduction, it appears that insulin is involved in the stimulation of the redox activity. However, insulin allows growth stimulation at higher concentrations of ferricyanide, which suggests an additional control on how the redox system functions.[31]

Normal and transformed cells are commonly used for studying growth regulation "in vitro" since the normal cells exhibit strong density-dependent inhibition of growth and the transformed cells have lost this property. NADP(H) pools appear to be generally similar in normal and transformed cells, being in a more reduced state and much less cell density-dependent in comparison to NAD(H) levels.

NAD(H) pools, however, seem to play a more important role in the control of animal cell division and growth. Although the mechanisms that regulate the levels of NAD(H) within the cells are not well understood, the levels appear to be lower in transformed than in nontransformed cells.[5,36-39] A variety of tumors,[15,18,40] fetal tissue and regenerating liver[20] contain lower NAD^+ levels than normal adult tissues. Additionally, carcinogens have been shown to cause a rapid reduction of cellular levels of NAD(H).[38,39,41]

The state of transformation seems to be reflected in a change in the $NAD^+/NADH$ ratio. $NAD^+/NADH$ seems to decrease by accumulation of the reduced form in normal cells as they approach confluency and stop dividing.[36,37] However, the lowering of the NADH pool and the increase of the oxidized form is the pattern found in transformed cells.[37] A high value of the $NAD^+/NADH$ ratio has also been reported in some tumors.[15]

It is possible that such an alteration of redox state is necessary for the processes involved in density-dependent inhibition of growth, and that transformed cells contain a defect in the mechanism that normally elevates NADH. The activity of the transplasma membrane dehydrogenase may contribute to control the redox state of the pyridine nucleotide pool. The decrease in this activity which is observed in transformed cells,[37] may reflect the decline in NADH within the cell so that the NADH concentration becomes rate-limiting for the dehydrogenase. Further studies of the factors implicated in the regulation of NAD(H) bio-

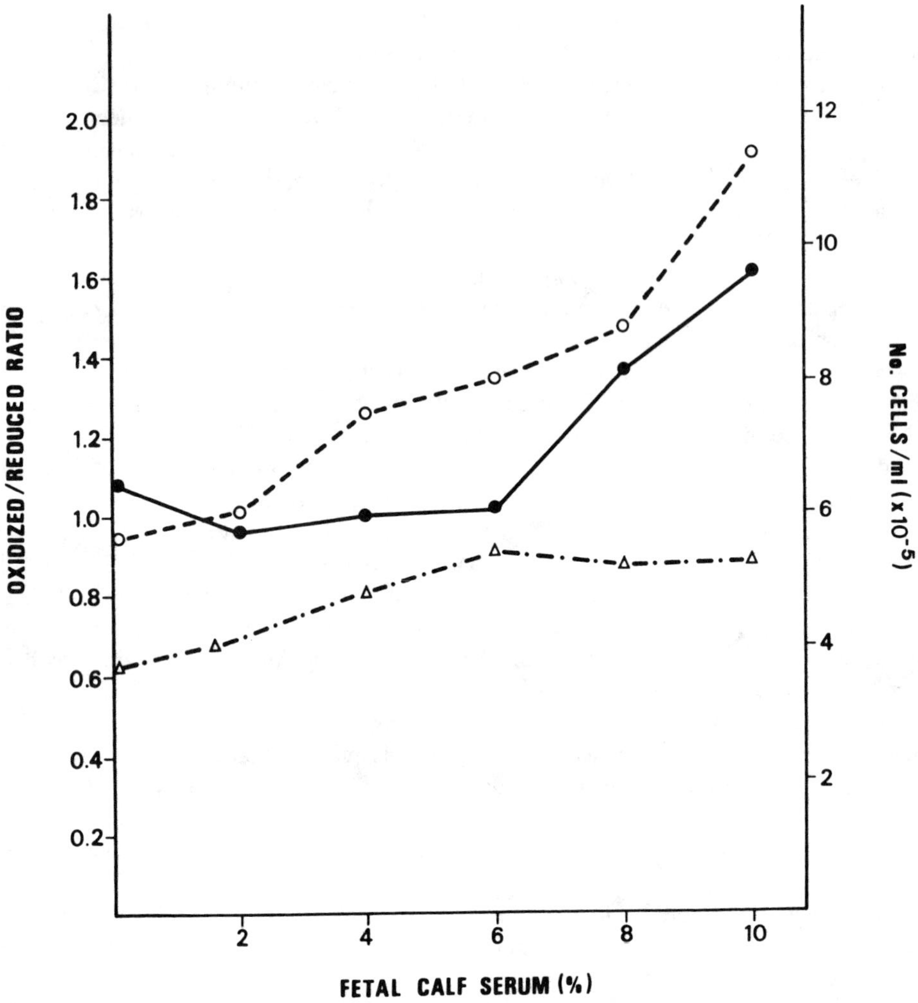

FIGURE 1. Fetal calf serum effect on redox state of pyridine nucleotides (●) and phosphopyridine nucleotides (○) of HeLa cells. Cell density at increasing percentages of fetal calf serum is also represented (△).

synthesis may aid in the elucidation of the mechanisms that regulate the redox state of the pyridine nucleotide pool in the normal and transformed cells.

III. CELL GROWTH STIMULATION BY TRANSPLASMA MEMBRANE REDOX SYSTEM

All eukaryotic cells have a transplasma membrane redox enzyme system.[42-45] In some cells the redox enzyme activity is sensitive to hormones.[46-48]

Activation of transplasma membrane electron transport with ferricyanide stimulates the growth of melanoma cells in serum-deficient media[35] and promotes HeLa cell proliferation in the absence of fetal calf serum and other growth factors.[31]

Diferric transferrin which acts as an electron acceptor for the transplasma membrane redox system,[49] has been shown to be essential for growth of many cultured cells in serum-free media[50,51] and stimulates HeLa cells growth in serum-limited media.[52]

A series of impermeable oxidants with redox potentials down to −125 mV was shown to have similar growth-promoting effects on animal cells.[53,54]

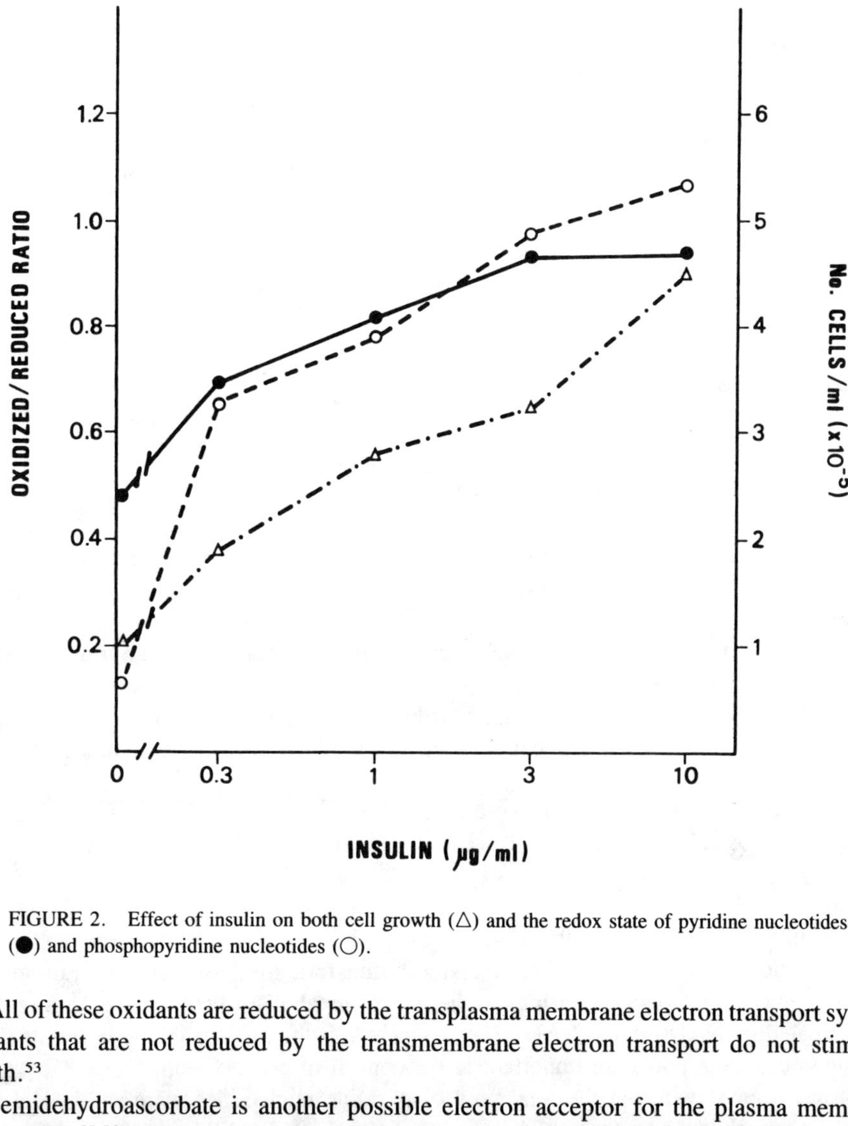

FIGURE 2. Effect of insulin on both cell growth (△) and the redox state of pyridine nucleotides (●) and phosphopyridine nucleotides (○).

All of these oxidants are reduced by the transplasma membrane electron transport system. Oxidants that are not reduced by the transmembrane electron transport do not stimulate growth.[53]

Semidehydroascorbate is another possible electron acceptor for the plasma membrane redox system[55,56] and also stimulates cell proliferation in serum-limited media.[57] The oxidized form of ascorbate, dehydroascorbate, does not stimulate cell growth.

Insulin, a growth-promoting hormone, stimulates growth of HeLa cells in the absence of serum (Figure 2) and also stimulates transmembrane redox activities.[9] An additive effect on the growth of HeLa cells by insulin and ferricyanide has also been demonstrated.[31]

The electron flow through the plasma membrane is accompanied by proton release from the cell.[58] Proton release due to ferricyanide or diferric transferrin is dependent on external sodium ions and it is inhibited by amiloride, which indicates that the proton release is mediated by the Na^+/H^+ antiport.[59,60] The inhibition of Na^+/H^+ antiport by amiloride also inhibits growth of both fibroblasts and hamster lung cells.[61]

Therefore, the transmembrane redox system might increase the pH of the cytoplasm, which is associated with cell proliferation.[62,63]

The increase of internal pH could be induced by external oxidants, either by energizing a proton pump of the plasma membrane, or by oxidation of internal NADH to remove protons. In the absence of buffering, oxidation of NADH could lead to an increased pH. Thus, a significant change in intracellular pH will depend upon the degree of pH buffering occurring in a given cell.[64]

TABLE 1
Intracellular $NAD^+/NADH$ and $NADP^+/NADPH$ Ratios of HeLa Cells after 48 h Growth in a Serum-Limiting Medium under the Action of External Oxidants

Addition	$NAD^+/NADH$	$NADP^+/NADPH$	No. cells/ml ($\times 10^{-5}$)
None	0.40	1.58	35
Ferricyanide	0.80	2.39	68
Ferrocyanide	0.42	1.40	30
Diferric transferrin	0.79	2.29	85
Apotransferrin	0.46	2.12	40
Semidehydroascorbate	0.75	2.46	70

Note: Ferricyanide and ferrocyanide were added at a final concentration of 0.033 m*M*, diferric transferrin and apotransferrin at 1.7 μ*M* and semidehydroascorbate at 2 m*M*.

It appears then that there is a direct connection between increased transplasma membrane redox enzyme activity and growth, probably due in part to an activation of the Na^+/H^+ antiport.

As the stimulation of growth is accompanied by an increase of total pyridine nucleotide pools[52,65] and their redox state at the logarithmic phase of growth (see above), intracellular pyridine nucleotide concentrations will be studied after both long-term and short-term oxidant stimulation to check the late and early events in the redox state of nucleotides.

IV. EFFECT OF TRANSPLASMA MEMBRANE REDOX ACTIVATION ON PYRIDINE NUCLEOTIDES

A. LATE EVENTS

Pyridine nucleotide levels are related to the proliferative state of cells under stimulation by conventional growth factors or the state of transformation (see above). Further, most dehydrogenase reactions and an important part of metabolism that changes during growth are mediated by NAD(H) and NADP(H) enzymes. It is then reasonable to postulate that the pyridine nucleotide pools are important in the control of cell division.

It has been shown that the total pool of pyridine nucleotides increases in HeLa cells after 48 h growth in the presence of diferric transferrin or ferricyanide in a medium containing a low concentration of serum.[52,65] Insulin increases the growth effect of ferricyanide[31] and stimulates the rise of the total pool of NAD(H) in HeLa cells under ferricyanide incubation.[52]

Although the total amount of nucleotides is induced to rise under growth conditions, the redox state of individual pyridine nucleotides appears to be more accurate for their relationship to the metabolism of growing cells.[11] Thus, $NAD^+/NADH$ and $NADP^+/NADPH$ ratios were determined in HeLa cells after 48 h growth in a serum-limiting medium supplemented with different oxidants (Table 1). Both $NAD^+/NADH$ and $NADP^+/NADPH$ ratios increase in HeLa cells growing in the presence of ferricyanide, diferric transferrin, or semidehydroascorbate in parallel to an increase of cell replication. Ferrocyanide and apotransferrin do not show a significant increase of both redox state of pyridine or phosphopyridine nucleotides and cell growth.

The pyridine nucleotides of cells induced to grow by the activation of the transplasma membrane electron transport are increased in both total concentration[52,65] and the ratio of reduced to oxidized forms. In this sense, late events on pyridine nucleotides in these cells appear to follow the same behavior as those stimulated by conventional mechanisms such as serum, growth factors, etc. Thus, increase of NAD^+ concentration could provide the substrate for enzymes acting in cell nuclei for DNA synthesis and repair mechanisms as was discussed above. A wide variety of DNA-damaging agents also cause a marked lowering of cellular NAD^+ levels.[66]

B. EARLY EVENTS

In this part we will discuss the change of redox states of intracellular pyridine nucleotides as a consequence of the initiation of transplasma membrane redox system activation by external oxidants. A short-term incubation of cells in the presence of external impermeable oxidants does not induce changes in the total concentration of pyridine nucleotides.[67]

It has been shown that all plasma membranes that have been obtained from different sources show NADH dehydrogenase activities,[45,49,56] but the major problem in the study of the transmembrane redox system in intact cells is to identify the electron donor.[58]

Glycolysis inhibitors such as iodoacetate and iodoacetamide inhibit ferricyanide reduction by erythrocytes and HeLa cells, indicating that the NADH produced by glycolysis can, therefore, be the natural substrate for these reductions.[68,69]

HeLa and HL60 cells were incubated for a short-term period in the presence of different oxidants, and the ratios of oxidized to reduced forms of individual pyridine nucleotides were determined. Both diferric transferrin and ferricyanide did not induce significant changes in the $NADP^+/NADPH$ ratio (Figure 3A) but clearly enhanced the $NAD^+/NADH$ ratio (Figure 3B). When the incubation was carried out in the presence of insulin, there was an increased effect of the external oxidants on the concentration of the NAD^+ oxidized form, but no changes were observed on phosphopyridine nucleotides.

Insulin does not itself produce any effect on pyridine nucleotide levels in HeLa cells after an incubation for 15 min.[67] Further, apotransferrin and ferrocyanide do not affect nucleotide concentrations after a short-term incubation.[65]

Semidehydroascorbate, which can be reduced by the transplasma membrane electron transport,[56] stimulates HL60 cell replication in a concentration that does not allow oxidant incorporation into the cell interior.[57,70,71] The $NAD^+/NADH$ ratio is enhanced during the first 20 min of HL60 cell stimulation by semidehydroascorbate (Figure 4). No significant changes in NADP(H) nucleotides were observed (data not shown).

All the above oxidants that induce the increase of oxidized over reduced NAD(H), enhance cell growth and are reduced by the transplasma membrane electron transport system. On the other hand, NADP(H) did not change, indicating that they are apparently not involved in the initial events of oxidation/reduction and/or growth induction.

The results presented here show a relative interchange of the reduced and oxidized forms of NAD(H) induced by the external electron acceptors diferric transferrin, ferricyanide, and semidehydroascorbate.

Ellem and Kay[35] postulated that the stimulation of growth by external oxidants may derive from oxidation of cytoplasmic NADH, and recently lactate, a source of cytosolic NADH,[60] was observed to increase transmembrane ferricyanide reduction in Ehrlich ascites tumor cells.[60]

Thus, not only does NADH serve as a primary intracellular electron donor for the transmembrane redox system of HeLa cells, but the stimulation of cell growth by external impermeable oxidant also may be correlated to the oxidation of the internal NADH.

V. CONCLUDING REMARKS

Pyridine nucleotides are clearly related to cell growth. Cells growing at log phase contain an increasing total amount of nucleotides, and oxidized forms are generally represented more than reduced forms. Cells stimulated by the activation of the transplasma membrane

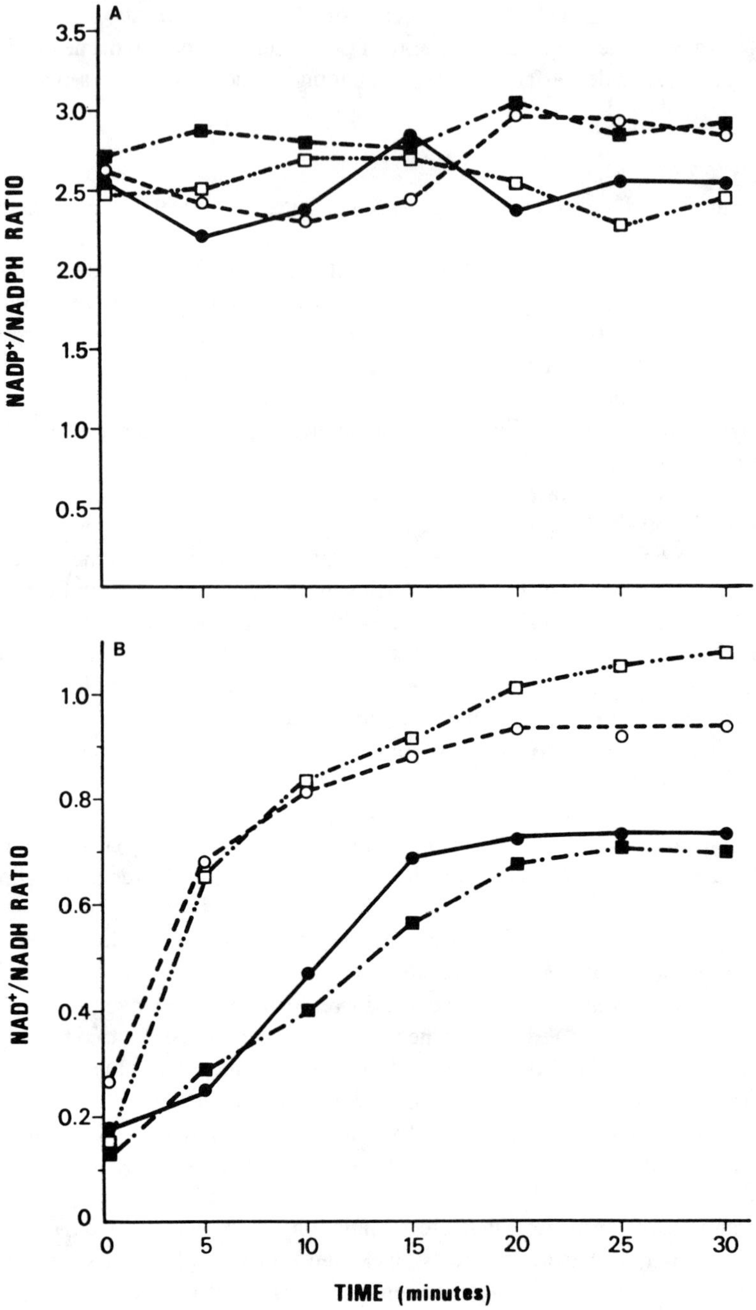

FIGURE 3. Early changes of redox states of phosphopyridine nucleotides (A) and pyridine nucleotides (B) in HeLa cells stimulated by external oxidants: (●) 1.7 μ*M* diferric transferrin; (○) diferric transferrin plus 3 μg/ml insulin; (■) 0.033 m*M* ferricyanide; (□) ferricyanide plus 3 μg/ml insulin.

redox system follow the same pattern for pyridine nucleotides as do mitogen- and growth factor-induced cells.

Initial events after the activation of the transmembrane electron transport system include the partial oxidation of intracellular NADH, but neither increase of the total pyridine nucleotide pool nor NADP(H) changes is observed. It is then an obvious assumption that

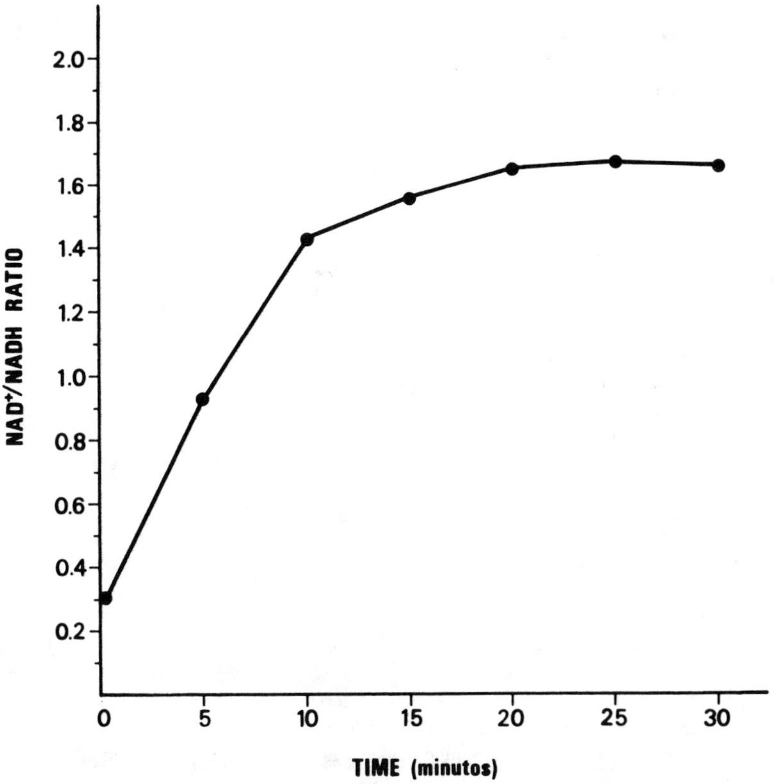

FIGURE 4. Short-term effect of 2 m*M* semidehydroascorbate on redox state of pyridine nucleotides in human promyelocytic leukemia cells (HL60).

NADH is the natural source of the intracellular electrons. This rapid interchange between oxidized and reduced forms during the first hour of treatment could induce the synthesis of new nucleotides observed after a long-term stimulation of cell growth.

ACKNOWLEDGMENTS

We are grateful to J. C. Rodriguez-Aguilera for his technical assistance. Partial financial support from the Spanish DGICYT is also acknowledged.

REFERENCES

1. **Kaplan, N. O.,** The role of pyridine nucleotides in regulating cellular metabolism, in *Current Topics in Cellular Regulation Modulation by Molecular Interactions,* Vol. 26, Levine, R. L. and Ginsburg, A., Eds., Academic Press, New York, 1985, 371.
2. **Kaplan, N. O., Colowick, S. P., and Neufeld, E. F.,** Pyridine nucleotide transhydrogenase, *Fed. Proc.,* 11, 238, 1952.
3. **Fisher, R. R. and Earle, S. R.,** Membrane-bound pyridine dinucleotide transhydrogenases, in *The Pyridine Nucleotide Coenzymes,* Everse, J., Anderson, B., and You, K.-S., Eds., Academic Press, New York, 1982, 279.
4. **Aström, A. and DePierre, J. W.,** Characterization of the induction of drug-metabolizing enzymes by 2-acetylaminofluorene, *Biochim. Biophys. Acta,* 673, 225, 1981.

5. **Sun, I., Mackellar, W. C., Crane, F. L., Barr, R., Elliot, W. L., Lem, N., Varnold, R. L., Heinstein, P. F., and Morré, D. J.,** Decreased NADH-oxidoreductase activities as an early response in rat liver to the carcinogen 2-acetylaminofluorene, *Cancer Res.,* 45, 157, 1985.
6. **Pekala, P. H. and Anderson, B. M.,** Non-oxidation-reduction reactions of pyridine nucleotides, in *The Pyridine Nucleotide Coenzymes,* Everse, J., Anderson, B., and You, K.-S., Eds., Academic Press, New York, 1982, 325.
7. **Navas, P., Minnifield, N., Sun, I. L., and Morré, D. J.,** NADP phosphatase as a marker in free-flow electrophoretic separations for cisternae of the Golgi apparatus midregion, *Biochim. Biophys. Acta,* 881, 1, 1986.
8. **Rechsteiner, M., Hillyard, D., and Olivera, B. M.,** Magnitude and significance of NAD turnover in human cell line D98/AH2, *Nature,* 259, 695, 1976.
9. **Hogebom, G. H. and Schneider, W. C.,** Cytochemical studies. VI. The synthesis of diphosphopyridine nucleotide by liver cell nuclei, *J. Biol. Chem.,* 197, 611, 1952.
10. **Rechsteiner, M. and Catanzarite, V.,** The biosynthesis and turnover of nicotinamide adenine dinucleotide in enucleated culture cells, *J. Cell. Physiol.,* 84, 409, 1974.
11. **Stollar, V. and Kaplan, N. O.,** Incorporation of isotopically labeled precursors into the pyridine nucleotide enzymes, *J. Biol. Chem.,* 236, 1863, 1961.
12. **Wang, T. P. and Kaplan, N. O.,** Kinases for the synthesis of coenzyme A and triphosphopyridine nucleotide, *J. Biol. Chem.,* 206, 311, 1954.
13. **Foster, J. W. and Baskowsky-Foster, A. M.,** Pyridine nucleotide cycle of *Salmonella typhimurium*: in vivo recycling of nicotinamide adenine dinucleotide, *J. Bacteriol.,* 142, 1032, 1980.
14. **White, H. B., III,** Biosynthetic and salvage pathways of pyridine nucleotide coenzymes, in *The Pyridine Nucleotide Coenzymes,* Everse, J., Anderson, B., and You, K.-S., Eds., Academic Press, New York, 1982, 225.
15. **Jedeikin, E. L. and Weinhouse, S.,** Metabolism of neoplastic tissue. VI. Assay of oxidized and reduced diphosphopyridine nucleotide in normal and neoplastic tissues, *J. Biol. Chem.,* 213, 271, 1955.
16. **Caiger, P. R., Morton, K., Filsell, O. H., and Jarret, J. G.,** A comparative study of nicotinamide coenzymes during growth of the sheep and rat, *Biochem. J.,* 85, 351, 1962.
17. **Nemeth, A. M. and Dickerman, H.,** Pyridine nucleotides and diphosphopyridine nucleotidase in developing mammalian tissues, *J. Biol. Chem.,* 249, 1980, 1960.
18. **Briggs, M. H.,** Vitamin and coenzyme content of hepatomas, *Nature,* 187, 249, 1960.
19. **Burch, H. B. and Von Dippe, P.,** Pyridine nucleotides in developing rat liver, *J. Biol. Chem.,* 239, 1898, 1964.
20. **Ferris, G. M. and Clark, J. B.,** The control of nucleic acid and nicotinamide nucleotide synthesis in regenerating rat liver, *Biochem. J.,* 128, 869, 1972.
21. **Berger, N. A., Berger, S. J., Sikorski, G. W., and Catino, D. M.,** Amplification of pyridine nucleotide pools in mitogen-stimulated human lymphocytes, *Exp. Cell Res.,* 137, 79, 1982.
22. **Lehmann, A. R., Kirk-Bell, S., Shall, S., and Whish, W. J. D.,** The relationship between cell growth, macromolecular synthesis and poly ADP-ribose polymerase in lymphoid cells, *Exp. Cell Res.,* 83, 63, 1974.
23. **Rochette-Egly, C., Ittel, M. E., Bilen, J., and Mandel, P.,** Effect of nicotinamide on RNA and DNA synthesis and poly (ADP-ribose) polymerase activity in normal and phytohemagglutinin stimulated human lymphocites, *FEBS Lett.,* 120, 7, 1980.
24. **Berger, N. A., Petzold, S. J., and Berger, S. J.,** Association of poly(ADP-rib) synthesis with cessation of DNA synthesis and DNA fragmentation, *Biochim. Biophys. Acta,* 564, 90, 1979.
25. **Miller, E. G.,** Stimulation of nuclear poly(adenosine diphosphate-ribose) polymerase activity from HeLa cells by endonucleases, *Biochim. Biophys. Acta,* 395, 191, 1975.
26. **Durkacz, B. W., Omidijii, O., Gray, D. A., and Shall, S.,** $(ADP\text{-}ribose)_n$ participates in DNA excision repair, *Nature,* 283, 593, 1980.
27. **Bertazzoni, U., Stefanini, M., Pedrali Noy, G., Giulotto, E., Nuzzo, F., Falaschi, A., and Spadari, S.,** Variations of DNA polymerases-α and -β during prolonged stimulation of human lymphocytes, *Proc. Natl. Acad. Sci. U.S.A.,* 73, 785, 1976.
28. **Gupta, P. K. and Sirover, M. A.,** Sequential stimulation of DNA repair and DNA replication in normal human cells, *Mutat. Res.,* 72, 273, 1980.
29. **Machaffie, R. A. and Wang, C. H.,** The effect of phytohemaglutinin upon glucose catabolism in lymphocytes, *J. Hematol. Blood,* 29, 640, 1967.
30. **Schwartz, J. P. and Passonneau, J. V.,** The effect of growth conditions on NAD^+ and NADH concentrations and the NAD^+: NADH ratio in normal and transformed fibroblasts, *J. Biol. Chem.,* 249, 4138, 1974.
31. **Sun, I. L., Crane, F. L., Grebing, C., and Löw, H.,** Transmembrane redox in control growth. Stimulation of HeLa cell growth by ferricyanide and insulin, *Exp. Cell Res.,* 156, 528, 1985.

32. **Breitman, T. R., Collins, S. J., and Keene, B. R.,** Induction of differentiation of the human promyelocytic leukemia cell line (HL60) by retinoic acid, *Exp. Cell Res.*, 126, 494, 1980.
33. **Bannai, S. and Sheppard, J. R.,** Cyclic AMP, ATP and cell contact, *Nature*, 250, 62, 1974.
34. **Miller, Z., Lovelace, E., Gallo, M., and Pastan, I.,** Cyclic guanosine monophosphate and cellular growth, *Science*, 190, 1213, 1975.
35. **Ellem, K. A. O. and Kay, G. F.,** Ferricyanide can replace pyruvate to stimulate growth and attachment of serum restricted human melanoma cells, *Biochem. Biophys. Res. Commun.*, 112, 183, 1983.
36. **Jacobson, E. L. and Jacobson, M. K.,** Pyridine nucleotide levels as a function of growth in normal and transformed 3T3 cells, *Arch. Biochem. Biophys.*, 175, 627, 1976.
37. **Sun, I. L., Navas, P., Crane, F. L., Chou, J. Y., and Löw, H.,** Transplasmalemma electron transport is changed in simian virus 40 transformed liver cells, *J. Bioenerg. Biomemb.*, 18, 471, 1986.
38. **Rakin, P. W., Jacobson, M. K., Mitchell, V. R., and Busbee, D. B.,** Reduction of nicotinamide adenine nucleotide levels by ultimate carcinogens in human lymphocytes, *Cancer Res.*, 40, 1803, 1980.
39. **Jacobson, M. K., Levi, V., Juarez-Salinas, H., Barton, R. A., and Jacobson, E. L.,** Effect of carcinogenic N-alkyl-N-nitroso compounds on nicotinamide adenine nucleotide metabolism, *Cancer Res.*, 40, 1797, 1980.
40. **Glock, G. E. and McLean, P.,** Levels of oxidized and reduced diphosphopyridine nucleotide and triphosphopyridine nucleotide in tumors, *Biochem. J.*, 65, 413, 1957.
41. **Gunnarson, R., Berne, C., and Hellerström, C.,** Cytotoxic effects of streptozotocin and N-nitrosomethylurea on the pancreatic B cell with special regard to the role of nicotinamide-adenine dinucleotide, *Biochem. J.*, 140, 487, 1974.
42. **Clark, M. G., Partick, E. J., Patten, G. S., Crane, F. L., Löw, H., and Grebing, C.,** Evidence for the extracellular reduction of ferricyanide by rat liver, *Biochem. J.*, 200, 565, 1981.
43. **Crane, F. L., Roberts, H., Linnane, A. W., and Löw, H.,** Transmembrane ferricyanide reduction by cells of the yeast *Saccharomyces cerevisiae*, *J. Bioenerg. Biomembr.*, 14, 191, 1982.
44. **Craig, T. A. and Crane, F. L.,** Evidence for a transplasma membrane electron transport system in plant cells, *Proc. Indiana Acad. Sci.*, 90, 150, 1981.
45. **Crane, F. L., Löw, H., and Clark, M. G.,** Plasma membrane redox enzymes, in *Enzymes of Biological Membranes*, Vol. 4, A. Martinosi, Ed., Plenum Press, New York, 1985.
46. **Clark, M. G., Patrick, E. J., and Crane, F. L.,** Properties and regulation of a transplasma membrane redox system in rat liver, *Biochem. J.*, 204, 795, 1982.
47. **Crane, F. L., Crane, H. E., Sun, I. L., MacKellar, W. C., Grebing, C., and Löw, H.,** Insulin control of a transplasma membrane NADH dehydrogenase in erythrocyte membranes, *J. Bioenerg. Biomembr.*, 14, 425, 1982.
48. **Goldenberg, H.,** Plasma membrane redox activities, *Biochim. Biophys. Acta*, 694, 203, 1982.
49. **Sun, I. L., Navas, P., Crane, F. L., Morré, D. J., and Löw, H.,** NADH diferric transferrin reductase in liver plasma membrane, *J. Biol. Chem.*, 262, 15915, 1987.
50. **Barnes, D. and Sato, G.,** Serum-free cell culture: a unifying approach, *Cell*, 22, 649, 1980.
51. **Tsao, M.-S., Sanders, G. H. S., and Grisham, J. W.,** Regulation of growth of cultured hepatic epithelial cells by transferrin, *Exp. Cell Res.*, 171, 52, 1987.
52. **Sun, I. L., Navas, P., Morré, D. J., and Crane, F. L.,** Modification of pyridine nucleotide pools on HeLa cells grown under growth promoting factors, *Proc. Indiana Acad. Sci.*, 96, 377, 1987.
53. **Sun, I. L., Crane, F. L., Löw, H., and Grebing, C.,** Transplasma membrane redox stimulates HeLa cell growth, *Biochem. Biophys. Res. Commun.*, 125, 649, 1984.
54. **Landschulz, W., Thesleff, J., and Ekblom, P.,** A lipophilic iron chelator can replace transferrin as a stimulator of cell proliferation and differentiation, *J. Cell Biol.*, 98, 596, 1984.
55. **Goldenberg, H., Grebing, C., and Löw, H.,** NADH monodehydroascorbate reductase in human erythrocyte membranes, *Biochem. Int.*, 6, 1, 1983.
56. **Navas, P., Estévez, A., Burón, M. I., Villalba, J. M., and Crane, F. L.,** Cell surface glycoconjugates control the activity of the NADH-ascorbate free radical reductase of rat liver plasma membrane, *Biochem. Biophys. Res. Commun.*, 154, 1029, 1988.
57. **Navas, P., Burón, M. I., and Alcain, F. J.,** Stimulation of HL60 cell growth by ascorbate free radical, *J. Cell Biol.*, 107, 503a, 1988.
58. **Crane, F. L., Sun, I. L., Clark, M. G., Grebing, C., and Löw, H.,** Transplasmamembrane redox systems in growth and development, *Biochim. Biophys. Acta*, 811, 233, 1985.
59. **Sun, I. L., Toole-Simms, W., Crane, F. L., Morré, D. J., Löw, H., and Chou, J. Y.,** Reduction of diferric transferrin by SV40 transformed pineal cells stimulates Na^+/H^+ antiport activity, *Biochim. Biophys. Acta*, 938, 17, 1988.
60. **Medina, M. A., Sánchez-Jiménez, F., Segura, J. A., and Núñez de Castro, I.,** Transmembrane ferricyanide reductase activity in Ehrlich ascites tumor cells, *Biochim. Biophys. Acta*, 946, 1, 1988.
61. **L'Allemain, G., Frauchi, A., Cragoe, E., Jr., and Pouyssegur, M.,** Blockade of the Na^+/H^+ antiport abolishes growth factor-induced DNA synthesis in fibroblasts, *J. Biol. Chem.*, 259, 4313, 1984.

62. **Moolenaar, W. H., Tsien, R. Y., van der Saag, P. T., and DeLeat, S. W.,** Na^+/H^+ exchange and cytoplasmic pH in the action of growth factors in human fibroblasts, *Nature,* 304, 645, 1983.
63. **Perona, R. and Serrano, R.,** Increased pH and tumorigenicity of fibroblasts expressing a yeast proton pump, *Nature,* 334, 438, 1988.
64. **Roth, Z., Chayen, N., and Dikstein, S.,** The involvement of the intracellular redox state and pH in the metabolic control of stimulus-response compling, *Int. Rev. Cytol.,* 85, 39, 1983.
65. **Navas, P., Sun, I. L., Crane, F. L., and Morré, D. J.,** Changes in the pyridine nucleotide pools of HeLa cells in response to growth promoting agents, in *Redox Functions of the Eukaryotic Plasma Membrane,* Ramírez, J. M., Ed., C.S.I.C., Madrid, 1987, 51.
66. **Skidmore, C. J., Davies, M. I., Goodwin, P. M., Halldorsson, H., Lewis, P., Shall, S., and Ziáee, A.-A.,** The involvement of poly(ADP-ribose) polymerase in the degradation of NAD caused by -radiaction and N-methyl-N-nitrosourea, *Eur. J. Biochem.,* 101, 135, 1979.
67. **Navas, P., Sun, I. L., Morré, D. J., and Crane, F. L.,** Decrease of NADH in HeLa cells in the presence of transferrin or ferricyanide, *Biochem. Biophys. Res. Commun.,* 135, 110, 1986.
68. **Sun, I. L., Crane, F. L., Löw, H., and Grebing, C.,** Inhibition of plasma membrane NADH dehydrogenase by adriamycin and related anthracycline antibiotic, *J. Bioenerg. Biomembr.,* 16, 209, 1984.
69. **Mishra, R. K. and Passow, H.,** Induction of intracellular ATP synthesis by extracellular ferricyanide in human red blood cells, *J. Membr. Biol.,* 1, 214, 1969.
70. **Raghoebar, M., Huisman, J. A. M., van der Berg, W. B., and van Ginneken, C. A. M.,** Characteristics of the transport of ascorbic acid into leucocytes, *Life Sci.,* 40, 499, 1987.
71. **Rose, R. C.,** Transport of ascorbic acid and other water-soluble vitamins, *Biochim. Biophys. Acta,* 947, 335, 1988.

Chapter 12

Na^+:H^+ ANTIPORT

Rafael García-Cañero

TABLE OF CONTENTS

I. INTRODUCTION

Cytoplasmic pH is strongly regulated in eukaryotic cells, clamped at values between 7.0 and 7.4, mainly by ion transport and the buffering capacity of the cytosol.[1] This is necessary because, due to a large inside-negative membrane potential of the typical cell, a passive distribution of H^+ across the plasma membrane is incompatible with the normal metabolic functions, such as enzyme activity, macromolecular synthesis, contractile activity, ion conductivities, and control of the cell cycle, H^+ concentration playing a major role in the above processes.[2]

Almost all eukaryotic cells appear to have two or more pH regulating mechanisms showing a rather complicated interaction among them; the pH-regulating anion antiport seems to exist in most cells.[3] Na-CO_3H symport systems have been described in a variety of cells, although their role in nonspecialized cells is still unclear.[4] H^+-translocating ATPases are involved in proton expulsion from the cytosol, when located in the plasma membrane and in the limiting membranes of the intracellular vesicles.[5]

It is generally accepted that the Na^+:H^+ antiport is present in all cells, playing an important role in pHi regulation after acid loads,[6] in volume regulation,[7] cell differentiation,[8] and in elevating pH_i as the result of the action of hormones and growth factors.[9]

II. THE Na^+:H^+ ANTIPORT OF EUKARYOTICS

The plasma membrane of eukaryotic cells of different origin and specialization contains a carrier-mediated transport system that drives the electroneutral exchange of Na^+ for H^+.[10] Under physiological, nonactivated conditions, this transport system carries out the uphill extrusion of H^+, coupled with the downhill flow of Na^+ into the cells. Its stoichiometry ranges between 0.6 and 1.2 in all the cell types studied, and the process is reversible depending on the direction of the concentration gradient for H^+ and Na^+. The exchange is independent of membrane potential and ATP content, and is electroneutral.[11]

Na^+ uptake by membrane vesicles from intestinal brush-border and kidney follows a single saturation kinetics with respect to external Na^+ concentration, suggesting a single external site for Na^+ binding.[12] However, other studies suggest sigmoidal characteristics when nonepithelial cells are used.[13]

The uptake of Na^+ through this pathway is stimulated by outwardly directed H^+ gradients. It is inhibited by Li^+ or NH_4^+, but not by K^+, Rb^+, or Cs^+, indicating its affinity for different monovalent cations. The K_m for external Na^+ ranges between 5 and 50 mM, depending on the cell being studied.[14]

By contrast, the dependence on internal H^+ shows a greater than first order kinetics for Na^+ uptake. Likewise, the net Na^+ influx via the Na^+:H^+ antiport shows a threefold increase as the internal pH decreases. The above facts suggest an interaction between internal H^+ and an allosteric site on the exchanger complex, other than the H^+-translocating site. This effect has been observed in plasma membrane vesicles derived from human epidermal carcinoma cells, muscle fibroblasts, and renal microvillus,[15] the K_m for internal H^+ being about 40 nM.[16]

The importance of an allosteric regulation site is shown by the quiescent state attained by the antiporter whenever pHi is above a certain value, which seems to coincide with the normal physiological pHi of cells, disregarding the combined value of the existing Na^+ plus H^+ gradient.[15]

Another interesting property of the antiport system is its sensitivity to a family of compounds from the pyrazinecarboxamide series,[17-19] the most well known being amiloride, which seems to compete for the external Na^+ site of the antiporter, as increasing external Na^+ concentration diminishes the inhibition produced. This inhibition is reversible, its extension depending on extracellular Na^+ concentration and, likewise, on external pH, as does the exchange activity.[20] Thus, at physiological Na^+ concentration, effective inhibition requires a rather high amiloride concentration. In addition, amiloride inhibits the antiport by binding to a second site independent of Na^+,[17] and for its analogs and itself the protonated state is required for inhibition.

Studies on covalent modification of the renal and placental Na^+:H^+ antiporter with dicyclohexylcarbodiimide indicate that amiloride binds to a 100-kDa protein,[21] due to its protection of Na^+:H^+ exchange from inactivation by dicyclohexylcarbodiimide near a catalytic site with an important carboxyl group. Histidyl groups appear to be equally important for antiporter operation,[22] seeming to form part of the external H^+ binding site on the exchanger.[23]

The transport protein has been detected in renal epithelial plasma membranes by means of radiation inactivation, rendering a molecular size of about 600 kDa.[24] Using amiloride-affinity chromatography, a Na^+-transporting protein has been purified from bovine adrenal papilla and cultured amphibian cells, with an apparent molecular mass of 670 kDa and composed of several subunits, one of them, of 150 kDa, able to bind covalently methylbromoamiloride.[25] Liposomes prepared with detergent extracts and partially purified transporter were sensitive to amiloride in inhibiting ^{22}Na uptake.[26] Attempts to raise monoclonal antibodies against the amiloride-sensitive, epithelial Na^+ channel, to improve its purification and recognition, have been reviewed.[11]

III. REGULATION OF Na^+:H^+ EXCHANGE, GROWTH, AND DIFFERENTIATION IN EUKARYOTICS

Aside from the intracellular pH regulation of Na^+:H^+ exchange in nonactivated cells, activation has been observed in a great variety of cells, of epithelial and nonepithelial origin, and of transformed and untransformed phenotypes,[15] induced in two different ways: fast responses, from seconds to minutes, induced by hormones and growth factors, Ca^{+2} ionophores, vanadate, chemotactic peptides, and serum; and slow responses, from hours to days, induced by acidosis, renal insufficiency, chronic potassium depletion, and glucocorticoids.[2,3,11,15,27-34]

The effect observed upon stimulation in most cases is coincident with a shift in the internal pH dependence of the exchanger,[15,29,35] increasing its affinity for internal H^+, which results in the enhancement of the Na^+ uptake and cytoplasmic alkalinization that controls the rate of initiation of DNA synthesis. Na^+ influx plays a greater role than the intracellular Na^+ level in promoting growth when accompanied by H^+ extrusion, amiloride and its analogs being more capable of inhibiting the growth of rapidly proliferating, transformed or untransformed cells than that of slowly proliferating, normal cells.[36-38]

Mitogen induced pHi variations range between 0.15 and 0.3 pH units, depending on cell type, mitogen or stimulus employed.[39] The requirements for this increase paralleled the same for Na^+:H^+ exchange activation, sharing similar properties.[39] This increase occurs rapidly and is complete after 10 min.[40]

The intracellular alkalinization controlled by Na:H antiport seems to play a crucial role in animal cells as well as in activation of cell proliferation,[41] being subject to genetic control, since a mutation abolishing the expression of the antiport impedes the growth of fibroblasts at physiological pH.[42,43]

In view of the data appearing in the literature over the past years, the molecular mechanisms by which Na:H antiport is activated are subject to speculation. The use of different cell types from different organs, including transformed cells, and the comparison of noncell specific mitogens and metabolic activators, as well as inhibition studies, argue against its ascription to a general or unique mechanism operating universally for Na^+:H^+ antiport activation.

Among the proposed mechanisms for mitogen activation of Na^+:H^+ antiport, the possible phosphorylation of the transport protein has been considered, which, in turn, could promote a conformational change, causing a modification in pHi sensitivity of the antiporter, induced by tyrosine-specific protein kinases detected in receptors for several growth factors.[44] This is supported by the finding that vanadate, an inhibitor of phosphotyrosine dephosphorylation, stimulates the Na^+:H^+ antiport.[45] However, this stimulation does not last as short a time as that induced by mitogens in the same cell type, indicating that the above effect could be ascribed to the inhibitory effect of vanadate rather than to direct phosphorylation of the antiporter.

Cell volume changes have been implicated in Na^+:H^+ antiporter activation; however, no volume changes have been observed upon mitogen addition, ruling this out as the activation mechanism.[39] In lymphocytes and epithelial cells, however, cell shrinking activates the antiporter,[46] although its role in growth remains unclear.

Another proposed mechanism deals with the requirement of steady levels of intracellular ATP for mitogen action; protein phosphorylation mediated by Ca^{2+}-phospholipid dependent protein kinase C (PKC) could be responsible for activation,[47] based on the effects of phorbol esters in increasing pHi and promoting protein phosphorylation,[48] as the result of a rise in cytoplasmic free calcium, and a coincident release of inositol (1,4,5)tris phosphate, and its role in activation of PKC.[49] However, a rise in Ca_i and/or phosphatidylinositol breakdown has not been shown to occur for all the mitogens in fibroblasts and lymphocytes. Moreover, although the pH_i increase induced by phorbol esters is both dependent on external Na^+ and amiloride-sensitive, other routes for Na^+:H^+ activation have been found in 3T3 fibroblasts,[50,51] in epidermal carcinoma cells,[52] and in Ehrlich ascites tumor cells.[53] In PKC-depleted cells, there is a possibility of further increasing pHi by adding growth factors or hormones, indicating alternative routes for Na^+:H^+ antiport activation.[54] Negative feedback control of antiport via PKC has been suggested as well.[55] Other protein kinases, such as the cAMP-dependent protein kinase A (PKA), fail to activate Na^+:H^+ exchange.[56]

Transmethylation of the antiporter[57] and phospholipase-dependent stimulation have been proposed as other possible mechanisms of Na^+:H^+ antiport activation. In cultured human fibroblasts treated with melittin, a phospholipase A2 activator, and mepacrine, an inhibitor, or after prolonged treatment with dexamethasone, which induces the appearance of an intrinsic inhibitor of the phospholipase, Na^+ uptake is either increased or inhibited,[58] the activation being due to a further activation of phospholipase C by products of phospholipase A2 activity.[59]

Finally, it has been observed in 3T3 fibroblasts that microinjection of the activated protein *ras* p21 induces a rapid cytoplasmic alkalinization,[60] as it does on the mitogenic signal transduction leading to DNA synthesis.[61] Recently, Maly et al. have reported the activation of Na:H antiport in NIH 3T3 cells by a PKC-independent mechanism.[62]

Thus, the possible mechanisms involved in the activation of the Na^+:H^+ antiport are covalent modification, gene expression, *de novo* synthesis, and allosteric regulation induced by transient intracellular acidification.

Concerning this last possibility for the mitogen action, there are some data in the literature which indicate the existence of a lag period of about 2 min after growth factor addition in which transient acidification is seen even in the presence of extracellular Na^+, prior to any increase in intracellular pH, and even before any activation of Na^+ uptake is measurable. For NR6 cells, platelet-derived growth factor (PDGF) and, to a lesser extent, serum, among others, induce a transient acidification lasting approximately 2 min immediately upon growth factor addition, prior to the pH_i increment, reaching values of about -0.07 pH_i.[63]

This phenomenon has been observed in other cells subjected to a variety of mitogenic and nonmitogenic agents, such as murine fibroblasts exposed to PDGF or bombesin;[64] in A431 cells activated by EGF;[65] in gastric cells;[66] in Ehrlich ascites tumor cells,[53] and in human neutrophils,[67] in this case implying redox effects in the cells.

The determination of the possible importance of the cellular redox state in growth factor activation of Na^+:H^+ antiport results from its role in hormone transduction mechanisms involving Ca^{2+} movements and regulation of adenylate- and guanylate-cyclase, and is related to pH_i, as reviewed by Roth et al.[68]

Pyridine nucleotide levels are different in proliferating cells than in nonproliferating cells, and when exogenous stimuli are applied, the NAD:NADH ratio increases in response to cell proliferation, whether or not hormones, such as insulin, or diferric transferrin or ferricyanide are used.[69]

On the other hand, transferrin is able to act as a progression factor in liver epithelial cells *in vitro*[70] and FeIII complexes with saturated coordination shells stimulate cell growth by a probable redox mechanism.[71]

IV. REGULATION OF GROWTH AND Na:H ANTIPORT BY TRANSMEMBRANE REDOX SYSTEMS

Electron transport systems are widespread in biological membranes from different cell organelles. Aside from mitochondria and endoplasmic reticulum, the redox activity has been detected and characterized in plasma membranes in all the cells studied.[72] The transmembranous NADH-FeIII reductase, able to employ diferric transferrin, is more abundant in most cell plasma membranes than in membranes from intracellular organelles, and is easily distinguishable from reductase activities contaminating plasma membrane preparations by its lack of sensitivity to inhibitors such as KCN, antimycin A, and azide, and by its sensitivity to hormones.[73]

Its operation is involved in control of membrane functions such as amino acid transport[74] and modulation of ionic channels.[75]

Diferric transferrin is required for cell growth, and ferricyanide stimulates the growth of HeLa cells alone, or synergistically with insulin and serum.[76] Another impermeable oxidant, ferricyanide, reduced by the transplasma membrane electron transport system, stimulates cell growth, and antitumor drugs that inhibit cell growth are able to inhibit the enzyme activity at the same doses. The reduction of the external oxidants paralleled a decrease in intracellular NADH levels in HeLa cells, indicating that NADH is the internal substrate for the enzyme.[77-79]

Ferricyanide has been involved in modulation of DNA synthesis and stimulation of protooncogene expression in fibroblasts, in the absence of growth factors.[80,81]

Transplasma membrane oxidoreductase activity is accompanied by proton extrusion to the medium[73] in erythrocytes, HeLa cells, rat adipocytes, and rat liver cells,[73,82,83] which seem to be insensitive to uncouplers, indicating that this proton movement goes against a concentration gradient and is not caused by a release of intracellular acids; it does not increase the membrane potential and requires a counter ion movement to equalize the extra proton movement or release of anions.[82,83]

In liver cells, amiloride-sensitive ^{22}Na uptake is activated by increasing concentrations of ferricyanide, the maximum activation observed coinciding with the maximal for ferricyanide reduction by the redox enzyme; ferricyanide reduction is inhibited by amiloride, with a K_i of about 150 μM. Ferricyanide reduction is dependent on external Na and can be activated by the specific hepatocyte mitogen, liver growth factor (LGF).[84,85] External Na deprivation and amiloride abolish the LGF-induced activation of the reductase.

Changes in ferricyanide reductase activity and its external Na dependence on growth has been shown in HeLa cells,[86] and stimulation of Na^+:H^+ antiport activity by diferric transferrin has been detected in HeLa and SV40-transformed pineal cells.[87,88]

In C3H/10T1/2 cells, ferricyanide modulates cytosolic PKC activity in a way sensitive to inhibition by amiloride, the redox activity being considered the priming event in transmembrane signaling.[89]

On the other hand, ferricyanide reductase is not activated by phorbol esters in hepatocytes, and this substrate does not promote changes in the levels of cytosolic PKC as does LGF; in these cells, diferric transferrin is able to increase DNA synthesis synergistically with submaximal doses of LGF, as does ferricyanide, whereas ferrocyanide or apotransferrin do not; phorbol dimyristoyl triacetate increases DNA synthesis as well. These results indicate different pathways for LGF and phorbol esters in activating rat liver cells.[87]

It has been observed that transretinoic acid, but no other analogs, is capable of inhibiting both diferric transferrin and ferricyanide oxidoreductase in HeLa cells and rat liver plasma membranes,[90,91] and that adriamycin inhibits HeLa cell redox activity;[79] this inhibition varies, depending upon the tumorigenic state of fetal rat liver cells infected with SV40 virus, in such a way that transformed cells are insensitive to inhibition by retinoic acid, and gain sensitivity to inhibition by adriamycin, reversing the relative sensitivity of their nontransformed counterparts[92] and changes in electron transport activity observed under carcinogen treatment,[86,93] indicating the importance of the redox activity in cellular growth control and differentiation, providing a new insight into the mechanisms of cell growth activation in a variety of cells.

V. CONCLUSIONS AND FUTURE PROSPECTS

Regulation of intracellular pH in eukaryotics is an important feature of cell growth. The regulation of the structural activity responsible for cytoplasmic alkalinization accompanying normal and tumoral cell growth has provided knowledge of many different modes of signal transduction in mitogen action, of vital importance in the understanding of mechanisms of growth control and, for instance, the future development of therapeutic modes for cancer treatment.

The understanding of the mechanisms by which the Na^+:H^+ antiport is regulated in cells seems to require the use of as many cell-specific mitogens as possible, even though cell growth is a pleiotropic phenomenon. From this point of view, it seems reasonable to accept the diversity in response shown by different cell types to different activating agents employed, and the impossibility of formulating a general mechanism for prereplicative activation involving antiport activation.

Although amiloride treatment has been questioned for its side effects, there is an increasing amount of literature on its use and the use of cellular alkalinization inhibitors as cancer chemotherapeutic agents, a fact that gives preponderance to the role of oncogene expression and transmembranous oxidoreductase enzymes in growth control.[94-96]

The need to obtain information about the structure and entity of the antiport protein, its real membrane localization and possible associated enzymatic activities should be kept in mind, in order to ascertain clearly the mechanism of activation in normal and abnormal growth.

REFERENCES

1. **Roos, A. and Boron, W. F.,** Intracellular pH, *Physiol. Rev.,* 61, 296, 1981.
2. **Seifter, J. L. and Aronson, P. E.,** Properties of the plasma membrane Na^+:H^+ exchanger, *J. Clin. Invest.,* 78, 859, 1986.
3. **Madshus, J. H.,** Regulation of intracellular pH in eukaryotic cells, *Biochem. J.,* 250, 1, 1988.
4. **Grassl, S. M. and Aronson, P. S.,** Na/CO_3H co-transport in basolateral membrane vesicles isolated from rabbit renal cortex, *J. Biol. Chem.,* 261, 8778, 1986.
5. **Aronson, P. S.,** Mechanisms of active H^+ secretion in the proximal tubules, *Am. J. Physiol.,* 245, F467, 1983.
6. **Boron, W. F.,** Approaches for studying intracellular pH regulation in mammalian renal cells, in *Current Topics in Membrane and Ion Transport,* Vol. 26, Aronson, P. and Boron, W. F., Eds., Academic Press, New York, 1986, 15.

7. **Grinstein, S., Cohen, S., Goetz, J. D., Rothstein, A., Mellors, A., and Gelfand, E. W.,** Activation of the Na^+-H^+ antiport by changes in cell volume and phorbol esters; possible role of protein kinase, in *Current Topics in Membrane and Ion Transport,* Vol. 26, Aronson, P. and Boron, W. F., Eds., Academic Press, New York, 1986, 115.
8. **Smith, L. L., Stanton, T. H., Calalb, H. H., and Bomsztyk, K.,** Recombinant murine interferon-gamma-induced differentiation of pre-B lymphocytes is associated with Na^+/H^+ exchange-dependent and -independent, cytoplasmic alkalinization, *J. Biol. Chem.,* 263, 7359, 1988.
9. **Pouysségur, J.,** The growth-factor activatable Na-H system, in *Oncogenes and Growth Factors,* Bradshaw, R. A. and Prentiss, S., Eds., Elsevier, New York, 1987, 292.
10. **Van Dyke, R. W. and Ives, H. E.,** Na^+/H^+ exchange: what, where and why?, *Hepatology,* 8, 960, 1988.
11. **Garty, H. and Benos, D.,** Characteristics and regulatory mechanisms of the amiloride-blockable Na^+ channel, *Physiol. Rev.,* 68, 309, 1988.
12. **Murer, H., Hopfer, V., and Kinne, R.,** Na/H antiport in brush-border membrane vesicles isolated from rat small intestine and kidney, *Biochem. J.,* 154, 597, 1976.
13. **Mosely, R. H., Meier, P. J., Aronson, P. S., and Boyer, J. L.,** Na-H exchange in rat liver basolateral but not canalicular plasma membrane vesicles, *Am. J. Physiol.,* 250, G35, 1986.
14. **Aronson, P. S.,** Kinetic properties of the plasma membrane Na-H exchanger, *Am. Rev. Physiol.,* 47, 545, 1985.
15. **Grinstein, S. and Rothstein, A.,** Mechanisms of regulation of the Na-H exchanger, *J. Membr. Biol.,* 90, 1, 1986.
16. **Aronson, P. S., Nee, J., and Suhm, M. A.,** Modifier role of internal H^+ in activating the Na-H exchanger in renal microvillus membrane vesicles, *Nature,* 299, 161, 1982.
17. **Kleinman, T. and Cragoe, E. J.,** Amiloride and its analogs as tools in the study of ion transport, *J. Membr. Biol.,* 105, 1, 1988.
18. **Cragoe, E. J., Woltersdorf, O. W., Bickling, J. B., Kwong, S. F., and Jones, J. H.,** Pyrazine diuretics. II. *N*-amidino-3-amino-5-substituted 6-halopyrazinecarboxamides, *J. Med. Chem.,* 10, 66, 1967.
19. **Zhuang, Y., Cragoe, E. J., Shaikewitz, T., Glaser, L., and Cassel, D.,** Characterization of potent Na/H exchange inhibitors from the amiloride series in A431 cells, *Biochemistry,* 23, 4481, 1984.
20. **Vigné, P., Frelin, C., and Laszdunski, M.,** Léchangeur Na^+/H^+ des cellules eucaryotes: propriétés biochimiques et pharmacologiques, rôle physiologique, *Biochimie,* 67, 129, 1985.
21. **Igarashi, P. and Aronson, P. S.,** Covalent modification of the renal Na/H exchanger by *N,N′*-dicyclohexylcarbodiimide, *J. Biol. Chem.,* 262, 860, 1987.
22. **Ganapathy, V., Balkovetz, D. F., Ganapathy, M. E., Mahesh, V. B., Devoe, L. D., and Leibach, F. H.,** Evidence for histidyl and carboxyl groups at the active site of the human placental Na^+:H^+ exchanger, *Biochem. J.,* 245, 473, 1987.
23. **Grillo, G. and Aronson, P. S.,** Inactivation of the renal microvillus membrane Na/H exchanger by histidine specific reagents, *J. Biol. Chem.,* 264, 1120, 1986.
24. **Béliveau, R., Demenle, M., and Potier, M.,** Molecular size of the Na-H antiport in renal brush-border membranes, as estimated by radiation inactivation, *Biochem. Biophys. Res. Commun.,* 152, 484, 1988.
25. **Benos, D. J., Saccomani, G., and Shariban-Sohraby, S.,** The epithelial sodium channel, subunit and location of the amiloride binding site, *J. Biol. Chem.,* 262, 10613, 1987.
26. **Sariban-Sohraby, S. and Benos, D. J.,** Detergent solubilization, functional reconstitution, and partial purification of epithelial amiloride-binding protein, *Biochemistry,* 25, 4639, 1986.
27. **Villereal, M. L.,** Na:H and Na:Ca exchange in activated cells, in *Current Topics in Membrane and Ion Transport,* Vol. 27, Aronson, P. and Boron, W. F., Eds., Academic Press, New York, 1986, 55.
28. **Leffert, H. L. and Koch, K. S.,** Growth regulation by Na ion fluxes, in *Control of Animal Cell Proliferation,* Vol. 1, Academic Press, New York, 1985, 386.
29. **Lagarde, A. E. and Pouysségur, J. M.,** The Na:H antiport in cancer, *Cancer Biochem. Biophys.,* 9, 1, 1986.
30. **Berridge, M. J. and Irvine, F. J.,** Inositol triphosphate: a novel second messenger in cellular signal transduction, *Nature (London),* 312, 315, 1984.
31. **Burns, C. P. and Rozengurt, E.,** Extracellular Na^+ and initiation of DNA synthesis: role of intracellular pH and K^+, *J. Cell. Biol.,* 98, 1082, 1084.
32. **Mettler, R. T. and Pollock, A. S.,** Modification of the internal pH sensitivity of the Na:H antiporter by PTH in a cultured renal cell line, *J. Biol. Chem.,* 262, 9115, 1987.
33. **Moolenaar, W. H., Defize, L. H. K., and de Laat, S. N.,** Ionic signaling by growth factor receptors, *J. Exp. Biol.,* 124, 359, 1986.
34. **Fine, L. G., Badie-Dezfooly, B., Lowe, A. G., Hanizeh, A., Wells, J., and Selehnogaddan, S.,** Stimulation of Na:H antiport is an early event in hypertrophy of renal proximal tubular cells, *Proc. Natl. Acad. Sci. U.S.A.,* 82, 1736, 1985.

35. **Pouysségur, J., Chambard, J.-C., Franchi, A., L'Allemain, G., Paris, S., and Van Obberghen-Schillig, E.,** Growth factor activation of Na:H antiporter controls growth of fibroblasts by regulating intracellular pH, in *Growth Factors and Transformation,* Feramisco, J., Ozanne, B., and Stiles, C., Eds., Cold Spring Harbor Laboratory, Cold Spring Harbor, N.Y., 1985, 409.
36. **Sparks, R. L., Pool, T. B., Smith, N. K. R., and Cameron, I. L.,** The role of ions, ion fluxes and Na-K ATPase in the control of proliferation, differentiation and transformation, in *Genetic Expression in the Cell Cycle,* Academic Press, New York, 1982, 363.
37. **Moolenaar, W. H., Tsien, R. Y., van der Saag, P. T., and de Laat, S. W.,** Na:H exchange and cytoplasmic pH in the action of growth factors in human fibroblasts, *Nature,* 304, 645, 1983.
38. **Doppler, W., Maly, W., and Grunicke, H.,** Role of the Na:H antiport in the regulation of the internal pH of Ehrlich ascites tumor cells in culture, *J. Membr. Biol.,* 91, 147, 1986.
39. **Reuss, L., Cassel, D., Rothenberg, P., Whitely, B., Mancuso, D., and Glaser, D.,** Mitogens and ion fluxes, in *Current Topics in Membrane and Ion Transport,* Vol. 27, Aronson, P. and Boron, W. F., Eds., Academic Press, New York, 1986, 3.
40. **Moolenaar, W. H.,** Effects of growth factors on intracellular pH regulation, *Annu. Rev. Physiol.,* 48, 363, 1986.
41. **Perona, R. and Serrano, R.,** Increased pH and tumorigenicity of fibroblasts expressing a yeast proton pump, *Nature,* 334, 438, 1988.
42. **Pouysségur, J., Sardet, C., Franchi, A., L'Allemain, G., and Paris, S.,** A specific mutation abolishing Na:H antiport activity in hamster fibroblasts precludes growth at neutral and acidic pH, *Proc. Natl. Acad. Sci. U.S.A.,* 81, 4833, 1984.
43. **Pouysségur, J.,** The growth factor-activatable Na^+/H^+ exchange system: a genetic approach, *Trends Biochem. Sci.,* 10, 453, 1985.
44. **Carpenter, G.,** Vanadate, EGF and stimulation of DNA synthesis, *Biochem. Biophys. Res. Commun.,* 102, 11115, 1981.
45. **Cassel, D., Zhuang, Y. X., and Glaser, L.,** Vanadate stimulates Na^+/H^+ exchange in A431 cells, *Biochem. Biophys. Res. Commun.,* 118, 675, 1984.
46. **Grinstein, S., Mack, E., and Mills, G. B.,** Osmotic activation of the Na:H antiport in protein kinase C depleted lymphocytes, *Biochem. Biophys. Res. Commun.,* 134, 8, 1986.
47. **Nishizuka, Y.,** Calcium, phospholipid turnover and transmembrane signaling, *Philos. Trans. R. Soc. London, Ser. B,* 302, 101, 1983.
48. **Moolenaar, W. H., Tertoolen, L. G. S., and De Laat, S. W.,** Phorbol esters and diacylglycerol mimic growth factor in raising cytoplasmic pH, *Nature,* 312, 371, 1984.
49. **Metcalfe, J. C., Moore, J. P., Smith, G. A., and Hesketh, T. R.,** Calcium and cell proliferation, *Br. Med. Bull.,* 42, 405, 1986.
50. **Daniel, T. O. and Ives, H. E.,** Cyclosporin A inhibits kinase C independent activation of the Na:H exchanger by PDGF and vanadate, *Biochem. Biophys. Res. Commun.,* 145, 111, 1987.
51. **Vara, F. and Rozengurt, E.,** Stimulation of Na:H antiport activity by EGF and insulin occurs without activation of protein kinase C, *Biochem. Biophys. Res. Commun.,* 130, 646, 1985.
52. **Macara, I. G.,** Activation of Ca^{2+} influx and Na:H exchange by EGF and vanadate in A431 cells is independent of phosphatidylinositol turnover and is inhibited by phorbol ester and diacylglycerol, *J. Biol. Chem.,* 261, 9321, 1986.
53. **Wiener, E., Dubyak, G., and Scarpa, A.,** Na:H exchange in Ehrlich ascites tumor cells: regulation by extracellular ATP and 12-*O*-tetradecanoylphorbol 13-acetate, *J. Biol. Chem.,* 261, 4529, 1986.
54. **Redondo, J. M., López-Rivas, A., Vila, V., Cragoe, E. J., and Fresno, M.,** The role of protein kinase C in T lymphocyte proliferation: existence of protein kinase C-dependent and independent pathways, *J. Biol. Chem.,* 263, 17467, 1988.
55. **Vicentini, L. M. and Villereal, M. L.,** Activation of Na:H exchange in cultured fibroblasts: synergism and antagonism between phorbol ester, Ca ionophore and growth factors, *Proc. Natl. Acad. Sci. U.S.A.,* 82, 8053, 1985.
56. **Rozengurt, E. and Mendoza, S. A.,** Early stimulation of Na:H antiport, Na-K pump activity and Ca fluxes in fibroblast mitogenesis, in *Current Topics in Membrane and Ion Transport,* Vol. 27, Aronson, P. and Boron, W. F., Eds., Academic Press, New York, 1986, 163.
57. **Sohraby, S. S., Burg, M., Weismann, W. P., Ciang, P. K., and Johnson, J. P.,** Transport into A6 apical membrane vesicles: possible mode of aldosterone action, *Science,* 225, 745, 1984.
58. **Vicentini, L. M., Miller, R. J., and Villereal, M. L.,** Evidence for a role of phospholipase activity in the serum stimulation of Na influx in human fibroblasts, *J. Biol. Chem.,* 259, 6912, 1984.
59. **Villereal, M. L., Mix-Meldon, L. L., Vicentini, L. M., Jamison, G. A., and Owen, N. E.,** Mechanisms of growth factor stimulation of Na-H exchange in cultured fibroblasts, in *Current Topics in Membrane and Ion Transport,* Vol. 26, Aronson, P. and Boron, W. F., Eds., Academic Press, New York, 1986, 175.
60. **Hagag, N., Lacal, J. C., Graber, M., Aronson, S., and Viola, M. V.,** Microinjection of ras p21 induces a rapid rise in intracellular pH, *Mol. Cell. Biol.,* 1984, 1987.

61. **Doppler, W., Jaggi, R., and Groner, B.,** Induction of v-mos and activated Ha-ras oncogene expression in quiescent NIH 3T3 cells causes intracellular alkalinization and cell-cycle progression, *Gene,* 54, 147, 1987.
62. **Maly, K., Überall, F., Loferer, H., Doppler, W., Oberhuber, H., Groner, B., and Grunicke, H. H.,** Ha-ras activates the Na:H antiporter by a protein kinase C-independent mechanism, *J. Biol. Chem.,* 264, 11839, 1989.
63. **Cassel, D., Rothenberg, P., Zhuang, Y. X., Deuel, T. F., and Glaser, L.,** Platelet-derived growth factor stimulates Na:H exchange and induces cytoplasmic alkalinization in NR6 cells, *Proc. Natl. Acad. Sci. U.S.A.,* 80, 6224, 1983.
64. **Ives, H. E. and Daniel, T. O.,** Interrelationship between growth factor-induced pH changes and intracellular Ca^{2+}, *Proc. Natl. Acad. Sci. U.S.A.,* 84, 1950, 1987.
65. **Rothenberg, P., Glaser, L., Schlesinger, P., and Cassel, D.,** Activation of Na:H exchange by EGF elevates intracellular pH in A431 cells, *J. Biol. Chem.,* 258, 12644, 1983.
66. **Paradiso, A. M., Tsien, R. Y., and Machen, T. E.,** Digital image processing of intracellular pH in gastric oxyntic and chief cells, *Nature,* 325, 447, 1987.
67. **Grinstein, S., Furuya, W., and Biggar, W. D.,** Cytoplasmic pH regulation in normal and abnormal neutrophils, *J. Biol. Chem.,* 26, 512, 1986.
68. **Roth, Z., Chayen, N., and Dikstein, S.,** The involvement of the intracellular redox state and pH in stimulus-response coupling, *Int. Rev. Cytol.,* 85, 39, 1983.
69. **Sun, I. L., Navas, P., Morré, D. J., and Crane, F. L.,** Modification of pyridine nucleotide pools in HeLa cells grown under growth promoting factors, *Proc. Indiana Acad. Sci.,* 96, 377, 1987.
70. **Tsao, M. S., Sanders, G. H. S., and Grisham, J. W.,** Regulation of growth of hepatic epithelial cells by transferrin, *Exp. Cell Res.,* 171, 52, 1987.
71. **Kay, G. F. and Ellem, K. A. O.,** Nonhaem complexes of FeIII stimulate cell attachment and growth by a mechanism different from that of serum, 2-oxocarboxylates and haemproteins, *J. Cell. Physiol.,* 126, 275, 1986.
72. **Crane, F. L., Goldenberg, H., and Morré, D. J.,** Dehydrogenases of the plasma membrane, in *Subcellular Biochemistry,* Vol. 6, Roodyn, D. B., Ed., Plenum Press, New York, 1983, 345.
73. **Crane, F. L., Sun, I. L., Clark, M. G., Grebing, C., and Löw, H.,** Transplasma membrane redox systems in growth and development, *Biochim. Biophys. Acta,* 811, 233, 1985.
74. **García Sancho, J., Sánchez, A., Handlogten, M. E., and Christensen, H. N.,** Unexpected additional mode of energization of amino acid transport into Ehrlich cells, *Proc. Natl. Acad. Sci. U.S.A.,* 74, 1488, 1977.
75. **García Sancho, J.,** Redox modulation of the sensitivity to Ca^{2+} of Ca-dependent K^+ channels, in *Redox Functions of Eukaryotic Plasma Membranes,* Ramírez, J. M., Ed., CSIC Office, Madrid, 1987, 133.
76. **Sun, I. L., Crane, F. L., Grebing, C., and Löw, H.,** Transmembrane redox in control of cell growth, *Exp. Cell Res.,* 156, 528, 1985.
77. **Navas, P., Sun, I. L., Morré, D. J., and Crane, F. L.,** Decreases of NADH levels in HeLa cells in the presence of transferrin or ferricyanide, *Biochem. Biophys. Res. Commun.,* 135, 110, 1986.
78. **Sun, I. L., Crane, F. L., Löw, H., and Grebing, C.,** Transplasma membrane redox stimulates HeLa cell growth, *Biochem. Biophys. Res. Commun.,* 125, 649, 1984.
79. **Sun, I. L., Navas, P., Crane, F. L., Morré, D. J., and Löw, H.,** Diferric transferrin reductase in the plasma membrane is inhibited by adriamycin, *Biochem. Int.,* 14, 119, 1987.
80. **Wenner, C. E., Cutry, A., Kinniburgh, A., Tomei, L. D., and Leister, K. J.,** Modulation of redox reactions involved in DNA synthesis by oxygen and artificial electron acceptors, in *Plasma Membrane Oxidoreductases in Control of Animal and Plant Growth,* Vol. 157, Crane, F. L., Morré, D. J., and Löw, H., Eds., NATO Series, 1988, 17.
81. **Cutry, A. F., Kinniburgh, A. J., and Wenner, C. E.,** Stimulation of c-fos and c-myc protooncogene expression by ferricyanide, an external electron acceptor, *J. Biol. Chem.,* in press.
82. **Sun, I. L., Crane, F. L., and Chou, J. Y.,** Modification of transmembrane electron transport activity in plasma membranes of simian virus 40 transformed pineal cells, *Biochim. Biophys. Acta,* 886, 327, 1986.
83. **García Cañero, R.,** Possible role of transmembrane ferricyanide reductase in mitogenic activation of rat liver cells, in *Plasma Membrane Oxidoreductases in Control of Animal and Plant Growth,* Vol. 157, Crane, F. L., Morré, D. J., and Löw, H., Eds., NATO Series, 1988, 27.
84. **García Cañero, R., Díaz Gil, J. J., and Guerra, M. A.,** Is transmembrane ferricyanide reductase connected with the early expression of mitogenic signals in hepatocytes?, in *Redox Functions of the Eukaryotic Plasma Membranes,* Ramírez, J. M., Ed., CSIC Office, Madrid, 1987, 41.
85. **Díaz Gil, J. J., Escartin, P., García Cañero, R., Veloso, J. J., Sánchez, G.,** Purification of a liver DNA synthesis promoter from plasma of partially hepatectomized rats, *Biochem. J.,* 235, 49, 1986.
86. **García Cañero, R. and Guerra, M.,** External Na dependency and variation of ferricyanide reductase activity during tumor cell growth, *Oncologia,* 11, 187, 1988.

87. **Sun, I. L., García-Cañero, R., Liu, W., Toole-Simms, W., Crane, F. L., Morré, D. J., and Löw, H.,** Diferric transferrin reduction stimulates the Na:H antiport in HeLa cells, *Biochem. Biophys. Res. Commun.*, 145, 467, 1987.
88. **Sun, I. L., Toole-Sims, W., Crane, F. L., Morré, D. J., Löw, H., and Chou, J. Y.,** Reduction of diferric transferrin by SV40 transformed pineal cells stimulates Na/H antiport activity, *Biochim. Biophys. Acta,* 938, 17, 1988.
89. **Malviya, A. N. and Anglard, P.,** Modulation of cytosolic protein kinase C activity by ferricyanide: priming event seems transmembrane redox signaling, *FEBS Lett.*, 200, 265, 1986.
90. **Sun, I. L., Toole-Simms, W., Crane, F. L., Golub, E. S., Díaz de Pagán, T., Morré, D. J., and Löw, H.,** Retinoic acid inhibition of transplasmalemma diferric transferrin reductase, *Biochem. Biophys. Res. Commun.*, 146, 976, 1987.
91. **Sun, I. L., Navas, P., Crane, F. L., Morré, D. J., and Löw, H.,** NADH diferric transferrin reductase in liver plasma membrane, *J. Biol. Chem.*, 262, 15915, 1987.
92. **Sun, I. L., Toole-Simms, W., Crane, F. L., Morré, D. J., Löw, H., and Chou, J. Y.,** Transformation with SV40 virus prevents retinoic acid inhibition of plasma membrane NADH diferric transferrin reductase in rat liver cells, *J. Bioenerg. Biomemb.*, 20, 383, 1988.
93. **Sun, I. L., MacKellar, W., Crane, F. L., Barr, R., Elliot, W. L., Lem, N., Varnold, R. L., Heinstein, P. F., and Morré, D. J.,** Decreased NADH-oxidoreductase activities as an early response in rat liver to carcinogen 2-acetylaminofluorene, *Cancer Res.*, 45, 157, 1985.
94. **Grunicke, H., Doppler, W., Hofmann, J., Lindner, H., Maly, K., Oberhuber, H., Puschendorf, B., and Ringsdorf, H.,** New strategies for the improvement of alkylating antitumor agents, in *Cancer Biology and Therapeutics,* Cory, J. G. and Szentivanyi, A., Eds., Plenum Press, New York, 1987, 127.
95. **Rotin, D., Wan, P., Grinstein, S., and Tannock, I.,** Cytotoxicity of compounds that interfere with regulation of intracellular pH: a potential new class of anticancer agents, *Cancer Res.*, 47, 1497, 1987.
96. **Chitambar, C. R. and Seligman, P. A.,** Effects of different transferrin forms on transferrin receptor expression iron uptake, and cellular proliferation of human leukemic HL 60 cells, *J. Clin. Invest.*, 78, 1538, 1986.

Chapter 13

PLASMA MEMBRANE OXIDOREDUCTION AND AMINO ACID TRANSPORT

Miguel Angel Medina and Ignacio Núñez de Castro

TABLE OF CONTENTS

I. INTRODUCTION

The study of transport systems across cellular membranes is a recurrent topic in biochemistry books. Transport systems have several important functions: the regulation of the ionic composition of the cells, the detoxification of toxic metabolites, and, finally, the supply of organic solutes to the cells. Many of these substances are transported against concentration gradients. Consequently, a coupled free energy input is required to transport them. Two main driving forces of active transport systems are described: a flux of protons or sodium ions, and hydrolysis of ATP. Nevertheless, the description of plasma membrane redox systems in prokaryotic and eukaryotic cells strongly suggests a linkage between the membrane redox system and solute transport, particularly that of amino acids. This first hypothesis was supported by further evidence obtained from the study of several amino acids transport systems. A comprehensive model of amino acid transport in eukaryotic cells which is coupled to a transplasma membrane redox system is proposed.

II. HOW IS AMINO ACID TRANSPORT ENERGIZED?

It is an established paradigm that the active uptake of amino acids is driven by cotransport by the H^+ or Na^+ inorganic ions when they migrate down their electrochemical potential gradients. To maintain this movement, H^+ or Na^+ must first be extruded from the cell; indeed, Ramos and Kaback[1] describe proton extrusion in prokaryotic cells, and Crane[2] describes the extrusion of Na^+ in eukaryotic cells. However, whether this mechanism alone is sufficient to explain the steady-state entrance of amino acids in eukaryotic cells has been questioned. Consequently, other energy sources have been postulated for amino acid transport.

Some amino acids can be taken up in the absence of extracellular Na^+. The model for this transport system is the norbornane amino acid 2-aminebicycle-(2,2,1)-heptan-2-carboxylic acid (BCH).[3] Although this transport system is described as Na^+-independent, some recent works suggest that it is most probably a H^+-gradient-dependent transport system.[4] After depleting endogenous intracellular amino acids, careful monitoring of amino acid loss from cells during BCH uptake shows that the active process is not explained by an exchange of extracellular BCH for intracellular endogenous amino acids.[5,6] Furthermore, even Na^+-dependent amino acids can be actively taken up when the transmembrane Na^+ concentration gradient is reversed.[7,8]

On the other hand, the transport of amino acid was shown to be greater in respiring or glycolyzing cells than in cells whose metabolic processes had been inhibited.[9] It has been suggested that the transport systems are energized by ATP, which has been previously formed by cellular metabolic processes and then cleaved in the transport process; however, the experimental evidence for this statement is inconclusive.[10] In fact, the uptake of aminoisobutyric acid (AIB) by Ehrlich cells is reported to be independent of the rate of ATP hydrolysis,[11] and moreover, the uptake of AIB is maintained after ATP depletion, even in the presence of unfavorable alkaline ionic gradients.[12]

Similarly, Philo and Eddy,[13] using carbocyanine dyes as voltage-sensitive probes that respond rapidly to changes in membrane potentials, discovered that the rate of methionine uptake by intact Ehrlich cells was much faster than that predicted from the observed Na^+-electrochemical gradient.

Johnstone[14] uses the ionophore, gramicidin, to eliminate transmembrane potentials and Na^+ concentration gradients in Ehrlich cells to establish a zero Na^+-electrochemical potential gradient. She reports that, under these conditions, the apparent K_m for glycine efflux was more than tenfold that of influx. This flux asymmetry cannot be explained by the Na^+-electrochemical potential gradient. Heinz et al.[15] use another ionophore, nigericin, to study the steady-state AIB uptake under conditions of zero Na^+-electrochemical potential gradient. AIB was accumulated to give an inside/outside distribution ratio of 5.4 in a low-Na^+ medium.

Because active cellular uptake of amino acids can occur in the absence of a Na^+-electrochemical gradient, other authors advance alternative hypotheses. García-Sancho et al.[16] suggest a different mode of energizing amino acid transport. Their theory is that a plasma membrane oxidoreductase system could contribute energy to both Na^+-independent and Na^+-dependent amino acid transport. Consequently, the reducing equivalents could supply the required energy for the transport. There are at least three different types of NADH dehydrogenases in eukaryotic plasma membranes, each with a different location relative to the membrane, i.e., extraplasma, intraplasma and transplasma membrane; the latter appears to be the most appropriate for transport studies.[17,18]

III. FIRST DIRECT DATA ON THE INVOLVEMENT OF PLASMA MEMBRANE REDOX SYSTEMS IN AMINO ACID TRANSPORT

Goldenberg[19] says that in the early 1970s there was a growing awareness that the carrier protein sulfydryl group redox state played a critical role in the transport of metabolites across the membrane. In fact, amino acid transport systems are most active in a state of limited oxidation, less active in the complete reduced state, and are inhibited by sulfhydryl blockers, and by prolonged exposure to oxidants or irradiation.[20,21]

Chemiosmotic coupling as a mechanism for oxidative phosphorylation in mitochondria and photophosphorylation in the chloroplast was initially suggested by Mitchell.[22] Later, the involvement of chemiosmotic coupling in active transport in bacteria was proposed.[23,24] Kaback's group, which adopted the chemiosmotic hypothesis, described bacterial plasma membrane systems which could use the energy obtained from the oxidation of metabolic substrates to build up an electrochemical proton gradient. This transmembrane gradient is the driving force for the uptake of several metabolites.[1,25]

However, the first direct experimental evidence of the involvement of eukaryotic plasma membrane redox systems in amino acid transport was reported in the now classical work of García-Sancho et al.[16] After 30 min of rotenone (100 ng/ml) treatment to deplete cellular ATP content, the uptake by Ehrlich cells of *N*-methyl amino-isobutyric acid (MeAIB) (transported by the Na^+-dependent system A), was inhibited by 75% and the uptake of BCH (transported by Na^+-independent system L) was inhibited by 52%. In that study, the use of an artificial donor system for electrons, phenazine methosulfate (0.1 m*M*) plus sodium ascorbate (20 m*M*) restored MeAIB and BCH uptake to 91% and 97% of their former levels, respectively. This restoration was not accompanied by an increase of cellular ATP levels. In all cases, the return to normal levels was prevented by 2 m*M* ouabain under conditions, in which an alkaline ionic gradient is not involved.

García-Sancho et al.[16] also demonstrated that Atebrin, an inhibitor of plasma membrane NADH dehydrogenase,[26] inhibits the 30-s uptake of MeAIB by 70%. On the other hand, they found that metabolic inhibitors such as rotenone, cyanide, or azide, do not inhibit this 30-s uptake of MeAIB.

On the basis of these results, these authors propose that the amino acid transport in eukaryotic cells may be energized by three different but compatible ways: (1) by an electrochemical Na^+ gradient; (2) by the cellular ATP generated in the cellular metabolic processes; or, (3) by reducing equivalents coupled to a transplasma membrane redox system, which generates an electrochemical proton gradient similar to that of the system described for the chemiosmotic hypothesis.

IV. FURTHER EVIDENCE FOR THE ROLE OF PLASMA MEMBRANE REDOX SYSTEMS IN AMINO ACID TRANSPORT

Kilberg and Christensen[27] isolated plasma membrane vesicles from Ehrlich cells and showed the existence of a very active transplasma membrane NADH: ferricyanide oxidoreductase. It is very probable that this system is involved in amino acid transport in these cells.

The data obtained with intact Ehrlich cells was confirmed by using isolated plasma membrane vesicles.[6,16,28] The 30-s uptake of MeAIB by plasma membrane vesicles increases 56% in the presence of 0.4 m*M* Atebrin in comparison with control vesicles. On the other hand, in the presence of 1.8 m*M* NADH, the 30-s uptake of MeAIB[16] is inhibited 50%. The inhibition decreases to 21% when only 1 m*M* NADH is used. When 0.1 m*M* phenazine methosulfate plus 20 m*M* sodium ascorbate, an artificial donor system for electron transport, is used, the inhibition is 31%.[6] These data are not inconsistent with the ones mentioned above for intact cells. As the authors point out,[28] the observed effects were in the right direction because most of the vesicles were inside-out sealed. The inhibitory effect of ouabain was also confirmed; in fact, the MeAIB uptake into rightside-out sealed vesicles is inhibited by 1 m*M* ouabain, even in the absence of K^+, and when an artificial, inwardly directed Na^+ gradient is imposed. Obviously, under these conditions, ouabain-produced inhibition is not due to its inhibition of Na^+,K^+-ATPase activity.[6]

Yamamoto and Kawasaki[29-31] confirm some of the results of Christensen's group. They observed a two to threefold increase in the initial rate of AIB uptake by Ehrlich cells incubated in the presence of 5 m*M* ascorbate and 0.09 m*M* phenazine methosulfate under low ATP conditions because the glycolysis was blocked by iodoacetate and the mitochondrial electron transport by rotenone. When they assayed the transport of other neutral amino acids transported via Na^+-dependent systems (alanine, glycine, proline, and cycloproline) by cells incubated in the presence of ascorbate and phenazine methosulfate, they also discovered a stimulatory effect similar to that obtained for AIB transport. However, under the same conditions there was no change in the rate of uptake of leucine and threonine. The fact that ascorbate can be replaced by externally added NADH suggests that the external NADH dehydrogenase described by Cherry et al.[32] is possibly involved in the energization of amino acid uptake by the Ehrlich cell. Phenazine methosulfate is the only electron acceptor that stimulated AIB uptake: ferricyanide was ineffective. The stimulation produced by ascorbate or NADH together with phenazine methosulfate is due to an increase in the V_{max} value for AIB transport and this effect requires the presence of a Na^+ gradient which in turn gives rise to an influx of Na^+. Yamamoto and Kawasaki[30,31] did not observe the ascorbate and phenazine methosulfate stimulatory effect on Na^+-independent amino acid transport systems. The plasma membrane NADH dehydrogenase inhibitor Atebrin inhibits the enhanced transport rates of both AIB and Na^+. They also confirm the stimulatory effect of ascorbate or NADH and phenazine methosulfate on AIB uptake in plasma membrane vesicles, but the observed 1.4-fold increase was dependent on the presence of an Na^+ gradient. In this case, too, the enhanced rate of AIB uptake was inhibited by atebrin.[31]

García et al.,[33] in a study of renal brush-border membrane vesicles, state that there is evidence that NADH oxidation energizes metabolic transport. When NADH is added to a brush border membrane vesicle suspension, the uptake of L-aspartate is greatly increased, so that at 6 min, the accumulation is increased sevenfold in comparison with that of control membranes without NADH. This effect is eliminated by rotenone, an inhibitor of the brush border membrane NADH oxidase activity.[34] On the other hand, the proton ionophere carbonylcyanide *m*-chlorophenylhydrazone (CCCP) inhibits the NADH-dependent L-aspartate uptake. Both nigericin, an ionophere which rapidly eliminates the pH gradient in the presence of K^+, and valinomycin, an ionophere which eliminates the transmembrane potential, inhibit the accumulation of L-aspartate but to a lower extent than CCCP. It is noteworthy that only negatively charged metabolites such as β-hydroxybutyrate, glutamate, or lactate are taken up by an NADH oxidation-dependent system. Neither neutral nor basic amino acids are taken up under the same conditions. These data and other complementary results led García et al.[33] to conclude that NADH oxidation could energize metabolic transport in renal brush border membrane vesicles. This energization appears to depend on the formation of an electrochemical proton gradient across the plasma membrane similar to that reported for the prokaryotic plasma membrane.[25] In physiological conditions, this proton gradient could be utilized to build up a Na^+ gradient through the Na^+/H^+ antiport previously described in brush border membranes.[35] Consequently, the electrochemical proton gradient created by the NADH oxidase could be used as the driving force for Na^+-dependent transport systems.[36]

Heinz et al.[15] showed for Ehrlich cells that when lactate, pyruvate, or glucose is added to an incubation medium low in sodium, the inside/outside AIB distribution ratio of 5.4 increased, respectively, to 11.9, 9.4, and 15.3. The effect of glucose can be explained partly by the fact that it increases the Na^+ electrochemical potential gradient. However, neither lactate nor pyruvate produce any change, either in membrane voltage, or in the intracellular Na^+ concentration. Heinz et al.[15] also observed a small but significant stimulation of AIB uptake by 0.1 m*M* phenazine methosulfate and 10 m*M* ascorbate under the same conditions used for the pyruvate or lactate experiments. They also report a 50% inhibition of AIB uptake in the presence of 0.5 m*M* Atebrin. The observations of similar experiments with pyruvate, phenazine methosulfate, plus ascorbate, and Atebrin support those previously reported by Christensen's group.[6,16] However, the result of lactate addition, the same as for pyruvate, is the opposite of that expected. This apparent contradiction is explained if the fact is taken into account that lactate is an immediate source of cytosolic NADH required by the oxidoreductase system.[37]

We have recently studied the plasma membrane system in intact Ehrlich cells[37] and now have preliminary data about the interaction on the plasma membrane redox system with amino acid transport systems. Both acivicin (5 μ*M*), and azaserine (10 μ*M*), two analogs of glutamine, and inhibitors of L-glutamine transport, inhibit ferricyanide reduction by 66% and 60%, respectively. On the other hand, we observed that in cellular suspensions incubated in the presence of 0.5 m*M* ferricyanide there is a significant increase in glutamine consumption which is accompanied by a significant decrease in glutamate production.[62] The 14% increase in glutamine consumption could be a consequence of an increase in the uptake of glutamine. The observed 20% decrease in glutamate production might be due to an increased flux via aspartate amino transferase.[38]

V. MEMBRANE POTENTIAL, PLASMA MEMBRANE REDOX SYSTEMS, AND AMINO ACID TRANSPORT

Many authors confirm that electron flow through the plasma membrane electron transport chain gives rise to unbalanced ionic flows and produces changes in the membrane potential.[18] Oxidation of external NADH causes a hyperpolarization of the plasma membrane. Zemkova et al.[39] show that extracellularly applied NADH increases the resting membrane potential from −74.1 to −76.6 mV (negative inside) in incubations of freshly isolated muscle in the presence of K^+ and from −64.6 to −72.9 m*M* in muscle equilibrated for 4 to 6 h in a K^+-free solution. This effect is not obtained if NAD^+ or NADPH is used instead of NADH. Furthermore, this hyperpolarization is blocked in muscle pretreated with ouabain or with any of the plasma membrane NADH dehydrogenase inhibitors adriamycin, Atebrin, azide, bleomycin, 4,4′diisothiocyano-2,2′-stilbenedisulfonate (DIDS) and *p*-chloromercuribenzoate (PCMB).

It has usually been assumed that the potential gradients across the plasma membrane of

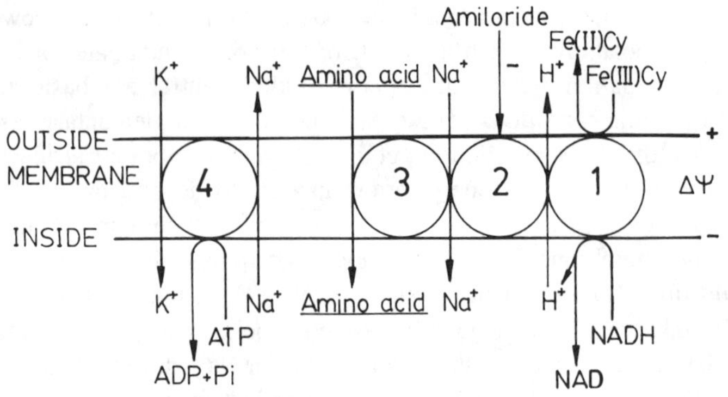

SCHEME 1. Model for the coupling between electron transport and amino acid transport across the plasma membrane. Details are given in the text.

most animal cells are generated by outward diffusion of K^+ ions down their chemical concentration gradients.[40] However, all animal cells produce H^+ ions and extrude them into the surrounding media; the media may then become significantly more acidic than the cytoplasm.[37,41] In addition, all animal cells have Na^+ pumps in their plasma membrane. Sodium pumps may contribute significantly to the generation of membrane potential in a variety of cells, particularly when the internal concentration of Na^+ is elevated.[42-44] Recently, Bashford and Pasternak[45,46] demonstrated that plasma membrane potential of Lettré cells does not depend on cation gradients but rather on pumps; this implies that in these cells the rate at which ions are pumped across the plasma membrane greatly exceeds the rate at which they can diffuse across it. On the other hand, the fact that the membrane potential in these cells is maintained in the absence of any net movement of Na^+ or K^+ also implies the existence of electroneutral return mechanisms for the pumped ions, such as Na^+/K^+ and/or Na^+/H^+ exchanges.[45]

Because the membrane potentials contribute substantially to the Na^+ electrochemical gradient, any change in those must affect amino acid transport.[47-51] Valdeolmillos et al.[52] showed that transmembrane potential strongly affects amino acid uptake through transport system A, but it does not alter transport through the ASC system. The initial rate of uptake of BCH (a model substrate of the Na^+-independent transport system L) is not modified by changes of the membrane potential; however, in hyperpolarized cells, the steady-state uptake of BCH increases. Moreover, the finding of Valdeolmillos et al. that changes of the extracellular K^+ concentration did not alter the plasma membrane potential in the Ehrlich cells supports the proposal of Bashford and Pasternak.[45,46]

Recently, Dawson and Smith[53] carried out membrane potential studies of Ehrlich cells using electrophysiological techniques. They concluded that electrochemical Na^+ gradients alone would be sufficient to serve as a source for AIB uptake. The apparent contradiction of Dawson and Smith's results with those of the above-mentioned works may be explained by the fact that they measure total cellular AIB uptake and, of course, this amino acid can be transported by several different systems. However, the fact that amino acid uptakes and transmembrane potentials are linked in many different types of cells does not imply that all different amino acid transport systems are energized in the same way.

VI. THE PROPOSED COMPREHENSIVE MODEL

On the basis of the several published experimental works quoted in the previous sections, we propose a comprehensive model (Scheme 1) in which electron transport is coupled to ionic and organic solute transport through plasma membrane.

The complete proposed system would comprise: (1) the transplasma membrane NADH dehydrogenase, (2) the Na^+/H^+ antiporter, (3) the Na^+-dependent amino acid transport system, and (4) the Na^+/K^+-ATPase. Plasma membrane NADH dehydrogenase activity is accompanied by proton extrusion.[18,37,54-57] Under physiological conditions, the proton gradient created in this way could be used to favor a net Na^+-influx into the cell through the Na^+/H^+ antiport. The interaction between the plasma membrane redox system and the Na^+/H^+ antiport appears to be reflected by the inhibition of the dehydrogenase by amiloride,[37,58,59] an inhibitor of the Na^+/H^+ antiporter.[60,61] This antiport, together with the proton gradient, could be an alternative mechanism to maintain the Na^+ influx required for all Na^+-dependent transport processes. The cycle would be closed by the activity of the Na^+/K^+-ATPase, or by a Na^+/K^+ antiport system.

The comprehensive model could possibly explain the involvement of plasma membrane redox systems in the energization of Na^+-dependent amino acid transport. The model would also explain the stimulatory effect on Na^+-independent amino acid transport reported by Christensen's group,[16,18] taking into account that the functioning of the plasma membrane redox systems create a proton gradient that would directly drive the accumulation of amino acid transported via an Na^+-independent transport system. Indeed, Na^+-independent amino acid transport systems have been proposed by other authors to be H^+-dependent systems.[4]

Much more work remains to be done to define and quantify clearly the interactions between plasma membrane redox systems and amino acid transport. If confirmed, this proposed comprehensive model of another driving force for amino acid transport across membranes would contribute to the understanding of many other similar systems.

REFERENCES

1. **Ramos, S. and Kaback, R.,** The relationship between the electrochemical proton gradient and active transport in *Escherichia coli* membrane vesicles, *Biochemistry,* 16, 854, 1977.
2. **Crane, R. K.,** Na^+-dependent transport in the intestine and other animal tissues, *Fed. Proc.,* 24, 1000, 1965.
3. **Christensen, H. N., Handlogten, M. E., Lam, I., Tager, H. S., and Zand, R.,** A bicyclic amino acid to improve discriminations among transport systems, *J. Biol. Chem.,* 244, 1510, 1969.
4. **Mitsumoto, Y., Sato, K., Ohyashiki, T., and Mohri, T.,** Leucine-proton cotransport system in Chang liver cells, *J. Biol. Chem.,* 261, 4549, 1986.
5. **Christensen, H. N., De Céspedes, C., Handlogten, M. E., and Ronquist, G.,** Energization of amino acid transport studied for the Ehrlich ascites tumor cell, *Biochim. Biophys. Acta,* 300, 487, 1973.
6. **Ohsawa, M., Kilberg, M. S., Kimmel, G., and Christensen, H. N.,** Energization of amino acid transport in energy-depleted Ehrlich cells and plasma membrane vesicles, *Biochim. Biophys. Acta,* 599, 175, 1980.
7. **Schafer, J. A. and Jacquez, J. A.,** Change in Na^+ uptake during amino acid transport, *Biochim. Biophys. Acta,* 135, 1081, 1967.
8. **Schafer, J. A. and Heinz, E.,** The effect of reversal of Na^+ and K^+ electrochemical potential gradients on the active transport of amino acids in Ehrlich ascites tumor cells, *Biochim. Biophys. Acta,* 249, 15, 1971.
9. **Morville, M., Reid, M., and Eddy, A. A.,** Amino acid absorption by mouse ascites tumor cells depleted of both endogenous amino acids and adenosine triphosphate, *Biochem. J.,* 134, 11, 1973.
10. **Johnstone, R. M.,** Role of ATP in the initial rate of amino acid uptake in Ehrlich ascites cells, *Biochim. Biophys. Acta,* 356, 319, 1974.
11. **Geck, P., Heinz, E., and Pfeiffer, B.,** Evidence against direct coupling between amino acid transport and ATP hydrolysis, *Biochim. Biophys. Acta,* 339, 419, 1974.
12. **Schafer, J. A.,** A reassessment of decreased amino acid accumulation by Ehrlich ascites tumor cells in the presence of metabolic inhibitors, *J. Gen. Physiol.,* 69, 681, 1977.
13. **Philo, R. D. and Eddy, A. A.,** Equilibrium and steady-state models of the coupling between the amino acid gradient and the sodium electrochemical gradient in mouse ascites-tumor cells, *Biochem. J.,* 174, 811, 1978.

14. **Johnstone, R. M.,** The basic asymmetry of Na^+-dependent glycine transport in Ehrlich cells, *Biochim. Biophys. Acta,* 512, 199, 1978.
15. **Heinz, A., Jackson, J. W., Richey, B. E., Sachs, G., and Schafer, J. A.,** Amino acid active transport and stimulation by substrates in the absence of a Na^+ electrochemical potential gradient, *J. Membrane Biol.,* 62, 149, 1981.
16. **García-Sancho, J., Sánchez, A., Handlogten, M. E., and Christensen, H. N.,** Unexpected additional mode of energization of amino-acid transport into Ehrlich cells, *Proc. Natl. Acad. Sci. U.S.A.,* 74, 1488, 1977.
17. **Crane, F. L., Löw, H., and Clark, M. G.,** Transport and trans-plasma-membrane redox systems, in *Membranes and Transport,* Vol. 2, Martonosi, A. N. Ed., Plenum Press, New York, 1982, 251.
18. **Crane, F. L., Sun, I. L., Clark, M. G., Grebing, C., and Löw, H.,** Transplasma-membrane redox systems in growth and development, *Biochim. Biophys. Acta,* 811, 233, 1985.
19. **Goldenberg, H.,** Plasma membrane redox activities, *Biochim. Biophys. Acta,* 694, 203, 1982.
20. **Kwock, L., Wallach, D. F. H., and Hefter, K.,** Involvement of sulphydryl groups in the action of insulin and radiation on thymocyte sodium-ion-dependent amino acid transport, *Biochim. Biophys. Acta,* 419, 93, 1976.
21. **Boumendil-Podevin, E. F. and Podevin, R. A.,** Stimulatory and inhibitory effects of sulphydryl reagents on *p*-aminohippuric acid transport by isolated renal tubules, *Biochim. Biophys. Acta,* 467, 364, 1977.
22. **Mitchell, P.,** Chemiosmotic coupling in oxidative and photosynthetic phosphorylation, *Biol. Rev. Cambridge Philos. Soc.,* 41, 445, 1966.
23. **Harold, F. M.,** Ion transport and electrogenesis in bacteria, *Bacteriol. Rev.,* 36, 172, 1972.
24. **Hamilton, W. A. and Niven, D. F.,** Transport driven by transmembrane gradients of chemical and electrical potential, *Adv. Microbiol. Physiol.,* 12, 1, 1975.
25. **Ramos, S., Schuldiner, S., and Kaback, H. R.,** The electrochemical gradient of protons and its relation to active transport in *Escherichia coli* membrane vesicles, *Proc. Natl. Acad. Sci. U.S.A.,* 73, 1892, 1976.
26. **Löw, H. and Crane, F. L.,** Hormone regulated redox function in plasma membranes, *FEBS Lett.,* 68, 157, 1976.
27. **Kilberg, M. S. and Christensen, H. N.,** Electron-transfering enzymes in the plasma membrane of the Ehrlich ascites tumor cells, *Biochemistry,* 18, 1525, 1979.
28. **Christensen, H. N.,** Amino acid transport, in *Redox Functions of the Eukaryotic Plasma Membrane,* Ramírez, J. M., Ed., CSIC Publication Office, Madrid, 1987, 119.
29. **Yamamoto, S.,** The mechanism of stimulation by the membrane-oxidoreduction system of amino acid uptake in Ehrlich cells, *Hiroshima Daigaku Igaku Zasshi,* 28, 697, 1980.
30. **Yamamoto, S. and Kawasaki, T.,** The involvement of the membrane oxidoreduction system in stimulating amino acid uptake in Ehrlich ascites tumor cells, *Biochim. Biophys. Acta,* 644, 192, 1981.
31. **Yamamoto, S. and Kawasaki, T.,** Evidence for the additional energy coupling to amino acid transport through a membrane redox system in Ehrlich cell membrane vesicles, *Biochem. Int.,* 3, 545, 1981.
32. **Cherry, J. M., Mackellar, W., Morré, D. J., Crane, F. L., Jacobsen, L. B., and Schirrmacher, V.,** Evidence for a plasma membrane redox system on intact ascites tumor cells with different metastatic capacity, *Biochim. Biophys. Acta,* 634, 11, 1981.
33. **García, M. L., Benavides, J., Giménez-Gallego, G., and Valdivieso, F.,** Coupling between reduced nicotinamide adenine dinucleotide oxidation and metabolite transport in renal brush border membrane vesicles, *Biochemistry,* 19, 4840, 1980.
34. **Giménez-Gallego, G., Benavides, J., García, M. L., and Valdivieso, F.,** Occurrence of a reduced nicotinamide adenine dinucleotide oxidase activity linked to a cytochrome system in renal brush-border-membranes, *Biochemistry,* 19, 4834, 1980.
35. **Murer, H., Hopfer, U., and Kinne, R.,** Sodium/proton antiport in brush-border-membrane vesicles isolated from rat small intestine and kidney, *Biochem. J.,* 154, 597, 1976.
36. **Giménez-Gallego, G. and García, M. L.,** Structure and function of an NADH oxidase system in renal brush-border membranes, in *Redox Functions of the Eukaryotic Plasma Membrane,* Ramírez, J. M., Ed., CSIC Publication Office, Madrid, 1987, 141.
37. **Medina, M. A., Sánchez-Jiménez, F., Segura, J. A., and Núñez de Castro, I.,** Transmembrane ferricyanide reductase activity in Ehrlich ascites tumor cells, *Biochim. Biophys. Acta,* 946, 1, 1988.
38. **Medina, M. A., Sánchez-Jiménez, F., Márquez, F. J., Pérez-Rodríguez, J., Quesada, A. R., and Núñez de Castro, I.,** Glutamine and glucose as energy substrates for Ehrlich ascites tumour cells, *Biochem. Int.,* 16, 339, 1988.
39. **Zemková, H., Teisinger, J., and Vyskocil, F.,** Hyperpolarization of mouse skeletal muscle plasma membrane induced by extracellular NADH, *Biochim. Biophys. Acta,* 775, 64, 1984.
40. **Williams, J. A.,** Origin of transmembrane potentials in non-excitable cells, *J. Theor. Biol.,* 28, 287, 1970.
41. **Bashford, C. L., Alder, G., Micklem, K. J., and Pasternak, C. A.,** A novel method for measuring intracellular pH and potassium concentration, *Biosci. Rep.,* 3, 631, 1983.

42. **Thomas, R. C.,** Electrogenic sodium pump in nerve and muscle cells, *Physiol. Rev.*, 52, 563, 1972.
43. **Pietrzyk, C., Geck, P., and Heinz, E.,** Regulation of the electrogenic (Na^+ + K^+)-pump of Ehrlich cells by intracellular cation levels, *Biochim. Biophys. Acta,* 513, 89, 1978.
44. **Lew, V. L., Ferreira, H. G., and Moura, T.,** The behaviour of transporting epithelial cells. I. Computer analysis of a basic model, *Proc. R. Soc. London Ser. B,* 206, 53, 1979.
45. **Bashford, C. L. and Pasternak, C. A.,** Plasma membrane potential of Lettré cells does not depend on cation gradients but on pumps, *J. Membrane Biol.*, 79, 275, 1984.
46. **Bashford, C. L. and Pasternak, C. A.,** Generation of plasma membrane potential by the Na^+-pump coupled to proton extrusion, *Eur. Biophys. J.*, 12, 229, 1985.
47. **Gibb, L. E. and Eddy, A. A.,** Electrogenic sodium pump as a possible factor leading to the concentration of amino acids by mouse ascites-tumor cells with reversed sodium ion concentration gradients, *Biochem. J.*, 129, 979, 1972.
48. **Reid, M., Gibb, L. E., and Eddy, A. A.,** Ionophore-mediated coupling between ion fluxes and amino acid absorption in mouse ascites tumor cells. Restoration of the physiological gradients of methionine by valinomycin in the absence of adenosine triphosphate, *Biochem. J.*, 140, 383, 1974.
49. **Colombini, M. and Johnstone, R. M.,** Na^+-gradient-stimualted AIB transport in membrane vesicles from Ehrlich ascites cells, *J. Membr. Biol.*, 18, 315, 1974.
50. **Johnstone, R. M. and Laris, P. C.,** Bimodal effects of cellular amino acids on sodium ion-dependent amino acid transport in Ehrlich cells, *Biochim. Biophys. Acta,* 599, 715, 1980.
51. **Valdeolmillos, M., García-Sancho, J., and Herreros, B.,** Stimulation of Na^+-dependent amino acid uptake by the activation of the Ca^{2+}-dependent K^+ channel in the Ehrlich ascites tumor cells, *Biochim. Biophys. Acta,* 689, 177, 1982.
52. **Valdeolmillos, M., García-Sancho, J., and Herreros, B.,** Differential effects of transmembrane potential on two Na^+-dependent transport systems for neutral amino acids, *Biochim. Biophys. Acta,* 858, 181, 1986.
53. **Dawson, W. D. and Smith, T. C.,** Energetics of Na^+-dependent amino acid co-transport in Ehrlich ascites tumor cells, *Biochim. Biophys. Acta,* 897, 5, 1987.
54. **Thom, M. and Maretzki, A.,** Evidence for a plasmalemma redox system in sugarcane, *Plant Physiol.*, 77, 873, 1985.
55. **Rubinstein, B. and Stern, A. I.,** Relationship of transplasmalemma redox activity to protein and solute transport by roots of *Zea mays, Plant Physiol.*, 80, 805, 1986.
56. **Sandelius, A. S., Barr, R., Crane, F. L., and Morré, D. J.,** Redox reactions of plasma membranes isolated from soybean hypocotyls by phase partition, *Plant Sci.*, 48, 1, 1986.
57. **Sun, I. L., Crane, F. L., and Chou, J. Y.,** Modification of transmembrane electron transport in plasma membrane of Simian virus 40 transformed pineal cells, *Biochim. Biophys. Acta,* 886, 327, 1986.
58. **Sun, I. L., García-Cañero, R., Liu, W., Toole-Simms, W., Crane, F. L., Morré, D. J., and Löw, H.,** Diferric transferrin reduction stimulates the Na^+/H^+ antiport of HeLa cells, *Biochem. Biophys. Res. Commun.*, 145, 467, 1987.
59. **García-Cañero, R., Díaz-Gil, J. J., and Guerra, M. A.,** Is transmembrane ferricyanide reductase connected with the early expression of mitogenic signals in hepatocytes?, in *Redox Functions of the Eukaryotic Plasma Membrane,* Ramírez, J. M., Ed., CSIC Publication Office, Madrid, 1987, 43.
60. **Kinsella, J. L. and Aronson, P. S.,** Properties of the sodium ion-hydrogen ion exchanger in renal microvillus membrane vesicles, *Am. J. Physiol.*, 238, F461, 1980.
61. **Kinsella, J. L. and Aronson, P. S.,** Amiloride inhibition of the sodium ion-proton exchanger in renal microvillus membrane vesicles, *Am. J. Physiol.*, 241, F374, 1981.
62. **Medina, M. A. and Núñez de Castro, I.,** unpublished results.

Chapter 14

INTERACTIONS OF ANTITUMOR DRUGS WITH PLASMA MEMBRANES

Iris L. Sun and F. L. Crane

TABLE OF CONTENTS

I. INTRODUCTION

Drugs used in cancer chemotherapy have been making rapid progress, but our awareness of the importance of antibiotic applied to this area has been slow in developing. Part of the difficulty was caused by the lack of screening systems that could detect activity in fermentation broths or crude isolates. However, in the last few years antibiotics have become firmly established as an important class of antitumor agents.

At the present time the anthracycline family provides the most exciting news in cancer chemotherapy. It has proven to be effective against a variety of human tumors despite its cardiotoxicity.

The bleomycins are remarkable in structure. In contrast to other anticancer agents, they do not depress bone marrow[1] and are therefore gaining increasing clinical acceptance.

Most people believe that the antitumor effect of the anthracycline antibiotics and the bleomycins is due to the effect on DNA.[2-7] However, there is growing evidence that the receptor sites of these drugs are at the plasma membrane.[8-10]

Actinomycin D has been well known for the inhibition of cell growth and mitosis.[11,12] There is, however, increasing consideration for an effect on membrane[13,14] and induced changes in protein conformation.[15] It also has been used against tumors since 1952.

Cis-diamminedichloroplatinum(II), cisplatin, a potent nonantibiotic antitumor drug also has a definite effect on cell surface.[16] This has been active against a number of malignant tumors in many experimental animals.[17] In contrast to cisplatin, transplatin has much less cytotoxicity.[18,19]

The importance of retinoid in the field of cancer studies has also been discussed.[20] One of the most exciting potential uses for the retinoids lies in the area of cancer prevention. The use of retinoids in combination with cytotoxic cancer chemotherapeutic drugs to inhibit tumor growth has also been reported.[21] The rationale for the use of retinoids in cancer prevention derives from their ability to control normal differentiation.

Transplasma membrane redox activity now can be related to several vital functions, which include control of cell growth,[22,23] facilitation of iron uptake,[24] and defense against bacteria.[25] In addition, there is good evidence that the action of hormones at the plasma membrane is also related to the redox activity.[26] There is further evidence that certain special transport functions may be stimulated or driven by the plasma membrane redox system.[27]

The mode of action of drugs listed-above has been proposed. Various concepts are widely scattered in the scientific literature. The purpose of this chapter is to bring together a broad range of chemical structure which show the same underlying biochemical effect — that is the inhibition of the transmembrane dehydrogenase, an enzyme which is involved in control of cellular proliferation.

II. TRANSPLASMA MEMBRANE REDOX ACTIVITY AND CELL GROWTH

A. IMPERMEABLE OXIDANTS AS GROWTH STIMULATORS

Stimulation of cell division in sea urchin eggs by impermeable indigosulfonates was shown 40 years ago.[28] Ferric sulfate has also been shown to stimulate growth of 3T3 cells.[29]

TABLE 1
Kinetic Constants for Ferricyanide Reduction by Transformed and Nontransformed Cells

Cells	K_m	V_{max}	Ref.
RPN209-11 (33°C) (transformed phenotype)	0.33 (fast rate)	1000 (fast rate)	39
	.074 (slow rate)	222 (slow rate)	39
RPN209-11 (40°C) (nontransformed phenotype)	0.18 (fast rate)	1086 (fast rate)	39
	.048 (slow rate)	333 (slow rate)	39
RLA209-15 (33°C) (transformed phenotype)	0.22 (fast rate)	500 (fast rate)	40
	0.07 (slow rate)	129 (slow rate)	40
RLA209-15 (40°C) (nontransformed phenotype)	0.086 (fast rate)	833 (fast rate)	40
	0.031 (slow rate)	294 (slow rate)	40
HeLa cells	0.125 (fast rate)	1250 (fast rate)	32
	0.24 (slow rate)	263 (slow rate)	32

Note: RLA209-15 are fetal rat cells infected with a temperature-sensitive SV40 virus that transforms cells at 30°C but not at 40°C. RPN-1 are rat pineal cells with the same temperature-sensitive SV40 virus.

However, the studies of Ellem and Kay,[23] which showed that external ferricyanide could stimulate the growth of melanoma cell when growth factors present in serum were limiting, opened a new area in control of cell function by transplasma membrane redox enzymes. The stimulating effect of ferricyanide on the growth of HeLa cells showed this even more dramatically[22,24] (Chapter 7), since this stimulation could be observed in the total absence of fetal calf serum. It is clear by now that the transmembrane redox system can be an important element in control of cell growth.[30-35]

Stimulation of cell growth by impermeable oxidants is not limited to iron compounds such as ferricyanide, transferrin or ferric complexes.[36] The growth of HeLa cells is also stimulated by hexamine ruthenium III, a trivalent cation, and by indigo tetrasulfonate[37,38] (Chapter 7). The use of a series of impermeable indigosulfonates with different standard redox potential shows that external oxidants with redox potential $E^{o\prime}_{7.0}$ above -125 mV can stimulate growth.[37] The external oxidants may increase growth by keeping redox components in the oxidized state in the plasma membrane,[23] or by increasing the concentration of oxidized pyridine nucleotide in the cytoplasm.[38]

B. MODIFICATION OF TRANSMEMBRANE REDOX ACTIVITY IN TUMOR CELLS

Tumor- or virus-transformed cells which have a defect in growth control modify the transplasmalemma redox system by changing the K_m and V_{max} of the enzyme[39,40] (Table 1). Transplasma membrane electron transport activity is noticeably depressed in tumor cells or transformed cells compared to nontransformed control cells[30,39,40] (Table 2). Similar evidence for a decrease in redox enzyme activity with transformation has also been shown by Sheinin and Onodera[24] that the ratio of NADH ferricyanide reductase to Na^+,K^+-ATPase activity in purified plasma membrane changes from 0.12 to 0.06 with SV40 and polyoma virus transformation of 3T3 cells. On the contrary, Warley and Cook[41] have reported much more rapid NADH ferricyanide reductase in plasma membranes from mouse leukemic lymphocytes than normal lymphocytes. Similarly, Powis et al.[42] showed that reduction of impermeable indigosulfonate was faster with transformed AKR mouse embryo cells than with nontransformed cells. A relation between transplasmalemma redox enzyme activity and the cell growth is clearly indicated.

TABLE 2
Comparison of Diferric Transferrin and Ferricyanide Reduction Rates of Cells between Transformed and Nontransformed Phenotype

Cell lines	Culture condition	Reduction rate ($nmol\ min^{-1}\ gww^{-1}$) Fast rate	Slow rate	Ref.
	Diferric Transferrin Reduction			
RLA209-15	33°C (Transformed phenotype)	3.4	1.1	64
RLA209-15	40°C (Nontransformed phenotype)	8.7	2.7	64
RPN209-1	33°C (Transformed phenotype)	4.7	1.7	39
RPN209-1	40°C (Nontransformed phenotype)	11.1	3.2	39
	Ferricyanide Reduction			
RLA209-15	33°C (Transformed phenotype)	260	99	30, 40
RLA209-15	40°C (Nontransformed phenotype)	709	320	30, 40
McA-RH7777	33°C (Hepatoma)	836	211	30
Primary fetal liver cells	33°C	1210	605	30
	Indigo Disulfonate Reduction			
Mouse AKR		350		42
Chemically transformed AKR		760		42

Note: RLA209-15 are fetal rat liver cells infected with temperature-sensitive SV40. RPN209-1 are ts SV40 rat pineal cells.

III. ANTHRACYCLINE ANTIBIOTICS

A. DEVELOPMENT OF ANTHRACYCLINE COMPOUNDS — MEMBRANE INTERACTION

Anthracycline antibiotics have been successfully used in the treatment of human cancer. It has long been thought that DNA is the primary target for the cytotoxic action of these drugs on susceptible cells.[5,6,8,43] The DNA receptor hypothesis is attractive, due to a reasonably high affinity between some of these drugs and nucleic acid. However, this hypothesis does not fully explain the antimitogenic effect of these drugs, since significant mitotic inhibition by daunomycin and Adriamycin are observed under conditions in which DNA synthesis is unaffected.[44,45] Part of the daunomycin added to membranes is sequestered deep in the membrane where it cannot react with agents at the surface.[46] The *N*-substituted derivative, such as *N*-acetyl daunomycin and *N*-trifluoro-acetyl Adriamycin-14-valerate (AD32) have very low affinity toward DNA, yet they are capable of inhibiting cell mitosis.[6,43,47] Therefore, there are reasons to suspect that additional targets other than DNA are involved in killing cancer cells with these drugs.

Effects of the drugs on the surface properties of sarcoma cells[9] and drug-cytoskeleton interaction[10] have been described. In addition, anthracycline antibiotics affect microsomal electron transport[48,49] and inactivate the last oxidation site of the respiratory chain, cyto-

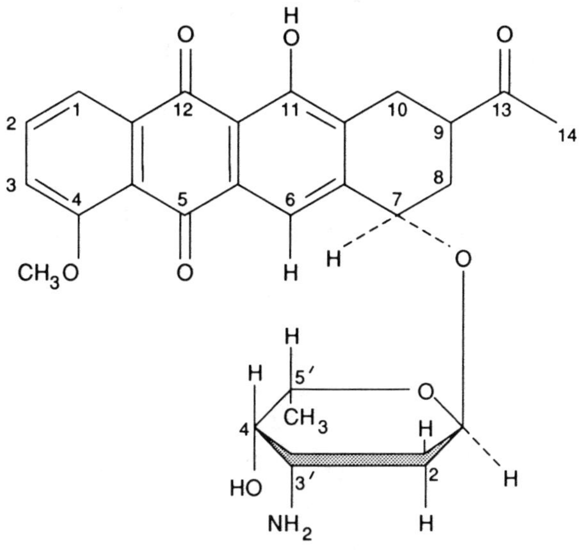

FIGURE 1. The basic structure of anthracyclines.

chrome oxidase.[50] Recently, Tritton and Yee have shown that Adriamycin bound to agarose beads, which cannot enter the cells, retains it cytotoxicity to sarcoma cells.[51,52,58,59] Likewise, the cytostatic action of Adriamycin coupled to polyglutaraldehyde microspheres has been demonstrated on human leukemia cells and rat hepatocytes.[53] A decrease in the fluidity of the plasma membrane is measured by ER spectroscopy in Sarcoma 180 cells[54] after Adriamycin treatment. One possible mechanism to explain the membrane rigidification is the formation of free radicals activated by the dehydrogenase as described in microsomes.[48,49] Modification of the composition[35] of the membrane glycoproteins and glycolipids[56,57] may explain a new susceptibility to immune attack[55,58] and to monoclonal antibody.[59] Unfortunately, it is possible that immune attack due to these surface property modifications can account for chronic cardiac toxicity. Finally, the effect anthracycline antibiotics on plasma membrane redox enzymes at concentrations equivalent to those used in therapy and inhibition of cell division has also been described.[1] Therefore, there is clear evidence that the killing effect of these drugs may be achieved by action at the cell surface.

B. STRUCTURE-ACTIVITY RELATIONSHIP

The basic structure of anthracycline antibiotics contains two main parts, anthraquinone and sugar. Figure 1 illustrates the structure, numbering system, and the constituents of anthracyclines which have been discussed in detail elsewhere.[60] The effect of some of the above-listed anthracyclines on HeLa cell growth and action against tumors has been determined.[60] Modification or substitution at position 4 seems to produce a stronger effect in contrast to loss of activity by change at position 11 and position 9. The inhibitory effect of these agents on plasma membrane electron transport appears to follow a similar pattern[61-64] (Table 3). Clinically used anthracyclines also have a strong effect on the activity of the transmembrane redox system (HeLa cells ferricyanide reduction)[61-63] (Figure 2). The *N*-substituted derivatives at the sugar moiety, such as AD32, also show a fairly strong inhibition. Dehydrogenase activity in purified plasma membrane has been shown to be inhibited in a similar manner[61,62] (Figure 3). The concentration of these agents which affect redox enzymes is correlated to their effect on the cell survival[61,63] (Table 4). The results indicate that HeLa cells are a little more resistant to AD32 than to Adriamycin. Similar results were previously reported[65] for human leukemia cells *in vitro*. AD32 appeared to be tenfold less potent than

TABLE 3
Comparison of Anthracycline Effect on Plasma Membrane Electron Transport Activities

Compound	Inhibition of Electron Transport: HeLa cell $Fe(CN)_6^{3-}$ reduction ID_{50} (μM)	HeLa cell Fe_2Tf reduction ID_{50} (μM)	Plasma membrane NADH → $Fe(CN)_6^{3-}$ ID_{50} (μM)
Doxorubicin	0.05	0.05	0.05
11-Demethoxydoxorubicin	0.5	0.5	>1.0
4′-Deoxy-4′-iodo DX	—	0.1	<0.05
4-Demethoxy DN	0.03	0.03	0.1
4′-Demethoxy 7R-9R DN	⩾1.0	⩾1.0	⩾1.0
4′-Amino-4′-deoxy 3′OH(A) deamino DN	0.5	0.05	<0.1
4′-Amino-4′-deoxy 3′OH(B) epi analog	⩾10	⩾10	⩾1.0
9-Deoxydoxorubicin	>10	>10	>1.0

Note: Fe_2Tf, diferric transferrin. Plasma membrane from pig erythrocyte. Data in part from Sun and Crane.[64] ID_{50} in concentration required for 50% inhibition. DX, doxorubicin; DN, daunorubicin. Anthracycline compounds were kindly supplied by Dr. F. Arcamone, Farmitalia, Milan.

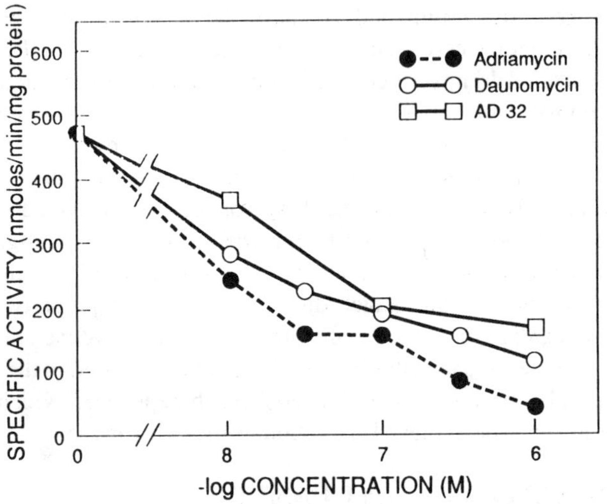

FIGURE 2. The effects of anthracycline compounds on the fast rate of ferricyanide reduction in HeLa cells. (Data from Sun and Crane.[63])

Adriamycin as measured by the effects on cell growth and survival. It is clear that plasma membrane redox enzymes are a promising target of the cytotoxic action for anthracycline compounds.

Coupling of proton release to the transmembrane redox activity has been found in HeLa cells.[32] The effect of anthracycline antibiotics on redox-induced proton extrusion by HeLa cells has therefore been examined[35,61] (Table 5). Under the influence of anthracycline antibiotics, a significant decrease (30 to 70%) in proton release from the cells is demonstrated at a concentration of 5×10^{-7} M. Daunomycin at 10^{-6} M has also been shown to inhibit ferricyanide reduction and proton release by Ehrlich ascites cells.[66]

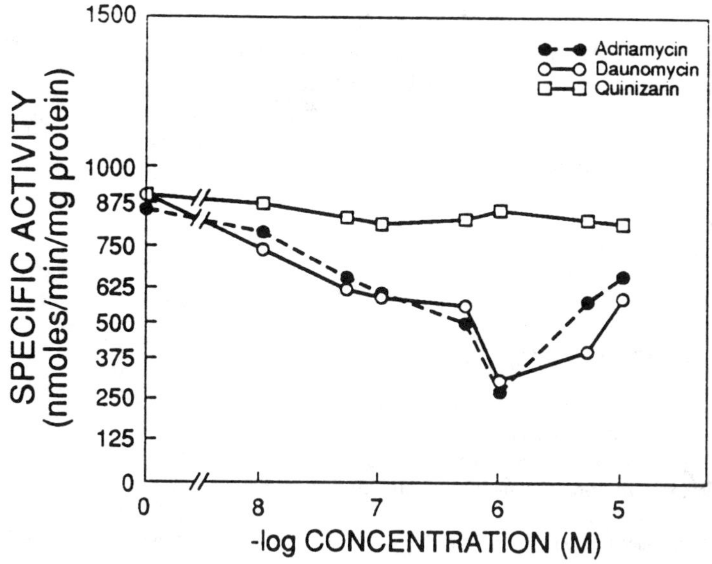

FIGURE 3. The effects of anthracycline antibiotics on NADH-ferricyanide reductase of purified mouse liver plasma membrane. (Data from Sun and Crane.[62])

TABLE 4
The Effect of Anthracyclines on Surviving Fractions of HeLa Cells

Dose of drugs	Surviving fraction (%)
Control	100
Adriamycin (10^{-6} *M*)	58
Adriamycin (10^{-5} *M*)	40
Aclacinomycin (10^{-6} *M*)	70
Aclacinomycin (10^{-5} *M*)	51
AD32 (10^{-6} *M*)	82
AD32 (10^{-5} *M*)	54

Note: AD32 = *N*-trifluoroacetyl adriamycin-14-valerate

Data from Sun and Crane.[63]

C. NORMAL AND TUMOR (OR TRANSFORMED) CELLS HAVE DIFFERENT RESPONSE TOWARD ANTHRACYCLINES

Adriamycin inhibits the ferricyanide reduction by transformed RLA209-15 liver cells and transformed RPNA209-1 pineal cells. However, the same concentration of drug gives much less inhibition of ferricyanide reduction by those cells exhibiting a nontransformed phenotype (Table 6). Cultured hepatoma cells, like transformed RLA209-15 cells and transformed RPNA209-1 cells, also show inhibition by Adriamycin. On the other hand, primary fetal liver cells, like nontransformed RLA209-15 cells and RPNA209-1 cells, show very little response to Adriamycin. These effects are seen with relatively short-term exposure (3 min) of the cells to Adriamycin. The sensitivity is increased by longer preincubation of cells with anthracycline antibiotics. For virus-transformed cells (33°C cultures), the drug has strong cytotoxicity. However, primary fetal hepatocytes like RLA209-15 (40°C cultures) and RPNA209-1 (40°C cultures) are not sensitive to killing.[30,39,40,61] Growth of these cells at 40°C does not permit expression of the temperature-sensitive SV40 virus with which they are infected.

TABLE 5
The Effect of Anthracycline Antibiotics on Ferricyanide Induced Proton Extrusion in HeLa Cells

Additions	Activity [nmoles H^+/min/g/cell (ww)]	% Inhibition
Control	416	—
Adriamycin (5 × 10^{-7} *M*)	240	43
Daunomycin (5 × 10^{-7} *M*)	168	60
Carminomycin (5 × 10^{-7} *M*)	96	77
Marcellomycin (5 × 10^{-7} *M*)	144	65
Aclacinomycin (5 × 10^{-7} *M*)	128	69
AD32 (7 × 10^{-6} *M*)	95	77
11-Deoxydoxorubucin (5 × 10^{-7} *M*)	116	72
4′-Deoxy-4′-iodo-doxorubucin (5 × 10^{-7} *M*)	104	75
4-Demethoxy-daunorubicin (5 × 10^{-7} *M*)	77	82
4′-Demethoxy-7*R*-9*R*-daunorubicin (5 × 10^{-7} *M*)	190	54

Note: Cell concentration was 0.1 gww/ml and $Fe(CN)_6$ was 0.3 m*M*; ww = wet weight.

Data from Sun and Crane.[63]

TABLE 6
Effect of Adriamycin and AD32 on Transmembrane Ferricyanide Reductase Activities of RLA209-15 Fetal Liver, McA-RH7777 Hepatoma, and Primary Fetal Liver Cells

	Percent inhibition by		
Cells, culture temp.	Adriamycin (10^{-6} *M*)	AD32 (10^{-6} *M*)	Ref.
Fast phase			
RLA209-15 (33°C) (transformed)	49±12	40	30,40
RLA209-15 (40°C) (nontransformed)	−2±4	−6	30,40
McA-RH7777 (33°C)	35	—	30,40
Primary fetal liver cells (33°C)	—	12	30
Slow phase			
RLA209-15 (33°C) (transformed)	55±16	36	30,40
RLA209-15 (40°C) (nontransformed)	3±4	−4	30,40
McA-RH7777 (33°C)	20	—	30
Primary fetal liver cells (33°C)	—	—	30
RPNA209-1 (33°C) (transformed)	52		30,39
RPNA209-1 (40°C) (nontransformed)	8		30,39

Note: (−) indicates stimulation instead of inhibition.

D. ANTHRACYCLINES AFFECT OTHER MEMBRANOUS ENZYME SYSTEMS

Besides transmembrane redox enzymes, adriamycin has been reported to inhibit human erythrocyte inositol metabolism. At concentrations of 5 × 10^{-4} *M* to 1 × 10^{-3} *M*, Adriamycin maintains a phosphatidyl-inositol-4,5 biphosphate level extracted from erythrocyte membrane at ~60% of control.[65] The drug also inhibits calmodulin-dependent Ca^{2+} uptake with a 50% inhibitory concentration of 5 × 10^{-4} *M*.[65] The specific effects of Adriamycin may be important in drug-induced cardiotoxicity. The inhibition of Na^+/K^+-stimulated ATPase[67] and Na^+/Ca^{2+}-antiport of heart sarcolemmal vesicles[68] could play a role in cardiotoxicity. Indeed, it has been found in heart cells that the Ca^{2+} fluxes are controlled by two plasma membrane (sarcolemma) pumping systems: a specific ATPase[69] and an Na^+/

Ca^{2+}-exchange system.[70,71] The Ca^{2+} disturbance induced by Adriamycin could conceivably be mediated by the impairment of either one of these two pumping systems or of both. Other effects of anthracycline glycosides on the plasma membrane enzymes such as an effect on the Na^+ transport,[72] on the isocitrate dehydrogenase,[73] on the rate of aggregation of lectin receptor,[9] and on the membrane potential[61,74] have been reported.

IV. BLEOMYCINS

A. BLEOMYCIN-CELL SURFACE INTERACTION

Bleomycins and tallysomycins, which are antibodies obtained from *Streptomyces verticillus*, are effective against several malignant diseases, such as squamous cell carcinoma and testicular neoplasm.[75-77] The receptor site for this drug action within the cell is generally believed to be DNA.[75,78] Evidence has been shown that damage to DNA by bleomycin *in vitro* requires the presence of iron II and oxygen, and the iron complex has been shown to generate hydroxyl radicals.[79,80] On the other hand, copper ions prevent damage to DNA by bleomycin.[79] The protective effect may be related to a conformational change in bleomycin as a result of copper chelate formation or to interference with electrostatic binding between bleomycin and DNA.[81] The copper chelate has also been reported to dismutate superoxide radicals produced by iron catalyzed oxidations.[79] Although copper-bleomycin complex does not damage DNA *in vitro*, it shows more cytotoxic effects than free bleomycin.[82] Therefore, antitumor action based only on the interaction with DNA has been questioned. Furthermore, tallysomycins are relatively more effective in antitumor activity than bleomycin but cause less breakdown of DNA.[83] These observations indicate that targets other than DNA are involved in the action of these drugs.

Recently, control of transplasma membrane redox activity and proton movement in HeLa cells by bleomycin has been reported.[33] Bleomycin stimulates both membrane Na^+/K^+-ATPase and electrogenic Na^+/K^+ pump, and partial removal of inhibition of these activities by vanadium ions has been observed.[84] An *in vitro* study of the effects of free and liposome-encapsulated bleomycin on a human hepatocellular carcinoma cell line has been described.[85] A higher percent of the cell population shows drug sensitivity with encapsulated bleomycin than free bleomycin. These results also suggest that the cell membrane plays a role in the action of bleomycin.

B. STRUCTURE-ACTIVITY RELATIONSHIP AND ITS DEVELOPMENT

The bleomycins are a remarkable family of antitumor antibiotics. The development of bleomycins are due largely to the outstanding ability of H. Umezawa and his collaborators at the Institute of Microbial Chemistry and the National Institute of Health in Japan. Umezawa has established a program for the screening of microbial culture (*Streptomyces verticillus*) filtrates against experimental tumors (squamous cell carcinoma). The novel approach led to the discovery of bleomycins in 1966. He also elucidated the structure of bleomycins[86] as indicated in Figure 4. The antibiotic has an unusual glycopeptide structure as the backbone and contains a carbonylated dissacharide moiety attached to the hydroxyl group of the β-hydroxyhistidine residue. Five years later, a new group of highly active compounds related to the bleomycins was described. This group, called tallysomycins, was discovered by Kanaguchi and co-workers at the Bristol-Banyu Research Institute.[87] The tallysomycins differ from the bleomycins in two of their amino acids and contain a unique amino sugar, 4-amino-4,6-dideoxy-L-talose. The presence of an additional amino acid, L-β-lysine, in tallysomycin A gives it a longer peptide chain than bleomycins.[88]

The tallysomycins were significantly more active than bleomycin against many strains of bacteria and fungi, including bleomycin-resistant strains, such as *Pseudomonas aeruginosa* and *Candida albicans*.[81] Tallysomycin A is more active than B against most bacteria, but

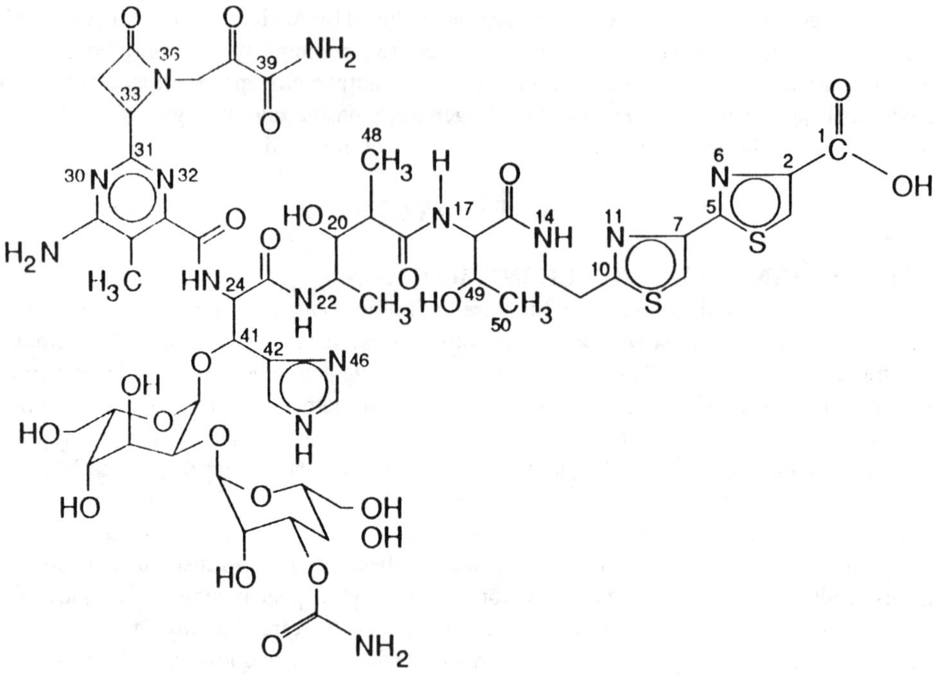

BLEOMYCINIC ACID

FIGURE 4. The structure of bleomycinic acid.

the converse is true against fungi. Prophage induction in the lysogenic bacterial *Escherichia coli* W1709 is caused by tallysomycin A at a minimum concentration of 0.0025 μg/ml. Equivalent induction requires about 32 times more bleomycin.[87] Tallysomycin A also shows potent activity in mice against experimental bacterial infections including *E. coli, P. aeruginosa*, and *Staphylococcus aureus*.[87]

Tallysomycins are active against experimental tumors in mice. Tallysomycin A was on the average four to five times more active than bleomycin. It is highly active on Walker 256 carcinoma, moderately active against P-388 leukemia and B16 melanoma, and slightly active against Lewis lung carcinoma. Against P-388 leukemia, tallysomycin A has a minimum effective dose 11.5 times lower than that of bleomycin on a dose schedule (daily for 9 d) at which it is 3.9 times more toxic than bleomycin. Tallysomycin A caused pneumonitis/fibrosis in only half as many mice as did bleomycin at equitoxic doses.[81]

C. BLEOMYCIN AND TALLYSOMYCIN CONTROL OF TRANSPLASMA MEMBRANE REDOX ACTIVITY AND PROTON MOVEMENT IN HeLa CELLS

There is a significant inhibition of ferricyanide reduction rates by bleomycins and tallysomycins[33,63] (Figure 5). Similar inhibition of proton release coupled to the redox system is also observed with bleomycins and tallysomycins[33] (Table 7). Tallysomycins and bleomycin show similar effects on all these reactions. The study with purified membranes[33] (Table 8)[33] clearly shows that a target for these drugs is on the plasma membrane.

The effect of bleomycin and its copper complex and tallysomycins on transmembrane redox activity is consistent with the cell growth inhibition effects of these compounds.[33,63] The inhibition of the redox-induced proton release by bleomycins, tallysomycins as well as

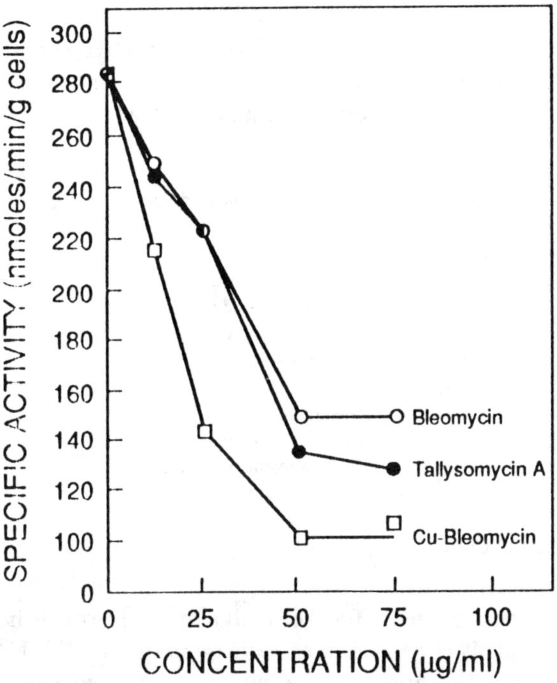

FIGURE 5. Effects of bleomycin and tallysomycin on the rate of ferricyanide reduction by HeLa cells. Cells were incubated with the drugs for 3 min before the addition of ferricyanide to start the reaction. The copper-bleomycin complex was prepared from equimolar bleomycin and cupric sulfate. (Data from Sun, I. L. and Crane, F. L., *Biochem. Pharmacol.*, 34, 617, 1985. With permission.)

TABLE 7
Effects of Bleomycin and Its Analogs on Ferricyanide-Induced Proton Extrusion in HeLa Cells[a]

Additions	Activity (nmol H^+/min/gww)	% Inhibition
None	403	
Bleomycin (25 μg/ml)	345	14
Bleomycin (50 μg/ml)	180	55
Copper-bleomycin (25 μg/ml)	315	22
Copper-bleomycin (50μg/ml)	150	63
Tallysomycin S_{10}b (25 μg/ml)	220	45
Tallysomycin S_{10}b (50 μg/ml)	70	83
Tallysomycin A (25μg/ml)	320	21
Tallysomycin A (50 μg/ml)	100	75

Note: [a]All assays were done with 100 mg cells and 0.3 m*M* potassium ferricyanide in sucrose-salts medium.

Data from Sun, I. L. and Crane, F. L., *Biochem. Pharmacol.*, 34, 617, 1985. With permission.

TABLE 8
Effects of Bleomycin and Its Analogs on NADH-Ferricyanide Reductase Activity of Mouse Liver Plasma Membrane and Pig Erythrocyte Plasma Membrane

Membrane source	Compounds added	Reductase activity, % control
Mouse liver plasma membrane	Control	100
	Bleomycin (50 μg/ml)	57
	Copper-bleomycin (50 μg/ml)	48
	Tallysomycin A (50 μg/ml)	72
	Tallysomycin S_{10}b (50 μg/ml)	69
Pig erythrocyte membrane	Control	100
	Bleomycin (75 μg/ml)	68
	Copper-bleomycin (75 μg/ml)	64
	Tallysomycin A (75 μg/ml)	69
	Tallysomycin S_{10}b (75 μg/ml)	63

Data from Sun and Crane.[33]

adriamycin, provides a possible basis for the inhibition of growth. It has been shown that an increase in cytosolic pH is correlated to an increase in mitosis.[89,90] If the transmembrane redox is stimulated, it would contribute to the movement of protons out of the cell. On the other hand, inhibition of electron transport by bleomycins inhibits proton release, which would tend to decrease intracellular pH and inhibit mitosis. Therefore, transmembrane electron transport activity coupled to an effect on cell mitosis may contribute to the cytotoxic effects of these drugs.

V. *CIS*-DICHLORODIAMMINE PLATINUM(II)

A. *CIS*-DICHLORODIAMMINE PLATINUM AFFECTS CELL SURFACE ARCHITECTURE

Antitumor activity of *cis*-dichlorodiammine platinum(II) (cisplatin) has been shown against a number of malignant tumors and cancer cells.[17,21,64,91,92] Sarna and Sodhi[91] observed the removal of mucopolysaccharides from the surface of tumor (fibrosarcoma) cells after cisplatin treatment and suggested that this removal of mucopolysaccharides from the surface of tumor cells can possibly expose some new or cryptic antigenic sites resulting in their increased antigenicity. It was suggested that platinum compounds may interact with the cells and change its antigenic properties by exposing new antigenic sites on the cell surface.[92] Additionally, Prasad and Sodhi[93] reported that the low agglutinability of normal cells (splenocytes) with concanavalin A (Con A) and wheat germ agglutinin (WGA) increases significantly after cisplatin treatment of the cells for 10 to 30 min at 37°C. On the other hand, the tumor (Dalton's lymphoma) cells which show a high degree of cell agglutination with Con A and WGA when treated with cisplatin, have an appreciable decrease in their degree of cell agglutination. Sodhi and Prasad[94] by ultrastructural studies showed that *in vivo* cisplatin treatment of tumor cells gradually removes the ruthenium red- and lanthanum nitrate-staining substances (extracellular mucopolysaccharides) from the surface of the cells and also leads to a redistribution, clustering, and complete disappearance of Con A binding sites (Con A-ferritin conjugates) on the surface of tumor cells. At the same time, the removal of sialic acid moieties from the surface of the cells after cisplatin treatment has been suggested to play an important role in the increase or decrease of cell agglutination of normal and tumor cells.[93] The effect of platinum complexes on some cellular enzymes has also been reported.[95]

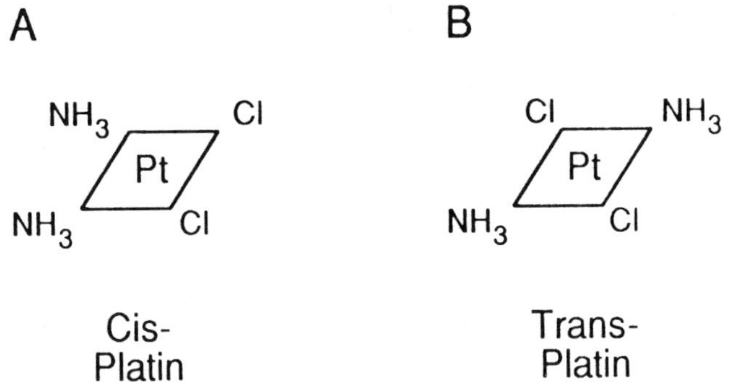

FIGURE 6. The structure of *cis* and *trans* Pt (II) diamminedichloride. (A) cisplatin; (B) transplatin.

B. THE SELECTIVE DRUG ACTION AND ISOMERS

Cis- and *trans*-Pt (II) diamminedichloride are two isomers with chemical structures shown in Figure 6. Pascoe and Roberts[21] have shown that higher doses of *trans*-Pt (II) diammine dichloride than the *cis* isomer have to be administered to cells to produce equal effects on cell survival. Determination of the amount of drug bound to cellular macromolecules shows that more of the *trans* compound was bound to DNA than *cis* when cell survival was reduced by the same amount. These suggest there is a real difference in the mechanism of action of the two isomers. Since both isomers bind to DNA equally well,[21,96] it is a rare event and possibly of little importance in determining that the specific cytotoxic effect of the *cis* compound is mainly due to the formation of DNA interstrand cross-links. The involvement of other targets is clearly indicated.

The drug specificity is also shown on the transmembrane redox system.[34] When cisplatin is preincubated with the cells for three min, the ferricyanide reduction in both slow and fast rates is inhibited at concentrations over 10^{-7} *M*. Incubation with transplatin under the same conditions shows very little inhibition and actually gives a stimulation of ferricyanide reduction above 10^{-6} *M* (Figure 7). The stimulation of ferricyanide reduction by the cis- and transplatin above 5×10^{-7} M is eliminated by superoxide dismutase consistent with evidence that higher concentrations of these compounds induce superoxide production.[97]

Isolated plasma membrane from various cells contain a transmembrane NADH dehydrogenase which transfers electrons from internal NADH to external ferricyanide and this enzyme is considered most likely to be the vehicle for ferricyanide reduction by whole cells. When membranes from erythrocytes or mouse liver are incubated with cisplatin for three min and assayed for NADH ferricyanide oxidoreductase activity, the enzyme is inhibited at concentrations similar to those required for inhibition of ferricyanide reduction by whole cells. Transplatin gives much less inhibition of the plasmalemma NADH ferricyanide reductase (Table 9). These results strongly indicate that the transmembrane enzyme is the site of specific inhibition by cisplatin.

C. PLATINUM COMPLEX AFFECTS OTHER MEMBRANE ENZYMES AND RESULTS IN CHANGING MEMBRANE PERMEABILITY

The effect of cisplatin on other enzyme activities of plasma membrane has been reported by Aggarwal and Niroomand.[98] When the rate of alkaline phosphatase, 5′-nucleotidase, Ca^{2+}-ATPase and Na^+,K^+-ATPase are checked, they show a gradual decrease in activity on the plasma membrane after the drug treatment. Quantitative analysis especially shows that alkaline phosphatase activity drops from 9.7 to 4.9 nmol in just 15 min and drops further to 0.79 nmol after 2 h of treatment of cultures. It is suggested that inactivation of various plasma membrane enzymes may result in permeability changes which are probably respon-

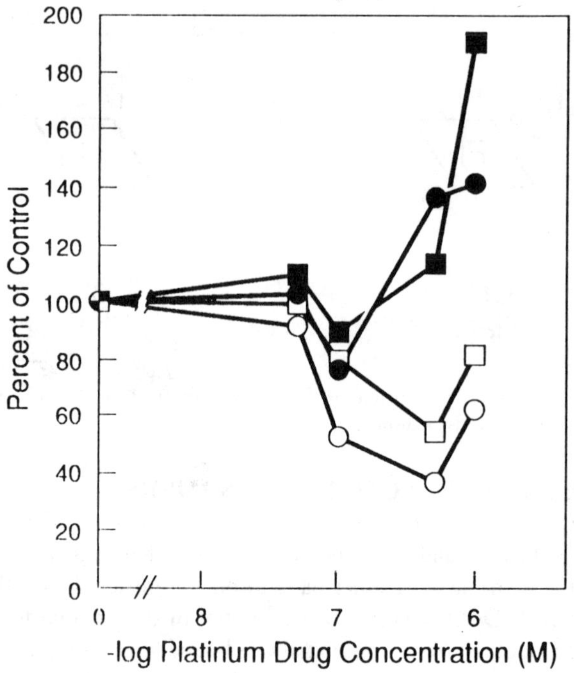

FIGURE 7. The effect of cisplatin and transplatin on ferricyanide reduction by HeLa cells. Control activity is initial fast rate of ferricyanide reduction 504 ± 37 nmol min^{-1} gww^{-1} (circles) and slow rate 160 ± 24 nmol min^{-1} gww^{-1} (triangles); open symbols, cisplatin; closed symbols transplatin. (Data from Sun and Crane.[34])

TABLE 9
Inhibition of NADH Ferricyanide Reductase Activity of Isolated Plasma Membranes by Platinum Drugs

	NADH Ferricyanide reductase	
Addition	**nmol/min mg protein**	**% inhibition**
Control	394	—
10^{-8} *M* cisplatin	388	2
10^{-7} *M* cisplatin	225	43
10^{-6} *M* cisplatin	308	22
10^{-7} *M* transplatin	390	1

Data in part from Sun and Crane[34] for pig erythrocyte plasma membranes.

sible for cell death. By using electrical discharges on cultures and rendering the plasma membranes transiently more permeable to drugs, Melvik et al.[99] show the cytotoxicity of cisplatin has been enhanced to an extent corresponding to at least a threefold increase in drug uptake relative to unpermeabilized cells. This result suggests that uptake across the plasma membrane may be the rate-limiting factor in the cytotoxic effect of platinum complex.

VI. ACTINOMYCIN D

A. DISCOVERY, ISOLATION, AND CHARACTERIZATION

The history of antitumor antibiotics begins with the discovery of actinomycins by Waksman and Woodruff in 1941.[100] The active principle has been crystallized and characterized.

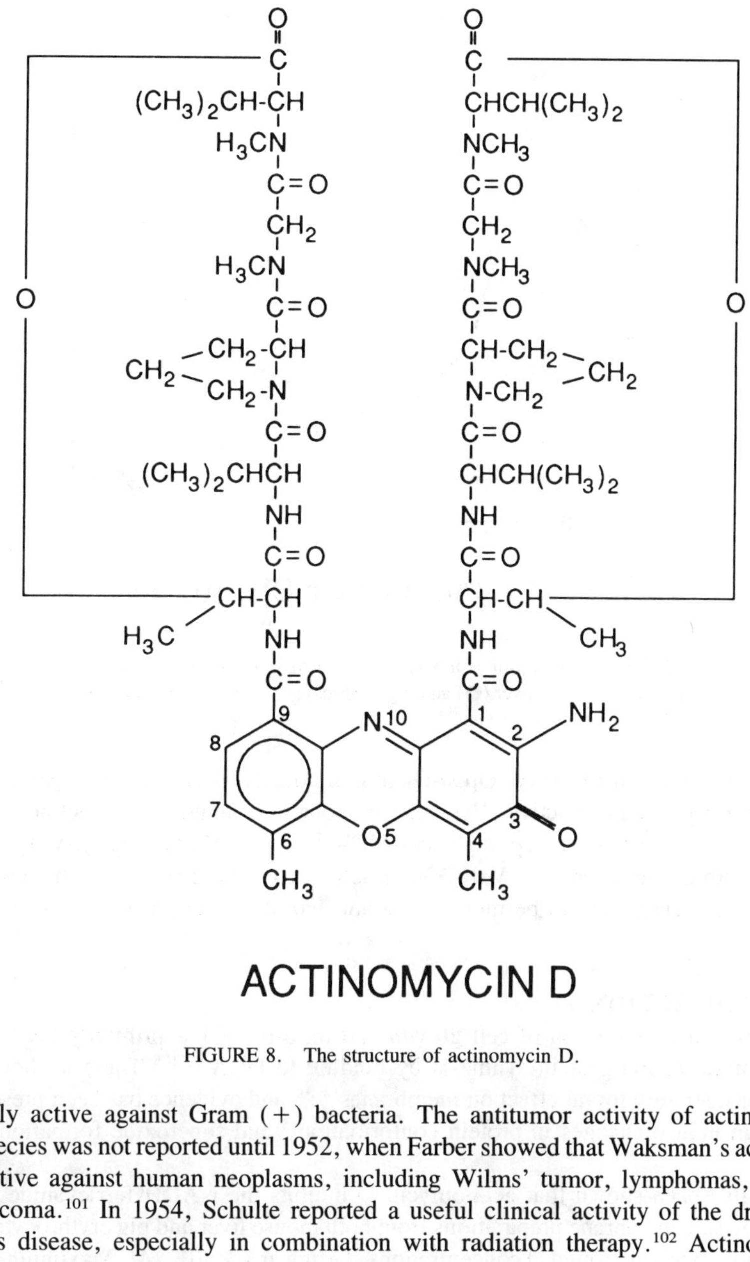

FIGURE 8. The structure of actinomycin D.

It is highly active against Gram (+) bacteria. The antitumor activity of actinomycin in animal species was not reported until 1952, when Farber showed that Waksman's actinomycin D was active against human neoplasms, including Wilms' tumor, lymphomas, and rhabdomyosarcoma.[101] In 1954, Schulte reported a useful clinical activity of the drug against Hodgkin's disease, especially in combination with radiation therapy.[102] Actinomycin D, which contains actinomycin C_1 almost exclusively is obtained from selected strains of *Streptomyces antibioticus* and *Streptomyces parvulus*. The complete and unambiguous structure of the chromophore was established in a series of acid degradations by Brockman, Gröne, and Muxfeldt.[102]

B. STRUCTURE-ACTIVITY RELATIONSHIPS

Actinomycins have structures composed of a chromophoric portion, 2-amino-4,5-dimethyl-3-phenoxazone-1,8-dicarboxylic acid, known as actinocin (Figure 8) and two pentapeptide lactones.[102] The structure-activity relationships were tested,[103] and are consistent for the inhibition of HeLa cells in culture, the inhibition of *B. subtilis* grown in culture, binding to DNA, and the inhibition of RNA polymerase isolated from HeLa cells and *E. coli*. The most obvious structure-activity factor is that both pentapeptide lactone rings must

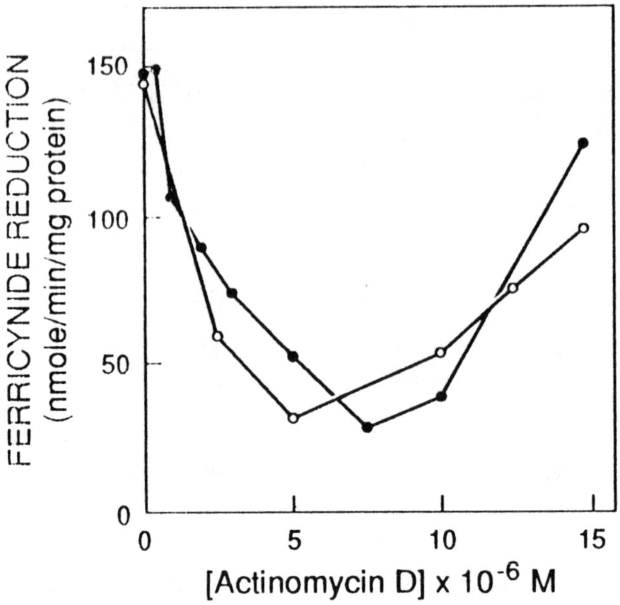

FIGURE 9. Effect of actinomycin D on NADH ferricyanide reductase activity of mouse liver (●) and pig erythrocyte (○) plasma membranes. (Data from Sun and Crane.[31])

maintain their structural integrity. Open-chain compounds such as actinomycinic acid and the secoactinomycins are inactive. Even the monolactone analogs (semiactinomycins) are inactive.[104] The isosteric pentapeptide lactams [Dbu^1]$_2$$Val^2$-AM and [$Dpr^1$]$_2$-$Val_2$-AM, show about one tenth of the potency of Val_2-AM of actinomycin D. This means that the shape of the pentapeptide chains might be more important than the nature of certain specific atoms in them.

C. MODE OF ACTION

Actinomycin D inhibition of cell growth and mitosis[11,12] has primarily been related to inhibition of RNA and protein synthesis by binding to DNA.[13,105] There is, however, increasing consideration for an effect on membranes[14,106] and evidence has been presented that the drug can induce changes in protein conformation[15] and superoxide formation in membranes.[107]

It has also been shown that actinomycin D inhibits the NADH ferricyanide reductase activity in plasma membrane preparations from both mouse liver and pig erythrocytes (Figure 9). Inhibitory effects are found at concentrations starting at $1 \times 10^{-6} M$. Maximum inhibition is seen in the range from 5 to $7.5 \times 10^{-6} M$ with both membranes. The inhibitory effects are shown with plasma membranes only,[31] but not with endoplasmic reticulum, as indicated in Table 10. The concentration of actinomycin D which affects transmembrane redox activities is in concentration ranges known to inhibit protein synthesis[11] and mitosis.[12]

VII. RETINOIC ACID

A. ANTITUMOR ACTION OF RETINOIC ACID

Retinoic acid is an unusual antitumor agent[20,21] which acts to induce differentiation of tumor cells into differentiated cells which have lost the capacity to proliferate freely. The basis for this action is unknown. It has been proposed that retinoic acid may enhance GTP levels in cells to activate genes involved in differentiation.[109]

TABLE 10
Effect of Antinomycin D on Cytochrome c Reductase Activity of Membranes from Mouse Liver and Pig Erythrocytes

	Cytochrome c reductase (nmol/min/mg protein)			
Membrane	**Substrate**	**Control**	**Actinomycin D (7.5×10^{-6} M)**	**% of control**
Erythrocyte	NADH	5.5	0.8	15
Liver plasma membrane	NADH	17	4	24
Liver endoplasmic reticulum	NADH	485	575	119
Liver endoplasmic reticulum	NADPH	40	43	108

Data from Sun, I. L. and Crane, F. L., *Biochem. Biophys. Res. Commun.*, 101, 68, 1981. With permission.

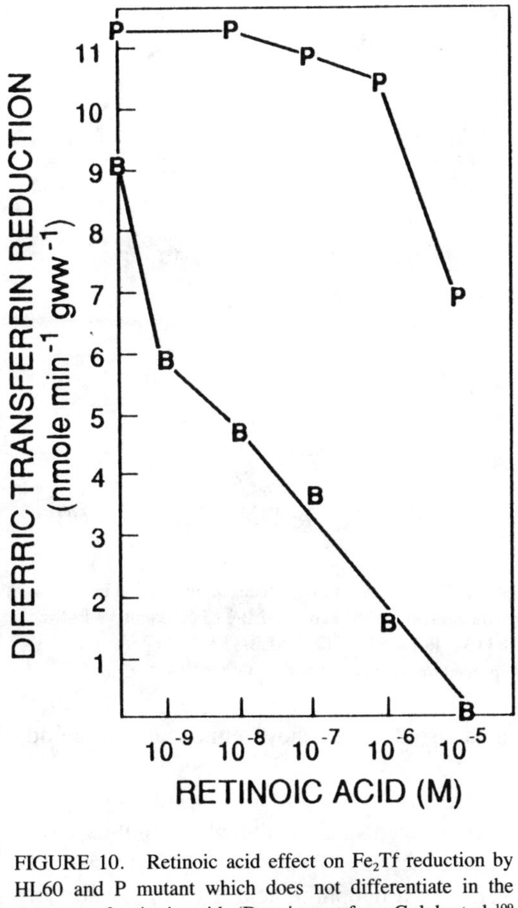

FIGURE 10. Retinoic acid effect on Fe_2Tf reduction by HL60 and P mutant which does not differentiate in the presence of retinoic acid. (Data in part from Golub et al.[109] [B-HL60, P-retinoic acid-resistant mutant HL60 cells].)

Retinoic acid is known to stimulate the Na^+/H^+ antiport in HL60 cells,[110] an effect that would normally increase cell division.[24]

B. RETINOIC ACID INHIBITION OF PLASMA MEMBRANE REDOX ACTIVITY

Retinoic acid is a very strong inhibitor of transplasma membrane ferricyanide reduction by HeLa[108] and HL60 cells. The concentrations which cause inhibition are the same as for inducing differentiation (Figure 10). A mutant strain of HL60 (P) that does not differentiate

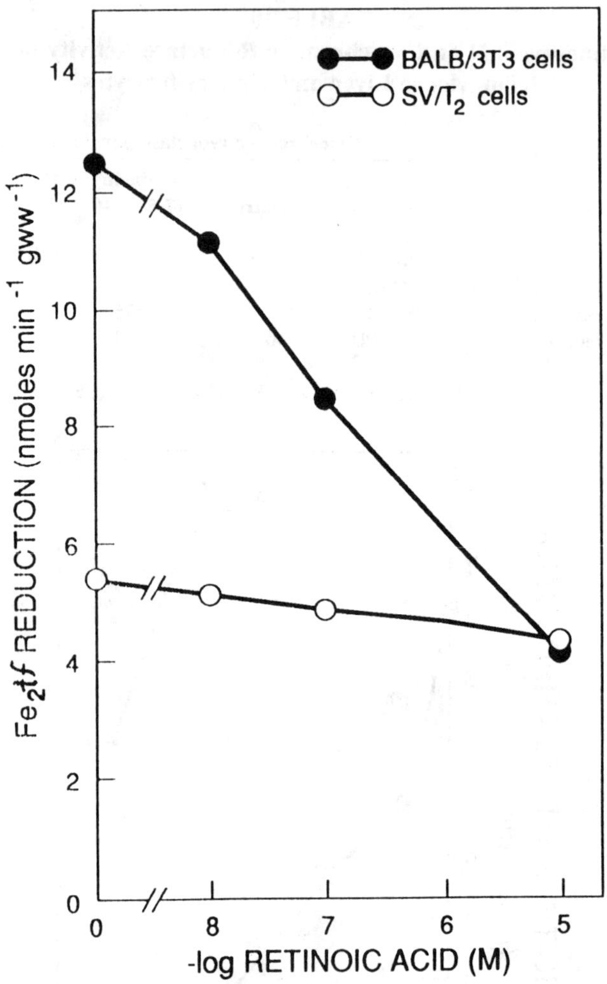

FIGURE 11. Comparison of effects of retinoic acid on the rate of diferric transferrin reduction by BALB/3T3 cells and SV40-transformed BALB/3T3 cells (SV/T_2). (●) BALB/3T3, (○) open circles SV40 3T3. Assay procedure as described in Reference 64.

in response to retinoic acid also does not show retinoic acid inhibition of the transmembrane electron transport.[109]

It should be noted that the retinoic acid has unusual properties that differentiate it from the other cytotosic antitumor agents. Even though it inhibits the transmembrane electron transport, it does not inhibit the Na^+/H^+ antiport. This effect can be based on the acidic nature of the retinoic acid. As a lipophilic acid it can pass through the plasma membrane and activate the Na^+/H^+ antiport at the allosteric H^+ binding site which has been identified with the activation of this antiport.[111,112]

C. PROTECTIVE EFFECT OF SV40 TRANSFORMATION AGAINST RETINOIC ACID EFFECTS

It has been shown that SV40 transformation of cells cannot be reversed by retinoic acid.[113] SV40-transformed cells also show very little inhibition of the transplasma membrane electron transport by retinoic acid[114,115,120] (Figure 11). Therefore, the insertion of the SV40 large T antigen into the plasma membrane must change a receptor site or the nature of the redox enzyme so that it cannot react with retinoic acid.

TABLE 11
Specificity of Retinoic Acid for Inhibition of Diferric Transferrin Reductase Activity of Fetal Rat Liver Cells Infected with a Temperature Sensitive SV40 Virus (RLA209-15 Cells)

	Diferric Transferrin Reduction (nmol min^{-1} gww^{-1})	
Additions	**Nontransformed cells 40°C**	**Transformed cells 33°C**
None	5.7	2.2
Retinoic acid 10^{-6} *M*	2.8	2.3
Retinoic acid 10^{-5} *M*	0.2	2.3
Retinol 10^{-6} *M*	4.4	2.2
Retinol 10^{-5} *M*	4.1	1.9
Retinyl acetate 10^{-6} *M*	4.8	—
Retinyl acetate 10^{-5} *M*	3.0	1.4

Note: Cells cultured at 40°C are not transformed, those cultured at 33°C are transformed as described in Reference 30. Data in part from Reference 120.

D. SPECIFICITY OF RETINOIC ACID EFFECTS

Retinoic acid is specific for many of its differentiation effects on cells in relation to similar concentrations of retinol or retinol esters. The effects on the plasma membrane electron transport show similar specificity. There is very little effect of retinol or its esters on the electron transport under conditions where retinoic acid shows strong inhibition. This specificity is clearly demonstrated on SV40 (nontransformed 40°C) cells (Table 11).[120]

VIII. MACROMOLECULAR ANTITUMOR AGENTS

Cesalin is a high molecular weight protein which reacts with glycoprotein receptors on the plasma membrane to inhibit DNA synthesis. It does not cause DNA scission or inhibit DNA synthesis in isolated nuclei.[114] Macromomycin is a similar polypeptide which inhibits cell growth but does not cause DNA scission.[117] Both cesalin and macromomycin inhibit oxygen uptake of cells[118,119] and there is evidence that the inhibition may not be at the mitochondria. Macromomycin inhibits ferricyanide reduction by HeLa cells. Six μg/ml gives a 40% inhibition. The inhibition of electron transport in the plasma membrane may be a basis for inhibition of oxygen uptake by these polypeptides and may be a basis for their inhibition of cell growth.

IX. SIGNIFICANCE

Transplasma-membrane redox systems have been observed to be related to many vital cellular functions.[122-125] It seems to be a general characteristic of cells because all cells examined do have these activities. The evidence of hormone action on these activities and the same response to inhibition (the inhibition of transmembrane dehydrogenase) by all antitumor drugs mentioned in this chapter, further strengthened the importance of the transmembrane redox system as a target for hormones and antitumor drugs.

The modification of transmembrane redox activity in tumors or virally transformed cells, as indicated by an alteration of V_{max} and K_m of enzyme kinetics,[30,40,41] may explain lack of control of growth activity in these cells. These cells also turn out to be more susceptible to antitumor drugs.[30] This may explain why antitumor drugs can be used clinically to treat cancers, for they have a greater inhibition of redox functions in cancer cells.

REFERENCES

1. **Remers, W. A.,** Adriamycin, in *The Chemistry of Antitumor Antibiotics,* John Wiley & Sons, New York, 1989, 144.
2. **Müller, W. E. G., Yamazaki, Z., and Zahn, R. K.,** Action of bleomycin on DNA and RNA, *Eur. J. Biochem.,* 31, 518, 1972.
3. **Barranco, S. C. and Humphrey, R. M.,** The effect of bleomycin on survival and cell progression, *Cancer Res.,* 31, 1218, 1971.
4. **Chandra, P.,** Role of chemical structure in biochemical activity of daunomycin, adriamycin and some structure analogs, *Cancer Chem. Rep.,* 59, 115, 1975.
5. **Dimarco, A.,** Adriamycin: mode and mechanism of action, *Cancer Chem. Rep.,* 59, 91, 1975.
6. **Kanter, P. M. and Schwartz, H. S.,** Quantitative models for growth inhibition of human leukemia cells by antitumor anthracycline derivatives, *Cancer Res.,* 39, 3661, 1979.
7. **Ross, W. E., Glaubizer, D., and Kohn, K. W.,** Qualitative and quantitative aspects of intercalator-induced DNA strand breaks, *Biochim. Biophys. Acta,* 90, 41, 1979.
8. **Berenblum, I. and Armuth, V.,** Bleomycin-resistant cells contain increased bleomycin-hydrolase activities, *Biochem. Biophys. Res. Commun.,* 101, 55, 1981.
9. **Murphee, S. A., Cunningham, L. S., and Hwang, K. H.,** Effect of adriamycin on surface properties of sarcoma 180 ascite cells, *Biochem. Pharmacol.,* 25, 1227, 1976.
10. **Na, C. and Timasheff, S. N.,** Physical-chemical study of daunomycin-tubulin interaction, *Arch. Biochem. Biophys.,* 182, 147, 1977.
11. **Hoober, J. K. and Cohen, S.,** Epidermal growth factor, the stimulation of protein on RNA synthesis in chick embryo epidermis, *Biochim. Biophys. Acta,* 138, 347, 1967.
12. **Kishimoto, S. and Lieberman, I.,** Synthesis of RNA and protein required for the mitosis of mammalian cells, *Exp. Cell. Res.,* 36, 92, 1964.
13. **Krugh, T. R.,** Association of actinomycin D and DNA as a model for binding of the drug to DNA, *Proc. Natl. Acad. Sci. U.S.A.,* 69, 1911, 1972.
14. **Pastan, I. and Friedman, R. M.,** Actinomycin D: inhibition of phospholipid synthesis in chick embryo cells, *Science,* 160, 316, 1968.
15. **Sinha, B. K. and Chignell, C. F.,** Interaction of antitumor drugs with human erythrocyte ghost membranes and mastocytoma P815: a spin label study, *Biochem. Biophys. Res. Commun.,* 86, 1051, 1979.
16. **Baliprasad, S. and Sodhi, A.,** Effect of *cis*-dichlorodiammine platinum (II) on surface of tumor and normal cells: biochemical, fluorescence, and electron microscopical studies, *Indian J. Exp. Biol.,* 20, 559, 1982.
17. **Rosenberg, B., Van Camp, L., Trosko, J. E., and Mansour, V. H.,** Platinum compounds: a new class of potent antitumour agents, *Nature,* 222, 385, 1969.
18. **Macquet, J. P. and Butour, J. L.,** Platinum-amine compounds: importance of labile and inert ligands for their pharmacological activity toward L1210 leukemia cells, *J. Natl. Cancer Inst.,* 70, 899, 1983.
19. **Pascoe, J. M. and Robert, J. J.,** Interactions between mammalian cell DNA and inorganic platinum compounds, *Biochem. Pharmacol.,* 23, 1345, 1974.
20. **Moon, R. C. and Itri, L. M.,** Retinoids and cancer, in *The Retinoids,* Vol. 2, 1984, 327.
21. **Hill, D. L. and Grubbs, C. J.,** The use of retinoids in combination with other chemotherapeutic agents against L1210 leukemia, *Anticancer Res.,* 2, 111, 1982.
22. **Sun, I. L., Crane, F. L., Grebing, C., and Löw, H.,** Transmembrane redox in control of cell growth: stimulation of HeLa cell growth by ferricyanide and insulin, *Exp. Cell. Res.,* 156, 528, 1985.
23. **Ellem, K. A. O. and Kay, G. F.,** Ferricyanide can replace pyruvate to stimulate growth and attachment of serum restricted human melanoma cells, *Biochem. Biophys. Res. Commun.,* 112, 183, 1983.
24. **Crane, F. L., Sun, I. L., Clark, M. G., Grebing, C., and Löw, H.,** Transplasma-membrane redox systems in growth and development, *Biochim. Biophys. Acta,* 811, 233, 1985.
25. **McLoughlin, P., Sun, I. L., and Crane, F. L.,** Membrane redox systems in porcine neutrophils, *Proc. Indiana Acad. Sci.,* 93, 333, 1984.
26. **Crane, F. L., Crane, H. E., Sun, I. L., MacKellar, W. C., Grebing, C., and Löw, H.,** Insulin control of a transplasma membrane NADH dehydrogenase in erythrocyte membranes, *J. Bioenerg. Biomembr.,* 14, 425, 1982.
27. **Christensen, H. N.,** in *Amino Acid Transport in Redox Functions of the Eukaryotic Plasma Membrane,* Ramirez, J., Ed., Consejo-Superior de Investigaciones Cientificas, Madrid, 1987, 119.
28. **Brooks, M. M.,** Activation of eggs by oxidation-reduction indicators, *Science,* 106, 320, 1947.
29. **Rudland, P. S., Durbin, H., Clingan, D., and De Asua, L. J.,** Iron salt and transferrin are specifically required for cell division of cultured 3T6 cells, *Biochem. Biophys. Res. Commun.,* 75, 556, 1977.
30. **Sun, I. L., Crane, F. L., Chou, J. Y., Löw, H., and Grebing, C.,** Transformed liver cells have modified transplasma redox activity which is sensitive to adriamycin, *Biochem. Biophys. Res. Commun.,* 116, 210, 1983.

31. **Sun, I. L. and Crane, F. L.,** Transplasmalemma NADH dehydrogenase is inhibited by actinomycin D, *Biochem. Biophys. Res. Commun.*, 101, 68, 1981.
32. **Sun, I. L., Crane, F. L., Löw, H., and Grebing, C.,** Properties of a transplasma membrane electron transport system in HeLa cells, *J. Bioenerg. Biomembr.*, 16, 583, 1984.
33. **Sun, I. L. and Crane, F. L.,** Bleomycin control of transplasma membrane redox activity and proton movement in HeLa cells, *Biochem. Pharmacol.*, 34, 617, 1985.
34. **Sun, I. L. and Crane, F. L.,** The antitumor drug, *cis*-platin, inhibits trans plasmalemma electron transport in HeLa cells, *Biochem. Int.*, 9, 299, 1984.
35. **Sun, I. L. and Crane, F. L.,** Antitumor drug inhibition of transplasma membrane redox function, *Fed. Proc.*, 41, 737, 1982.
36. **Ellem, K. A. O., Kay, G. F., Dunstan, A. J., and Stengel, D. J.,** Redox agents which modulate the growth of melanoma cells, in *Plasma Membrane Oxidoreductases in Control of Animal and Plant Growth*, Crane, F. L., Morré, D. J., and Löw, H., Plenum Press, New York, 1988, 17.
37. **Sun, I. L., Crane, F. L., Grebing, C., and Löw, H.,** Growth stimulation by impermeable oxidants through plasma membrane redox, *J. Cell. Biol.*, 99, 293a, 1984.
38. **Sun, I. L., Crane, F. L., Löw, H., and Grebing, C.,** Transplasma membrane redox stimulates HeLa cell growth, *Biochem. Biophys. Res. Commun.*, 125, 649, 1984.
39. **Sun, I. L., Crane, F. L., and Chou, J. Y.,** Modification of transmembrane electron transport activity in plasma membranes of SV40 transformed pineal cells, *Biochim. Biophys. Acta*, 886, 327, 1986.
40. **Sun, I. L., Crane, F. L., Chou, J. Y., and Löw, H.,** Transplasmalemma electron transport is changed in SV40 transformed liver cells, *J. Bioenerg. Biomembr.*, 18, 471, 1986.
41. **Warley, A. and Cook, G. M. W.,** The isolation and characterization of plasma membranes from normal and leukaemic cells of mice, *Biochim. Biophys. Acta*, 323, 55, 1973.
42. **Powis, G., Svingen, B. A., and Appel, P.,** Quinone-stimulated superoxide formation by subcellular fractions, isolated hepatocytes and other cells, *Mol. Pharmacol.*, 20, 387, 1981.
43. **Kanter, P. M. and Schwartz, H. S.,** Quantitative models for growth inhibition of human leukemia cells by anthracycline derivatives, *Cancer Res.*, 39, 3661, 1979.
44. **Silvertrini, R., DiMarco, A., and Dasdia, T.,** Interference of daunomycin with metabolic events of the cell cycle synchronized cultures of rat fibroblasts, *Cancer Res.*, 30, 966, 1970.
45. **Silvertrini, R., Lenaz, L., and DiFronzo, C.,** Corrections between cytotoxicity, biochemical effects, drug level, and therapeutic effectiveness of daunomycin and adriamycin on scarcome 180 ascites in mice, *Cancer Res.*, 33, 2954, 1973.
46. **Ferrier-Mouteäl, A. V., Gousales-Ros, J. M., and Ferragut, J. A.,** *Biochim. Biophys. Acta*, 937, 379, 1988.
47. **DiMarco, A., Silvertrini, R., DiMarco, S. D., and Dasdia, T.,** Inhibiting effect of the new cytotoxic antibiotic daunomycin on nucleic acid and mitotic activity of HeLa cells, *J. Cell. Biol.*, 27, 545, 1965.
48. **Bachur, N. R., Gordon, S. L., and Halcolm, V. G.,** Anthracycline antibiotic augmentation of microsomal electron transport and free radical formation, *Molec. Pharmacol.*, 13, 901, 1977.
49. **Bachur, N. R., Gordon, S. L., Gee, M. V., and Kon, H.,** NADPH cytochrome P-450 reductase activation of quinone anticancer agent to free radicals, *Proc. Natl. Acad. Sci. U.S.A.*, 76, 954, 1979.
50. **Goormaghtigh, E., Brasseur, R., and Ruysschaert, J. M.,** Adriamycin inactivates cytochrome c oxidase by exclusion of the enzyme from its cardiolipin essential environment, *Biochem. Biophys. Res. Commun.*, 104, 314, 1982.
51. **Tritton, T. R. and Yee, G.,** The anticancer agent adriamycin can be actively cytotoxic without entering cells, *Science*, 217, 248, 1982.
52. **Tritton, T. R., Yee, G., and Wing, L. B., Jr.,** Immobilized adriamycin: a tool for separating cell surface from intracellular mechanisms, *Fed. Proc.*, 42, 284, 1983.
53. **Tokes, Z. A., Rogers, K. E., and Rembaum, A.,** Synthesis of adriamycin-coupled polyglutaraldehyde microspheres and evaluation of their cytotoxic activity, *Proc. Natl. Acad. Sci. U.S.A.*, 79, 2026, 1982.
54. **Murphree, S. A., Tritton, T. R., Smith, P. L., and Sartorelli, A. C.,** Adriamycin-induced changes in the surface membrane of sarcoma 180 ascites cells, *Biochim. Biophys. Acta*, 649, 317, 1981.
55. **Schlager, S. I. and Ohanian, S. H.,** A role for fatty acid composition of complex cellular lipids in the susceptibility of tumor cells to humoral immune killing, *J. Immunol.*, 123, 146, 1980.
56. **Peterson, R. H. F., Meyers, M. B., Splenger, M. B. A., and Biedler, J. L.,** Alteration of plasma membrane glycopeptides and gangliosides of Chinese hamster cells accompanying development of resistance to daunorubicin and vincristine, *Cancer Res.*, 93, 222, 1983.
57. **Kessel, D.,** Enhanced glycosylation induced by adriamycin, *Mol. Pharmacol.*, 16, 306, 1979.
58. **Schlager, S. I.,** Relationship between cell-mediated and humoral immune attack on tumor cells, *Cell. Immunol.*, 66, 300, 1982.
59. **Sugimoto, Y., Suzuki, H., and Tanaka, N.,** Alteration of plasma membrane of drug-resistant tumor cells: 230-kilodalton protein identified by monoclonal antibody, *Biochem. Biophys. Res. Commun.*, 114, 969, 1983.

60. **Acamone, F.,** Properties of antitumor anthracyclines and new developments in their application, *Cancer Res.,* 45, 5995, 1985.
61. **Sun, I. L. and Navas, P.,** in *Redox Functions of the Eukaryotic Plasma Membrane,* Ramirez, J., Ed., Graficas Urpe, S. A., Rufino Gonzales, Madrid, 1987, 65.
62. **Sun, I. L., Crane, F. L., Löw, H., and Grebing, C.,** Inhibition of plasma membrane NADH dehydrogenase by adriamycin and related anthracycline antibiotics, *J. Bioenerg. Biomembr.,* 16, 209, 1984.
63. **Sun, I. L. and Crane, F. L.,** Effect of anthracycline compounds on transmembrane redox function of cultured HeLa cells, *Proc. Indiana Acad. Sci.,* 93, 267, 1984.
64. **Sun, I. L. and Crane, F. L.,** Control of transplasma membrane diferric transferrin reductase by antitumor drugs, in *Plasma Membrane Oxidoreductases in Control of Animal and Plant Growth,* Crane, F. L., Morré, D. J., and Löw, H., Eds., Plenum Press, New York, 1988, 181.
65. **Thompson, M. G., Chabwala, S. B., and Hickman, J. A.,** Inhibition of human erythrocyte inositol lipid metabolism by adriamycin, *Cancer Res.,* 47, 2799, 1987.
66. **Medina, M. A., Sanchez Jiminez, F., Segura, J. A., and Nüñez de Castro, I.,** Transmembrane ferricyanide reductase activity in Ehrlich ascite tumor cells, *Biochim. Biophys. Acta,* 946, 1, 1988.
67. **Van Rossum, G. D. V. and Gosalvez, M.,** Inhibition of ion-transport by the antitumor antibiotic adriamycin, *Fed. Proc.,* 35, 787, 1976.
68. **Caroni, P., Villani, F., and Carafoli, E.,** The cardiotoxic antibiotic doxorubicin inhibits the Na^+/Ca^{2+} exchange of dog heart sarcolemmal vesicles, *FEBS Lett.,* 130, 184, 1981.
69. **Caroni, P. and Carafoli, E.,** An ATP-dependent Ca^{2+}-pumping system in dog heart sarcolemma, *Nature (London),* 283, 756, 1980.
70. **Reeves, J. P. and Sutko, J. L.,** Sodium-calcium ion exchange in cardiac membrane vesicles, *Proc. Natl. Acad. Sci. U.S.A.,* 76, 590, 1979.
71. **Villani, F., Piccinini, F., Merelli, P., and Favalli, L.,** Influence of adriamycin on calcium exchangeability in cardiac muscle and its modification by ouabain, *Biochem. Pharmacol.,* 27, 985, 1978.
72. **Solie, T. N. and Yuncker, C.,** Adriamycin induced changes in translocation of sodium ions in transporting epithelial cells, *Life Sci.,* 22, 1907, 1978.
73. **Yasumi, M., Minaga, T., Nakamura, K., Kisu, A., and Igichi, H.,** Inhibition of cardiac NADP-linked isocitrate dehydrogenase by adriamycin, *Biochem. Biophys. Res. Commun.,* 93, 631, 1980.
74. **Weiner, L., Averbich, S., and Singer, D.,** Adriamycin: cellular electrophysiologic changes, *Fed. Proc.,* 38, 987, 1979.
75. **Shirakama, I., Azegami, M., Ishii, S., and Umezawa, H.,** Reaction of bleomycin with DNA strand scission of DNA in the absence of sulfhydric on peroxide compounds, *J. Antibiot. (Tokyo),* 24, 761, 1971.
76. **Blum, R. H., Carter, S. K., and Agee, K. A.,** A clinical review of bleomycin — a new antineoplastic agent, *Cancer N.Y.,* 31, 903, 1973.
77. **Haas, C. D., Coltman, C. A., Gottlieb, J. A., Hunt, A., Luce, J. K., Talley, R. W., Samal, B., and Wilson, H. E.,** Phase II evaluations of bleomycin, *Cancer N.Y.,* 38, 8, 1976.
78. **Müller, W. E. G., Yamazaki, Z., and Zahn, R. K.,** Action of bleomycin on DNA and RNA, *Eur. J. Biochem.,* 31, 518, 1972.
79. **Gutteridge, J. M. C. and Chang, F. X.,** Protection of iron-catalysed free radical damage to DNA and lipids by copper (ii) - bleomycin, *Biochem. Biophys. Res. Commun.,* 99, 1354, 1981.
80. **Burger, R. M., Deisach, J., and Horwitz, S. B.,** Effect of O_2 on the reactions of activated bleomycin, *J. Biol. Chem.,* 257, 3372, 1982.
81. **Remers, W. A.,** Bleomycins and phleomycins, in *The Chemistry of Antitumor Antibiotics,* Vol. 1, John Wiley & Sons, New York, 1979, 176.
82. **Umezawa, H.,** Chemistry and mechanism of action of bleomycin, *Fed. Proc.,* 33, 2296, 1974.
83. **Strong, J. E. and Crooke, S. T.,** DNA breakage by Tallysomycin, *Cancer Res.,* 38, 3322, 1978.
84. **Vyskocil, F., Pilar, J., Zemkova, H., Svofoda, P., Vitek, V., and Teisinger, J.,** Bleomycin stimulates both membrane ATPase and electrogenic pump and partially removes the inhibition on vanadium ions, *Biochem. Biophys. Res. Commun.,* 116, 783, 1983.
85. **Saunders, R. M. and Alexander, J. J.,** The effect of free and liposome-encapsulated bleomycin on a human hepatocellular carcinoma cell line, *S. Afr. Med. J.,* 69, 103, 1986.
86. **Takita, T., Muraoka, Y., Yoshioka, T., Fujii, A., Maeda, K., and Umezawa, H.,** The chemistry of bleomycin, the structure of bleomycin and phleomycin, *J. Antibiot. (Tokyo),* 25, 755, 1972.
87. **Kawaguchi, H., Tsukiura, H., Tomita, K., Konishi, M., Saita, K., Kobaru, S., Numata, K., Fujisawa, K., Miyaki, T., Hatori, M., and Koshiyama, H.,** Tallysomycin, a new antitumor antibiotic complex related to bleomycin. I. Production, isolation and properties, *J. Antibiot. (Tokyo),* 30, 779, 1977.
88. **Konishi, M., Saito, K., Numata, K., Tsuno, T., Asana, K., Tsukiura, H., Naito, T., and Kawaguchi, H.,** Tallysomycin, a new antitumor antibiotic complex related to bleomycin. II. Structure determination of tallysomycins A and B, *J. Antibiot. (Tokyo),* 30, 789, 1977.
89. **Moolenaar, W. H., Tsien, R. Y., Van den Saag, P. T., and de Loat, S. W.,** Na^+/H^+ exchange and cytoplasmic pH in the action of growth factors in human fibroblasts, *Nature (London),* 304, 645, 1983.

90. **Gerson, D. F., Kiefer, H., and Eufe, W.,** Intracellular pH of mitogen-stimulated lymphocytes, *Science,* 216, 1009, 1982.
91. **Sarna, S. and Sodhi, A.,** Removal of mucopolysaccharides from the surface of fibrosarcoma by *cis*-platin, *Indian J. Exp. Biol.,* 16, 1236, 1978.
92. **Sodhi, A.,** The effect of cell surface antigenicity by *cis*-platin, *Indian J. Exp. Biol.,* 14, 383, 1976.
93. **Prasad, S. B. and Sodhi, A.,** Chemico-biological interactions, 36, 355, 1981.
94. **Sodhi, A. and Prasad, S. B.,** Ultrastructural and fluorescence microscopical observations on the effect of *cis*-platin on the surface of tumor and normal cells, *Indian J. Exp. Biol.,* 19, 328, 1981.
95. **Aull, J. L., Allen, R. L., Bapat, A. R., Daron, H. H., Friedman, M. E., and Wilson, J. F.,** The effects of platinum complexes on seven enzymes, *Biochim. Biophys. Acta,* 571, 352, 1979.
96. **Salles, B., Butour, J. L., Lesca, C., and Macquet, J. P.,** Cis-$Pt(NH_3)_2Cl_2$ and trans-$Pt(NH_3)_2Cl_2$ inhibit DNA synthesis in cultured L1210 leukemia cells, *Biochem. Biophys. Res. Commun.,* 112, 555, 1983.
97. **McGinness, J. E., Procter, P. D., Demopoulis, H. B., Hokanson, J. A., and Kirkpatrick, D. S.,** Superoxide production by *cis* and *trans*-Pt(II) diamine dichloride, *Physiol. Chem. Phys.,* 10, 267, 1978.
98. **Aggarwal, S. K. and Niroomand, R. I.,** Effect of *cis*-platin on plasma membrane phosphatase activities in ascite sarcoma-180 cells, *J. Histochem. Cytochem.,* 31, 307, 1983.
99. **Melvik, J. E., Pettersen, E. O., Gordon, P. B., and Seglen, P. O.,** Electrical discharges on cultures affect the cytotoxicity of *cis*-platin, *Eur. J. Cancer Clin. Oncol.,* 22, 1523-1530.
100. **Waksman, S. A. and Woodruff, H. B.,** Actinomyces antibioticus, a new soil organism antagonistic to pathogenic and non-pathogenic bacteria, *J. Bacteriol.,* 42, 231, 1941.
101. **Farber, S.,** Ciba Foundation Symp. on Amino Acids and Peptides with Antimetabolic Activity, Ciba Foundation, London, 138.
102. **Remers, W. A.,** Actinomycin, in *The Chemistry of Antitumor Antibiotics,* Vol. 1, John Wiley & Sons, New York, 1979, 3.
103. **Reich, E., Goldberg, I. H., and Rabinowitz, M.,** Structure-activity correlations of actinomycins and their derivatives, *Nature (London),* 196, 743.
104. **Lackner, H.,** in *Chemistry and Biology of Peptide,* Heienhofer, J., Ed., Ann Arbor Science Publishers, Ann Arbor, MI, 1972, 147.
105. **Goldberg, I. H., Rabinowitz, M., and Reich, E.,** Basis of actinomycin action I. DNA binding and inhibition of RNA-polymerase synthetic reactions by actinomycin, *Proc. Natl. Acad. Sci. U.S.A.,* 48, 2094, 1962.
106. **Fico, R. M., Chen, T. K., and Canallakis, E. S.,** Biofunctional intercalators: relationship of antitumor activity of diacridines to the cell membrane, *Science,* 198, 53, 1977.
107. **Sinha, B. K. and Cox, M. G.,** Stimulation of superoxide formation by actinomycin D and its H^2-substituted spin-labeled derivatives, *Mol. Pharmacol.,* 17, 432, 1980.
108. **Sun, I. L., Toole-Simms, W., Crane, F. L., Golub, E. S., Diaz de Pagan, T., Morré, D. J., and Löw, H.,** Retinoic acid inhibition of transplasmalemma diferric transferrin reduction, *Biochim. Biophys. Acta,* 146, 976, 1987.
109. **Golub, E. S., Diaz de Pagan, T., Sun, I. L., and Crane, F. L.,** The retinoic acid-induced cascade of events leading to granulocyte differentiation in HL60 through a postulated GTP-regulated inducing molecule, in *Plasma Membrane Oxidoreductases in Control of Animal and Plant Growth,* Crane, F. L., Morré, D. J., and Löw, H., Eds., Plenum Press, New York, 1988, 313.
110. **Ladoux, A., Abita, J. P., and Gerry, B.,** Retinoic acid-induced differentiation of HL60 cells is associated with biphasic activation of the Na^+/H^+ pump, *Differentiation,* 33, 142, 1986.
111. **Grinstein, S., Rotin, D., and Mason, M. J.,** Na^+/H^+ exchange and growth factor induced cytosolic pH changes: role in cellular proliferation, *Biochim. Biophys. Acta,* 988, 73, 1989.
112. **Ladoux, A., Cragoe, E. J., Gerry, B., and Arbita, J. P.,** Retinoic acid activates the Na^+/H^+ antiport in HL60 cells, *J. Biol. Chem.,* 262, 811, 1987.
113. **Jetten, A. M., Jetten, H. E. R., Shapiro, S. S., and Poun, J. P.,** Characterization of the action of retinoids on mouse fibroblast cell lines, *Exp. Cell Res.,* 119, 289, 1979.
114. **Sun, E. E., Sun, I. L., Crane, F. L., and Chou, J. Y.,** Retinoic acid effects on plasma membrane diferric transferrin reductase in relation to control of differentiation, *FASEB J.,* 2, 1096A, 1987.
115. **Sun, E. E., Crane, F. L., Sun, I. L., and Chou, J. Y.,** Modification of retinoic acid binding efficiency in the plasma membranes of SV40 viral transformed cells, *J. Cell Biol.,* 109, 254A, 1989.
116. **Shepherd, V. L., Vandre, D. D., Elting, D., and Montgomery, R.,** Effects of cesalin on the ultrastructure and biological properties of cultured mammalian cells, *Cancer Res.,* 40, 225, 1980.
117. **Vandre, D. D., Shepherd, V. L., and Montgomery, R.,** The effect of macromomycin on the ultrastructural and biological properties of cultured mammalian cells, *Cancer Res.,* 39, 409, 1979.
118. **Elting, D. and Montgomery, R.,** Biological effects of cesalin, an antitumor protein, on cultured mammalian cells, *J. Biol. Chem.,* 254, 453, 1979.

119. **Thomas, K. R.,** Inhibition of O_2 consumption in cultured mammalian cells by macromomycin, *Fed. Proc.*, 40, 1578A, 1981.
120. **Sun, I. L., Toole-Simms, W., Crane, F. L., Morré, D. J., Löw, H., and Chou, J. Y.,** Transformation with SV40 virus prevents retinoic acid inhibition of plasma membrane NADH diferric transferrin reductase in rat liver cells, *J. Bioenerg. Biomemb.*, 20, 383, 1988.

Chapter 15

REDOX CONTROL OF NEUROTRANSMITTER ACTION

J.-L. Dreyer, T. Treichler, F. Gueggi, and C. Buillard

TABLE OF CONTENTS

I. INTRODUCTION

Transplasma-membrane redox systems have received increasing attention in the last years as they have been found in all cells examined and appear to be a general characteristic of cells. The basic evidence for plasma-membrane redox systems and some of the properties of the dehydrogenase enzymes have been discussed by Crane et al.[1-4] These redox activities can now be related to several vital functions which include control of cell growth,[5] facilitation of iron-uptake,[6] defense against bacteria,[7] stimulation of certain special transport function, and modulation of the action of some hormones and adenylate cyclase.[8,9] These redox activities in cells appear therefore to be related to postreceptor signal transduction and membrane (de-)polarization, and to promote ion or amino acid transport, or proton transfer across plasma membranes.

Transplasma-membrane redox activities have not been characterized in plasma membranes of neuronal tissue. These membranes, however, are typically endowed with different types of ion channels and carriers providing the molecular basis of excitability. Synaptic transmission involves the stimulation of cation flux by neurotransmitters or neuroactive agents to bring about a change in the membrane electrical potential via an electrogenic process.[10,11] The conductivity change reflects permeability changes mobilizing the steady state gradients of ions such as H^+, Na^+, K^+, and Ca^{2+}. Yet the molecular mechanism underlying the permeability changes in excitable membranes is still largely unsolved. Recent studies show the occurrence of transplasma-membrane redox systems in synaptic membranes and their function in relation to synaptic transmission. The characterization of redox components associated with synaptic membranes and the evidence for the presence of NADH-dehydrogenase activity in these preparations stresses the sensitivity of NADH-dehydrogenases toward neurotransmitters and neuroactive agents, particularly towards catecholamines and acidic amino acids.

II. NADH-DEHYDROGENASES IN SYNAPTIC PLASMA MEMBRANES

A. FRACTION PURITY

The assessment for redox compounds and metals in synaptic membranes requires critical investigations of the preparations for contamination, particularly with respect to mitochondrial or microsomal membranes. The common procedures for the preparation of nerve terminals yield a complex mixture of pre- and postsynaptic membrane vesicles which generally contain other cell organites, mainly mitochondria and synaptic vesicles. Further treatment of these "synaptosomes vesicles", e.g., by means of an osmotic shock, yields a mixture of intracellular membranes (from intracellular synaptic vesicles) and mostly plasma membranes. The classical tissue fractionation procedure by Cotman et al.[12] yields highly purified synaptic membranes (i.e., a mixture of pre- and postsynaptic plasma membranes and membranes from synaptic vesicles), devoid of mitochondrial and microsomal enzymatic activities, which exhibit no activity for, e.g., succinate dehydrogenase, cytochrome c oxidase, or NADPH-cytochrome c reductase. Another procedure for the preparation of synaptic plasma membranes in high yield and of high purity involves phase separation on polyethylene-glycol-dextran.[13]

B. OCCURRENCE

NADH-acceptor oxidoreductase activities in synaptic membrane fractions are given in Table 1. A variety of oxidizing agents have been tested as substrates and the membranes tested exhibit various NADH-dehydrogenases depending upon the redox potential of the electron acceptor, unless it reflects the presence of several enzymes, and heterogeneity of

TABLE 1
Occurrence of Plasma Membrane Dehydrogenases in Purified Rat Brain Synaptic Membranes

Acceptor	Specific activity
Ferricyanide	150—170
DCIP	13—17
Glyoxylate	100
Juglone	29—35
Cytochrome c	4—6
Nitroblue-tetrazolium	4

Note: 200 μg protein of purified synaptic membranes was incubated with 1 m*M* NADH and assayed for NADH-acceptor dehydrogenase activity in 100 m*M* phosphate buffer pH 7.0 (except for DCIP, measured at pH 8.0 in the same buffer) in the presence of 0.14 m*M* acceptor. Appropriate blanks and controls were run in parallel to exclude unspecific oxidation of NADH. Specific activities are in nmol of acceptor, reduced per min per mg of protein of synaptic membranes.

the membrane preparations (mainly containing a mixture of neuronal pre- and postsynaptic plasma membranes, besides glial plasma membrane). A fairly high concentration of NADH-dehydrogenases is found in these plasma membrane preparations. The activities are not accompanied by mitochondrial or microsomal enzymes and the activities with substrates such as ferricyanide and DCIP are clearly too high to be accounted for by contamination with either endoplasmic reticulum, mitochondrial membranes, or a combination of the two. Preparations of the synaptic membrane fractions yield enrichment in NADH-ferricyanide oxidoreductase activity parallel with the increase of receptor-binding activity for neurotransmitters (e.g., for L-glutamic acid) or acetylcholine-esterase activity.

C. LIMITED PROTEOLYSIS AND SOLUBILIZATION

Synaptic plasma membrane preparations exposed to limited proteolysis with trypsin or pepsin are readily inactivated and all the NADH-acceptor oxidoreductase activities vanish upon 1 h incubation at 37°C. Furthermore, the NADH-acceptor oxidoreductases can be solubilized under suitable conditions, without loss of activity. It is known that some proteins which copurify with plasma membrane fractions are not integral membrane proteins and can be removed by treatment of the membranes with high ionic strength buffers or with chaotropic salt solutions. However, treatment of intact synaptic membranes with high ionic strength media or with chaotropic agents yielded either low solubilization of the NADH-oxidoreductase activities or inactive material. Only with the use of detergents can significant amounts of dehydrogenase activity be solubilized. Na^{+}-cholate or Triton X-100 (both 0.5% w/v) solubilized 40 to 45% of the membrane-associated proteins, and about as much dehydrogenase activity. Nevertheless, complete solubilization of both membrane proteins and activity can be achieved upon treatment with Triton X-114. Treatment of the synaptic membrane preparations with 0.1% Triton X-114 followed by ultracentrifugation results in solubilization of the NADH-dehydrogenase activity with 40 to 60% recovery. The remaining NADH-dehydrogenase activity in the pellet is sensitive to dopamine, exhibiting complete inhibition in the presence of 1 m*M* dopamine. From these data, it is apparent that NADH-acceptor oxidoreductases are integral membrane proteins.

D. OXYGEN RADICALS

Oxygen radicals play an important role in the nervous system, and their importance as "mediators" of the action of certain neurotoxins or of antioxidants and vitamin E has been stressed (see Reference 14 for a review). Furthermore, NADH-dehydrogenases have been shown to play a protective role against oxygen radicals in erythrocytes[15] and neutrophil leukocytes.[16] Under similar conditions, however, NADH-acceptor oxidoreductases in synaptic membranes are not sensitive to either catalase, lipoxygenase, superoxide dismutase, or oxygen radical scavengers such as, e.g., vitamin E, or ascorbate.

E. REDOX COMPONENTS OF THE SYNAPTIC MEMBRANE PREPARATIONS

The membranes are devoid of trace amounts of copper but exhibit significant amounts of iron. It is noteworthy that contamination by transition metals in preparations of biological samples is difficult to avoid at these low levels; a common observation has shown, however, that when iron contamination is observed in such preparations, it is generally accompanied by copper contamination as well. The absence of the latter in these samples indicates that the presence of iron in these synaptic membrane preparations is not accidental and cannot be exclusively attributed to adventitious iron. The preparations contain also small amounts of acid-labile sulfur (which cannot be generated adventitiously). Upon cholate treatment a four- to tenfold increase in detectable acid-labile sulfide is observed in the pellet, accompanied by a concomitant increase of detectable iron (Table 2). The synaptic membrane preparations contain low amounts of flavin and ubiquinone, accounting for about 20 to 50 times less than mitochondrial membranes. If these amounts were due to trace contamination, e.g., by mitochondrial membranes, one would expect to detect significant enzymatic activities for mitochondrial marker enzymes (e.g., cytochrome c oxidase, succinate dehydrogenase) and copper, which is not the case.

F. EPR SPECTROSCOPY

Synaptic membrane preparations tested by EPR spectroscopy at $>$13 K reveal no EPR-detectable signal for either mitochondrial complex I or complex II, in samples either oxidized or reduced by succinate, NADH, or dithionite (in agreement with the absence of mitochondrial contamination found in such preparations by means of activity tests for mitochondrial marker enzymes such as cytochrome oxydase and succinate dehydrogenase), and therefore any observed EPR signal to be described below cannot be originated from contaminating species of known origin, e.g., center S-3 of succinate dehydrogenase. When rat brain synaptic membranes are analyzed by EPR spectroscopy, no signal can be detected, either in untreated samples, or in samples oxidized with ferricyanide, except for a small amount of h.s. Fe^{3+} exhibiting a minor signal at $g = 4.3$. No signal in the $g = 2.0$ region can be observed upon reduction with either NADH, dithionite, or other reducing agents (e.g, succinate). Upon reduction with dithionite, the faint signal at $g = 4.3$ (h.s. Fe^{3+}) develops by a factor of almost two.

Upon treatment of the synaptic membranes with sodium cholate, an isotropic EPR signal centered at $g = 2.012$ appears (Figure 1). The signal exhibits a line-width of 32.5 G and is not EPR-detectable above about 25 K, because of line broadening due to relaxation. Power saturation studies (Figure 1, insert), together with these observations, indicate that the signal is typical for an iron-sulfur cluster of the (3Fe-XS) type. In parallel with the development of this signal upon treatment of the synaptic membranes with cholate, an increase of the signal at $g = 4.3$ is observed, which then accounts for about 25 to 30% of the spin concentration of the $g = 2.01$ species. Upon reduction with sodium dithionite, all signals would vanish from cholate-treated preparations. Attempts to titrate the (Fe-S) center, in order to determine its approximate oxidation-reduction potential, are not satisfactory and show a great lability of the (Fe-S) cluster.

TABLE 2
Chemical Analysis of Purified Rat Brain Synaptic Membrane Preparations and Unmasking of the Fe-S Cluster and Binding Activity upon Sodium Cholate Treatment

Sytnaptic membranes	Fe (nmol/mg protein)	S-labile (nmol/mg protein)	Cu (nmol/mg protein)	Ubiquinone (nmol/mg protein)	Flavin (nmol/mg protein)	Spec. Glu-binding activity (pmol/mg protein)	Spin intensity of the g = 2.01 signal (spin/mg)
Untreated	0.4 ± 0.2	0.62 ± 0.04	0.0	0.120	0.050	5.2 ± 1.0	Undetectable
After cholate treatment	1.9 ± 0.3	3.5 ± 0.8				24.0 ± 0.3	0.95
After cholate treatment and oxidized w. 5 m*M* K_3 (Fe(CN)$_6$)		2.4 ± 0.5				18.0 ± 2	0.68

Note: Purified synaptic plasma membranes were suspended in 50 m*M* tris-HCl pH 7.4, and analyzed for metals, labile sulfide, flavin, and ubiquinone. The values are expressed in nmol per mg protein. Glutamate-binding activity was measured (expressed in nmol bound L-glutamate per mg of protein), together with the spin intensity of the g = 2.01 signal, measured by EPR-spectroscopy at 13 K, standardized with 1 m*M* Cu(ClO_4), run under the same conditions. The membranes were treated with 0.5% sodium cholate and measurements were repeated before and after oxidation with 5 m*M* ferricyanide.

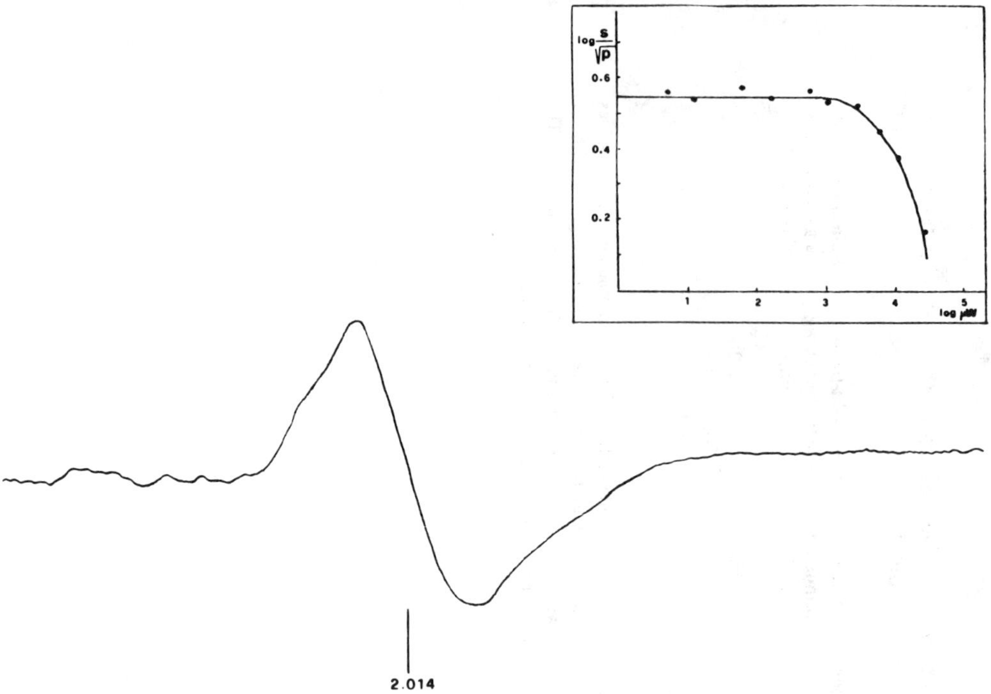

FIGURE 1. EPR spectrum at 12.5 K of cholate-treated synaptic membranes. Purified synaptic membranes were treated with 0.5% cholate in 50 m*M* Tris-HCl pH 8.0 for 45 min, at room temperature, and centrifuged for 30 min at 100,000 × g at 4°C. The pellet, resuspended in 50 m*M* Tris pH 7.4 (15 mg/ml), was placed into an EPR tube and frozen immediately in liquid N_2. Instrument settings were: modulation frequency 110 kHz; modulation amplitude: 2 G; microwave power: 2 mW; field center: 3340 G; scan range: 200 G; temperature: 12.5 K; time constant: 0.3 sec. Insert shows a power saturation curve at 12.5 K. s = signal amplitude at g = 2.01; P = microwave power.

TABLE 3
Effect of Fe Chelators on the Glu-Receptor in Purified Synaptic Membranes

	Na-Azide (m*M*)					*o*-Phenanthroline (m*M*)			
Treatment	**0.0**	**0.08**	**0.4**	**0.8**	**8.0**	**0.0**	**0.08**	**0.4**	**0.8**
Untreated synaptic membranes	100.0	100.0	100.0	67.0	—	100.0	79.0	46.0	42.0
Treated with cholate	100.0	100.0	56.0	12.0	0.0	100.0	31.0	23.0	17.0

Note: Untreated membranes (5 mg protein/ml) or membranes treated with Na cholate (1 mg protein/ml) were incubated in 10 m*M* phosphate buffer pH 7.4 containing 2.5% sucrose and various amounts of *o*-phenanthroline or sodium azide, as indicated. The membranes were tested for their binding activity for L-Glu. Activities are expressed as percentage residual L-Glu-binding activity.

G. ASSOCIATION OF REDOX COMPONENTS TO RECEPTOR PROTEINS

Glutamate-binding activity is detected in synaptic plasma membranes. The membranes exhibit two binding sites for glutamic acid, one with low affinity (K_D = μ*M*) and a second one with high affinity (K_D = 0.675 μ*M*), in excellent agreement with data from the literature.[17,18] Upon cholate treatment, a fivefold increase in the binding activity for L-glutamic acid is observed. This increase parallels the development of the g = 2.01 signal detected by EPR spectroscopy (Table 2). Furthermore, the receptor activity for glutamic acid is sensitive to iron chelators such as *o*-phenanthroline or sodium azide (Table 3). These chelators inhibit the binding capacity of the high affinity receptor in a dose-dependent manner. Upon treatment of the synaptic membranes with sodium cholate, the sensitivity towards metal chelators is even more pronounced and full inhibition of the receptor activity can be achieved at about < 1 m*M* inhibitor. From the above data, an association of, e.g., the glutamine-receptor-like glycoprotein and the Fe-S protein (as has been proposed by Michaelis et al.[19-21]) is quite possible. However, at this stage, no claim for a specific interaction can be made: as a matter of fact it is highly probable that receptor proteins for other neurotransmitters (e.g., L-aspartic acid, dopamine, etc.) would behave alike with respect to solubilization with detergents and inhibition by chelators. We may assume that the observed Fe-S center in synaptic membranes is associated in some manner with an electron transport system and is therefore part of a membraneous NADH-dehydrogenase complex. In such case, these data bring supporting evidence for the close association of the NADH-dehydrogenases and neurotransmitter-receptors (e.g., the receptor for L-glutamic acid).

III. MODULATION OF NADH-DEHYDROGENASES

A. INHIBITION STUDIES

The NADH-acceptor oxidoreductase activity is differently affected by inhibitors, depending upon the acceptor used. Azide has no effects on DCIP-reductase, but significantly inhibits both ferricyanide-reductase or glyoxylate-reductase activity. By contrast, Adriamycin is able to inhibit DCIP-reductase completely but has no effects on other NADH-acceptor oxidoreductase activities, even at high concentrations (Table 4). Fluoride, an inhibitor of NADH-dehydrogenase in liver plasma membranes[2,4] has no effects in synaptic membrane preparations. These membrane dehydrogenases are poorly affected by Atebrin alone (see, however, the more specific effects of Atebrin on NADH-dehydrogenases in the presence of neuroactive agents).

Thiol reagents such as *N*-ethylmaleimide or iodoacetate have no effect on the NADH-acceptor oxidoreductases. Nevertheless, the system is readily inhibited by very low concentrations of *p*-chloromercuribenzoate (PCMB). Glyoxylate-reductase is significantly insensitive towards PCMB compared to the other NADH-acceptor oxidoreductases.

Metal and metal chelators — The importance of metal coordination phenomena in the vesicular binding and storage of catecholamines in synaptosomes has been stressed.[22] Furthermore, transition metals have been used as synaptic blocking agents and interfere with the propagation of action potentials.[23] On the other hand, membrane dehydrogenases are often metalloproteins. This prompted us to investigate the effects of different transition metals on synaptic membrane NADH-acceptor oxidoreductases. However, most bivalent metals tested (Zn^{2+}, Mg^{2+}, Mn^{2+}, Fe^{2+}, and Ca^{2+}) have no effect on the electron transfer rate of NADH-acceptor oxidoreductases. Metal chelators such as EDTA, *o*-phenanthroline, α-picolinate, and nitrilo-triacetate (all tested up to 1 m*M*) are also without effect.

B. CATECHOLAMINES

Phenylalanine metabolites inhibit the electron transfer rates of NADH-acceptor oxidoreductases. As shown in Table 5, the inhibition is mostly observed with DOPA or isopropylnoradrenaline, followed by noradrenaline and dopamine. The inhibition by catecholamines is concentration dependent. On some preparations of synaptic membranes the inhibition is complete with 1 m*M* dopamine. The effects of dopamine antagonists (spiperone) or agonists (bromocriptine) as well as the dopamine-receptor blocker haloperidol have been tested but these agents induce no inhibition on the dehydrogenases. However, apomorphine, a dopamine agonist which labels both the D-2 and D-3 sites of dopamine receptors, effectively inhibits the NADH-dehydrogenases. Catecholaminergic innervation involves α- and β-adrenergic receptors or dopamine receptors.[20] α-receptor-active agents such as phenylephrine, methoxyamine, yohimbine (an α_2-antagonist), or typical α_2-receptor agonists, such as clonidine,

TABLE 4
Selective Modulation of Synaptic Membrane Dehydrogenases by Various Inhibitors (A) and Different Neurotransmitters (B)

(A) Inhibitors

	Concentration (*M*)	Activity NADH-$K_3(Fe(CN)_6)$ reductase	NADH-DCIP reductase	NADH-glyoxylate reductase
Azide	10^{-1}	64.0 ± 8 (40%)	16.0 ± 2 (100%)	70.0 ± 6 (70%)
Atebrin	3×10^{-3}	160.0 ± 10 (100%)	1.5 ± 0.1 (10%)	100.0 ± 9 (100%)
Adriamycin	10^{-5}	160.0 ± 10 (100%)	1.2 ± 0.1 (8%)	100.0 ± 10 (100%)
Adriamycin	10^{-4}	160.0 ± 10 (100%)	3.8 ± 0.2 (25%)	99.0 ± 0.5 (99%)
Fluoride	10^{-3}	160.0 ± 10 (100%)	16.0 ± 2 (100%)	90.0 ± 2 (90%)
p-Chloromercuri-benzoate	5×10^{-6}	80.0 ± 5 (50%)	6.0 ± 1 (48%)	97.0 ± 1 (97%)
Antimycin	10^{-4}	160.0 ± 10 (100%)	13.0 ± 2 (80%)	100.0 ± 8 (100%)
Rotenone	10^{-5}	160.0 ± 10 (100%)	16.0 ± 2 (100%)	100.0 ± 9 (100%)
o-Phenanthroline	10^{-3}	160.0 ± 10 (100%)	16.0 ± 2 (100%)	100.0 ± 9 (100%)

(B) Neurotransmitters

	Concentration (*M*)	Activity NADH-Glyoxylate reductase	NADH-DCIP reductase	NADH-Juglone reductase
Noradrenaline	0.250	95.0 ± 8 (95%)	11.0 ± 1 (60%)	29.0 ± 2 (90%)
Noradrenaline	1.	95.0 ± 8 (95%)	6.5 ± 1 (43%)	22.0 ± 2 (67%)
Dopamine	1.	95.0 ± 8 (95%)	7.0 ± 1 (46%)	22.0 ± 2 (67%)
Glutamate	20.	100.0 ± 9 (100%)	21.0 ± 2 (140%)	38.0 ± 3 (120%)

Note: Purified synaptic membranes were prepared and assayed for the different NADH-dehydrogenase activities in the presence of effectors at various concentration, as indicated. In (B), strong interference of ferricyanide with phenol derivatives prevented assaying for NADH-ferricyanide reductase. Enzymatic activity is expressed in nmol of acceptor reduced per min per mg of protein of synaptic membranes. Inhibition or activation is expressed in brackets as % residual activity (with 100% = activity without effector present).

have no effect (see Table 5). Amphetamine and ''false'' neurotransmitters or catabolites of catecholamines are also without effect on NADH-acceptor oxidoreductases.

C. AMINO ACIDS

NADH-acceptor oxidoreductase activity in synaptic membranes is markedly activated by the acidic amino acids and neurotransmitters, L-glutamic and L-aspartic acids (Figure 2). The effects are concentration dependent. These amino acids are known to produce rapid and reversible depolarization of central neurons resembling the production of excitatory post-synaptic potentials. Antagonists of L-glutamic acid such as kainic acid or 2-amino-5-phosphonovaleric acid or antagonists of L-aspartic acid such as 2-phosphono-4-butyric acid significantly activate the synaptic NADH-dehydrogenases (Table 5). The activation observed with 2-amino-5-phosphonovalerate (among the most potent and specific antagonist for *N*-methyl-D-aspartate-sensitive receptors[24] is relatively low; however, these effects have been tested in the absence of Mg^{2+} ions, while it is well known that Mg^{2+} ions can markedly affect the sensitivity of certain excitatory aminoacid receptors (particularly ''*N*-methyl-D-aspartate-sensitive receptors'').[25]

Other amino acids with action as neurotransmitters such as glycine, taurine, cysteic

TABLE 5
Effect of Neurotransmitters on Synaptic Membrane NADH-Dehydrogenase

Excitatory acidic amino acids 20 m*M*

Aspartic acid	24.0 ± 1 (150%)
Glutamic acid	22.4 ± 1 (140%)
2-Amino-5-phosphonovaleric acid	18.9 ± 1 (118%)
2-Amino-5-phosphonobutyric acid	22.5 ± 1 (140%)
Kainic acid	19.2 ± 1 (120%)

Other Aminoacids 1 m*M*

Glycine	16.0 ± 1 (100%)
Strychnine	16.0 ± 1 (100%)
Muscinol	16.0 ± 1 (100%)
Cysteic acid	16.0 ± 1 (100%)
Homocysteic acid	16.0 ± 1 (100%)

Ion-channel agents 1 m*M*

Veratridine	16.0 ± 1 (100%)
Verapamil	17.3 ± 1 (110%)
Diltiazem	18.5 ± 1 (116%)
Nifedipine	16.8 ± 1 (105%)

Tyrosine metabolites 1 m*M*

None	16.0 ± 1 (100%)
Phenylalanine	15.8 ± 1 (93%)
Tyrosine	14.9 ± 1 (93%)
Tyramine	15.5 ± 1 (98%)
DOPA	3.2 ± 1 (20%)
Dopamine	7.4 ± 1 (46%)
Noradrenaline	6.9 ± 1 (43%)
Adrenaline	10.2 ± 1 (64%)
Isopropyl-noradrenaline	3.2 ± 1 (20%)
3,4-Dihydroxy-mandelic acid	7.3 ± 1 (46%)
3,4-Dihydroxyphenylglycol	9.6 ± 1 (60%)
4-Hydroxy-3-methoxy-mandelic acid	13.6 ± 1 (85%)
Nor-metanephrine	16.0 ± 1 (100%)
3-Methoxy-4-hydroxyphenylglycol	16.0 ± 1 (100%)
3,4-Dimethoxynoradrenaline	16.0 ± 1 (100%)
3-Methoxy-4-hydroxynoradrenaline	16.0 ± 1 (100%)

Cholinergic agents 1 m*M*

Acetylcholine	14.9 ± 1 (93%)
Nicotinic acid	16.8 ± 1 (98%)
Atropine	16.0 ± 1 (100%)
Muscarine	10.6 ± 1 (65%)
Pilocarpine	21.6 ± 1 (135%)
Oxotrenorine	16.0 ± 1 (100%)
Quinuclydinylbenzoate	15.3 ± 1 (96%)

Dopaminergic agents 1 m*M*

Bromocriptine 250 µ*M*	12.5 ± 1 (78%)
Apomorphine	7.0 ± 1 (45%)
Spiperone 250 µ*M*	16.0 ± 1 (100%)
Haloperidol	16.0 ± 1 (100%)
Ephedrine	16.0 ± 1 (100%)
ψ-Ephedrine	15.4 ± 1 (96%)
Methyl-DOPA	16.0 ± 1 (100%)

Peptides

Leu-enkephalin 0.1 m*M*	16.0 ± 1 (100%)
Met-enkephalin 0.1 m*M*	16.0 ± 1 (100%)
Glucagon 0.01 m*M*	16.0 ± 1 (100%)
Insulin 0.01 m*M*	15.2 ± 1 (95%)
Substance P 0.001 m*M*	16.0 ± 1 (100%)
ACTH 0.1 m*M*	16.0 ± 1 (100%)
Triiodo-thyronine 0.1 m*M*	16.0 ± 1 (100%)

Benzodiazepine receptor agents 1 m*M*

Diazepam	16.0 ± 1 (100%)
GABA	16.0 ± 1 (100%)

α- and β-adrenergic agents 1 m*M*

Noradrenaline	6.9 ± 1 (43%)
Clonidine	16.0 ± 1 (100%)
Methoxyamine	14.9 ± 1 (93%)
Phenylephrine	16.0 ± 1 (100%)
Yohimbine	16.0 ± 1 (100%)
Alprenolol	16.0 ± 1 (100%)
Pindolol	16.0 ± 1 (100%)

Various agents 1 m*M*

Adriamycin 0.1 m*M*	12.0 ± 1 (75%)
D-Amphetamine	16.0 ± 1 (100%)
Iproniazid	10.0 ± 1 (56%)
Nitrofurantoin	16.0 ± 1 (100%)
Naloxone	16.0 ± 1 (100%)

Amines and antidepressants 1 m*M*

Serotonin	13.6 ± 1 (86%)
Histamine	16.0 ± 1 (100%)
Tryptamine	16.0 ± 1 (100%)
Desipramine	27.2 ± 1 (170%)
Imipramine	25.6 ± 1 (160%)

Adenosine receptor agents 1 m*M*

Adenosine	16 ± 1 (100%)
Caffeine	16 ± 1 (100%)
Theophylline	16 ± 1 (100%)

Note: All effectors tested were 1 m*M* in the assay mix, unless specified, except for amino acids (20 m*M*). The assay was performed on NADH-DCIP reductase activity. Appropriate controls and blanks (without enzyme) were conducted to exclude unspecific oxidation of the metabolites. Activity is expressed in nmol of acceptor reduced per min per mg of protein and in bracket () as % with reference to NADH-DCIP activity performed in the absence of effector (= 100% activity).

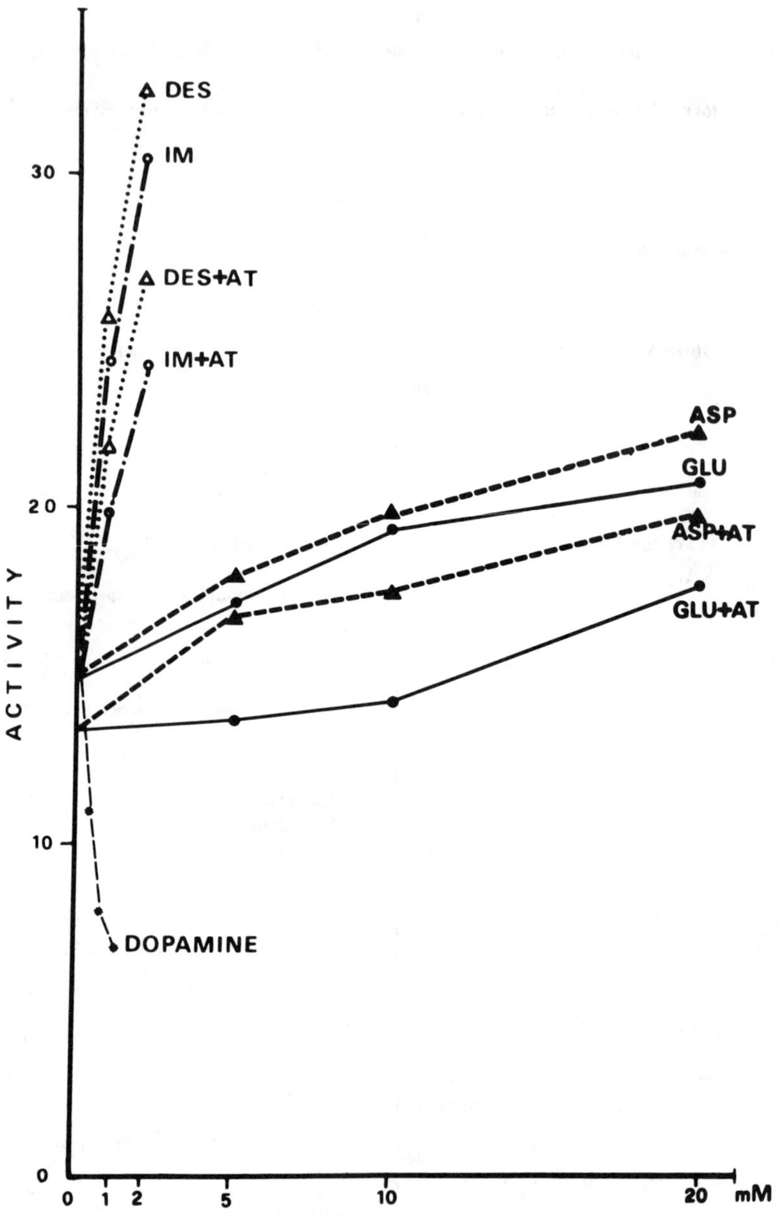

FIGURE 2. Effect of neurotransmitters on NADH-dehydrogenase in purified rat brain synaptic membranes. 200 mg purified synaptic membranes were incubated in 0.1 *M* phosphate buffer containing varying concentrations of the neurotransmitters (dopamine, Asp or Glu) or the antidepressants (imipramine or desipramine) in the presence of Atebrin (3 m*M*) where indicated and the NADH-DCIP reductase activity was measured. Controls to exclude interferences were conducted under identical conditions. All measurements were performed at least in triplicate. The activity is expressed in specific activity per mg of membrane protein per min. The curves correspond to the following agents: (a) (·····△·····), Des: desipramine; (b) (−○−○−), Im: imipramine; (c) (−○−○−), Im + AT: imipramine in the presence of 3 m*M* Atebrin; (d) (·····△·····), Des + AT: desipramine in the presence of 3 m*M* Atebrin; (4) (−●−●−), Glu: L-glutamic acid; (f) (----▲----), Asp: L-aspartic acid; (g) Asp + AT: L-aspartic acid with 3 m*M* Atebrin; (h) Glu + AT: L-glutamic acid with 3 m*M* Atebrin; (i) (− * −), DOPAMINE.

acid, and even γ-aminobutyric acid are without effect on the NADH-acceptor oxidoreductases, even at very high concentrations (up to 20 m*M*). Also diazepam, an agonist of the benzodiazepine receptor, is ineffective.

D. VARIOUS NEUROACTIVE AGENTS

Tricyclic antidepressants such as desipramine and imipramine, markedly stimulate the rates of electron transfer in NADH-acceptor oxidoreductases (Table 5 and Figure 2). The stimulation is comparable in intensity to the effects observed with excitatory amino acids. These agents are also considered as serotonin agonists and compete with the presynaptic serotonin uptake complex.[26] Serotonin, but not other amines (histamine and tyramine), has a small inhibiting effect.

Adenosine which has pronounced effects on nervous tissue function,[27] or antagonists such as alkylxanthines (caffeine or theophylline) are ineffective against NADH-acceptor oxidoreductases, even at very high concentrations. Similarly, NADH-acceptor oxidoreductases are insensitive towards acetylcholine, muscarinic acid, and nicotinic acid, or agonists (oxotremorine), or peptides (enkephalin and derivatives, also agonists of the opiate receptor) or peptide antagonists (such as naloxone). Peptide hormones such as insulin,[28] ACTH,[29] and glucagon,[30] or triiodothyronine have been shown to modulate NADH reduction in liver plasma membranes,[2,3] but have a negligible influence on synaptic NADH-dehydrogenases under comparable experimental conditions. In the same lines, ion channel reagents (veratridine, verapamil, diltiazem, and nifedipine) could not modulate the electron transfer rates of NADH-acceptor oxidoreductases, even at high concentration (up to 1 m*M*).

E. OTHER AGENTS: ATEBRIN

As pointed out above, NADH-acceptor oxidoreductases in synaptic membrane preparations are sensitive to the flavin antagonist Atebrin, by analogy with observations by Crane et al. on other plasma membrane dehydrogenases.[2,3] Atebrin, however, inhibits the stimulatory effects on NADH-acceptor oxidoreductases by either L-Asp or L-Glu or by the antidepressants, imipramine or desipramine (Figure 2).

F. MULTIPLICITY OF NADH-ACCEPTOR OXIDOREDUCTASES

Depending upon the acceptors and the experimental conditions used, variations in the effects of neurotransmitters on NADH-dehydrogenases are observed (Table 4). Glyoxylate-oxidoreductase is markedly less sensitive to acidic amino acids — even at high concentrations — and catecholamines than ferricyanide or DCIP-oxidoreductases. In addition, the juglone-oxidoreductase activity is effectively modulated by neurotransmitters but it is markedly less sensitive than either ferricyanide- or DCIP-oxidoreductases. It is thus very possible that the observed differences reflect the presence of diverse NADH-acceptor oxidoreductases in synaptic membranes.

This notion is supported by several other data. Five different NADH-dehydrogenase activities can be separated by electrophoresis of solubilized synaptic membrane preparations (Figure 3). The purified enzymes display molecular weights of 240,000 to 280,000 Da in their native state. The different enzymes may not all be involved in similar cell processes and this multiplicity may also reflect the heterogeneity of the plasma membrane preparations used (made of a mixture of pre- and postsynaptic membrane vesicles and membranes of synaptosomal origin as well as possibly glial plasma membrane vesicles).

IV. MODULATION OF ADENYLATE CYCLASE

A. INTRODUCTORY REMARKS

Since the discovery by Rall et al.[31] that several putative neurotransmitters stimulate the synthesis of cyclic adenosine monophosphate (cAMP) in cell-containing preparations of

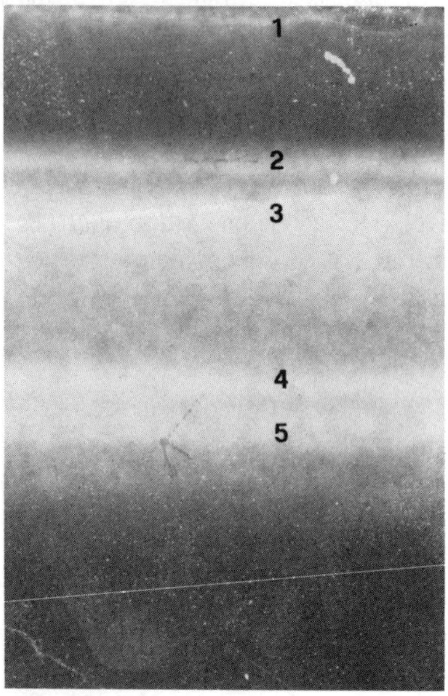

FIGURE 3. Nondissociating polyacrylamide-gel electrophoresis of solubilized synaptic plasma membrane. Purified synaptic membranes were solubilized as in Figure 1 and the extract submitted to electrophoresis and stained for NADH-dehydrogenases activity in 100 m*M* NADH and tetranitroblue tetrazolium in 50 m*M* phosphate buffer, pH 7.4.

nervous tissue, the interest in adenylate cyclase in neural tissues has grown and the significance of the ''second messenger'' concept in synaptic transmission has been stressed.[32] This development has been brought forth by the early observation that the brain is the richest source of adenylate cyclase.[33] Subsequent work has shown that most of the hormone-stimulated cyclase in the brain is located in the postsynaptic membrane[34] making it clear that in many instances cAMP mediates some if not all of the actions of neurotransmitters on their target neurons. Recent studies have shown that cAMP is not only involved in postsynaptic events, but may well be involved in many other neural processes, such as the release of transmitters from presynaptic terminals,[35] the secretion of hypothalamic and pituitary hormones,[36] the control of glial cell metabolism[37] or microtubular function,[38] and sensory transduction in the retina.[39] Noradrenaline, dopamine, and some other neurotransmitters activate brain adenylate cyclase.[40-42] The sensitivity of the adenylate cyclase from various brain regions parallels in general the distribution of the transmitters in those regions. The coupling of the neurotransmitters to adenylate cyclase requires guanosine triphosphate and the transsynaptic formation of cAMP depends on calmodulin.[43] Brain tissue contains both Ca^{2+}-dependent, and Ca^{2+}-independent forms of adenylate cyclase and of the phosphodiesterases that cleave cAMP and cGMP.[44] Protein phosphorylation can effect a number of events in synaptic transmission, particularly the Ca^{2+}-dependent release of noradrenaline from synaptic vesicules,[45] or the expression of desensitization of the nicotinic acetylcholine receptor.[46] Many pieces of evidence suggest a role for cAMP in enhancing presynaptic efficacy.[47] Activation of most receptors negatively coupled to adenylate cyclase inhibits the release of neurotransmitters, and hence inhibition of adenylate cyclase is presumably involved in the inhibitory modulation of neurotransmitter release,[48] unless it reflects other processes.

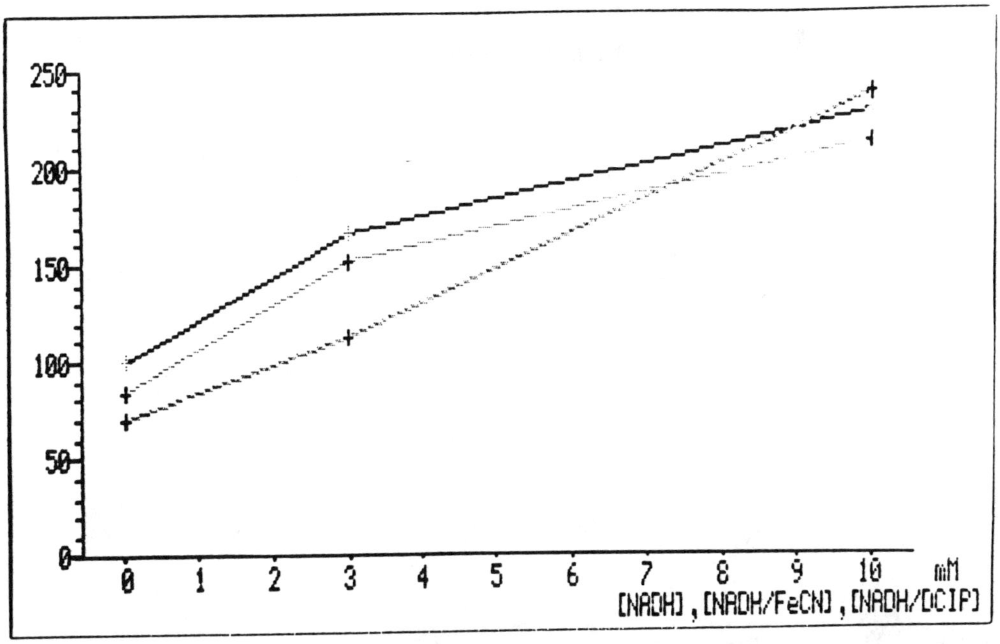

FIGURE 4. Effect of NADH-dehydrogenases on the production of cAMP in purified rat brain synaptic membranes. Purified membrane preparations were incubated in the presence of various concentrations of substrates for NADH-dehydrogenases as indicated (upper line, NADH; middle line, NADH with 0.3 m*M* FeCN; lower line, NADH with 0.3 m*M* DCIP).

B. REDUCING AGENTS AND THE PRODUCTION OF cAMP IN SYNAPTIC MEMBRANES

Synaptic plasma membranes exhibit high levels of adenylate cyclase activity in the absence of stimulation by either drugs or other agents. In the presence of reduced pyridine nucleotides, NADH or NADPH, an increase in the production of cAMP compared to basic standard conditions is observed (Figure 4). The effect is concentration-dependent and results in a 2.4, respectively, 3.6-fold increase in cAMP production in the presence of 10 m*M* NADH or NADPH, respectively. By contrast, NAD or NADP yield only small activation, attributable to the presence of partially reduced products in the chemicals used in these experiments.

Electron donors, oxidizing agents, or substrates for the synaptic membrane NADH-dehydrogenases, such as DCIP or ferricyanide, cause a significant inhibition of cyclase activity in the absence of reduced pyridine nucleotides (Tables 6 and 7). The inhibition is concentration-dependent and is more pronounced with DCIP. As shown, 0.3 m*M* DCIP or ferricyanide result, respectively, in a 30% or 16% inhibition of adenylate cyclase. However, when NADH is added to the same incubation mixture, the observed activation of cyclase activity is restored (Table 7). It is therefore apparent that adenylate cyclase in synaptic membranes is sensitive to the redox environment and that activation of the NADH-dehydrogenase results in a significant increase in the production of cAMP.

NAD added to the incubation in 10 m*M* concentration (Table 6) does not affect the enzyme activity, which stresses the fact that the reduced state of the nucleotide is of importance for the effect. Finally, a drop in the basal activity in the presence of fluoride is observed, which also is overcome by the addition of NADH.

The effect described in Table 6 for synaptic membranes is opposite to observations made under comparable experimental conditions by Loew et al. in adipocyte plasma membranes.[49] These authors have stressed the importance of the oxidation-reduction state of the cell as a possible regulator of the adenylate cyclase activity and have demonstrated an inhibitory

TABLE 6
Influence of NAD(P)$^+$ and NAD(P)H on Adenylate Cyclase Activity in Rat Brain Synaptic Membranes

Pyridine nucleotide	Conc (m*M*)	No agent
—	—	100.0
NADH	1	137.5
	3	171.9
	5	200.0
	10	237.8
NADPH	1	135.6
	3	176.9
	5	269.2
	10	365.3
NAD$^+$	1	102.2
	3	115.6
	10	148.2
NADP$^+$	1	123.9
	3	163.5
	10	262.5

Note: The amount of cAMP is expressed in percent relative to no activation as reference.

TABLE 7
Influence of Oxidation-Reduction Agents on Adenylate Cyclase Activity in Rat Brain Synaptic Membranes

Pyridine nucleotide	Conc (m*M*)	No agent	FeCN (0.3 m*M*)	DCIP (0.5 m*M*)
—	—	100.0	84.8	70.1
NADH	3	166.7	151.8	112.7
	10	227.3	212.4	236.7

Note: The amount of cAMP is expressed in % relative to no activation as reference.

effect of the reduced pyridine nicotinamide on the cyclase activity in these tissues, whereas reduced nicotinamide strongly activate brain adenylate cyclase in synaptic membrane preparations.

C. EFFECTS OF NEUROTRANSMITTERS ON THE DEHYDROGENASE-MEDIATED STIMULATION OF ADENYLATE CYCLASE

The NADH-dehydrogenases in synaptic membranes are modulated by a number of neurotransmitters and agonists. We thus became interested in the effects on the adenylate cyclase achieved by modulating NADH-dehydrogenase activities in the presence of such agents. As these agents are known in some cases to modulate the cyclase themselves, proper controls are required to exclude unspecific effects. The results from these studies are depicted in Tables 8 to 10 and in Figures 5 and 6.

1. Neuroexcitatory Amino Acids (Table 8, Figure 5)

Synaptic membrane NADH-dehydrogenases are mostly activated by the excitatory amino acids Glu and Asp. At lower concentrations, Glu and Asp affect neither basal adenylate cyclase activity nor adenylate cyclase activity under conditions of NADH-dehydrogenase activation. Nevertheless, at higher concentrations of the excitatory amino acids, an inhibition

TABLE 8
Influence of Neurotransmitters on Adenylate Cyclase upon Activation by NADH-Dehydrogenases

Transmitter	Conc (m*M*)	Basal activity	NADH/FeCN	NADH/DC
Glutamic acid	0	100	183.8	191.5
	0.01	104.5		
	0.1	105.7		
	1	110.2	181.2	172.7
	10	75	159.3	135.4
Aspartic acid	0	100	183.8	194.7
	1	108	186.9	184.7
	10	71	145.6	143
Dopamine	0	100	169.3	169.3
	0.1	97	152.2	167.9
	1	100	182.8	163.4
DOPA	0	100	169.3	169.3
	0.1	120.4	156.7	170.1
	1	130.t	161.9	179

Note: 2 mg synaptic membranes were incubated at 37°C for 5 min in 80 m*M* tris-HCl, pH 7.4, 8 m*M* $MgCl_2$, 4 m*M* ATP, 10 m*M* theophylline plus test substances as indicated, and cAMP was assayed after terminating the assay in a boiling water bath for 2 min, followed by centrifugation. cAMP assay of aliquots from the supernatant was performed by means of a double antibody radioimmunoassay (RIANEN™, New England Nuclear Corporation). Stimulation of NADH-dehydrogenase was achieved in the presence of 1 m*M* NADH and either 0.3 m*M* ferricyanide or 0.5 m*M* DCIP, unless otherwise stated. The amount of cAMP is expressed in % relative to no activation as reference.

of basal adenylate cyclase is observed. A similar influence appears upon activation of the NADH-dehydrogenases where both Glu and Asp inhibit the stimulation of adenylate cyclase activity by NADH-dehydrogenases. These effects are observed under various conditions and are reproducible for both ferricyanide-reductase and indophenol-reductase.

2. Catecholamines (Table 8, Figure 6)

Dopamine and DOPA markedly inhibit NADH-dehydrogenase activities. On the other hand, dopamine-sensitive adenylate cyclase has been (largely) described.[50,39] It is noteworthy however that dopamine-sensitivity of adenylate cyclase is easily lost upon purification of brain homogenates.[35,39,50] Preparations of purified synaptic membranes from rat brain cortex display practically no activation of adenylate cyclase. Upon stimulation of adenylate cyclase by the NADH-dehydrogenases, no indication for a significant and clear-cut effect is found. Similarly, DOPA, which induces an enhanced production of cAMP under basal conditions, has no significant effects on the stimulation of adenylate cyclase by NADH-dehydrogenases: the stimulatory effect induced on adenylate cyclase by ferricyanide-reductase is slightly and dose dependently decreased, whereas on the other hand a small increase occurs at high concentrations of DOPA in the case of activation of adenylate cyclase by indophenol-reductase.

3. Dopamine Antagonists and Agonists

A number of observations of the activation of adenylate cyclase, achieved by the activation of NADH-dehydrogenases in the presence of various agonists or antagonists, are summarized in Table 9 and Figures 7 and 8.

Apomorphine, a dopamine agonist, dramatically inhibits NADH-dehydrogenase activity in synaptic membrane preparations. In the absence of substrates for NADH-dehydrogenases,

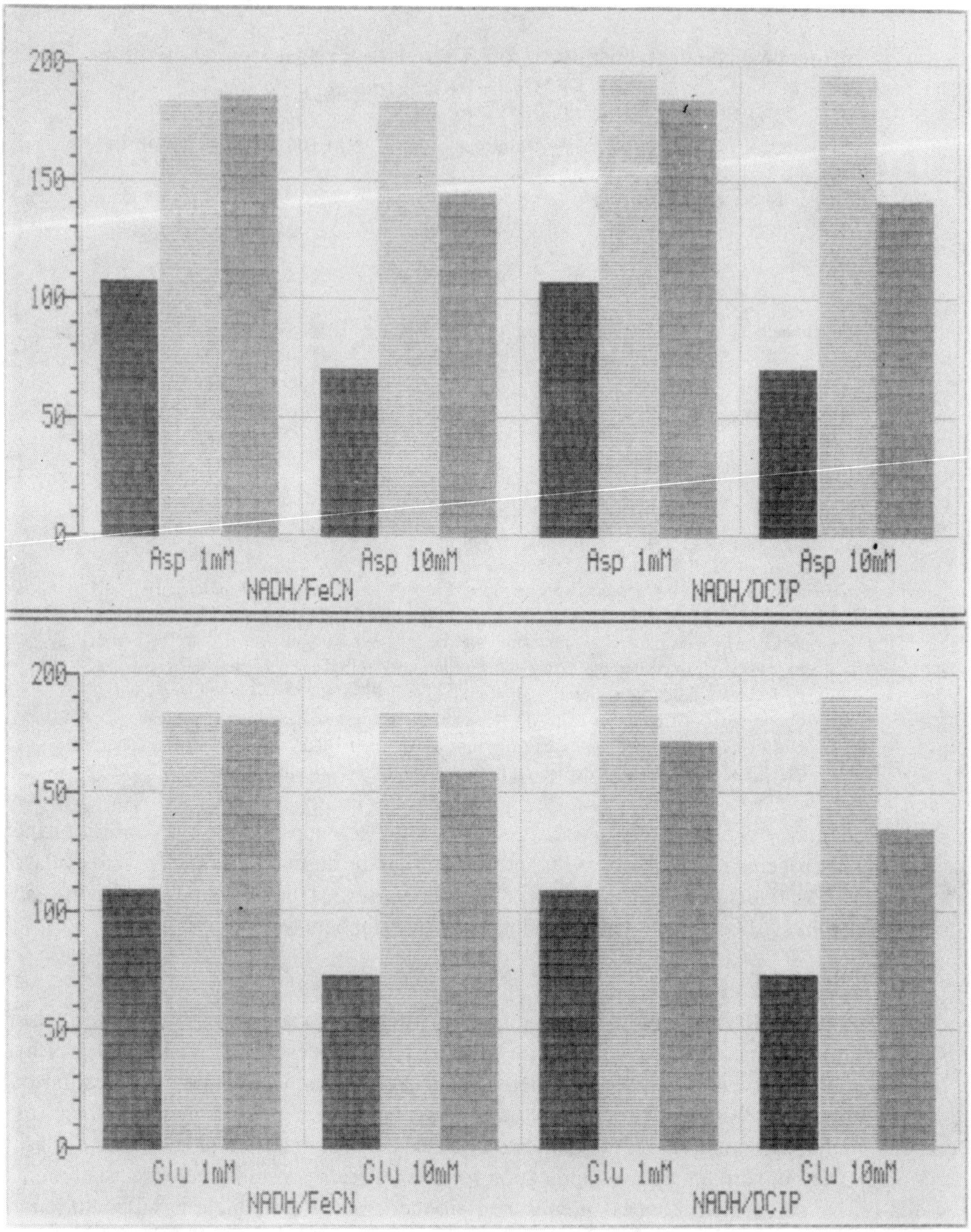

FIGURE 5. Influence of excitatory aminoacids on the activation by NADH-dehydrogenases of synaptic adenylate cyclase. Adenylate cyclase was assayed by incubating synaptic membranes in the presence of substrates for NADH-dehydrogenase (either 1 m*M* NADH and 0.3 m*M* ferricyanide for the activation of the ferricyanide-reductase activity or 1 m*M* NADH and 0.5 m*M* DCIP for the activation of the indophenol-dehydrogenase activity). For each group, the incubations were performed with (right) or without (center) the respective aminoacid upon activation of the NADH-dehydrogenases, whereas control experiments (left) were run in the presence of the respective aminoacid without substrates for the NADH-dehydrogenases. Activities are expressed as percent activation over the control (= 100%).

similarly an inhibition of adenylate cyclase is observed at very low concentrations of the drug ($<10^{-6}$), but the effect is removed by increasing the concentrations of apomorphine. Under conditions where the NADH-ferricyanide reductase is activated, apomorphine potentializes at low concentrations the stimulation of adenylate cyclase by the dehydrogenase.

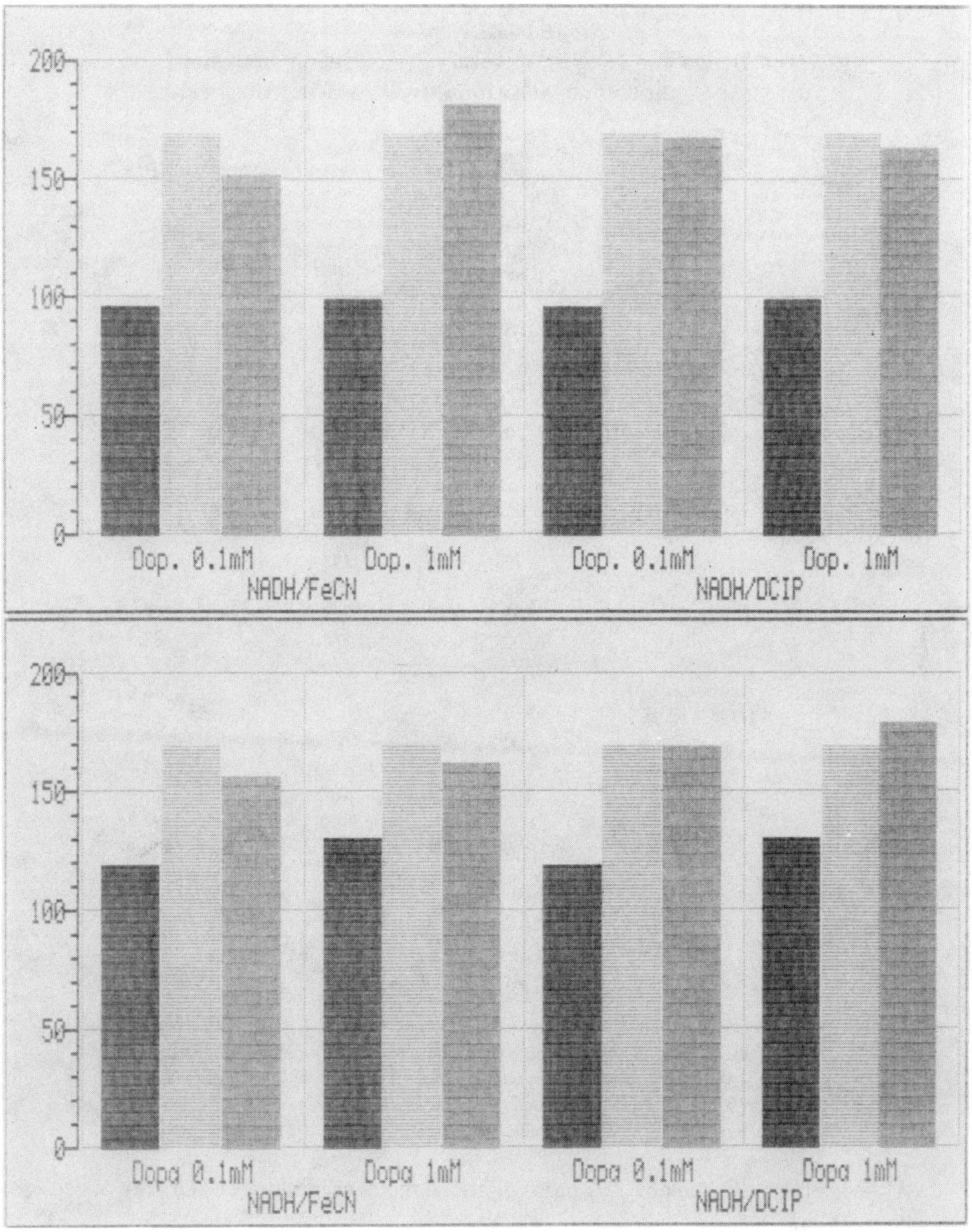

FIGURE 6. Effects of DOPA and Dopamin on the dehydrogenase-mediated activation of synaptic adenylate cyclase. The incubations included for each group either test substance alone (left — concentration as indicated) or substrates for the activation of NADH-dehydrogenases (without test substance) (center) or both test substances and substrates of NADH-dehydrogenases (right). Substrates for dehydrogenases when added included 1 m*M* NADH and 0.3 m*M* ferricyanide, resp. 0.5 m*M* DCIP.

However, no synergistic effect is observed when NADH-indophenol reductase is stimulated; in the latter case, the addition of apomorphine does not further enhance the dehydrogenase-dependent stimulation of adenylate cyclase.

By contrast, the dopamine antagonist bromocryptine stimulates adenylate cyclase (see References 50, 39) also under conditions where the dehydrogenases are not activated. Upon activation of the latter in the presence of bromocryptine, a marked increase in the production

TABLE 9
Effect of Drugs and Neurotransmitters Agonists or Antagonists on Adenylate Cyclase upon Activation by NADH-Dehydrogenases

Agents	Conc (<m*M*>)	Basal activity %	w. NADH/FeCN	w. NADH/DCIP
Pindolol	0	100	185.5	160.5
	0.5	98	218.2	360.5
	5	93.2	198	136.5
	50	95.2	185	130.7
Spiperone	0	100	209.1	189.8
	0.5	79.5	183.6	153
	5	102	193.6	175.4
	50	121.4	199.9	204.5
Apomorphine	0	100	119.6	126.3
	0.5	88.7	149.6	133.9
	5	89.5	160.9	120.3
	50	94.8	131.6	133.9
Bromocryptine	0	100	146.9	156.5
	0.5	170.4	179.1	178.2
	5	168.6	200	182.5
	50	153	217.4	224.3
Clonidine	0	100	183.6	180.1
	1000	99.4	156.2	156
Yohimbine	0	100	186.4	148.6
	10	98.8		
	100	100		
	1000	92.9	162.1	135.1
Chlorpromazine	0	100	180.8	211.6
	100	96.2	166.7	193.7
Naloxone	0	100	253.7	156.6
	0.5	95.6	253.7	156.6
	50	97.8	152.2	133.1
Atebrin	0	100	171.1	139
	1000	78.9		
	3000	58.3	118.1	108.7

Note: The assay conditions were as described in Figure 8. Basal activity refers to control experiments performed in the absence of NADH and either ferricyanide or DCIP. Experiments performed with activated synaptic redox enzymes were performed in the presence of 1 m*M* NADH with either 0.3 m*M* ferricyanide or 0.5 m*M* DCIP. The amount of cAMP is expressed in % relative to no activation as reference.

of cAMP is observed. The effect is proportional to the concentration of the drug added up to about 100 μ*M*.

The effects of pindolol, another antagonist of dopamine and a dopamine-receptor blocker, are more complex. The drug exhibits strong potentializing effects of the dehydrogenase-dependent stimulation of adenylate cyclase at very low concentrations ($< 10^{-6}$ *M*), but inhibits this effect at higher concentrations. The effect is more pronounced on indophenol-reductase compared to ferricyanide-reductase, an observation which may reflect the presence of a variety of dehydrogenases in synaptic membrane preparations. Spiperone, also a known antagonist of dopamine, inhibits adenylate cyclase in the absence of activation of NADH-dehydrogenase and at low concentrations ($<10^{-6}$ *M*), but exhibits no effects at higher concentrations. When the incubation is performed under conditions where NADH-dehydrogenases are activated, an inhibition is observed also at very low concentrations of the drug, but the effect disappears as the concentration of spiperone is raised. The effects observed in these cases appear to be additive, rather than synergistic.

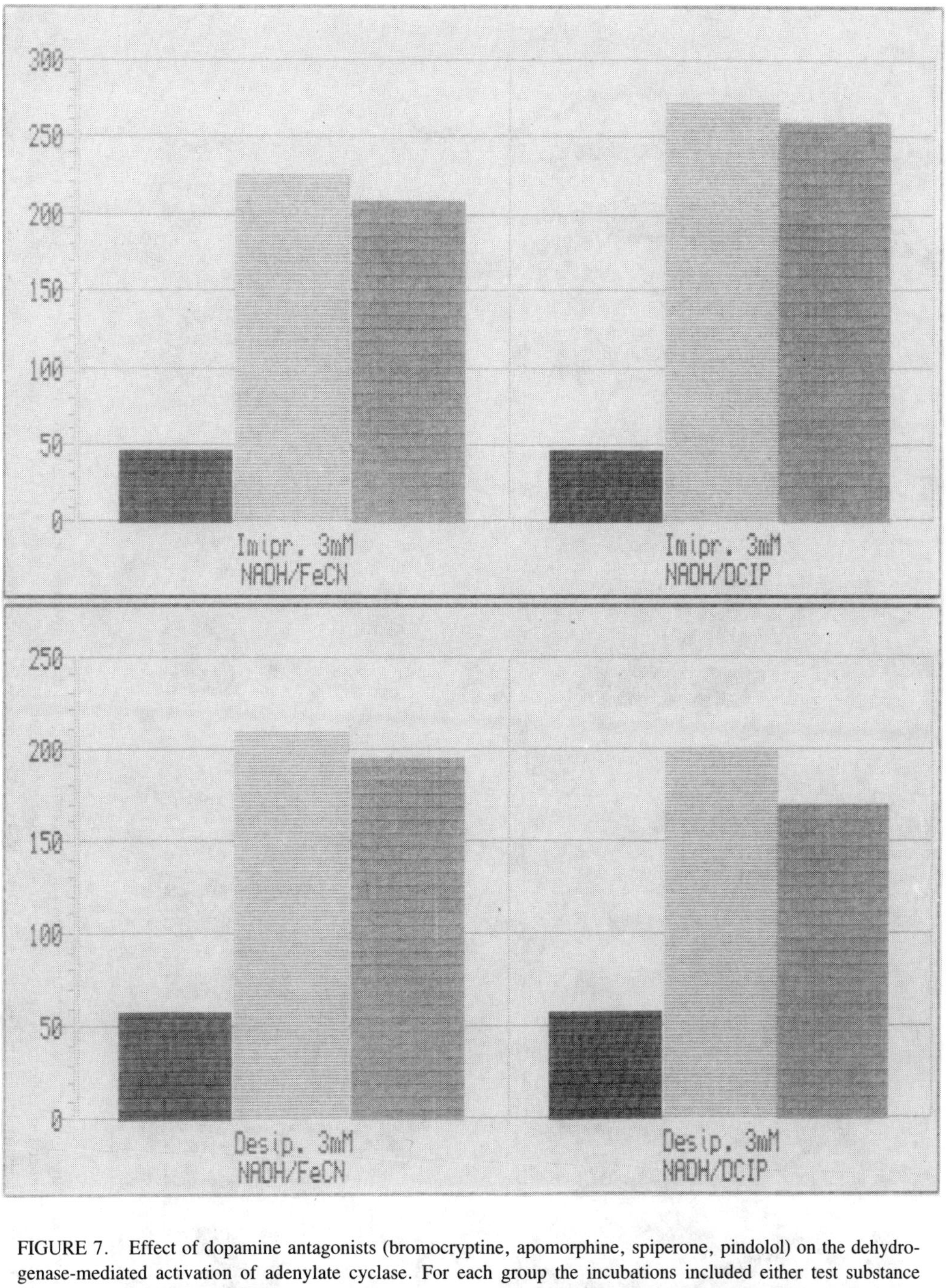

FIGURE 7. Effect of dopamine antagonists (bromocryptine, apomorphine, spiperone, pindolol) on the dehydrogenase-mediated activation of adenylate cyclase. For each group the incubations included either test substance alone (right) or substrates alone (center) for the activation of NADH-dehydrogenases (without test substance) or both substrates and test substances (right). Activation of the NADH-dehydrogenases proceeded in the presence of 1 m*M* NADH and 0.3 m*M* ferricyanide, resp. 0.5 m*M* DCIP. The test substances were included at the concentrations indicated in the figure.

D. OTHER AGENTS

Clonidine and yohimbine, respectively, α_1- and α_2-adrenergic agents which display no direct action on NADH-dehydrogenases, markedly inhibit the stimulation of adenylate cyclase by the NADH-dehydrogenases (Figure 8 and Table 9) although they have practically no effects on the adenylate cyclase in the absence of NADH and electron acceptors. Chlor-

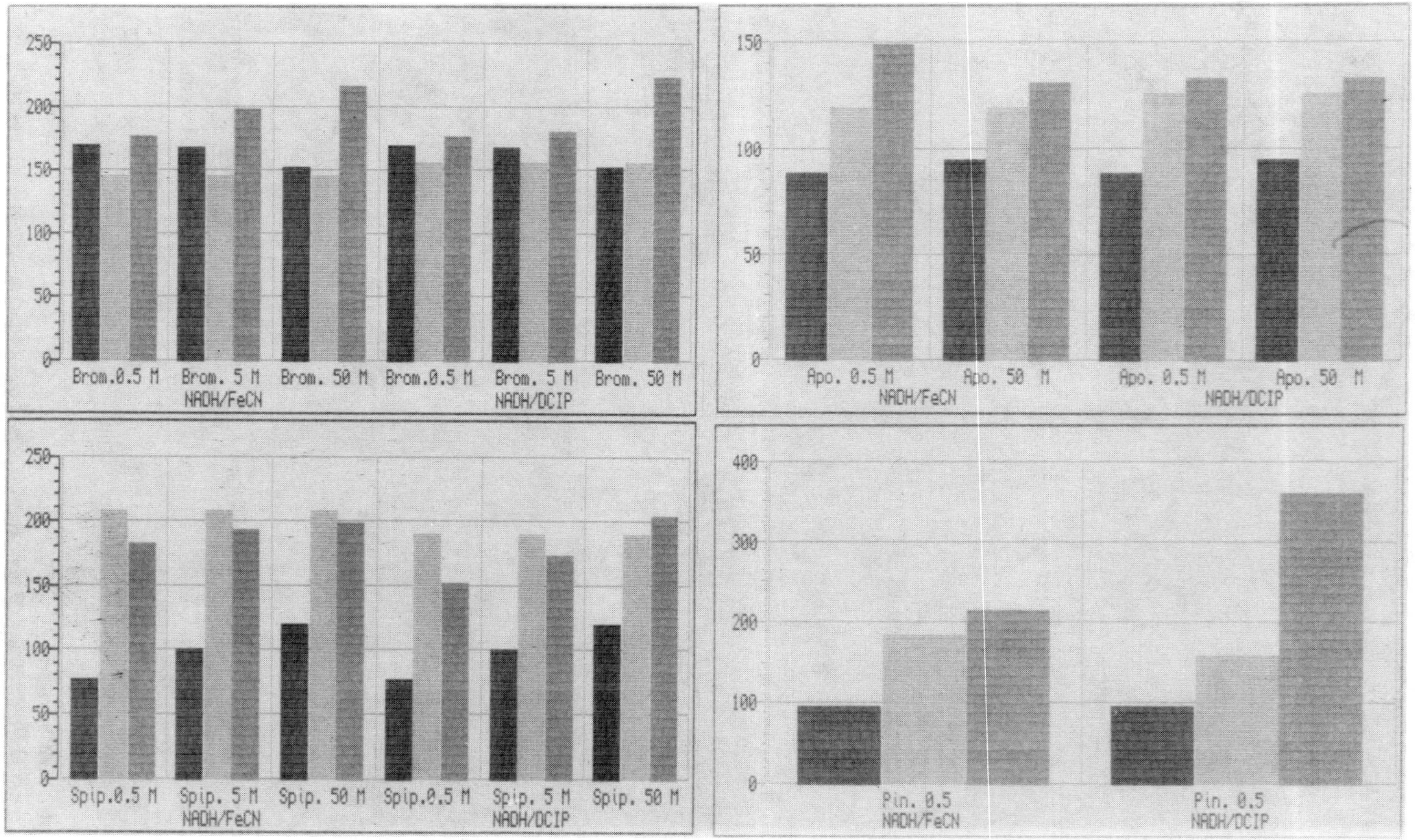

FIGURE 8. Effect of various drugs (clonidine, yohimbine, Atebrin and chlorpromazine) on the dehydrogenase-mediated activation of adenylate cyclase. For each group, the incubations included either test substances alone (right) or substrates alone (center) for the activation of NADH-dehydrogenases (without test substance) or both substrates and test substances (right). Activation of the NADH-dehydrogenases proceeded in the presence of 1 m*M* NADH and 0.3 m*M* ferricyanide, resp. 0.5 m*M* DCIP. The test substances were included at the concentrations indicated in the figure.

TABLE 10
Influence on Adenylate Cyclase of the Inhibition of NADH-Dehydrogenases by Tricyclic Antidepressants

Agent	Basal activity		w. NADH/FeCN	w. NADH/DCIP
Imipramine	0	100	227.4	274.5
	1 m*M*	80.4		
	3 m*M*	48.6	209.7	259.9
Desipramine	0	100	209.6	199.9
	1 m*M*	68.2		
	3 m*M*	59.1	196.5	171.7

Note: The assay of adenylate cyclase was performed as in Figure 8. Basal activity refers to control experiments performed in the absence of NADH and either ferricyanide or DCIP. Experiments performed with activated synaptic redox enzymes were performed in the presence of 1 m*M* NADH with either 0.3 m*M* ferricyanide or 0.5 m*M* DCIP. The amount of cAMP is expressed in percent relative to no activation as reference.

promazine, typically a calmodulin antagonist (but also an inhibitor of protein kinase c and a dopamine antagonist), displays a similar pattern but the effect is less pronounced (Table 9 and Figure 8). A similar effect is found with alprenolol and neurotoxins (results not shown).

Naloxone, a cholinergic agent which is also without influence on the NADH-dehydrogenase activity, strongly inhibits the activation of adenylate cyclase by synaptic NADH-dehydrogenases. The effect is concentration dependent and fully blocks the stimulatory effects by NADH-dehydrogenase on adenylate cyclase at concentrations $>5.10^{-5}$ *M*.

Atebrin, a flavin antagonist, markedly influences the rate of electron transport in synaptic membrane NADH dehydrogenase. Atebrin exhibits a dose-dependent inhibition of adenylate cyclase and fully blocks the dehydrogenase-dependent stimulation of adenylate cyclase (Figure 8 and Table 9).

E. ANTIDEPRESSANTS

The tricyclic antidepressants imipramine and desipramine markedly affect NADH-dehydrogenases in synaptic membranes. These agents inhibit in a dose-dependent way the basal activity of adenylate cyclase, yielding up to about 50% inhibition at 3 m*M* (Table 10, and Figure 9). This effect is overcome upon activating the NADH-dehydrogenases, which yields only a partial decrease in cAMP production. Under these conditions the stimulatory effect of NADH-dehydrogenases on adenylate cyclase is unchanged (or in some cases slightly reduced), and results in a twofold activation of adenylate cyclase.

F. NADH-DEHYDROGENASE AND FORSKOLIN-INDUCED STIMULATION OF ADENYLATE CYCLASE

The hormone-sensitive adenylate cyclase system is composed of several interacting subunits embedded in the lipid bilayer of the plasma membrane; stimulatory and inhibitory hormone or neurotransmitter receptors, acting as external discriminators, the adenylate cyclase catalytic moiety, catalyzing the formation of cAMP from MgATP, and two guanine nucleotide-binding regulatory components N_s and N_i, transducing the signal from the hormone-activated stimulatory and inhibitory receptors, respectively, to the adenylate cyclase.[51,52] Modulation of adenylate cyclase activity by N_s and N_i is associated with changes in the affinity of the adenylate cyclase for its substrate or for its activating cation Mg^{2+}.[53] The adenylate cyclase can furthermore be stimulated by the diterpene forskolin,[54] which can potentiate adenylate cyclase stimulation by hormones, and it has been proposed that the

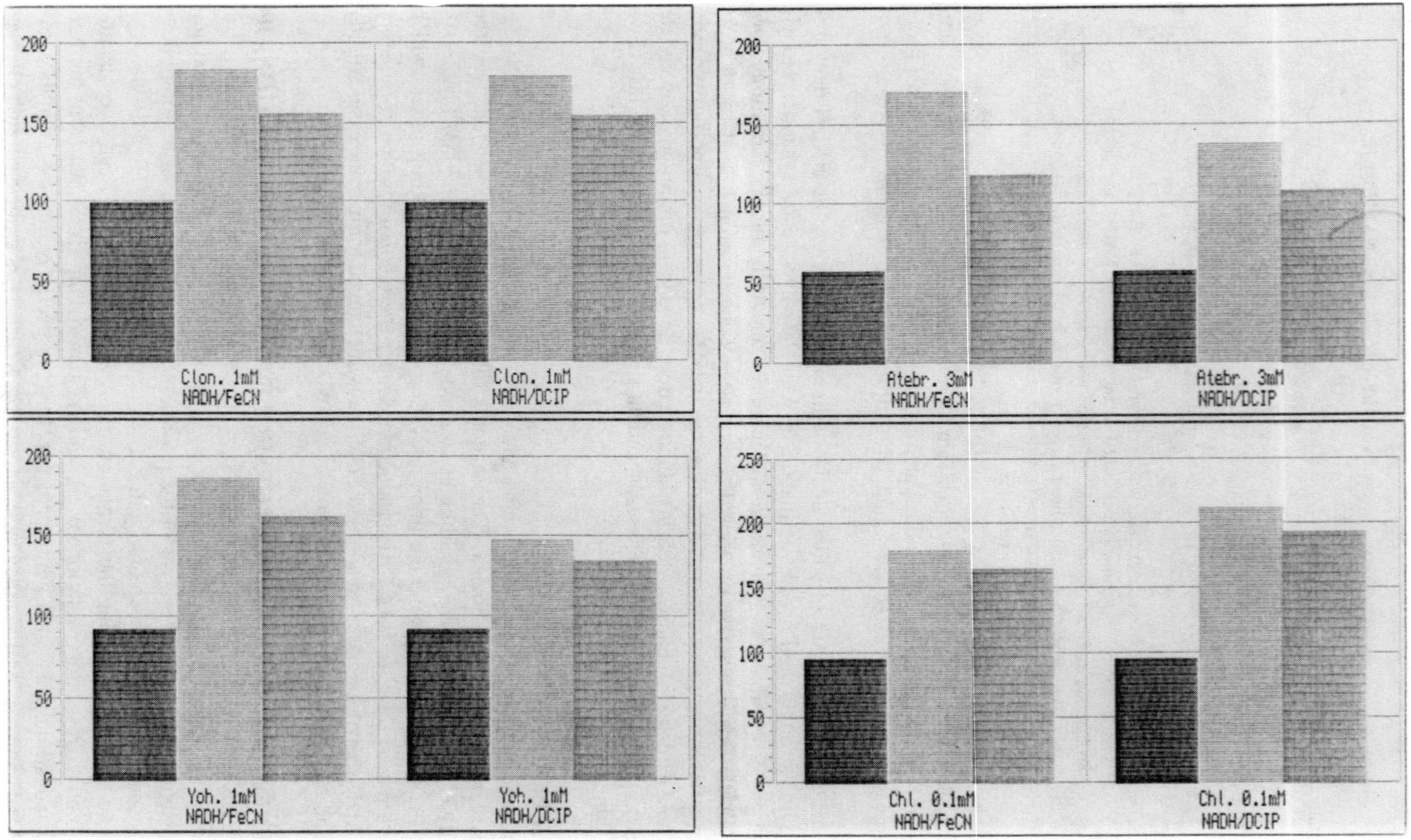

FIGURE 9. Effects of tricyclic anti-depressants on the dehydrogenase-mediated activation of synaptic adenylate cyclase. For each group the incubations included either test substances alone (right) or substrates alone (center) for the activation of NADH-dehydrogenases (without test substance) or both substrates and test substances (right). The activation of the NADH-dehydrogenases proceeded in the presence of 1 m*M* NADH and 0.3 m*M* ferricyanide, resp. 0.5 m*M* DCIP. The test substances were included at the concentrations indicated in the figure.

diterpene has a direct activating effect on the catalytic unit of adenylate cyclase, since the stimulation is observed in the absence of a functional N_s.[55] Forskolin has also been shown to effect the interaction between N_s and the catalytic unit of adenylate cyclase, and appears to stabilize the N_s-activated state of adenylate cyclase.[56] Figure 10A shows the effects of forskolin on adenylate cyclase in the absence and in the presence of stimulation by the NADH-dehydrogenases. In the absence of NADH/ferricyanide or NADH/DCIP, forskolin (at 0.1 m*M*, respectively, 1 m*M*) induces a 2.1, respectively, 2.4-fold activation of adenylate cyclase. However, when adenylate cyclase is stimulated by the NADH-dehydrogenases, i.e., by incubating the synaptic membrane preparations with their respective substrates and in the presence of added forskolin (either 0.1 m*M* or 1 m*M*), a synergistic production of cAMP is observed. This results in a 3.1, respectively, 4.5-fold increase in cAMP production when the ferricyanide-reductase has been activated and in a 2.5, respectively, 3.9-fold increase, when the DCIP-reductase is stimulated. Figure 10B shows similar effects observed when GTP is added to the incubation mixture: GTP which alone is ineffective on adenylate cyclase, enhances the stimulation by forskolin; furthermore, the activation of NADH-dehydrogenases markedly potentializes the latter effects (Figure 10B, b and c).

G. ADENYLATE CYCLASE AND THE REDOX STATE

The finding that the adenylate cyclase is sensitive to the redox state of the environment opens a new line of investigation on the control mechanism of neurotransmitter action. The effect is specific and enzyme mediated and not an unspecific artifact. Although the NADH concentration used might seem to be high, the concentration of NADH + NAD^+ which has been observed in the cytosol is in the same milimolar range. As a control responsive to the metabolic state of the cell, mediation by NADH seems to be ideal.

Efforts to ascribe particular brain functions to individual neurotransmitters have been hampered by the existence of 50 to 100 putative transmitters, each of which acts through several receptor subtypes.[57] One group of neurotransmitters as exemplified by the excitatory and inhibitory amino acids, acts directly to open or close specific ion channels with rapid synaptic effects whereas by contrast biogenic amines and neuropeptides act through biochemical second messenger systems to produce more subtle synaptic modulation.[58,59] Two major messenger systems, the adenylate cyclase and phosphatidyl-inositol systems, mediate actions of these transmitters. Information of the modulation may clarify the role of a whole class of receptors linked to each second messenger system. cAMP accumulation in rat brain cerebral cortex is involved in signal response processes for several receptors, e.g., α-adrenergic receptors,[60] dopamine-receptors,[50] and nicotinic acetylcholine receptors.[61]

Multiple interactions between these second messengers have been reported,[62-64] and mutual regulation is inferred. On the other hand, there are many examples of the modulation of a receptor response in a cell by activation of a second receptor.[65-70] Such mechanisms are of obvious importance in the integrative and discriminative function of the nervous system. Most of the interactions that have been described have been found in whole cell systems and involve an interaction between the second (or subsequent) messenger of one receptor system and the receptor-response of a second system.

V. MODULATION OF Ca^{2+} TRANSPORT

In synaptic plasma membranes the ATP-dependent Mg-dependent Ca^{2+} transport located at the presynaptic plasma membrane has been investigated in more detail. The transport system is very sensitive to reducing agents and can be blocked by NADH (IC_{50} about 8 m*M*) or NADPH (IC_{50} about 3 m*M*). The Ca^{2+} transport system is also inhibited by oxidized NAD and NADP, indicating that the effect is more due to the nicotinamide adenine dinucleotide molecule self, rather than to its redox state (Table 11). Oxidizing agents such as

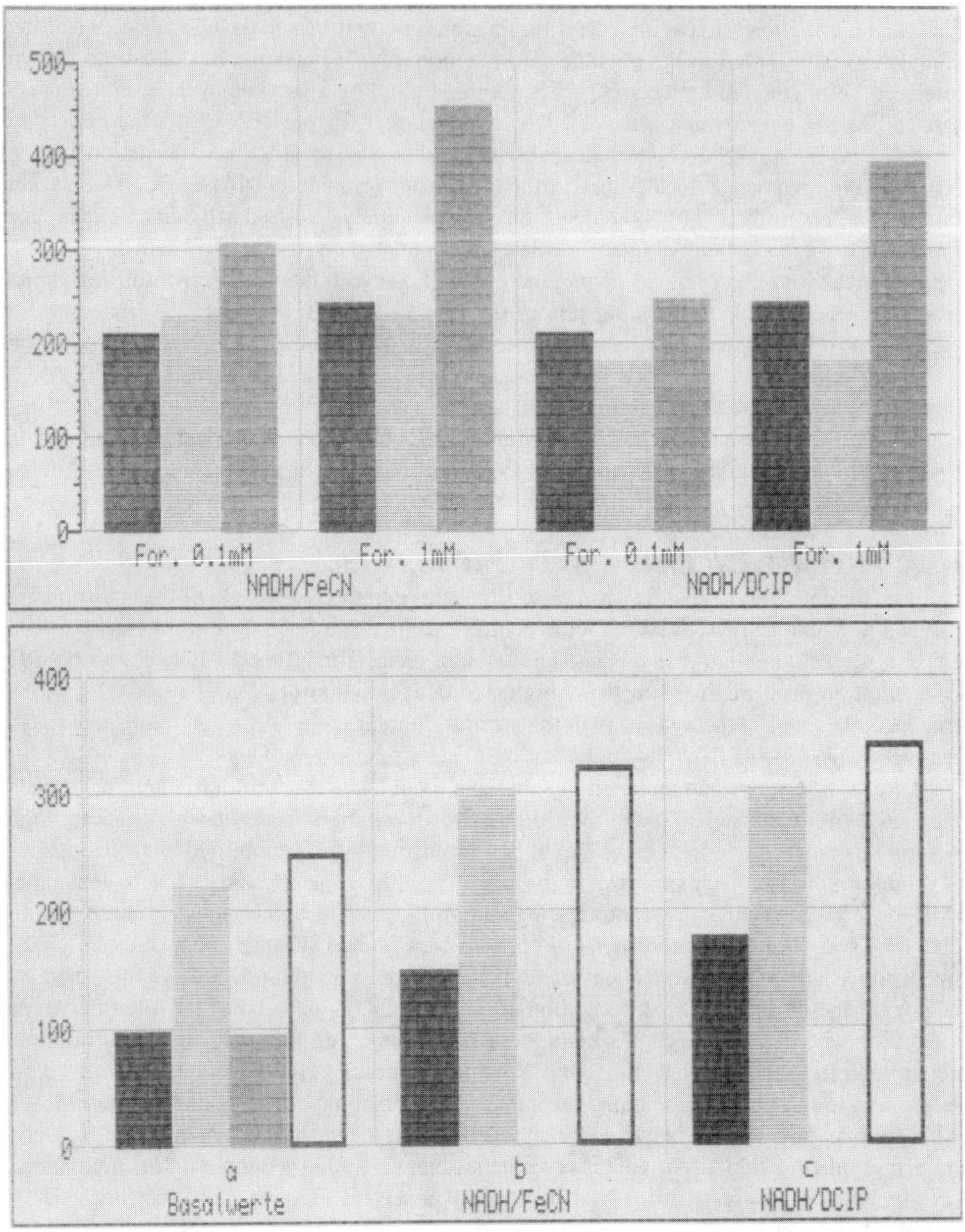

FIGURE 10. Effect of NADH-dehydrogenase on the forskolin-mediated activation of adenylate cyclase. (A, top) Synaptic membranes were incubated in the presence of either 0.1 m*M* or 1 m*M* forskolin without further additions (left) or under addition (right) of substrates for NADH-dehydrogenases (i.e., 1 m*M* NADH and either 0.3 m*M* ferricyanide or 0.5 m*M* DCIP). After incubation, the production of cAMP was measured. In addition, for each group, control experiments (center) were run in the absence of forskolin, but upon activation of NADH-dehydrogenases. The activation of the NADH-dehydrogenases proceeded with 1 m*M* NADH and 0.3 m*M* ferricyanide, resp. 0.5 m*M* DCIP in the incubation medium, where necessary. (B, bottom) For each group of assays, synaptic membranes were incubated either alone (left), in the presence of forskolin (center-left), or in the presence of 0.1 m*M* GTP (center-right) or with both forskolin and GTP (right). These experiments were performed either in the absence of activation of NADH-dehydrogenase (a) or under activating conditions in the presence of substrates for NADH-ferricyanide reductase, i.e., 1 m*M* NADH and 0.3 m*M* ferricyanide (b) or substrates for NADH-indophenol reductase, i.e., 1 m*M* NADH and 0.5 m*M* DCIP (c).

TABLE 11
Effect of NAD(P)$^+$ and NAD(P)H and DCIP on the Mg$^+$- and ATP-Dependent Ca^{2+} Transport in Synaptic Membranes

	Conc (m*M*)	% Residual Ca^{2+} transported
NADH	0	100
	1	88
	5	62
	10	48
	20	27
NAD	0	100
	1	75
	5	10
	10	7
NADPH	0	100
	5	24
	10	7
NADP	0	100
	5	5
	10	0
DCIP	0	100
	0.1	79
	0.5	14

Note: Synaptic plasma membranes were incubated in 5 m*M* tris, pH 7.4 with 0.1 *M* KCl, 25 μg oligomycin, 0.1 m*M* ouabain, 0.1 m*M* $MgCl_2$, 1 μ*M* isotopically diluted $^{45}CaCl_2$ (free), 0.2 *M* *trans*-cyclo-1,2-diamine-*N,N,N′,N′*-tetraacetic acid and 0.1 m*M* ATP. Transport activity was initiated by addition of ATP and allowed to proceed for 120 s. The activity was terminated by addition of an ice-cold solution containing 150 m*M* KCl, 10 m*M* tris, 1 m*M* $LaCl_3$ to inhibit Ca^{2+} efflux, and filtration through cellulose nitrate filters. The data indicate % residual Ca^{2+} transport activity as compared to 100% activity measured in the absence of test substance.

DCIP also completely inhibit the ATP-dependent Ca^{2+} transport system, whereas the effects by ferricyanide are less conclusive.

The dramatic changes induced by the substrates of plasma membrane dehydrogenases on Ca^{2+} transport may be of great physiological significance, considering the function of Ca^{2+} in cell activation processes in general and particularly with respect of transmitter release. Nerve cells have become specialized to conduct signals in the form of electrical impulses. The basis for this is ion currents through the cell membrane driven by a periodically formed electric potential difference between the inside and the outside of the cell. The role of Ca^{2+} on the activation of specific K channels of the plasma membrane has been extensively documented. It has been often postulated that other factors related to the cell metabolic state could modify the sensitivity to Ca^{2+} of Ca^{2+}-dependent K^+-channels, but direct evidence was never obtained.[71] In the last few years, evidence has been produced that the sensitivity to Ca^{2+} of the K^+-channels can be modified by changes of the redox state of a membrane component.[72] Artificial electron donor systems, such as ascorbate plus phenazine methosulfate, can increase the sensitivity of Ca^{2+} of these Ca^{2+}-dependent K^+-channels in a variety of animal cells. In erythrocytes, the shift from the low to the high-affinity state depends on the reduction of a membrane component accepting two electrons and with a standard redox potential of about 47 mV.[73] The individual K^+-channels show, in the reduced state, both a decreased threshold for Ca^{2+} activation and an increased apparent permeability for Rb. In intact cells, whose $NAD^+/NADH$ quotient was modified by incubation with pyruvate, lactate and xylitol, the apparent sensitivity to Ca^{2+} of Ca-dependent K^+-transport is increased in the reduced state, suggesting that this redox modulation can be relevant under physiological conditions.[74] Similar effects have been reported in the Ehrlich cell,[75,76] human lymphocytes, rat thymocytes,[77] and guinea pig hepatocytes.[77]

VI. MODULATION OF RECEPTOR STATES BY OXIDATION-REDUCTION PROCESSES OF THE SYNAPTIC PLASMA MEMBRANES

The importance of oxidation-reduction processes at the level of the synaptic plasma membrane is well documented. Strong perturbations of postsynaptic responses at cholinergic synapses upon incubation with reducing agents such as thiol agents are known. The importance of these processes in postsynaptic responses raises the question of the implication of a NAD(P)-protein disulfide reductase in the activation-deactivation of cellular signaling in excitable membranes. It is known that the reduction of disulfide bond in general brings about alterations in the physiological response of excitable membranes.

A. CHOLINERGIC SYNAPSES

As an example, reduction of the living electroplax with DTT inhibits the permeability response of acetylcholine.[78] Subsequent alkylation of the reduced membranes renders the inhibition irreversible, while oxidation restores the initial sensitivity to agonists.[79] This reflects the complex regulation of the redox balance in the living excitable membrane.[80] Such oxidation-reduction processes influence various receptor functional characteristics, among which one can specify the state of receptor aggregation or receptor clustering, surface charge properties, and ligand binding phenomena.[81]

Exposure of the membrane-bound acetylcholine receptor to certain cholinergic ligands, most notably with typical agonists of the nicotinic system, is known to cause a concentration- and time-dependent increase in the affinity of the receptor for such ligands.[82-91] Oxidation or reduction gives rise to changes in receptor affinities, in agonist-induced transitions and in the physical state of the receptor itself. The changes are dependent on the nature of the ligand and on the thiol modification involved. It has been suggested that in the acetylcholine receptor, the disulfide bond in the vicinity of the receptor site is affected by oxidation-reduction treatments and is mainly responsible for the altered ligand discrimination and the formation of modified receptor states.[82]

In a number of cases, the cooperativity of ligand binding is lost upon reduction. Furthermore the Na^+-efflux from receptor-rich membrane vesicles remains normal after treatment with reducing agents, but the dissociation constant for acetylcholine increases tenfold.[92]

In addition to the alterations of affinity states of the acetylcholine receptor and transitions thereof, the reductive cleavage of certain disulfide bonds in the receptor results in specific changes of its recognition capacity toward ligands, and therefore oxidation-reduction of the acetylcholine receptor is a key factor in the cholinergic ligand discrimination.[93]

Interconversion of high- and low-affinity sites by means of reduction has also been demonstrated in the case of muscarinic acetylcholine receptors for agonist and antagonist binding. The changes in affinity for muscarinic ligands does not result from interaction between receptors and GTP-binding proteins. In the presence of reducing agents the affinity for muscarinic ligands is decreased whereas preincubation with DTNB resulted in an increase in the proportion of high-affinity binding sites for muscarinic agonists to 75% of the total number of ligand-binding sites. Thus the redox state of thiol groups in the receptor molecules is relevant to their affinities for ligands.[94]

B. BETA-ADRENERGIC RECEPTORS

Oxidation-reduction of beta-adrenergic receptors causes intramolecular modification and association-disassociation which cause functional activation of the receptor. Agonists such as isoproterenol as well as other reducing agents, including thiol reagents increase the receptor-promoted activation of stimulatory GTP-binding protein.[95]

Similar reports have been made for other receptors (α_1-receptors,[96] excitatory amino acids,[97] opiate receptors,[98,99] etc.

C. RELATED REDOX PROCESSES

The effects of micro-electrophoretic application of oxidizing and reducing agents at both the endplate region and extrasynaptic sites of neuromuscular preparation have been studied, by means of iontophoretically applying reducing and oxidizing sulfhydryl agents in situ.[100] Both oxidizing and reducing agents elicit depolarizations only when applied at the endplate region. These studies which suggests the presence of reactive sites within the receptor complex, could particularly be involved with the activation of the α-subunit of the acetylcholine receptor.

A number of food dyes are known to inhibit neurotransmitters or neurotransmitter precursors, by means of oxidation-reduction in brain membranes. These indications of operating oxidation-reduction processes at the level of the synaptic plasma membrane have been recently presented by elucidation of the mechanisms of action of Erythrosine B, a dye implicated in hyperkinesis and learning disabilities among children and an inhibitor of dopamine transport into rat caudate synaptosomes.[101] Erythrosine B propagates oxidation-reduction events occurring outside brain membranes into the interior of the membrane bilayer. Spin-label experiments demonstrated that the nitroxide group of membrane spin-labels are reduced via an Erythrosine B transient species permeable into the membrane bilayer, most likely a free radical of the dye. The dye interacts with neuromal membranes and neuromuscular membranes, increasing the resting membrane potential by increasing the potassium conductance.

VII. CONCLUDING REMARKS

Purified rat brain synaptic membrane preparations display NADH-dehydrogenase activity presumably associated with redox components in the plasma membrane. This redox activity is part of an electron transport system in the plasma membrane which is inhibited by azide, but not by antimycin at low concentration or by rotenone, so it would not be attributed to mitochondrial inner membranes. Several NADH-dehydrogenases appear to be operative at the synaptic plasma membranes since three to five different enzyme activities can be detected and separated electrophoretically, depending upon the procedure used for the preparation of the synaptic membranes. These observations are corroborated by inhibition studies which speak in favor of the presence of a variety of oxidoreductases in the same preparations, and also by comparative solubilization studies of synaptic membrane proteins where a close association of NADH-dehydrogenases and other receptor proteins is demonstrated. The NADH-dehydrogenases and, e.g., the high affinity binding protein-receptor for glutamic acid are inhibited or solubilized under identical conditions. Electrophoretic data also support this notion.

The real function of these transplasma-membrane NADH-dehydrogenases remains to be established. It is already clear that this redox activity in synaptic plasma membranes is linked to postreceptor signal transduction phenomena. It dramatically modulates adenylate-cyclase-dependent processes and is influenced by a number of neurotransmitters and analogs. It is clearly related to dopamine, noradrenaline, and excitatory amino acid receptor functions.

It is widely accepted that neurotransmission causes transient membrane conductance for

Na^+, K^+ and Ca^{2+} in vertebrate neurons which lead to electrical depolarization of these cells. The stimulation of cation efflux by neurotransmitters (e.g., L-Glu) is assumed to be an electrogenic process bringing about a change in the membrane electrical potential. The involvement of NADH-dehydrogenases in electrogenic processes associated with the function of neurotransmitters (e.g., dopamine or excitatory amino acids, such as L-glutamic acid) in neurotransmission is possible but requires further investigations. For many cellular processes, mainly cellular signaling, a control by the redox state of NADH seems ideal, but does not exclude other control mechanisms. At this stage it is presumed that NADH-acceptor oxidoreductases in synaptic membranes are involved in some manner in the mechanism of postreceptor signal transduction. They appear to be selectively associated with catecholamine- and excitatory amino acid-receptors. Finally, some function in membrane depolarization processes (e.g., linked to H^+ transport systems) should also be considered.

However, it does not seem possible at this stage to assign the NADH-dehydrogenase activities to specific synaptic receptors. For example the sensitivity of NADH-acceptor oxidoreductase towards dopamine and norepinephrine is about the same, whereas dopamine receptors coupled to adenylate cyclase usually respond to dopamine better than to norepinephrine. The situation is also complicated by the fact that dopamine-receptor blockers of the butyrophenone type (haloperidol) have no effect, in contrast to the usual behavior of adenylate-cyclase-dependent dopamine receptors.

The question also remains whether these dehydrogenases are pre- or postsynaptic membrane proteins. The following drugs involve presynaptic receptors: (1) tricyclic antidepressants, (blockers of presynaptic α_2-adrenoceptors in the central nervous system); (2) apomorphine (preferentially presynaptic dopamine receptor agonist); (3) methyl-DOPA. This would speak for the possibility that presynaptic receptors are involved, at least partly, in the mechanism of action of NADH-acceptor oxidoreductase. On the other hand, an interaction of NADH-acceptor oxidoreductase with D-2 receptors is also likely, and it is known that D-2 receptors in the striatum are postsynaptic to dopamine terminals. The nature of the preparations, mainly the relative contamination by plasma membrane of glial cells, also complicates the investigations relative to the functions of the dehydrogenases in nerve cells.

REFERENCES

1. **Loew, H. and Crane, F. L.,** Redox functions in plasma membranes, *Biochim. Biophys. Acta,* 515, 141, 1978.
2. **Crane, F. L., Goldenberg, H., Morré, J. D., and Loew, H.,** Dehydrogenases of the plasma membranes, in *Subcellular Biochemistry,* Vol. 6, Roodyn, D. B., Ed., Plenum Press, New York, 1979, 345.
3. **Crane, F. L., Sun, I. L., Clark, M. G., Grebing, C., and Loew, H.,** Transplasma membrane redox systems in growth and development, *Biochim. Biophys. Acta,* 811, 233, 1985.
4. **Barr, R., Craig, T. A., and Crane, F. A.,** Transmembrane Ferricyanide reduction in carrot cells, *Biochim. Biophys. Acta,* 812, 49, 1985.
5. **Sun, I. L., Crane, F. L., Grebing, C., and Loew, H.,** Transmembrane redox in control of cell growth, *Exp. Cell Res.,* 156, 528, 1985.
6. **Bienfait, H. F.,** Regulated redox processes at the plasmalemma of plant root cells and their function in iron uptake, *J. Bioenerg. Biomemb.,* 17, 73, 1985.
7. **Segal, A. W., Croos, A. R., Garcia, R. C., Borregaard, N., Valerius, N. H., Soothill, J. F., and Jones, O. T. G.,** Absence of cytochrome b_{-245} in chronic granulomatous disease, *N. Engl. J. Med.,* 308, 245, 1983.
8. **Loew, H. and Werner, S.,** Effects of reducing and oxidizing agents on the adenylate cyclase activity in adipocyte plasma membrane, *FEBS Lett.,* 65, 96, 1976.
9. **Mukherjee, S. P. and Mukherjee, C.,** Role of sulfhydryl oxidation in adipocyte plasma membrane surface in the response of adenylate cyclase to isoproterenol and glucagon, *Biochim. Biophys. Acta,* 677, 339, 1981.

10. **Neumann, E. and Bernhardt, J.,** Physiological chemistry of excitable membranes, *Annu. Rev. Biochem.*, 46, 117, 1977.
11. **Schmalzing, G.,** Mechanisms of depolarization of rat cortical synaptosomes at submicrosomal external Ca^{2+} activity, *Biochem. J.*, 225, 671, 1985.
12. **Cotman, C. W. and Mattews, D. A.,** Synaptic plasma membranes from rat brain synaptosomes. Isolation and partial characterization, *Biochim. Biophys. Acta,* 249, 380, 1971.
13. **Albertsson, P. A., Andersson, B., Larsson, C., and Akerlund, H. E.,** Phase partition — a method for purification and analysis of cell organelles and membrane vesicles, *Meth. Biochem. Anal.*, 28, 115, 1982.
14. **Halliwell, B. and Gutteridge, M. C.,** Oxygen radicals and the nervous system, *Trends Neurosci.*, 8, 22, 1985.
15. **Zamudio, I., Cellino, N., and Canessa, F. M.,** The relation between membrane structure and NADH:(acceptor) oxidoreductase activity in erythrocyte ghosts, *Arch. Biochem. Biophys.*, 129, 336, 1969.
16. **Nisimoto, Y., Wilson, E., Heyl, B. L., and Lambeth, J. D.,** NADH-dehydrogenase from bovine neutrophil membranes. Purification and properties, *J. Biol. Chem.*, 261, 285, 1986.
17. **Kanner, B. I. and Sharon, I.,** Active transport of L-glutamate by membrane vesicles isolated from rat brain, *Biochemistry,* 17, 3949, 1978.
18. **Logan, W. J. and Snyder, S. H.,** High affinity uptake systems for glycine, glutamic and aspartic acid in synaptosomes of rat central nervous tissue, *Brain Res.*, 42, 413, 1972.
19. **Michaelis, E. K.,** The glutamate receptor-like protein of brain synaptic membranes is a metalloprotein, *Biochem. Biophys. Res. Commun.*, 7, 106, 1979.
20. **Michaelis, E. K., Michaelis, N. N., and Boyarski, L. L.,** High affinity glutamic acid binding to brain synaptic membranes, *Biochim. Biophys. Acta,* 367, 338, 1974.
21. **Michaelis, E. K., Michaelis, M. L., Chang, H. H., Grubbs, R. D., and Kuonen, D. R.,** Molecular characterization of glutamate receptors in the mammalian brain, *Mol. Cell. Biochem.*, 38, 163, 1981.
22. **Birdsall, N. J. M.,** Can different receptors interact directly with each other?, *Trends Neurosci.*, 5, 137, 1982.
23. **Cotman, C. W. and Taylor, D.,** Isolation and structural studies on synaptic complexes from rat brain, *J. Cell. Biol.*, 55, 696, 1972.
24. **Creese, I.,** Dopamine receptors explained, *Trends Neurosci.*, 5, 40, 1982.
25. **DeFeudis, F. V.,** Aminoacid neurotransmitters, *Trends Neurosci.*, 4, XIII, 1981.
26. **Langer, S. Z. and Briley, H.,** High affinity ^3H-imipramine binding: a new tool for studies in depression, *Trends Neurosci.*, 4, 28, 1981.
27. **Williams, M.,** Adenosine — A selective neuromodulator in the mammalian CNS?, *Trends Neurosci.*, 7, 164, 1984.
28. **Loew, H., Crane, F. L., Grebing, C., Tally, M., and Hall, K.,** Hormone control of plasma membrane oxidation-reduction reactions, *FEBS Lett.*, 91, 166, 1978.
29. **Loew, H. and Crane, F. L.,** Hormone regulated redox function in plasma membranes, *FEBS Lett.*, 68, 157, 1986.
30. **Goldenberg, H., Crane, F. L., and Morré, D. J.,** Influence of hormones on NADH-dehydrogenases in mouse liver plasma membranes, *Biochem. Biophys. Res. Commun.*, 83, 234, 1978.
31. **Rall, T. W.,** Role of adenosine 3′,5′-monophosphate (cyclic AMP) in actions of catecholamines, *Pharmacol. Rev.*, 24, 399, 1972.
32. **Rodnight, R. and Perrett, C.,** Protein phosphorylation and synaptic transmission: receptor mediated modulation of protein kinase C in a brain fraction enriched in synaptosomes, *J. Physiol.*, 81, 340, 1986.
33. **Klainer, L. M., Chi, Y. M., Friedberg, S. L., Rall, T. W., and Sutherland, E. W.,** Adenylate cyclase. The effects of neurohormones on the formation of adenosyl-3′-5′-phosphate on preparation from rat brain and other tissues, *J. Biol. Chem.*, 237, 1239, 1962.
34. **Weller, M. and Morgan, I.,** Distribution of protein kinase activities in subcellular fractions of rat brain, *Biochem. Biophys. Acta,* 436, 675, 1978.
35. **Erulkar, S. D.,** The modulation of neurotransmitter release at synaptic junctions, *Rev. of Physiol. Biochem. Pharmacol.*, 98, 62, 1983.
36. **Daly, J.,** in *Cyclic Nucleotides in the Nervous System,* Plenum Press, New York, 1977, 98.
37. **Schubert, D., Tarikas, H., and Lacobière, M.,** Neurotransmitter regulation of adenosine-3′-5′-monophosphate in clonal nerve, glia and muscle cell lines, *Science,* 192, 471, 1976.
38. **Rodnight, R.,** Aspects of protein phosphorylation in the nervous system with particular reference to synaptic transmission, *Progr. Brain Res.*, 56, 1, 1982.
39. **Brown, J. H. and Mackman, M. H.,** Stimulation by dopamine of adenylate cyclase in retinal homogenates and of adenosine-3′-5′-cyclic monophosphate formation in intact retina, *Proc. Natl. Acad. Sci. U.S.A.*, 69, 539, 1972.
40. **Sieghart, W., Leitich, H., and Karobath, M.,** Subcellular distribution of dopamine sensitive adenylate cyclase, *Brain Res.*, 109, 418, 1976.

41. **Sulzer, F. and Vetulani, J.,** Psychotropic drugs and cyclic AMP in the central nervous system (with particular reference to striatal and mesolimbic structures, in *Cyclic Nucleotides: Mechanisms of Action,* Cramer, H. and Schultz, J., Eds., John Wiley & Sons, London, 1977, 337.
42. **Blumberg, J. B., Vetulani, J., Stawarz, R. J., and Sulser, F.,** The noradrenergic cyclic AMP generation system in the limbic forebrain: pharmacological characterization *in vitro* and possible role of limbic noradrenergic mechanisms in the mode of action of antipsychotics, *Eur. J. Pharmacol.,* 37, 357, 1976.
43. **Beam, K. G. and Greengard, P.,** Cyclic nucleotides, protein phosphorylation and synaptic function, *Cold Spring Harbour Symp. Quant. Biol.,* 10, 157, 1976.
44. **Lynch, T. J., Tallant, E. N., and Cheung, W. Z.,** Rat brain adenylate cyclase. Further studies on its stimulation by a calcium-binding protein, *Arch. Biochem. Biophys.,* 182, 124, 1977.
45. **Greengard, P.,** Possible role for cyclic nucleotides and phosphorylated membrane proteins in postsynaptic actions of neurotransmitters, *Nature,* 260, 101, 1976.
46. **Kuba, K., Kato, E., Kumamoto, E., Koketsu, K., and Hirai, K.,** Sustained potentiation of transmitter release by adrenaline and dibutyryl cyclic AMP in sympathetic ganglia, *Nature (London),* 291, 654, 1981.
47. **Dolphin, A. C. and Archer, E. R.,** 2-Chloroadenosine blocks the K^+-stimulated release of glutamate but not GABA from dendate gyrus slices, *Neurosci. Lett.,* 43, 596, 1984.
48. **Dolphin, A. C.,** $GABA_B$ receptors: has adenylate cyclase inhibition any functional relevance?, *Trends Neurosci.,* 7, 363, 1984.
49. **Loew, H. and Crane, F. L.,** Hormone regulated redox function in plasma membrane, *FEBS Lett.,* 68, 157, 1976.
50. **Kebakian, J. W., Petzold, G. L., and Greengard, P.,** Dopamine-sensitive adenylate cyclase in caudate nucleus of rat brain and its similarity to the "dopamine receptor", *Proc. Natl. Acad. Sci. U.S.A.,* 69, 2145, 1972.
51. **Gilman, A. G.,** G-proteins and dual control of adenylate cyclase, *Cell,* 36, 577, 1984.
52. **Birnbaumer, L., Codina, J., Mattera, R., Cerione, R. A., Hildebrandt, J. D., Sunyer, T., Rojas, F. J., Caron, M. G., Lefkowitz, R. G., and Iyengar, R.,** Regulation of hormone receptors and adenylate cyclase by guanine nucleotide binding N-proteins, *Recent Prog. Horm. Res.,* 41, 41, 1985.
53. **Watanabe, Y. and Jakobs, K. H.,** Forskolin sensitizes human platelet adenylate cyclase to modulation of substrate (MgATP) affinity by hormones, *Biochem. J.,* 237, 273, 1986.
54. **Daly, J. W.,** Forskolin, adenylate cyclase, and cell physiology: an overview, *Adv. Cycl. Nucl. Prot. Phosphor. Res.,* 17, 81, 1984.
55. **Codina, J., Hildebrandt, J. D., Birnbaumer, L., and Sekura, R. D.,** Effects of guanidine nucleotides and Mg on human erythrocytes N_i and N_s, the regulatory components of adenylate cyclase, *J. Biol. Chem.,* 259, 11408, 1984.
56. **Yamashita, A., Kurokawa, T., Higashi, K., Dan'ura, T., and Ishibashi, S.,** Forskolin stabilizes a functionally coupled state between activated guanine nucleotide-binding stimulatory regulatory protein N_s and catalytic protein of adenylate cyclase system in rat erythrocytes, *Biochem. Biophys. Res. Commun.,* 137, 190, 1986.
57. **Snyder, S. H.,** Drug and neurotransmitter receptor in the brain, *Science,* 224, 22, 1979.
58. **Kupfermann, I.,** Modulatory actions of neurotransmitters, *Annu. Rev. Neurosci.,* 2, 447, 1979.
59. **Worley, P. F., Baraban, J.-M., DeSouza, E. R., and Snyder, S. H.,** Mapping second messengers systems in the brain: differential localization of adenylate cyclase and protein kinase C, *Proc. Natl. Acad. Sci. U.S.A.,* 83, 4053, 1986.
60. **Pilc, A. and Enna, S. J.,** Activation of alpha-2 adrenergic receptors augments neurotransmitter-stimulated cyclic AMP accumulation in rat brain cerebral cortical slices, *J. Pharmacol. Exper. Ther.,* 237, 725, 1986.
61. **Albuquerque, E. X., Deshpande, S. S., Aracava, Y., Alkondon, M., and Daly, J. W.,** A possible involvement of cyclic AMP in the expression of desensitization of the nicotinic acetylcholine receptor, *FEBS Lett.,* 199, 113, 1986.
62. **Watson, S. P., McConnell, R. T., and Lapetina, E. G.,** The rapid formation of inositol phosphate in human platelets by thrombin is inhibited by prostacyclin, *J. Biol. Chem.,* 259, 13199, 1984.
63. **Bell, J. D., Buxton, I. L. O., and Brunton, L. L.,** Enhancement of adenylate cyclase activity in S49 lymphoma cell by phorbol esters, *J. Biol. Chem.,* 260, 2625, 1985.
64. **Nabika, T., Nara, Y., Yamori, Y., Lovenberg, W., and Endo, J.,** Angiotensin II and phorbol ester enhance isoproterenol and vasoactive intestinal (VIP)- induced cyclic AMP accumulation in vascular smooth muscle cells, *Biochim. Biophys. Res. Commun.,* 131, 30, 1985.
65. **Watanabe, A. M., McConnaughey, M. M., Strawbridge, R. A., Fleming, J. W., Jones, L. R., and Besch, H. R.,** Muscarinic cholinergic receptor modulation of β-adrenergic receptor affinity for catecholamines, *J. Biol. Chem.,* 253, 4833, 1978.
66. **Lundberg, J. M., Anggard, A., Fahrenkrug, J., Hökfelt, T., and Mutt, V.,** Vasoactive intestinal polypeptide in cholinergic neurons of exocrine glands: functional significance of existing transmitters for vasodilatation and secretion, *Proc. Natl. Acad. Sci. U.S.A.,* 77, 1651, 1980.

67. **Lundberg, J. M., Hedlund, B., and Bartfai, T.,** Vasoactive intestinal polypeptide enhances muscarinic ligand binding in cat submandibular salivary gland, *Nature (London),* 295, 147, 1982.
68. **Martin, I. L. and Candy, J. M.,** Facilitation of benzodiazepine binding by sodium chloride and GABA, *Neuropharmacology,* 17, 993, 1978.
69. **Olsen, R. W.,** GABA-drug receptor interactions, *J. Neurochem.,* 37, 1, 1981.
70. **Dunn, S. M., Blanchard, S. G., and Raftery, M. A.,** Effects of local anesthetics and histrionicotoxin on the binding of carbamoylcholine to membrane-bound acetylcholine receptor, *Biochemistry,* 20, 5617, 1981.
71. **Sanchez, A., Garcia-Sancho, J., and Herreros, B.,** Effects of several inhibitors on the K^+-efflux induced by activation of the Ca^{2+}-dependent channel and by valinomycin in the human red cell, *FEBS Lett.,* 110, 65, 1980.
72. **Garcia-Sancho, J., Sanchez, A., and Herreros, B.,** Stimulation of monovalent cation fluxes by electron donors in the human red cell membrane, *Biochim. Biophys. Acta,* 556, 118, 1979.
73. **Garcia-Sancho, J.,** Redox modulation of the sensitivity to Ca^{2+} of Ca-dependent K channels, in *Redox Functions of the Eukaryotic Plasma Membrane,* Ramirez, J. M., Ed., Rufino Gonzales, Madrid, 1988, 131.
74. **Alvarez, J., Camaleno, J. M., Garcia-Sancho, J., and Herreros, B.,** Modulation of Ca^{2+}-dependent K^+ transport by modifications of the NAD^+/NADH ratio in intact human red cells, *Biochim. Biophys. Acta,* 856, 408, 1986.
75. **Valdeolmillos, M., Garcia-Sancho, J., and Herreros, B.,** Ca^{2+}-dependent K^+ transport in the Ehrlich ascites tumor cell, *Biochim. Biophys. Acta,* 685, 273, 1982.
76. **Valdeolmillos, M., Garcia-Sancho, J., and Herreros, B.,** Stimulation of Na^+-dependent amino acid uptake by activation of the Ca^{2+}-dependent K^+ channel in the Ehrlich ascites tumor cell, *Biochim. Biophys. Acta,* 689, 177, 1982.
77. **Miner, C., Lopez-Burillo, S., Garcia-Sancho, J., and Herreros, B.,** Plasma membrane NADH-dehydrogenase and Ca^{2+}-dependent potassium transport in erythrocytes of several animal species, *Biochim. Biophys. Acta,* 727, 266, 1983.
78. **Karlin, A. and Bartels, E.,** Effects of blocking sulfhydryl groups and of reducing disulfide bonds on the acetylcholine-activated permeability system of the electroplax, *Biochim. Biophys. Acta,* 126, 525, 1966.
79. **Karlin, A.,** Nicotinic acetylcholine-receptor, *Methods Enzymol.,* 46, 582, 1977.
80. **Barrantes, F. J.,** Modulation of acetylcholine receptor states, *Annu. Rev. Biophys. Bioeng.,* 8, 287, 1979.
81. **Barrantes, F. J.,** Modulation of acetylcholine receptor states by thiol modification, *Biochemistry,* 19, 2957, 1980.
82. **Weber, M., David-Pfeuty, J.-L., and Changeux, J. P.,** Regulation of binding properties of the nicotinic receptor protein by cholinergic ligands in membrane fragments of *Torpedo marmorata, Proc. Natl. Acad. Sci. U.S.A.,* 72, 3443, 1975.
83. **Barrantes, F. J.,** Intrinsic fluorescence of the membrane-bound acetylcholine receptor: its quenching by suberyldicholine, *Biochem. Biophys. Res. Commun.,* 72, 479, 1976.
84. **Barrantes, F. J.,** Agonist-mediated changes of the acetylcholine receptor in its membrane environment, *J. Mol. Biol.,* 124, 1, 1978.
85. **Colquhoun, D. and Rang, H. P.,** Effects of inhibitors on the binding of iodinated α-bungarotoxin to acetylcholine receptors in rat muscle, *Mol. Pharmacol.,* 12, 519, 1976.
86. **Weiland, G., Georgia, B., Wee, V. T., Chignell, C. F., and Taylor, P.,** Ligand-interactions with cholinergic receptor-enriched membranes from *Torpedo*: influence of agonist exposure on receptor properties, *Mol. Pharmacol.,* 12, 1091, 1976.
87. **Lee, T., Witzemann, V., Schimerlik, M., and Raftery, M. A.,** Cholinergic ligand-induced affinity changes in *Torpedo californica* acetylcholine receptor, *Arch. Biochem. Biophys.,* 183, 57, 1977.
88. **Quast, U., Schimerlik, M., Lee, T., Witzemann, V., Blanchard, S., and Raftery, M.,** Ligand-induced conformation changes in *Torpedo californica* membrane-bound acetylcholine receptor, *Biochemistry,* 17, 2405, 1978.
89. **Lukas, R. J., Morimoto, H., and Bennett, E. L.,** Effects of thiol-group modification and Ca^{2+} on agonist-specific state transition of a central acetylcholine receptor, *Biochemistry,* 18, 2384, 1979.
90. **Moreau, M. and Changeux, J. P.,** Studies on the electrogenic action of acetylcholine with *Torpedo marmorata* electric organ, *J. Mol. Biol.,* 106, 457, 1979.
91. **Karlin, A. and Cowburn, D.,** The affinity-labeling of partially purified acetylcholine receptor from electric tissue of *Electrophorus, Proc. Natl. Acad. Sci. U.S.A.,* 70, 3636, 1973.
92. **Donaldson, J. and LaBella, F. S.,** The effects of manganese on the cholinergic receptor *in vivo* and *in vitro* may be mediated through modulation of free radicals, *Neurotoxicology,* 5, 105, 1984.
93. **Hamilton, S. L., Pratt, D. R., and Eaton, D. C.,** Arrangement of the subunits of the nicotinic acetylcholine receptor of *Torpedo californica* as determined by α-neurotoxin cross-linking, *Biochemistry,* 24, 2210, 1985.
94. **Berstein, G., Haga, K., Haga, T., and Ischiyama, A.,** Agonists and antagonists binding of muscarinic acetylcholine receptors purified from porcine brain: interconversion of high- and low-affinity sites by sulfhydryl reagents, *J. Neurochem.,* 50, 1687, 1988.

95. **Davies, A. D.,** Coupling of human beta-2-adrenergic receptors: relationships to redox potential, *J. Endocrinol. Invest.,* 11, 239, 1988.
95a. **Davies, A. D.,** Effects of endogenous redox-active compounds on coupling of human beta-2-adrenergic receptors, *Am. J. Med. Sci.,* 292, 257, 1986.
96. **Curro, F. A. and Greenberg, S.,** Characteristics of postsynaptic alpha-1 and alpha-2 adrenergic receptors in canine vascular smooth muscle, *Can. J. Physiol. Pharmacol.,* 61, 893, 1983.
97. **Foster, A. C. and Fagg, G. E.,** Acidic amino acids binding sites in mammalian neuronal membranes: their characterization and relationship to synaptic receptors, *Brain Res.,* 319, 103, 1984.
98. **Marzullio, B. and Hine, B.,** Opiate receptor function may be mediated through an oxidation-reduction mechanism, *Science,* 208, 1171, 1980.
99. **Kolb, V. M.,** New insights in the clastic binding hypothesis for opiate-receptor interactions. I. Electron-transfer mechanisms, *J. Pharm. Sci.,* 73, 715, 1984.
100. **Bravo, C., Garcia-Baro, J. A., and Sobrino, J. A.,** Activation of mammalian adenylate receptor by iontophoretic application of reducing and oxidizing sulfhydryl reagents, *Rev. Esp. Fisiol.,* 41, 243, 1985.
101. **Floyd, R. A.,** Erythrosine B (Red dye Nr. 3)-mediated oxidation-reduction in brain membranes, *Biochem. Biophys. Res. Commun.,* 96, 1305, 1980.

INDEX

A

B

C

G

H

I

K

L

M

N